A Social History of Science and Technology in Contemporary Japan

JAPANESE SOCIETY SERIES

General Editor: Yoshio Sugimoto

Lives of Young Koreans in Japan
Yasunori Fukuoka

Globalization and Social Change in Contemporary Japan
J.S. Eades Tom Gill Harumi Befu

Coming Out in Japan:The Story of Satoru and Ryuta
Satoru Ito and Ryuta Yanase

Japan and Its Others:
Globalization, Difference and the Critique of Modernity
John Clammer

Hegemony of Homogeneity:
An Anthropological Analysis of *Nihonjinron*
Harumi Befu

Foreign Migrants in Contemporary Japan
Hiroshi Komai

A Social History of Science and Technology in Contemporary Japan
Volume I: The Occupation Period 1945–1952
Shigeru Nakayama

A Social History of Science and Technology in Contemporary Japan

Volume I

The Occupation Period
1945–1952

Edited by Shigeru Nakayama

with
Kunio Gotō and Hitoshi Yoshioka

Trans Pacific Press
Melbourne

First published in Japanese in 1995 by Gakuyō Shobō (Tokyo) as *Tsūshi Nihon no kagaku gijutsu, Dai 1-kan.*
This English edition first published in 2001 by
Trans Pacific Press, PO Box 120, Rosanna, Melbourne, Victoria 3084, Australia
Telephone: +61 3 9459 3021 Fax: +61 3 9457 5923
E-mail: info@transpacificpress.com
Website: http://www.transpacificpress.com

Designed and set by digital environs Melbourne
enquiries@digitalenvirons.com

Printed in Melbourne by Brown Prior Anderson

Distributors

Australia
Bushbooks
PO Box 1958, Gosford, NSW
2250
Telephone: (02) 4323-3274
Fax: (02) 9212-2468
Email:
bushbook@ozemail.com.au

Japan
Kyoto University Press
Kyodai Kaikan
15-9 Yoshida Kawara-cho
Sakyo-ku, Kyoto 606-8305
Telephone: (075) 761-6182
Fax: (075) 761-6190
Email: sales@kyoto-up.gr.jp
Web: http://www.kyoto-up.gr.jp

USA and Canada
International Specialized Book
Services (ISBS)
5824 N. E. Hassalo Street
Portland, Oregon 97213-3644
USA
Telephone: (800) 944-6190
Fax: (503) 280-8832
Email: orders@isbs.com
Web: http://www.isbs.com

ISSN 1443–9670 (Japanese Society Series)
ISBN 1–8768–4310–1

National Library of Australia Cataloging in Publication Data

[Tsūshi Nihon no kagaku gijutsu]
A social history of science and technology in contemporary Japan.

Bibliography.
Includes index.
ISBN 1 876843 10 1.

1. Science - Japan - History. 2. Technology - Japan - History. 3. Japan - History - 1945-. I. Nakayama, Shigeru, 1928-. (Series : Japanese society series).

509.52

Contents

Table of Tables

Table of Figures

Preface to the English Edition

When I visited China as the leader of the First Japanese History of Science Visiting-China Team, the Chinese Academy of Science asked me to give a talk addressing their question 'Why did the Japanese succeed in developing science and technology despite their defeat in World War II?' This was in 1980, about the time that the particular style of Japanese development in science and technology began to receive worldwide recognition. Hence, this was a legitimate and important question.

My ready-made answer at the time was that, while the prewar paradigm of Japanese science and technology was military-oriented research and development, after the war the focus was almost totally economic. The actual course of development was, of course, not so simple, consisting of a continuous series of trial and error, of ups and downs. While Japan has not produced many Nobel laureates, the postwar development was characteristically Japanese, quite in contrast to the Western path in the Cold War environment.

Thus, in the 1980s, many countries, both developed and developing, sent observation teams to Japan to find out the 'cause' of the Japanese 'miracle' – and whether or not it really was a miracle. There are no simple or easy answers to such questions, and this work makes no pretense at providing any such thing. Instead, we have attempted to chart the course of development, both its merits and its demerits – from the 'inside' the shortcomings are often more visible – to enable readers to discover this unique course, with both its positive and negative lessons, for themselves.

In the 1990s, we Japanese came to a turning point in history after half a century of this particular course of development. Thus, it is a good time to examine the course of development and to consider the possible future of development with the wisdom of hindsight, without being distracted by the short-term ups and downs of the economic cycle.

In 1999 we completed the seventh volume of this project in Japanese and turned our attention to making this work available in English, so that our lessons can be shared with those who do not read Japanese. In the process of translation we encountered numerous problems that the translators and editors struggled to make comprehensible and, hence, some further explanation became necessary. The Japanese authors had originally addressed their message to a Japanese audience, so they had included contexts and inferences that were internationally unintelligible, while at the same time omitting some information necessary for an international audience's understanding. To overcome these difficulties Professor Sugimoto, working in Australia, and I, in Japan, tried to fill the lacuna between the Japanese and world contexts.

Our thanks must therefore include email and internet technology; we have had daily exchanges of electronic files back and forth between Australian translators and Japanese authors. Without such technology, international editing like ours is virtually impossible.

I am most obliged to Professor Yoshio Sugimoto of La Trobe University, who not only organized translators and editors in Australia – which I consider to have the highest level of proficiency in Japanese language among English-speaking countries, perhaps higher than that of the US and UK – but also provided competent editorial care on this difficult, many-authored work. Thanks are also due to Karl Smith, style editor of the volume, who has worked tirelessly, paid meticulous attention to the details of each chapter and contributed to the production of a manuscript more coherent and integrated than the Japanese version. I am also grateful to a team of translators, including Francis Conlan, Mike Danaher and Naohiko Shimizu. Finally, we acknowledge the Japan Society for the Promotion of Science for their 2000–2001 Grant-in-Aid for Publication of Scientific Research Results for the translation and publication of such a large work, unprecedented in their experience.

Shigeru Nakayama
Melbourne, Australia
18 February 2001

Preface

Shigeru Nakayama

Historical Introduction

There is a widespread belief that the Second World War (WWII) was brought to an end by the atomic bombs. From such a perspective, history can be periodized according to the science and technology discovered, invented and employed during any given era. Thus, the postwar era is known as the atomic age. The atomic age was brought forth by scientists – physicists, in particular – whom many regard as possessing the only valid perspective on this new age. Hence the spokesmen of the scientific community were in great demand by the mass media.

WWII was, however, by no means terminated with the atomic bombs. Rather, the bomb marked the advent of the new and more enduring 'Cold War.' The Manhattan Project that developed and manufactured the A-Bombs was not only the prelude to the atomic age but also provided a paradigmatic model for postwar research and development (R&D) due to its strategic success; science became a 'big' project, supported by big powers and closely connected with technology. The 'little science' practiced in prewar academia became simply antiquarian. A dominant belief developed among postwar scientists that the best and easiest way to advance the research frontier was simply to expand the scale of their research facilities and organizations far beyond the prewar standard.

The USA and the USSR created military-industrial-academic complexes that were self-perpetuating and continually growing. The 'complex' played a leading role in R&D throughout the postwar period, with minor fluctuations in its power and influence over the years. The science policy of the European nations was generally modeled after these complexes, too, although they were not always militarily oriented. The Cold War lasted for half a century, an unexpectedly long period. So too, the military-industrial complex endured and prospered, mutually excited and enlarged. It appeared to be an everlastingly perpetuated mechanism.

Now that era has passed. In the last decade of the twentieth century, the military-industrial complex began to decline and was wholly restructured. At the present point in history, it is the duty of historians of science to recapitulate the scientific activities of the Cold War era and develop a prospective for future generations.

Uniqueness of Japanese Development

Japanese science and technology (S&T), however, followed a unique course of development, considerably dissimilar to the major Western nations.

S&T in prewar Japan was heavily inclined towards military application. Japan shares the same experience as other invaded countries in having been awakened to the necessity of modernization through the threat of the Western powers in the nineteenth century. The difference from other non-Western countries lies in the 'successful' transformation from an invaded to an invading nation in the course of development. At the time of transformation the only thing needed was more advanced military science and technology.

The departure from militant imperialism was rapid after 1945, following Japan's military defeat in WWII. The pace of change is partially explainable by the pressure from the Occupation Forces to demilitarize, but endogenous motivation provides a better explanation for the overall changes in direction. In the Japanese scientific community, as well as the political and business communities, a pacifist mood prevailed. Following their bitter lessons, they admitted the irrationality of the militaristic 'progress' that had ultimately invited the catastrophic misery of 1945 and the immediate postwar period. Consequently, they chose to forge a new path, a path that led to postwar Japan being a military-political dwarf but an economic giant.

This path was diametrically opposed to the predominant trend of the military-industrial complex established by the USA and the USSR. The only significant parallel to the Japanese experience is West Germany. Hence, it is a unique path and is worth recording for posterity.

The STS (Science, Technology and Society) Perspective

'Science, or S&T, is not an idea but a social institution.' The truth of this statement becomes increasingly clear as history unfolds. Modern science began in the seventeenth century as an intellectual revolution. Much of the history of science appears to be a history of intellectual elites. Today, however, it is apparent that science is simply a social infrastructure.

It was the Marxist historians of science in the 1930s that first noted the social and institutional aspects of science. At the second International Conference of the History of Science in 1931, Boris Hessen of the Soviet Union presented the socio-economic basis of Newtonian mechanics. Following the Hessen paradigm, a number of materialist historians of science appeared. Hessen's work was translated into Japanese in two different versions and was very influential amongst Japanese historians of science before WWII.

The materialist historians had disputes with conventional intellectuals who believed that the history of science was driven purely by the achievements of a limited number of geniuses. The materialists argued, instead, that science is influenced by the social infrastructure. In the prewar period, in fact, some materialist historians made an even stronger claim, arguing that science is socially determined. However, in the days of 'small science,' science played only a minor role in determining the course of social events. Science was like many other cultural phenomena, remaining in the 'superstructure,' with little influence on the social base. In other words, science was socially determined, but did not, in turn, determine society. The social origin of science is now accepted even amongst academics who previously maintained that science is self-determined; that is, that scientific matters (questions, validity, methods, etc.) are determined internally by the logic of each discipline. Still, this concession has not ended the disputes between the social determinists and the internalists.

During the Soviet regime – under Stalinist intimidation – it was a dangerous taboo for a historian of science to deal with the contemporary (post-revolution) history of science. On the other hand, in the capitalist regimes up to the 1950's, particularly in the US, scholars kept their distance from the social history of science under the threat of McCarthyist purges and, instead, practiced the safer internalist approach, i.e., the intellectual history of science.

Western capitalist society experienced a watershed in the 1960s, culminating in widespread protests in 1968 (some people call it a 'revolution'). Previously academic historians of science focused on the classical period and/or the seventeenth century Scientific Revolution. Subsequently, the younger generations extended their interests to the eighteenth, nineteenth and twentieth centuries. It became apparent that science had been increasingly institutionalized over this period of history. Hence, rather than considering science to be an isolated activity, they regarded scientific activity as tightly integrated into the scientific community. This approach is called the sociology of science, and was highly influenced by Thomas S. Kuhn's *The Structure of Scientific Revolutions*.[1] The sociology of science is a social constructivist approach that, at its extreme manifests as social determinism. Social determinists maintain that all scientific activity is determined by the group structure of the scientific community. While the extreme is still debated, more moderate social constructivist views are now almost universally accepted.

In the 1970s a group of researchers and educators, following the constructivist trend, began to advocate the Science, Technology and Society (STS) approach. STS is not a specific method or discipline but, rather, a problem-area. In the postwar period, as S&T's power appeared everywhere, it generated a great deal of problems vis-à-vis society at large.

Solutions would not be found within traditional disciplines such as the history or sociology of science. For instance, 'the social acceptability of nuclear reactors in Japan' is a problem area in the sociology of science. Such problem-areas need to be approached from a variety of angles, transcending disciplinary boundaries.

Japan is not the birthplace of modern science. Hence, in the historical recollections of the average Japanese, science does not appear to be as significant as it is for Europeans. Japan's first encounter with 'modernity' was when it awoke one morning in 1853 to discover Western gunboats anchored in Tokyo Bay. Thus, for the Japanese, 'modern' originally referred to the military superiority of the Western powers rather than Western science. They soon realized that behind the military threat lay the Industrial Revolutions. Thus, for modern Japanese historians it made more sense to investigate the technological achievements of the nineteenth century than the Scientific Revolution of the seventeenth century. Their historical experience generated a keen interest in the institutional and social history of technology transfer processes. The Japanese thus readily accepted the STS problematics, and members of the STS-oriented generation were keen to participate in the project of researching and recording *A Social History of Science and Technology in Contemporary Japan*.

Four-Sector Approach

I have elsewhere criticized the naiveté of applying Marxism's base-superstructure model to the history of science[2] and have argued for the insertion of a middle structure consisting of the institutional and sociological aspects of the scientific community between science and society.[3] Thus, I have been sympathetic towards the STS approach from the beginning. Both economic base-determinism and scientific community (social) determinism are appealing for their theoretical simplicity; problems are simply reduced to one single factor. The reality, however, is much more complex. Most things cannot be explained in such simple terms.

According to the opinion of those who were once deeply committed to base-determinism or base-reductionism and later deeply disappointed by them, complex problems must be treated as complex phenomena and approached from many viewpoints. Hence, I propose here a 'four-sector' approach – public (governmental), private (corporate), academic (university) and citizenry – to provide a historical account of the phenomena at the interstices of the interests of the different social sectors. In postwar Japan, S&T is where all of these different sectarian interests meet. The balance of power between the four sectors has determined the direction that S&T has developed.

I proposed that the four-sector approach should provide the 'official' perspective for our Social History Project, expecting it to be acceptable to all members. There seems to have been no particular objections: some participants – particularly the sociologists – were quite willing to explore these phenomena from the perspective of the citizenry. Other participants in our project are public or private sector professionals.

However, most of the contributors to our project were neither bureaucrats nor corporate executives, but rather university staff. Hence their standpoint does not represent public or private interests; it is academic. Historians of science could readily assume the perspective of the citizenry or the academic scientist while experiencing more difficulty in sharing the public sector or corporate standpoints, even where the latter is essential. In other words, it seems somehow easier to share the perspective of the citizenry who had no vested interest in the promotion of S&T.

A mass operation such as ours should not be conducted under a rigid official ideology and methodology that restricts members' personal freedom and is often alienating. Rather, a loosely outlined approach, such as the four-sector perspective, is more liberating and therefore, perhaps, more fruitful.

Classification of Scientific activity according to the Sociology of Science

We deem it essential that S&T be classified according to the 'social actors' who have the power to assess scientific activity, the people to whom scientists address their work. In other words, the question 'To whom is the scientific research addressed?' is the most important factor in shaping and defining the character of each scientific activity. Other distinctions such as research location, form of presentation and sources of funds are not necessarily essential and often remain derivative. The following sections explain our classification of science according to the social sector of the assessors.

Academic Sector and Academic Science

Within the scientific community, members present their research for debate, discussion and criticism by their colleagues. This form of science is based on the principle of publishing or otherwise making the results of research available to the scientific community. I refer to this sort of scientific activity as 'academic sector science' or simply 'academic science.' In the case of 'academic science,' fellow scientists conduct professional assessments.

Research is initiated out of individual interest and pursued for reasons of personal honor and distinction, which is meted out through a referee system. For example, it is simply an academic concern of engineers

Figure A: Circular diagram illustrating various scientific activities

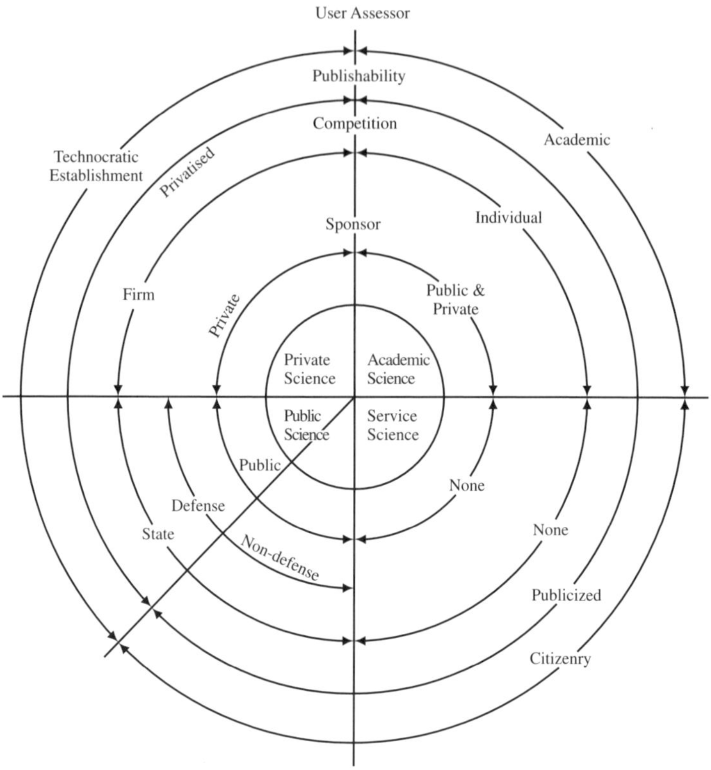

whether Japanese technology is more advanced than the rest of the world; a matter of pride for individual engineers but nothing more.

Academic science as practiced in the academic sector can be called 'science' rather than S&T (this distinction is clarified below). In academic science, the academic freedom of choosing one's own research topic is well respected. It leads to the idea of scientism (science for its own sake), but for that reason is often less directly significant for society at large.

Universities, where the traditional ideal of academic freedom is supposed to be maintained, are generally thought to be the best locations for conducting academic science. We can call it academic science when it is practiced in various professional schools – such as the school of medicine, technology or agriculture – as long as the researchers have freedom to choose their own research topics and are peer-reviewed via established procedures.

This mechanism has guaranteed and strengthened the universality and objectivity of modern science. Academic science retains its prestige because of its educational role. Nevertheless, with the advent of big science with its big budgets and large numbers of personnel, the value judgments of sponsors became more important and industrialized science became the mainstream of the R&D community. It is here where tensions emerge between the mechanisms for assessment and sponsorship.

Public Sector and Public Science

Such academic scientists alone have not characterized the main current of postwar S&T. Few would disagree that, apart from the academic sector, the outstanding feature of postwar S&T has been the way in which the politics of bureaucrats – the public sector – have largely determined the direction that S&T has progressed, e.g., for strengthening national defense and welfare, or simply strengthening bureaucratic power.

Scientists in the public sector organize and plan mission-oriented research. Throughout history, the government's contact point with science has been primarily focused on defense-related research. Other scientific activities can also be found in modern nation-states; in nationwide surveys of resources, the maintenance of national standards and, more recently, public healthcare and managing the global environment. These activities are mostly performed in the in-house public laboratories and hence we shall define them as 'public sector science' or 'publicly-performed science.' In the postwar period, however, government sponsorship was extended to the academic and private sectors; thus it could be called 'publicly-sponsored science.' For example, the US system of scientific R&D since WWII is the product of an amalgamation of the governmental and industrial sectors – whose two largest branches are nuclear and space technologies. Because the government is the largest purchaser (consumer) of the technology produced (i.e., nuclear weapons, spacecraft, communications equipment, etc.) this approach can be called 'publicly-sponsored science' or 'public science.' The major direction of 'publicly-sponsored science' is decided by, and its output assessed and appraised by, bureaucrats. Since we tend to place more emphasis on assessors than performers, we call it 'public science.'

Private Sector and Private Science

The R&D carried out in the private laboratory of a corporation is conducted in the interests of the industrialists and is hence called 'industrial science' in a narrow sense. While academic science is primarily pursued in universities, industrial science is mainly practiced in private enterprises. A distinction based on the type of institution where the research work is conducted might be more readily understood, in which case we would speak of 'university science' and 'corporate science.'

The broader term 'industrial science' refers to sponsored research including research at a think-tank. Similar to 'publicly-sponsored science,' 'privately-sponsored science' as practiced in the private sector is appraised by private enterprises in order to increase profits. As we place emphasis on sponsors and assessors rather than the site of performance, we name it 'private science.'

Depending on whether the sponsor of R&D belongs to the public or private sector, the nature of scientific activity differs. The US system of scientific R&D since WWII is the product of an amalgamation of the governmental and industrial sectors – whose two largest branches are nuclear and space technologies. In contrast to the US, postwar Japan did not develop a military-industrial complex; the military was not the major sponsor or consumer of science. The hallmark of postwar Japanese R&D is, as was internationally recognized in the 1980s, the predominance of 'private science' over 'public science,' for the sake of maximum corporate profit.

Citizen Sector and Service Science

The citizens may be defined as people who do not belong to either the academic, public or private sectors described in the above categorization; in other words, ordinary people who have no vested interest in and derive no direct profit from the advance of science and technology. Hence, there is no place in a sectorial analysis based on public, private and academic for the S&T activities that aim directly at the interests of the citizenry or for the benefit of mankind. For these practices we require another science sector. Those scientific activities that directly contribute to the welfare of the citizenry, I refer to as 'citizen-sector science' or 'service science.'[4] A typical example of service science, that is, science that is directly assessed by the citizen sector, is health care. In spite of the recent bureaucratized trend, a general physician's medical service is even today directly assessed by his patients.

In spite of the devastating effects of nuclear bombs and the continual, cumulative destruction of the natural environment, it is still firmly believed that S&T will contribute to, if not actually bring about human happiness. Various questionnaires confirm that this belief is widespread.

Medical research and services are reportedly the fields most anticipated to provide benefits. We repeat the definition from the sociology of S&T that the citizenry is the social sector that does not benefit directly from the promotion of science and technology. Academic science, public and private sciences all have internally instituted promotion and assessment mechanisms that exist to protect their own interests. Therefore, the fourth sector also requires its own promotion and assessment mechanism by the citizenry. Here, I am more proscriptive than descriptive, attributing it with characteristics that it 'should have' rather than what it does have at the present time (we will discuss citizenry activism in more detail below).

Public Service Science

Narrowly defined, service science must be public service science conducted in public laboratories and financed by tax money for a return[5] to the taxpayers. In addition to medical research, prime examples of public service science are: weather forecasting, environmental protection, nutritional research and information and library services. Such government activity constitutes a gray area between 'service science' and 'public science.'

Of the Japanese public science that is funded by tax money, defense R&D accounts for an insignificant part. Most public service science seems to be conducted in various national laboratories, which should be in the interests of the taxpaying citizenry. However, this type of institution lacks the competitive mechanisms that are integral to every other sector, academics, nation-states and corporations. It also lacks mechanisms for responding to the opinions and judgments of the citizenry. Thus, scientists in these public laboratories have no avenue for professional assessment other than through publishing articles in scientific journals, just like academic scientists.

There are also R&D services that are not considered to be public service science. For example, many public laboratories belonging to MITI (Ministry of International Trade and Industry) serve the industrial sector and often aim to patent their works. Likewise, the laboratories of each government ministry provide services to the sector of society for which their ministry is responsible.

The knowledge produced by public service science should be public knowledge that remains openly available to the general public. In contrast, military science, which is also funded by tax money, is very secretive. Nuclear research is also highly secretive as its military applications prohibit the disclosure of information even within the nuclear science community, thus creating tensions between the public and citizenry sectors.

Other Criteria for Classifying Science

Science and Technology are Indistinguishable

The adoption of the term 'Japanese S&T' in our project title generally does not distinguish between science and technology. Any such distinction between science and technology is a Western conception. At the beginning of Japanese modernization in the late nineteenth century, samurai bureaucrats did not understand a clear-cut distinction. For them, Western science and Western technology, combined, were the tools for civilization, industrialization and modern armament.

In Europe, there was an aristocratic tradition of 'high science,' which was maintained by the 'gentlemanly class' at universities. Technology was nurtured within the apprenticeship system and later taught at polytechnics to the newly emerging lower and middle class city dwellers. In Japan there was also a class system carried over from the premodern Edo period with four categories: samurai, peasantry, artisan and tradesman. In the 1870s the Meiji government recognized technology as a legitimate subject to be studied within the university system. Yet, although from 1886 the Imperial University distinguished the Faculty of Science from the Faculty of Technology, the practitioners – mainly samurai descendants – did not recognize any clear distinction and engaged in both. The social status of engineers was therefore higher in Japan than it was in Europe.

While the samurai-class university educated engineers were responsible for disseminating Western S&T in the public sector, the traditional artisan class developed technology for private profit. While the samurai engineers often failed in their attempts to apply technology to business because of the failure to properly consider the economic aspects, success was often achieved when that technology was transferred from the public sector samurai to the private sector artisans, as in the case of Toyota, where Western technology took root before WWII. The privatization of technology continued into the postwar period when the structure of privately led S&T and R&D neared perfection.

In the interwar period, a tradition of promoting 'pure science' was founded by the second generation of modern Japanese scientists. They accepted the same distinction between science and technology as that made in the West. Outside of this particular community, though, even in the social science community, such a distinction was not so obvious. The in-distinction intensified with the general trends in the postwar world where science was no longer 'free enquiry' but a techno-nationalistic endeavor. This in-distinction still holds true today in postwar Japanese society.

This has also been true internationally since the success of the Manhattan Project, wherein science and technology were indistinguishably intertwined. In postwar Japan S&T is characterized by the national slogan 'science and technology for economic recovery' and the private dominance of R&D. Overshadowed by the private sector success of industrial science, the relative status of pure science declined. Internationally this situation is often regarded as a deficiency in Japan's basic science research, a claim 'supported' by the observation, made since the 1980s, that in contrast to the economic success, there are few Nobel laureates in Japan.

Some sectors of Japanese society maintain that in comparison to the good old European tradition, Japan is shamefully backward and is still underdeveloped in promoting science. However, from the STS perspective it is apparent that the transfer of science from one cultural environment to another invariably generates some modification to the nature of science, if not wholesale transformation. In other words, the prominent nineteenth century German *Wissenschaft* differs substantially from the 'big science' of twentieth century America. Likewise, the personal character of researchers changes from a savant or a *Wissenschaftler* to a scientific professional. Hence, it was inevitable that postwar Japanese science-and-technology would undergo transformation in an environment entirely different from the West.

It appears that what Japan has developed in the postwar period is not quite a science but rather a production technology. Japanese *kagakugijutsu* (literally translated as sciencetechnology) is somewhat different from what has developed in the West. The Chinese coined a new word for it: Ke-I (Sci-Tech is a literal translation). Thus this failure to distinguish between science and technology is a Japanese modern tradition that is shared by the Chinese.

Publicized and Privatized Science

The most basic criterion of postwar science is whether or not it is open to the public. The former is academic science as practiced in the academic sector, namely universities, including applied science fields such as medicine, engineering and agricultural study as long as the research findings are publicized. This same value of openness is shared by the citizenry sector.

Private science as practiced in the private sector, however, does not share this value; its research findings are the private property of corporations. From this perspective, scientific information that has been produced by research is worthless unless it is privately owned. The information, if publicized, can be used by anyone, even competitors, and will come to have little economic value. Public science as practiced by the military sector also shares the value of closedness, and the secrecy of scientific findings. Non-military public science is in a gray zone on this issue, although nominally research findings should be open to the public since it is financed by tax-money. These differences are essential to our discussion of postwar Japanese science.

Competition Mechanism

Competition is the incentive for such intellectual activities as scientific research and development. In academic science, competition is between individual researchers, but in sponsored science competition between nation-states and competition between corporations is the main motivation for funding R&D. When competition becomes intense, governments and corporations tend to place their own needs over those of the citizenry. In order to prevent such situations occurring, it may be desirable that the citizenry be empowered to make assessments of both public and private sector R&D. However, the citizenry can never become a powerful regulating force since it does not incorporate any competitive incentives.

Since the fourth sector does not possess the professional drive and profit incentive of the first three sectors and, as it is often on the receiving end of the activities of the other three sectors, citizens' movements, such as the anti-pollution campaign, have to depend on participants' motivation and must utilize the mass media if they are to influence the tide of technocratic trends so that the damage inflicted on civilians does not exceed acceptable limits.

Technocracy vs. the Viewpoint of the Citizenry Sector

In recent years, the public, private and academic sectors have become tightly integrated in a technocratic structure, in which no critical assessment function is installed. Japan is no exception. Under these conditions a critical function can only exist in the citizenry sector. Hence, the citizenry sector's positive contribution to S&T lies in its critical competence exceeding self-interest and the regulatory power arising from consensus conferences. As such, 'critical science' – as J. Ravetz called it – is the major role of service science. But of course, to speak of this as a 'positive' contribution is to assume a particular position on the issue – it is certainly seen as a 'negative' contribution by those who are criticized, at least to the extent that they are disadvantaged in the pursuit of their self-interests. The citizenry sector activists are therefore frequently victimized by the other, generally more powerful sectors.

The citizenry sector's active avant-garde includes environmental action groups and non-government organizations (NGOs). These social actors utilize mass-communications and the mass media to speak on behalf of the citizenry sector. The voice of the citizenry is thus often represented as a consensus opinion.

Technocracy and Democracy: Polarized Concepts

In the following discussion, we treat technocracy and democracy as polarized concepts. We rather simplistically define technocracy – an amalgam of 'technology' and (hierarchized) 'bureaucracy' – as a top-down communication mechanism in pursuit of efficiency while democracy is a bottom-up mechanism of majority rule. 'Technocratic structure'

Table A: Characteristics of the four science sectors

Science	Academic	Private	Public	Service
Assessor	peer	sponsor	sponsor	citizenry
Publishability	publicized	private	private	publicized
Competition	individual	firm	nation-state	none
Performer	university	private laboratory	public laboratory	media
Sponsor	public or private	private	public	none
Utilizer	academic	technocratic establishment	technocratic establishment	citizenry

denotes a complex consortium, the incorporated structures of the public, private and academic sectors (i.e., government, industry and universities). Historically, the dominant power in a society – usually the government – assessed, controlled and protected the S&T that emerged outside of the power structure. In recent years, however, the integration of three sectors in most developed nations gave rise to technocratic structures that incorporated and promoted S&T as its main function. Japan is one (obvious) instance of such processes.

The modernization of Japan was largely samurai-spirited (samurai: warrior-bureaucrat). Under imperialist competition, modernization was encapsulated by the celebrated slogan 'Fukoku Kyōhei (make the nation wealthy and strengthen the army).' In spite of a major turn after the surrender in 1945, the same spirit has been embedded in the postwar period in a nationalistic slogan 'nation-building through science and technology' – or 'techno-nationalism.' The techno-nationalist vocabulary is rife with terms of warlike strategy and competition, such as 'the struggle for survival,' which is not entirely comfortable for, or compatible with, the consumerist citizenry seeking peaceful coexistence. Consequently, the function of assessment should be transferred to the fourth sector, the citizenry.

Technocratic literature is peppered with clichés such as 'under the close cooperation of the public, private and academic sectors, the research and development of science and technology is rapidly and efficiently conducted.' Here, the citizenry sector is significantly avoided. The truth of this cannot be judged until we look closely at what science means to the citizenry sector.

In reality, the three technocratic sectors together have controlled and directed S&T so far. But something is missing: unless the citizens' perspective is integrated, the 'society' part of the 'Science, Technology and Society' approach cannot be well represented.

The Citizenry's Concern with Science Arose in the 1970s

'The pattern of industrialization in Japan and other Asian countries is technocratic developmental dictatorship.' A small number of elite bureaucrats produced a highly centralized plan of industrialization and executed it through the absolute power of the newly emerged nation-state. This approach has been called the Asian model of capitalism. Modern Japan set the precedent for this model which the other Asian nations later followed.

Langdon Winner claims that the US, West Germany and Japan attained a peak of technocratic achievement in the 1960s.[6] At the same time, though, the citizenry sector began to speak-out about STS problems, the environmental (anti-pollution) and anti-war movements appeared and eventually the course of development of S&T was significantly altered. These events are impossible to explain without reference to the citizenry sector. For instance, when planning the Narita International Airport in the 1960s, technocrats did not anticipate the impact of opposition from the citizenry, but by the 1970s they realized that they had to seriously consider citizens' claims on every significant development.

Up until the 1960s, the consumer citizenry had no direct interest or concern with S&T, and there was no discussion about its social meaning. It was not until the 1970s, that a pollution-concerned citizenry began to recognize that they were directly connected with the S&T activity of corporations and began to question these activities. Subsequently, the Toyota Foundation was created, primarily to address criticisms of technocratic ways of thinking.

The Toyota Project

This 'Japanese Science and Technology' project began when I met Yoshinori Yamaoka, a program officer of the Toyota Foundation, in 1983. He was looking for a new project to replace the 'Minamata Project' which was nearing completion. He was also a supporter of the 'service science' concept and had created an 'environment competition' program to encourage the citizenry to participate in environmental research at the Toyota Foundation

Yamaoka suggested that I should coordinate a major project for the Foundation on the postwar history of Japanese science and technology. I revised my life-long research plan and decided to accept the offer. I calculated that the project would take ten years to complete.

I recalled my experience from a quarter of a century earlier when I participated in a project with the History of Science Society of Japan to produce *Nihon kagaku gijutsu-shi taikei* (Source book on the history of Japanese science and technology, 25 vols.)[7] which had taken ten years. That project was organized and completed with a tremendous contribution from Professor Mitsutomo Yuasa. I have often tried to model myself after him.

While I was enjoying the benefits of working on this large-scale, fully funded project, the younger generation of Japanese historians of science urged me to plan some similar project for their training.

When the Toyota Project began I was managing 'discipline formation' seminars with the Ministry of Education's Grant-in-aids program and the 'Science and Society' Forum at the Toyota Foundation. The staff members of these organizations in their twenties and thirties were the most eager workers for my project. Twenty-five years earlier, Prof. Yuasa was in his fifties and we active workers were in our thirties. So, for the Toyota Project I made a strategic decision to invite older historians of science to be resource persons, those in their fifties – my generation – would be organizers and the younger generation would be the real workers.

In the beginning I invited our older predecessors to the 'Science and Society' Forum and asked them: 'If you write the history of Japanese postwar science and technology, how do you write, what arrangement do you use' and so on. We tape-recorded our discussions with Yūjirō Hayashi, Yoshirō Hoshino, Junnosuke Kishida and others. Some of these were edited and appeared, entitled *Nihon no gijutsuryoku: Sengoshi to tenbō* (Japan's technological power: Postwar history and perspective) in 1986 in the *Asahi Shimbun*. This provided a preview for the whole project.

The headquarters of our project, the 'Science and Society' Forum, is located in Tokyo. I invited Kunio Gotō (of my generation) to organize a similar forum in Osaka where we had workshops twice a year.

There were less than ten members to the core group who sat together for days to construct a preliminary outline and agree on our editorial principles. Our model, of course, was the *Nihon kagaku gijutsu-shi taikei*. Following its construction, we produced four *tsūshi* (Outline history) and twenty various *kakuron* (volumes on special topics) about the postwar period.

The Tsūshi team began by collecting and analyzing major bibliographies and journal catalogues. Their aim was to systematically search not only academic journals but also sources from all sectors of society.

Besides the Tsūshi team, several Kakuron teams were created for particular projects, such as 'Nuclear Research and Reactors,' 'Food and Agriculture,' 'Grassroots Attitudes' and 'Scientific Manpower.' There were some overlaps in membership between the Tsūshi and Kakuron teams. At the half way point we decided to concentrate on completing the Tsūshi first and shelved the Kakuron projects for later publication.

Tekunohisutorii (Technohistory)

By this stage, we had already published the texts so far completed for the critical examination of the readers of the magazine *Shūkan ekonomisuto* (Weekly economist) published by Mainichi Shimbun. The series was entitled 'Tekunohisutorii hanseiki (Half a century of techno-history)' and printed between April 1990 and September 1991. These articles were then edited in a single volume of *Sengo kagaku gijutsu no shakai-shi* (A Social history of postwar science and technology).[8] It is an intermediary product as well as another preview of the larger project.

Compiling the Four Outline Volumes

The characteristic point of our editing is the arrangement of historical events in a topical-module system rather than the ordinary style of chapter-section subdivision. The topical-module approach was preferable to a simple time sequence of historical occurrences because our subject matters are highly specialized and each author has their own specialist topic that can stand alone and work as a module. We organized monthly meetings at which each article was subjected to a group discussion as well as being peer reviewed. During the university vacation period, the core members stayed together to examine and coordinate articles on various themes and topics to avoid overlap and fill lacuna. As a result, we often demanded rewriting up to three times. Hence, each chapter of this work is a refereed article of academic standing.

There is, as yet, no established viewpoint from which to write a contemporary history. Hence, it is nearly impossible for editors to impose a homogeneous unity on the problematics, viewpoints and writing styles of the various authors. Thus, the final responsibility for each chapter lies with the author, rather than the editors.

Periodization and Structure

The choice between the topical-module approach and the time sequence approach arises as part of a larger problem inherent to writing history. Historians are perpetually confronted with the decision of whether to adopt a synchronic view by exploring every event and aspect horizontally and arranging them in chronological order, or a diachronic view by vertically following a theme and topic from the specific perspective of each social sector, or to strike some sort of compromise between these two approaches.

We chose to, first, divide the volumes in chronological order as follows:

Volume 1: Occupation Period: 1945–52
Volume 2: Peace Treaty to the end of 1950s: 1952–59
Volume 3: 60s: 1960–1969
Volume 4: 70s: 1970–1979

It was in the 1980s when we initially planned our project to end up with four volumes of Japanese history from WWII to the 1970s. The first four volumes were published in 1995. We then spent another four years preparing Volume Five to present as much of the developments that had occurred during the writing of the first four volumes as possible; i.e., from 1980 to 1995 and, where possible, even later.

Thus, *Volume 5: Internationalization Period: 1980–95* (two volumes) was published in 1999.

Next, we prepared an 'Outline' for each chapter to provide a synchronic overview of the period covered. Each volume, however, is a diachronic presentation of topic-module-like chapters. The construction is modeled after Charles Singer (ed.), *The History of Technology* (five volumes). Of course we often found it difficult to achieve this compromise between synchronic and diachronic; the topics of each module-chapter do not necessarily end concurrently with the periodization of our volumes; for instance, some topics begin in the 1960s but extend into the 1970s. Such a module would be found in Volume 3 where the story begins chronologically.

Whilst we have adopted a four-sector approach it is not our intention to simply divide each topic and problematic into separate analyses from four independent and orthogonal social sectors. Our true intention is to look at any event from at least these four different angles to discover the points of contact and conflict between the various sectors, thereby revealing the dynamic moments of the history.

It is rather a casual analogy to use in a scholarly publication, but I shall use the quadruple rhythms of ancient Chinese poetry – the outset, the sequence, the turning and conclusion – to represent the development of postwar Japanese science and technology. From the Allied Occupation to the 1950s is the 'outset': the high growth of the 1960s is the 'sequence'; the anti-pollution movements and oil shock mark the 'turning' in the 1970s from heavy-thick-long-big business to light-thin-short-small industry. We 'conclude' this particular stage of Japanese history with Japan's 'arrival' as a 'world-class' S&T nation/society.

The 'outset' period lasts from J-day to the end of the 1950s, which has been divided into two volumes; Volume 1 discusses the Occupation Period and Volume 2 explores the development of endogenous policy. There is a turning point within Volume 1 when the Occupation policy shifts from demilitarization to economic reconstruction in the 'American way.' The period covered by Volume 2 is characterized by the revision and demolition of the Occupation Forces' legacy and establishing the direction of Japanese initiatives that would determine the high economic growth of the subsequent period. Thus there were two outsets to the history of postwar Japan; they were of different qualities and had different impacts on the subsequent period. We shall clarify these differences in Volume 2.

The 1960s – the 'sequence' phase of Volume 3 – saw the formation of technocratic structures in Japan, France and the US (amongst others). Toward the end of the 1960s, there was a watershed demarked by the anti-

Vietnam-war movement, campus unrest and other social 'uprisings' that occurred in most of the developed countries and affected a 'turning' in the high economic growth paradigm. This 'turn' is the subject of Volume 4. Volume 5 is the 'conclusion,' characterized by the arrival of Japanese science and technology at the international frontier, and the global demise of the Cold-War system.

Source-Materials

During the course of our Project, all of the Occupation Forces documents preserved at the National Recording Center, Maryland, USA, were microfilmed, sent to Japan and compiled by archivists of the National Diet Library at Tokyo. In editing Volume 1, in particular, access to these documents provided us with a great advantage over the earlier work *Nihon kagaku gijutsu-shi taikei, Tsūshi 5*. These records were supplemented by the microfilmed diplomatic records preserved at Gaimushō, Gaikō Shiryōkan (Diplomatic Sources Archive, the Ministry of Foreign Affairs).

Compared to the US, the Japanese have a poor – in fact, almost no – archivist tradition. Hence, we suffer from a scarcity of available public documents. There is no public archive that preserves 'gray' literature such as internal sources to show the process of policy formulation and the proceedings of various committee meetings. The Japanese National Archive was created for this purpose but they have no authority to command that any administration hand over their records. The National Diet Library contains all of the published White Papers and other government publications. The libraries of the Ministry of Foreign Affairs and the Science Council of Japan were useful for our Project. The Science and Technology Agency and the Ministry of International Trade and Industry (MITI) also preserve some useful sources in their libraries. The Ministry of Education makes no such preservation effort and its library is not open to the public. Hence, most of the 'gray' literature we had access to was 'accidentally' preserved by individual bureaucrats and found in second-hand bookshops.

Sources from the private industrial sector are totally inadequate. Official company histories are readily available but with the exception of some few works that were edited by professional business historians, the quality is not satisfactory. Company secrecy policies prohibit free access to the basic source materials. Even commercial advertisements are not well preserved.

At one stage of our Project, Mr. Satoshi Kōju proposed to construct a database of public relations movies and videos. There is no doubt about the value of these sources or the power of visual media over the written text. But the collection was so voluminous that our efforts remained experimental and far from completion.

Resources from the academic and citizenry sectors are all open to the public. These sectors have tried to create a database while our Project has been in progress. The situation should be improved in the future.

Oral History

Many people who were involved in the historical events recorded here were still alive during our research project, so we attempted to employ 'oral histories' in our research. This method 'creates' historical documents by interviewing participants of events and then formally authorizing transcripts of the interviews with the interviewee's signed agreement to publish. Such an interview is thereby enhanced in status as a publishable primary source. This methodology was, however, not very familiar to either the interviewers or the interviewees. Thus, in most cases, interviews were conducted informally and quoted without formal procedure.

Sources of the Citizenry Sector

The methodology of historical study should not be evaluated for its theoretical consistency *per se* but must rather be appreciated in terms of its capacity to discover a new angle and incorporate new findings. Though the viewpoint of the citizenry may seem irrelevant to S&T activity, approaching S&T from this perspective may bring something fresh into our historical picture.

In practice, however, it is virtually impossible to uncover historical sources and documents of the citizenry that might be directly related to S&T activity. The primary sources are – and will probably remain – government reports, private sector documents written from the industrialist perspective and academic scientific papers. These are all written, collected and published by individuals and institutions that are clearly committed to the promotion of S&T. In a sense, they all amount to the collected advertising of the S&T sectors. Even scientific papers must be considered to be advertising for the academic sector in the broadest sense.

Hence, to develop the citizenry view of S&T from readily available sources, we are dependent upon the sources that exclusively represent the public, private and academic perspectives. For instance, the Science and Technology Agency issues a Science and Technology White Paper each year, which lists all sources and data for each nation-state. If we rely too heavily on them, we get only the public sector's perspective on science and technology. For example, the White Papers list an abundant literature about the technological conflict and competition between Japanese and US corporations; but there are few arguments presented from the perspective of consumers or the citizenry, who's interests must be different from a nation-state's or a corporation's. In fact, the interests of the citizenry are often diametrically opposed to the competition among nation-states, corporations and scientists, e.g., governmental protection of particular industries in international trade wars is frequently disadvantageous for consumers.

Unsuccessful Non-academic Approach

A number of works have appeared in the international arena of the history of science since the 1970s that have tried to reveal something new by adopting a feminine perspective. Paralleling such attempts, we tried to establish the everyday life standpoint of consumers, from which to examine basic science and industry, a reversal of the linear approach from basic to complex reality. We invited contributions from those who are closest to the consumers' perspective such as home economists and interior designers and we organized workshops for planning our project. Regrettably, it was too early to bear fruit.

We also invited non-academic people like science journalists, think-tankers, free-lance video collectors and bureaucrats to join our team, but by the final stage most of them had dropped out and the responsibility for final drafting was left to academics who could do such work for the academic achievement rather than for their livelihood. We also tried out a variety of approaches but the only one consistently adopted was the four-sector approach.

On the Whiggish View of History

In the midst of a revolution, participants often feel that everything that is happening is worth recording. Then, as time passes, some things are lost, some are destroyed and even the things left out undergo a transformation. If, on the one hand, we only collect the surviving documents and evaluate them as historically worthwhile, we will produce a distorted history. On the other hand, if we apply contemporary criteria to historical records and write history accordingly, we are also producing a Whiggish view of history that should be considered useless among professional historians. The history of science and technology, where progress is most prominent, often falls into such a trap.

When we look to our past experience, postwar democracy for my generation, the radical student movement of the late 1960s for the next generation, we find differing assessments.

For instance, the influence of the Minshu Shugi Kagakusha Kyōkai (Association of Democratic Scientists) during the postwar period of democratic activism is now almost totally forgotten: the role that the

Science Council of Japan (JSC) played around the time of its creation is under-appreciated on the basis of its present diminished status. Various ideas that appeared in the late 1960s and early 1970s – e.g., the influence of the Great Cultural Revolution in China and the radical student movement around the world – risk being totally forgotten without proper assessment.

It is the task of a historian to record forgotten events for posterity. In order to develop an historical perspective, it is essential that we attempt to move our viewpoint to the time period under consideration. To reevaluate forgotten events is meaningful in revealing alternative possibilities to younger generations, but it is almost impossible to recover the lost passion of a revolutionary movement.

Conversely, it is both the privilege and a desire of contemporary generations to approach a past event through present-day problematics. It is doubtful that contemporary history can be written in the same way as it would have been, say, ten or twenty years ago. I wrote a contemporary history of the late 1960s and 1970s in *Kagaku to shakai no gendai-shi* to record the decade after the '1968 revolution.'[9] If I were to write it now, though, the style would undoubtedly be different. The former contemporary history was a record of conspicuous events that had appeared on the journalistic surface. It was the kind of history that can be written only contemporaneously with the events – 'in the moment' as it were. If we write it later, we write it from a different historical perspective. Making judgments based on the present knowledge about a trend that, at the time, was promising to develop further is utterly unavoidable after the trend has played itself out or some resolution has been reached. However, if we neglect contemporary judgments, we are then guilty of being too caught-up in the trends of the time.

There is another way of writing history, as Benedetto Croce claimed, to describe the past events in ascending order of time from the present-day problematics. Writing contemporary history is made possible at the point where two perspectives cut across; the perspective from the past point of historical events and another perspective ascending from the present-day problematics. The historical course is selected by people at each point in history. At a historical moment, the point-of-view before and after the selection should be different. Hence in the following, we intentionally distinguish two perspectives: one descending from the past and another ascending from the present. By so doing, we can create a historical dynamism; a conversation of past and present.

Therefore, in addition to approaching our topical problematic from the four sectorial angles, and organizing our material both synchronically and diachronically, we have also attempted to reveal the dynamic moments in history by developing a dialogue between the past and present perspectives.

Notes

1 Second edition published in 1970.
2 Shigeru Nakayama, 'Joseph Needham, an Organic Philosopher,' in S. Nakayama and N. Sivin (eds.), *Chinese Science*, MIT, 1973.
3 Shigeru Nakayama, *Academic and Scientific Traditions in China, Japan and the West*, Tokyo University Press , 1984. Originally published in Japanese in 1974 as *Rekishi to shiteno gakumon* (Intellectual activity as history).
4 Shigeru Nakayama, 'The Future of Research: A Call for A Service Science,' in *Fundamenta Scientiae* vol.2, no.1, 1981, pp. 85–97.
5 This 'return' may be in the form of a 'service' or 'public benefit' – i.e., a contribution to the 'social-' or 'common-good' – rather than a narrowly conceived financial return as required by the industrial-private sectors.
6 Langdon Winner, *Autonomous Technology*, MIT Press, 1977, p. 135.
7 Daiichi Hōki, 1964–70.
8 Shigeru Nakayama and Hitoshi Yoshioka (eds.), Asahi Sensho, 1994.
9 Iwanami, 1981.

Introduction: Occupation Period 1945–'52

Shigeru Nakayama

Prologue

Wartime S&T Activity Evaluated

It is widely accepted that the development of science and technology (S&T) in Japan from the early nineteenth century until World War II (WWII) was unbalanced. What is unbalanced and what natural is arguable but the fact that in comparison to Western development, modern Japan prioritized military science and technology in order to 'catch up' with the Western powers – as quickly as possible and under the pressure of imperialistic competition – is indisputable.

As a result, and in spite of the fact that Japan was far behind the USA in the general production capacity of machineries at the time of the outset of the Pacific War, there were certain technologies in which Japan's achievements – for example, the Zero-Fighter and the construction of large warships (Yamato and Musashi) – surpassed the world frontier.

J-day, 15 August 1945, was the day that the Japanese surrendered to the Allied Powers and accepted their defeat in WWII. This was an historically unprecedented turning point for the Japanese. Every value-system and institution was overturned. Science and technology was no exception.

During the democracy movements immediately following WWII, many argued that the wartime activity of science and technology was terribly distorted by militarism and ultra-nationalist ideology and therefore unable to rationally pursue its quest. Many of the war generation – such as Yoshirō Hoshino, a noted historian of technology – held the institutions of science and technology in low esteem because of what they had witnessed during the war.[1]

This conventional view was challenged by counter-evidence, however. A survey of technologists concluded that the wartime period counted as their 'happiest days.' For the first time in history, the military establishment fully appreciated their expertise.

During the war, scientists and engineers were isolated from the West. New information became scarce. But this shortage of information provided technologists with the best opportunity to demonstrate their abilities. The wartime mobilization system radically altered a tradition that regarded Western science and technology as superior to its own. It is also worth noting that substantial grant-in-aids funding was introduced for the first time since the Manchurian Incident (1931).

The first postwar generation of historians of technology (i.e., the generation following Hoshino), having examined and criticized the war generation's view of the poor performance of the science and technology sector during wartime activity, then overestimated the contribution of the scientists and engineers of the war generation to the high economic growth of the late 1950s and '60s.[2] Similarly, wartime experience and the cultural influence of Nazi Germany provided an enhanced social status for scientists and engineers, a trend that continued into the postwar period.

Of course, different standards and different perspectives produce different evaluations of Japanese S&T, both during and after the war. While, at the time of surrender, the Japanese destroyed most of the documents produced during their wartime activity, it was SCAP (Supreme Commander of the Allied Powers) who thoroughly investigated Japanese wartime S&T activity. Their evaluation was based in a comparison to the US wartime Science Mobilization. By applying the US standard, SCAP determined that Japanese S&T was good in limited fields such as particle physics, textiles and ceramics. Generally speaking, there were many excellent individual researchers but the coordination of research was so poor that they were unable to produce significant outcomes. This evaluation was so firmly accepted that it was incorporated into SCAP's science policy.

International Background

J-day marks the advent of a new epoch in world history. Nazi Germany had already been crushed in May of the same year and it was obvious which side would eventually win the war. But J-day is nevertheless memorable for demarcating the end of World War II.

For S&T, J-day symbolizes the end of the wartime mobilization structure. The primary aim of the prewar system of science was the pursuit of universal truth. During WWII, however, the international scientific community had been divided into opposing political systems: Allied and Axis. Scientists were subordinated to their own political system and involved in the mobilization of research for victory. After J-day academic scientists should, in principle, have been freed from their duty and returned to the prewar system. Industrial scientists also should have returned to peacetime research and development.

J-day, however, also marked the turning point from the Allied-Axis opposition to the US-USSR opposition. With the advent of the Cold War, the scientific community was again involved in militarist strategy. While this new strategic situation was never as urgently pressed as the wartime mobilization had been, it eventually lasted for more than fifty years, with unexpected and unprecedented impact. This has come to be known as the 'military-industrial complex.'

Japanese Situation

J-Day also marks the beginning of the US Occupation of Japan. The Occupation lasted from the end of August 1945 when SCAP marched into Japan, to 1952, when the San Francisco Peace Treaty became effective.

During the Occupation the focus of Japanese S&T shifted from war to the recovery of the economy. In the beginning of the Occupation, SCAP's assignment was to destroy Japanese industry as well as its S&T and demilitarize them. Wartime scientists had been treated as if they were on a special mission for the state. In the postwar period, the entire industrial complex was restructured. Even the big military manufacturers turned to manufacturing kitchen utensils. Aeronautical and military technology engineers had to look elsewhere for employment. Science students who had enjoyed a special exemption from compulsory military service changed their majors to humanities. Engineering graduates were no longer in demand in the labor market. Curiously (because contradictorily) though, desperate students rushed into the basic science disciplines during this period. The generalized economic crisis was even more devastating for the scientific professions than for 'ordinary' people as their once highly valued skills were no longer in demand. This state of affairs lasted for the first half of the Occupation: 1945–48.

In due course, SCAP found that the Japanese were not willing to restore militarism. This realization – combined with a general feeling that the harsh reparations imposed on Germany after World War I had fueled the German militarization that had led to WWII – inspired a rethinking of the Occupation policies towards Japanese industry. From about 1949 SCAP's strategy was to re-construct the Japanese economy as a local Asian base for fighting the Cold War. The suppression of left-wing ideology and activity was an integral component of this strategy.

The isolation from world-news and technological developments that Japan had experienced during the war continued during the Occupation period through the restrictions of the Occupation press code. Throughout the Cold War the Japanese scientific community was not involved in any militarily directed R&D. Even during one of the 'hottest' phases of the Cold War – i.e., the Korean War – Japan's contribution was always supplemental and indirect.

Occupation Policies: Demilitarization and Democratization

The seven years of the Occupation were an unprecedented experience for the Japanese. One thing that was unique was that during this period all scientists and engineers were controlled by SCAP. The experience of the Allied Forces in the German Occupation is strikingly similar to the experience in Japan. There are, however, two major differences.

1 The German Occupation was managed by four major powers, the USA, the USSR, the UK and France. US policies therefore applied only to the US-occupied region.

The US did not want to duplicate this situation in the Occupation of Japan. The American Force brought 400,000 personnel, who *de facto* managed it exclusively. Other powers checked the US's exclusive Occupation policy whenever possible through the Far Eastern Commission (established in December 1945) and tried to pursue their own interests. MacArthur and his GHQ (General Headquarters of SCAP), however, successfully avoided their interference. Consequently, GHQ could enforce the US Occupation policy systematically and uncompromisingly.

The status of S&T policy was rather exceptional and somewhat different to other areas of the economy, since the ESS/ST (Science and Technical Division of the Economic and Scientific Section) was formally headed by an Australian Brigadier. Substantially, however, Harry C. Kelly and his American associates did most of the decision-making.

2 There are also substantial differences between the two nations being occupied. While Germany was devastated by invasion and the Nazi government had been torn down, the Japanese government continued to exist and its function remained intact. Accordingly, while German S&T was completely destroyed and/or confiscated by invading forces, Japanese facilities and equipment were preserved. Furthermore SCAP adopted a policy of controlling Japan indirectly through the existing Japanese bureaucracy. SCAP appointed civilian members along with military members to the Special Staff Section. As a result, many of the ESS/ST staff were civilians.

Indirect Control Policies

SCAP's indirect approach to Occupational governance resulted in an effective *de facto* government with absolute authority. Under these conditions the Occupation Forces could implement social and industrial changes that no domestic government could have achieved. In fact, the US bureaucrats were able to implement idealist policies in Japan that they could never have put in place in their own country.

By the end of the Occupation, however, the idealism of the reform policy was compromised by the realpolitik of the Cold War, and Japan was caught up in America's Far East strategy. The Japanese native reform plan was also under the scrutiny of SCAP. Of course there was also native institutional resistance to the Occupation policies which inevitably ensured that the outcomes of reform strategies never turned out precisely as projected.

In many respects, it seems that SCAP's indirect control was better than direct military control. However, for native reformists, the postwar reform ended up rather unsatisfactorily, since most of the old bureaucracy like the Monbushō (Ministry of Education) continued to function in accordance with their prewar traditions. The Department of Army and Navy was the only government bureau to be totally dismantled. It soon became apparent that the surviving native Japanese institutions had no intention of radical change on the basis of reflections on their past failure.

The Organization of SCAP

On 5 August 1945, five days before the Japanese government accepted the Potsdam Declaration, the Office of Military Administration was established within the US Pacific Army Command in Manila. In Manila, Karl T. Compton and Edward L. Moreland, both high officers of the Office of Scientific Research and Development (OSRD: the wartime science mobilization office), made a blueprint of the Occupation science policy for postwar Japan. When Japan suddenly surrendered sooner than expected they found they had no time to go home. Instead they took a direct route to Japan, investigated the Japanese wartime research mobilization, and then went home. Subsequently, all Japanese S&T activity was placed under the control of SCAP for the duration of the Occupation.

On 30 August 1945, SCAP entered Japan. On 15 September the Economic and Scientific Section (ESS) was created out of the Military Command Section. The formation of the ESS had been ordained already in the 'Early basic directives right after the surrender of Japan by SCAP for the Occupation and Control of Japan' even before the Japanese had surrendered. Among various sections and divisions of SCAP, the ESS was the largest. They were known as the office of economic democratization (which included reforming labor-relations).

It is still unknown why 'Economy' and 'Science' were put together. Within the ESS there were many followers of President Franklin D. Roosevelt's New Deal policy. It is so named perhaps because their ideology might have closely connected economics with scientific rationality.

The 'Science Division' was created within the ESS in early 1946, later changing its name to the 'Science and Technical Division (ESS/ST).' The central tenet of ESS/ST was to implement the economic policy of the ESS.

On 22 September 1945, the Civil Information and Education Section (CIE) was created. Its charge was educational reform to assimilate the American school system into the Japanese system. Both ESS/ST and CIE were involved in the provision of higher education, the ESS/ST in research and CIE on education policy. The policy overlap often led to quarrels and territorial disputes.

Other scientific sections include Public Health and Welfare (PHW) and the Natural Resource Section (NRS). PHW was in charge of hygiene in the Occupied area. They are infamous today for their efforts to improve the hygiene of the postwar Japanese by spraying them with DDT. The absolute power of the NRS was crucial to the achievement of SCAP's land reform policies, arguably SCAP's most important achievement.

The Investigation of Wartime Research and Demilitarization

Between J-day and SCAP's arrival, rumors were rife. But with no experience of surrender, the Japanese had no idea what would happen under the Occupation Forces. Apathy prevailed in the scientific community. There seemed little choice but to adopt a wait and see attitude until SCAP announced their science policy. But those who had engaged in wartime military research prepared themselves for the demolition and prohibition of their work.

SCAP's highest priority was to demilitarize Japan. Weapon production was categorically prohibited. The Bureau of Technology (Gijutsuin), which had been in charge of wartime R&D, was disbanded and SCAP ordered that all military research facilities and equipment was to be preserved until their investigations were completed.

WWII was ultimately a science war, a physics war. American scientists moved quickly after J-day, not only to confiscate facilities for ordinary weapons but also to investigate Japan's wartime R&D on atomic bombs, killer-rays, poison gas and balloon bombs. They followed the precedent of the Alsos investigations of Nazi German S&T (cf. ch. 1). As previously mentioned, though, the US's absolute authority as the sole occupier of Japan removed the competitive tension that existed between the US and the USSR for confiscating German S&T resources. Unlike Germany, the Japanese administrative organization and universities remained intact. Hence the investigation was easier.

While the military personnel of SCAP were facing the enemy of yesterday, and focusing on issues such as the possibility of hidden military research facilities, resistance and counterattack, in the academic sector a cosmopolitan value of science prevailed. The scientists of the investigation team were open to informal academic communication with Japanese scientists. The scientific investigation team immediately recognized that the Japanese scientists were not dangerous; they were not willing to continue the war. Furthermore, the Americans evaluated the Japanese wartime mobilization of S&T less highly than they had the Germans.

Amongst the various investigation teams in ESS/ST was the Atomic Bomb Casualty Survey Team. Their report on this historic experiment was not only highly valuable to the scientific community, it was also of the highest priority for the US government who wished to control the postwar world through a Pax Americana ensured by the atomic bombs.

The Japanese scientific community was asked to cooperate with the US scientists and physicians in the Survey Team. In July 1947, the US Atomic Bomb Casualty Commission (ABCC) was established to measure the long-time effects of radiation hazards. While observing the effects of radiation on the victims, their objective was not to care for the victims, but rather, to acquire scientific data. Needless to say there is a lingering dissatisfaction amongst the subjects of their surveillance.

Ethically, the scientists argued that they first had to collect accurate objective information; only afterward could they develop effective treatments. Politically, however, under US strategic policy, all research findings were classified (cf. ch. 2).

The Occupation press code maintained strict media control on information pertaining to atomic bombs. Hence, there was no place for bomb-victims to express their complaints and any nascent anti-nuclear movement was suppressed.

Prohibition of Research: Nuclear and Aeronautical

One of SCAP's highest priorities in the demilitarization of Japan was the prohibition of military-oriented research. This prohibition was not limited only to weapons-related research but extended categorically to most nuclear and aeronautical research. In spite of petitions by the Japanese scientific community, the prohibition lasted to the end of the Occupation period. In some areas, though, such as radar and microwave research, the prohibition (which was never anticipated by the Japanese) had been relaxed as early as 1946.

The prohibition of nuclear and aeronautical research during the Occupation inevitably led to Japan 'falling behind' in comparison to the major powers of the Cold War. But it also forced the Japanese S&T community to focus their efforts on areas where they could compete: the commercialization of technology for the market. Hence, the demilitarization of Japan is central to the formation of the paradigm of economically oriented industrialized science.

Kelly's Arrival: the Conversion from Demilitarization to Democratization and Economic Recovery

Harry Kelly is pivotal to the shift in Occupation science policy from demilitarization to democratization, and then to the formation of economically oriented industrialized science. Arriving in January 1946 in response to a telegram requesting expert advice about cyclotrons, Kelly was to become a key figure in the formation of SCAP's S&T policy.

He was initially assigned to investigating military research facilities with the aim of enforcing the demilitarization policies of the Occupation. But through his contact with the scientific community, Kelly quickly concluded there was no need to fear any militarist intent. So he assigned himself the job of encouraging the S&T community to focus on getting out of the postwar economic misery as early as the spring of 1946. But in the process he tried to radically restructure the scientific community.

Demilitarization was, of course, not limited to the scientific community. On a broader political scale the Occupation Forces set out to fundamentally eradicate the ideological basis for militarism. Aiming to destroy the existing hierarchy, they appropriated Marxist formulae and disbanded the Zaibatsu (financial/industrial concerns controlled by powerful families).

Kelly adopted the same approach to democratizing the scientific community. His immediate target was to dissolve the three 'old-school' organizations: Nihon Gakushiin (Japan Academy), Gakujutsu Kenkyū Kaigi (the National Research Council) and Gakujutsu Shinkōkai (the Japan Society for the Promotion of Science: JSPS). His ambition was to create a new democratic organization of scientists, the likes of which had never existed anywhere in the world. It was to be a parliament of scientists where members were elected from the entire body of researchers and scientific matters would be managed by the consensus of all scientists.

He began by approaching the Japanese physics and basic science community where he found like-minded friends. They formed a kernel of democratization within the Japanese scientific community. After overcoming many complications, Kelly's effort was finally rewarded by the creation of the Nihon Gakujutsu Kaigi (the Science Council of Japan: JSC) in January 1949.

Of course, the JSC did not turn out precisely as Kelly and his associates at ESS/ST had envisaged it. For example, Kelly wanted to exclude the old establishment of the Ministry of Education and the Imperial Universities, especially Tokyo Imperial. But in the process of transforming his plan into an institution, he was obliged to compromise with them. Differences from Kelly's original intention include the following:

1 While Kelly modeled the JSC on the American Academy of Science which was comprised of natural scientists, the JSC also included social scientists and Minshushugi Kagakusha Kyōkai (Minka: the Association of Democratic Scientists; a gathering of Marxist scientists).
2 The JSC was dominated by academic scientists and, therefore, was not the right organization to promote the kinds of R&D necessary for an economic recovery.

Kelly's aim was to create a democratic organization of scientists and engineers and contribute to the economic recovery. In practice, though, democratization occurred only in the academic sector (which, here, includes the JSC) while economic recovery was affected by the technocratic regimes of the bureaucracy and the industrial sector. Hence, rather than creating a cohesive and consensual S&T community governed by a united Council, the JSC, elected from the total association of researchers, found itself in conflict with both the Kagaku Gijutsu Kaigi (the Science and Technology Council) appointed by the government and Science and Technology Agency (STA).

Turnover of Occupation Policy

The radical reform program implemented by GHQ in the early stages of the Occupation underwent a series of changes. Significant amongst these was the reparations policy. In December 1945, while fending off Soviet demands for reparations, GHQ established a Reparations Mission headed by Edwin Pauley. GHQ was less certain about how to extract reparations than they had been about the establishment of the ESS/ST and other policies such as land reform, disbanding the Zaibatsu and encouraging the creation of labor unions. The experience of the harsh reparations extracted from Germany following the First World War (WWI) made them think hard about the best policy for Japan.

Again, the focus of the reparations mission was not limited to weapons-production facilities, but extended to all industry, including steel, chemical, machinery, shipbuilding and light metal industries, power stations and railways. Pauley's mission investigated these industries and drew up an interim policy for making reparations to the Asian countries damaged during the war. The policy was formed around the idea that the defeated Japanese should not enjoy a better standard of living than their victims. This was an obvious threat to the Japanese economy.

But this policy proved to be short-lived. In practice, the Pauley Mission began by red-tagging weapons-related industrial equipment for reparations. Although, the industrial facilities being dismantled for reparations were deemed to be too outdated to be of direct value for the

Americans, from the perspective of China, the Philippines, other Asian countries and Australia, the Japanese wartime facilities had sufficient value to demand as reparations. Army arsenals, naval dockyards and all airplane factories were designated. In addition, perhaps as much as half of all machine tools, shipbuilding facilities, power stations, sulfuric acid, and sodium manufacturing facilities were appropriated as reparations. Thus, the reparations programs had far more devastating short-term effects for Japanese industry than did the prohibitions on S&T research. It was also far more threatening to any chances of an economic recovery.

Surprised by this unexpectedly harsh treatment, the Japanese industrial sector petitioned GHQ for leniency and was rewarded with a slightly mitigated Interim Reparation Plan being adopted by the Pauley Mission. In practice, the focus of GHQ soon turned more towards the economic recovery of Japan, a recovery that demanded the resurrection of the Japanese industrial sector. In 1947 the Strike Mission arrived in Japan and soon cancelled all reparations except for purely military installations. It is undeniable that these policy changes were crucial to the recovery of Japanese industry and the development of S&T (cf. ch. 14).

Japanese Reaction

Economic Difficulties

To understand these policy shifts, changes to institutions and the urgency with which the state of the economy was viewed we must locate them in the context of the human scale of demilitarization and reparations. We have already noted that demilitarization was GHQ's highest priority. In the process of demilitarization, the first priority was to disarm and disband four million domestic and two and half million overseas Japanese military personnel, confiscate several thousand aircraft and several million weapons, destroy arsenals and then maintain control of the confiscated equipment. The Japanese government was assigned the task of maintaining railway and ship transportation, telecommunications and electric power for SCAP. Other industrial facilities were left untouched.

In defeat, Japan had to support a population of 72 million, including 6.5 million returned servicemen. During the final stages of the War, the government's controlling power weakened; resources were mal-distributed and black marketing was widespread. After the War, the material scarcity was even more disastrous than before.

In December 1945, it was estimated that there were 13.24 million people unemployed due to the demobilization of the military and the contraction of industry. Unemployment in the educated class reached 14,000 due to the curtailment of executive jobs in government and industry.

Engineers lost jobs due to the closure of military arsenals. Many jobless and returnee university students tried to survive by working in the black market. The situation was at its worst in the early part of 1946.

No exceptions were made for scientists or engineers, who found themselves in the miserable reality of postwar Japan with no protection, no employment, and no privileges. Those whose homes and facilities had been burned by the US air raids and the returnees from former colonies and occupied lands had no place to stay. They often slept in their laboratories. Normal research life was not possible.

Recovery of the Freedom of Expression

Those scientists who were formerly assigned work in the government and industrial sectors had nothing to do during the transition period immediately following the war. Most private laboratories were closed. Military laboratories were all disbanded. Other government laboratories had no research focus or direction during the transition from war to peace. The academic sector was less affected, as they were accustomed to selecting their own research topics freely. Hence, their morale was higher.

In spite of material hardships, the academic and citizenry sectors had recovered their freedom of expression. In spite of the shortage of paper for printing, scientific magazine publication experienced an unprecedented boom (cf. ch. 27). Academics influenced public opinion. Physicists, in particular, were welcomed in the media as the people most qualified to predict the future of the atomic age. The resurrection of Marxist ideas in the S&T sectors is particularly noteworthy (cf. ch. 28), having been suppressed during the war. It was widely accepted by Japanese society that the reconstruction of Japanese society could only be achieved through science and technology.

Visions of Postwar Reconstruction

The Japanese value system underwent radical changes both during and after WWII. Such change predictably generated new conceptions of Japanese society, such as the Cultural Nation, Democratic Nation, Scientific Nation and Marxist notion of a revolutionary society. In spite of economic difficulties, the people wanted a fresh vision. Debates were happening everywhere.

Amongst these diverse conceptions it was generally agreed that science was a value worth pursuing. The spokesmen of science proclaimed, 'We lost the war because of inadequate science. Hence, we should promote science from now on...' While this strongly resembles a nationalist plea for strengthening the country for war, they were in fact trying to promote S&T for the economic recovery of a peaceful society.

A food shortage was the most urgent problem to be resolved. Proposed solutions were heatedly disputed. While some advocated the creation of an agrarian society on the Danish model, most people thought this impossible for an industrialized society like Japan. A more widely accepted view was expressed by Kiyoshi Tomizuka who claimed, 'heavy industry was destroyed by the War. We should abandon the approach of "strengthening the nation by strengthening armament". We should instead restructure Japan for fine industry, like Swiss watch manufacturing.' Weapons manufacturers restructured themselves as sewing machine manufacturers and began exporting in 1949. During the Korean War, the Japanese camera, 'Nikon' was highly appreciated by American photojournalists. It found international markets, eventually superseding West Germany in units of production. By the 1950s, sewing machine, camera and watch industries were flourishing. In the 1960s Japan became the world's second largest watch manufacturer after Switzerland (cf. vol. 2, ch. 31).

These are very significant accomplishments considering the almost total destruction of Japanese industry in 1945. Although the reparations policy was short-lived and never fully implemented during its existence, Japanese manufacturing hardware had already been severely damaged by US air raids during the war. The channels – both official and unofficial – for gathering information (software) were not functioning properly because of the wartime seclusion (cf. ch. 12). Massive unemployment, food shortages, disbanding the Zaibatsu and prohibitions on various areas of S&T research generated what can be considered as a 'scrap heap' economy from which to build a new one.

In March 1946, Saburō Ōkita drafted 'Basic Problems of the Economic Recovery (Nihon keizai hukkō no kihon mondai),' in which the technocrats' role in planning was fully developed and the importance of technology was emphasized. It was at this time that the policy of 'economic recovery by means of science and technology' was adopted by the ESS/ST.

Priority Production

The Economic Stabilization Board (Keizai Antei Honbu; Anpon) was created in August 1946 to address the economic crisis. Though criticized as a return to a wartime-controlled economy, Anpon marked the beginning of technocratic long-term planning for economic development. On 27 December 1946, economists working at Anpon drew up a concrete blue print for postwar economic recovery and named it priority production (keisha seisan).

Anpon decided that the first priority for economic recovery was the production of coal and steel. By investing the limited resources available primarily in coal and steel production, coal and steel would stimulate demand for each other, cyclically increasing both productions. Precious imported resources such as crude oil were allocated to the steel industry. The steel thus produced was then put into coal production, and the coal produced, in turn, was invested back into the steel industry. Price-difference subsidies to the steel industry, electric power, fertilizer production and other strategic industries were also employed to stimulate production.

These measures were an attempt to introduce a planned economy, rather than a wartime controlled economy, in the postwar confusion. It was not quite a socialist policy but it was nevertheless welcomed by the Socialist government that took office on 1 June 1947 and played an important role in the process of economic recovery in 1947–48. It is important to note that this was endogenous planning, independent of GHQ.

The planning of priority production took a long-term perspective starting with basic industry to establish a foundation from which later to reach the consumer market. This approach tended to protect basic industry with price difference subsidies. Governmental protection and interference were all abolished after Joseph M. Dodge, a special envoy sent to Japan in early 1949, criticized the priority production measures and denied access to the subsidies designated for the heavy and chemical industry in the recovery measures planned by the Economic Recovery Committee in 1947.[3] Japanese industry was thus forced to find a way to survive in a free market economy.

The End of Wartime Seclusion and the Adoption of a US Model

Before the War, there were few chances for researchers in the private sector to study overseas. However, university scientists – who were all public servants – were sent abroad by the Ministry of Education. Most of them headed for Germany. Thus universities played a leading role in providing prewar and wartime scientific information from abroad. S&T in both the public and private sectors had been in a long slump while free individual basic research in the academic sector was less affected by postwar restructuring. Their efforts began to fill the lacuna of scientific information created by the wartime seclusion.

During WWII, some important S&T information was transmitted to Japan by submarines from Nazi Germany but most scientific journals were not available. At the time of surrender in 1945, the only Western S&T information the Japanese had access to was from the 1930s. The Western patent information Japan had had been confiscated at the outbreak of the Pacific War by a special action based on industrial property law.

During the Occupation, GHQ ordered the abandonment and disclosure of the secret patents that the Japanese military possessed (cf. ch. 19). The enforcement of foreign patents denied Japanese engineers access to information that they had freely used during the War. In contrast, pirate publications of foreign books and translations became available to a research community hungry for the latest information (cf. ch. 27). The Occupation Forces frequently warned Japanese publishers for violating copyrights.

In the early part of the Occupation the Japanese were not permitted to travel overseas. Scientists had to make great efforts to overcome barriers of every sort. They managed to obtain cutting edge S&T information through all sorts of channels (cf. ch. 12). For example, the information of a transistor was transmitted informally by contact with Occupation scientists providing the genesis of the Japanese study of transistors (cf. vol. 2, ch. 34). The most up-to-date S&T information was hand copied during frequent visits to the CIE libraries. PB Reports, the database of wartime technological information, were also accessed (cf. ch. 12).

Only in the latter half of the Occupation were Japanese scientists permitted to travel overseas, and even then it was to a limited extent (cf. ch. 13). The first Japanese scientist to visit oversea was Hitoshi Kihara in 1947. Hideki Yukawa followed in the same year with the assistance (political rather than financial) of Robert Oppenheimer, then the director of the Princeton Institute of Advanced Studies, who put pressure on GHQ to get Yukawa's visa. An Occupation Forces project was established to invite Japanese leaders to the US. There were a number of scientists of various disciplines who joined this program. In 1950, an official channel for studying abroad was created with the GARIOA Fund. The Fulbright Program later superseded it.

Gradually the Japanese scientific community returned to international society. In the meantime, the US university model replaced the German model. This shift is most apparent in medical education (cf. ch. 16). Under the guidance of PHW, an intern system and nation-wide examination system was introduced, but the old Ikyoku (literally: medical bureau/office) system survived intact. Most prewar Japanese universities had employed the German language for medical terms. It was well after the end of the Occupation before English superseded it.

CIE's Educational Reforms

In higher education, the wartime demand had produced an expanded cohort of science students. The change back to the study of humanities reversed the wartime ratio of humanities: science = 4:6 and returned to a more normal ratio of 6:4.

The four principal slogans of the CIE were elements from US models: 'Democratization,' 'Equity,' 'Decentralization' and 'Lay Control.' In 1947–48 the CIE proposed to delegate supervisory power of national universities to local prefectural educational commissions similar to the system of the state universities in the US. Thus, 'Decentralization' was intended to transfer power from the Ministry of Education to the local education commissions. The Japanese central bureaucracy resisted this threat to their power, arguing that local governments were not accustomed to handling matters of higher education and that they lacked appropriate resources. 'Lay Control' was also presented by the CIE in 1950 with the introduction of the US-style 'Board of Trustees' as the governing body of Japanese national universities. The university authorities, afraid of the loss of academic freedom, resisted both bills, while student activism opposed the capitalist control of academia. The opposition was so strong that even GHQ, with its almost absolute power, could not resist it. Thus the bills were eventually repealed.

The US-style governance of national universities was modeled after business practices while the traditional Japanese national universities had developed as a part of the government bureaucracy. While the US system was run according to a top-down technocratic efficiency principle, the Japanese universities ran on the principle of professorial democracy – i.e., self-governance by faculty vote – following the German university system. There were also different interpretations of democracy by GHQ and the Japanese faculty members. Democracy for GHQ was merely an ideology against authoritarian nationalism. In the Japanese practice of faculty democracy, though, all faculty members participate in the election of the president and deans. It is a formalistic democracy of the administration system. Most faculty members find this democratic system highly satisfying, but the inefficiencies of such a system are also apparent. One significant inefficiency is found in the conservatism which silences all progressive and minority voices and stifles innovation.[4]

CIE ordered the Japanese government to establish a national university in each prefecture along the lines of the US state university system. It was financially difficult to establish so many new universities in the postwar economy. The history of universities has traditionally consisted of the development of institutions from lower grade college or professional schools to university status. Every educational institution's final target was historically geared to attaining university status. Before the war, the Japanese had established a new university approximately every ten years, usually in commemoration of a war victory or something similar. Each new university was established with its own vision, which guided the selection of lecture chairs, as well as the creation of administrative and research systems that had not previously existed.

In contrast, the Occupation was an era of cheap and rapid university creation. Normal schools, professional schools and private schools all scrambled for the privilege of university status. The postwar boom in the creation of new institutions meant that the schools were being built with no institutional innovation, as replicas of existing successful institutions, but without the research funding and facilities required to maintain the standards of the old Imperial universities.

CIE created the University Accreditation Association (Daigaku Kijun Kyōkai) in 1947 to screen applications for university status. In the beginning a number of applications were rejected; but the CIE decree inevitably created an unprecedented expansion of the Japanese higher education system. Within a few years, universities, as well as their teachers and students, numbered more than ten times the prewar figures.

When compared with the prewar standards of universities, the quality of the new universities was undoubtedly lower. However, its 'cheap-ness' was beneficial in the sense that higher education came within the reach of the common citizen. It changed the Japanese view of higher education from elitist to egalitarian. Arguably, this rapid expansion in the university sector provided mass higher education to the Japanese populace rather preemptively. By following American precedents, Japanese universities expanded much more rapidly than those in the European countries. In fact, according to a comparative study conducted by the OECD, *Towards Mass Higher Education* (1974), Japan is the only country with a radical increase in the number of university students throughout the 1950s.

Naturally, the established academic community based in the Imperial universities expressed fears about lowering the standards of university education. They went to great lengths to differentiate themselves from the newly created universities and thereby defend their privileges. But the Occupation Forces, who preferred democratization and equity of higher education, parried their attempts. Following the San Francisco Peace Treaty, when the education system regained its independence, the old universities recovered their privileged position to some extent. But the overall effect of the massive expansion of the education system undoubtedly weakened academic science research, at least temporarily.

Certainly, though, the expansion of the higher education system could not have happened if the populace had not been so eager to provide advanced education for their children. The Occupation Forces' policy only influenced the direction of parents' efforts. The development of mass higher education provided the expanded manpower demanded by the S&T sector in the ensuing decades of high economic growth. Whereas the traditional university system was geared to producing an elite few highly specialized scientists and researchers, the new system mass-produced middle-class engineers who could support the high growth in industrial technology in the latter half of the twentieth century.

New Graduate Schools

The Education Mission, established by the CIE, followed the policies of demilitarization, democratization and general education for the citizenry. With the exception of demilitarization policy, the Education Mission had little concern about S&T. However, the Science Mission, established by ESS/ST, was very concerned about S&T research. The concerns of the CIE and ESS/ST overlapped with regard to graduate education. This was a source of much conflict. To begin with, the Science Mission's recommendation for graduate schools was strongly resisted by the CIE (cf. ch. 8, 10 and vol.2 ch. 15).

Nevertheless, SCAP's efforts to reform the education sector in accordance with the US model of education had a dramatic impact on the Japanese education community. The reform reached step by step from the lowest grades of grammar school right through to middle and higher education. By the time the reform program had reached graduate school education, though, the San Francisco Peace Treaty was concluded and the Occupation Forces quit Japan. Hence, the reform of graduate school education remained incomplete.

Although the University Accreditation Association had issued the requirements for graduate school accreditation in 1949 it was effective only after the Peace Treaty in 1953 when a new system of graduate schools in national universities had their first intake of science-majors.

Postwar Democracy and S&T

Left-Wing Background of Postwar Democracy

Looking back at Japanese history from the 1990s, there is a certain Whiggish perspective from which the 'miracle of Japan' appears as a straight linear trajectory from postwar economic crisis to the leading economic power in the world by technocratically employing capitalist economic principles.

Those who maintain this view, however, are either ignorant of, or choose to ignore the fact that leftist thought dominated Japanese thinking immediately following the War. This seems to be completely forgotten by the post-Cold War generation of Japanese. This leftist orientation is apparent in the ideology of the Minshushugi Kagakusha Kyōkai (Minka: the Association of Democratic Scientists), the professional association of average Japanese scientists and engineers.[5]

Many scientists regarded science from an entirely different perspective than the technocratic way of thinking typical of both the public and private sectors. Some were actively engaged as opinion-leaders for the academic and citizenry sectors. During the early postwar period, especially during the Occupation, scientific activity in both the public and private sectors was rather inconspicuous. In contrast, leftist or progressive men of culture were dominating the media.

Intellectuals and students supported left-wing political parties at election to a much higher degree than the general population. Many believed the creation of a socialistic society was an historical necessity, though they disagreed over the timetable for its materialization. Some believed that the social model of S&T differs between capitalist and socialist societies. Some even advocated a strategy for a radical communist revolution rather than incremental reform of the existing system. Perhaps many general intellectuals, students and scientists thought that what the Communist Party demanded was an ideal while the strategy of the Socialist Party was more realistic. Regardless of the reasoning and beliefs, for politically minded scientists and engineers, the conservative position on S&T was unclear while the Communist Party's Science and Technology Department provided a clearly articulated and therefore more attractive science policy.[6]

Among general intellectuals, the Japanese Communist Party was highly regarded, as they had fought uncompromisingly to stop Japanese militarism before the war, and, indeed, many of them had been imprisoned during the war.

Significantly, these leftist leanings were supported and fostered by the Occupation Forces. There were a number of 'New Dealers' in GHQ who were eager to implement more progressive social policies than they could achieve at home. They supported the enactment of a Trade Union Law that enabled the rapid expansion of the labor movement in the Japanese workforce. Lifetime employment and the seniority wage systems are among the achievements of this strong labor movement. It certainly encouraged democratization in the work place. Unions also were leading the democratization of universities and research institutes, where collective bargaining was in practice.

Science and Democracy as Liberators

Coincidentally, while there were GHQ officers sympathetic to the ambitions of left-wing politics, the Japanese Communist Party was also sympathetic to the Occupation Forces. The Kuninomiya Cabinet created during the transition period did not free imprisoned communists, not even after the end of the War. It was the Occupation Forces who released them, as late as 10 October 1945, when the Kuninomiya Cabinet was disbanded.[7]

Hence, the Japanese communists in the early postwar days regarded the US Army as a liberation army. At the convention of the Japanese Communist Party from 1–3, December 1945, its leader, Kyūichi Tokuda, while advocating the overthrow of the Imperial System, stated that 'GHQ is not our enemy. They are powerful supporters of democratic revolution. They are our liberators.'

For most Japanese scientists and engineers, the postwar period was a time of liberation from wartime militarism. As a part of the intellectual class they were willing to accept democracy right away. Their prewar higher education had familiarized them with Western culture and thought, and democracy seemed much easier to deal with than militarism. Hence, GHQ found ready support for their democratization programs. There was no need for sermons; they would have been preaching to the converted.

Happy Marriage of Science and Democracy: Historical and International Background

Some may interpret the fact that both science and democracy originated in classical Athenian culture, and share a uniquely Greek debating tradition as suggesting a common-heritage and common tradition for them. Yet, evidence indicates that the two traditions have developed independently throughout history. Science is an elitist culture, somewhat incompatible with the bottom-up thinking of democracy. Already in Ancient Greece, Hellenist science can be seen as separated from the democratic culture as it seeks sponsorship amongst the powerful elite. Similarly, modern European science was rooted in academies and sought the patronage of rulers and the aristocracy. This association with patronage, aristocracy and monarchy led the early nineteenth century US Congress to neglect science altogether.

In the East, too, science was supported and sponsored by an absolute monarchy and its bureaucracy. They had little interest in passing on the fruits of scientific knowledge to the ordinary people. It was only in the twentieth century, when 'democracy and science' was promoted as a symbol of modernization and a political slogan in the Chinese May Fourth movement in 1919, that science came to be associated with the progress and emancipation of ordinary people. In this new context, science and democracy were viewed synonymously and became a symbol of opposition against the Confucian tradition of their common enemy.

It is unlikely that the Japanese scientists after World War II were especially conscious of the Chinese precedent; yet, one common feature is evident. Japanese intellectuals and scientists saw science and democracy as a symbol of modernization against a feudal regime, against the common enemies of militarism and ultra-nationalism.

Laboratory Democracy

For the average rank-and-file scientists, democracy meant the eradication of a feudal environment in the laboratory. Within the Japanese scientific community, feudalism was interpreted as an ethical code of oppressive, paternalistic professors who did not respect the democratic claims of junior academics. Primarily the younger members of the scientific community who sought to overturn the feudal system promoted laboratory democracy. One of their tactics was the ostracism of ineffective and rightwing professors. Minka was one of the left-wing organizations created in this environment (cf. ch. 4–2).

In January 1946 when Minka first formed, the participants were not conscious of any conflict between science and democracy. Most active members were Marxist, while practicing liberal scientists and engineers had all participated in the discussion of reforms to the research system. Minka was a purely volunteer organization consisting of a transdisciplinary forum for discussing a common reform strategy. At the end of 1949, the peak of its activity, Minka had approximately 10,000 members across 110 local branch organizations.

There were democracy movements in a number of scientific laboratories right after the War. Interpretations of democracy vary. In the basic science institutions of postwar Japan, democracy meant the democratization of laboratories and opposition to the paternalistic chair system. The first formalized democracy movement in a scientific laboratory was at the Research Council (Kenkyū Kaigi) of the Toshiba Electronic Works (Tōshiba Denshi Kōgyōsho). Started in November 1945, the movement proposed replacing the old top-down system with a laboratory managed through the participation of all the researchers. Toshiba's Research Council soon became a model for other laboratories to follow and exerted considerable influence.

Another influential institutional paradigm was the Laboratory Constitution (Kyōshitsu Kenshō) initiated by Shōichi Sakata, a distinguished physicist at Nagoya Imperial University. Sakata established a Laboratory Council (Kyōshitsu Kaigi) system in June 1946, following the principles outlined in J. D. Bernal's *Social Function of Science*. The aim was to overthrow the hierarchical chair system and to establish departmental management (including recruiting and appointing new members) via a decision-making mechanism based on the egalitarian participation of all members of staff (i.e., professors as well as assistants and graduate students having an equal right to vote) (cf. ch. 28).

This system was welcomed and followed nationwide by those who sought a new organization for research. Support was especially strong amongst young researchers in the physical sciences. In the late 1960s when campus rioting spread all over, the demands for laboratory democracy resumed in medicine and agricultural studies, areas where democratization had not happened earlier.

It is important to recognize that these movements were limited to the scientific community and failed to formulate any links with citizens' movements or the wider populace. They did, however, maintain close contact with the labor-union movements. Scientists and engineers often participated in the union organization within a corporation, sometimes playing leading roles as happened at Toshiba. At the time of the establishment of the Science Council of Japan, Minka's platform was adopted with the participation of private sector scientists and engineers. In geology, the Association for Geological Collaboration Japan (Chigaku Dantai Kenkyūkai, abbreviated as Chidanken) was organized integrating teachers and students of primary and secondary education under its democratic principle (cf. ch. 26). However, for social scientists, democracy means parliamentary democracy. To them science means a particular rationalism opposed to irrational wartime ultra-nationalism; or it means 'scientific socialism' as coined by Friedrich Engels, rather than research activity. The social scientists were less interested in the democratization of the university system.

Laboratory democracy was much more thoroughly established in Japan than in the USA. The unionization of universities and research laboratories is widespread in Japan; salary scale is determined by age, regardless of one's ability. This is the achievement of the postwar union movement, seeking salaries for survival rather than as meritocratic rewards. The criticism against the democratic atmosphere of Japanese egalitarianism includes arguments to the effect that it is not conducive to the competitive and elitist culture of science. There is an egalitarian intellectual and institutional environment in which cooperative research of individual participation is made possible by Japanese laboratory democracy, but big science projects requiring a hierarchical structure are difficult to complete.

Conflict between Science and Democracy

In the intellectual milieu at the time that Minka was established, an inherent conflict between science and democracy was not recognized. The JSC was the first in the world to attempt a 'parliament of scholars' consisting of elected members from the scientific community. Beginning with different expectations from the various sectors of the community, their voting was nevertheless effective in overthrowing the old regime. But as time passed, conflicts appeared between science and democracy.

First of all, academic science seems to require a certain elitism that inherently conflicts with the principles of democracy. Evidence of this can be found in considering the fact that those who had been elected to the JSC were not necessarily outstanding scientists. In other words, those scientists most adept at marshalling votes in an election are not necessarily those who excel in basic research skills.

As the Cold War deepened, the Japanese Communist Party and the cadres of Minka began political action within the JSC that divided the scientific community. If the JSC were purely a political organization, according to the rules of parliamentary democracy it would be legitimate to oppose it by founding another party. In the scientific community however, there is certain apathy towards politics. Instead of actively opposing politicized members, most of the scientists and engineers simply lost interest in the JSC. But this meant that politicized party members, even though constituting far less than a majority, could make strategic political decisions with a limited number of organized votes.

The top-down directive system of the Japanese Communist Party drew accusations of political maneuvering to deny scholars the liberty to speak freely. Liberal scientists were repulsive to the Minka cadres, who deemed it most efficient to follow the Party directives despite the denial of bottom-up democratic processes. But this un-democratic position, by the 1960s, led to the diminution of Minka's influence. When the Japan Scientists' Conference (Nihon Kagakusha Kaigi), considered to be the successor of Minka, was established as a mouthpiece for the anti-Vietnam War campaign by the Japanese Communist Party, the Conference was careful not to be confused with Minka, who had become too unpopular within the scientific community.

Table B: Contrast between academic/citizenry and public/private sectors

Academic and citizenry sectors	Public and private sectors
postwar democracy of science	technocracy
breakdown of old feudalistic regime	top-down directives
laboratory democracy	scientific planning
laboratory conference	corporate organization
J. D. Bernal's *Social Function of Science*	US business management
voice of young people	voice of leaders
egalitarianism	meritocracy
anti-militarism	wartime was best time; tolerance of militarism
inclination to socialism	indifference to ideology
Science Council of Japan	Science and Technology Council

Democracy vs. Technocracy in Science

Although academic science is basically individualistic activity, it can be conducted democratically. Project-oriented research, however, cannot. Especially in militarily oriented research such as the Manhattan Project, secrecy and security are high priorities and a top-down directive management style is generally considered to be the best suited and most efficient for achieving specifically defined purposes. Postwar science projects were primarily modeled on the Manhattan Project. There was a predisposition towards big (budget) science under public sector initiatives. In fact, it became normal to request funding for R&D budgets from the government. Science and technology effectively became a national enterprise. Academic freedom, such as the freedom to select one's own research topic, was weakened as organizations and facilities became larger. Parliamentary-type democracy is incapable of managing such enterprises. Direct democracy – as practiced by the JSC – does not function efficiently. Thus the management of governments and corporations is essentially top-down – technocratic not democratic.

Technocracy had a positive image in prewar Japan, supported by a widespread belief in the capacity of scientists and engineers to rationally design and manage society. But, technocracy is the antinomy of democracy. It was not until the 1970s that technocracy and democracy came into conflict, but the origin of the conflict is to be found immediately after WWII. Proponents of democracy in the academic and citizens sectors of Minka, journalism, labor unions and laboratory conferences were all diametrically opposed to the establishment of technocratic governance in the public and private sectors; particularly in the Science and Technology Agency, MITI, the Science and Technology Council, the Federation of Economic Organizations (Keidanren) and big corporations. Table B highlightings the contrasts between the two sides in this debate.

From Democracy to Technocracy

What we define here as technocracy is the system that employs technical bureaucrats to design and implement rationalized plans for industry, science and society in a top-down fashion. In this sense, the New Dealers were technocratic, Harry Kelly included. The way they conducted themselves was somewhat democratic but not quite liberal. They have been criticized for having too much confidence in social planning.[8] There was conflict between the New Dealers and proponents of a more liberal free-market type of economic system. Eventually a compromise was struck between the New Dealers and the conservatives. At the end of 1948, GHQ introduced its Economic Stabilization Program, to liberalize industry from the planning of priority production and encourage instead market-oriented free enterprise.

In the official history of the ESS/ST, *Reorganization of Science and Technology in Japan* (1952), democracy is proudly presented as one of the ESS/ST's central achievements. The primary narrative of this history describes the role of the organization in transforming the JSC from a nascent liaison-group into a fully democratic organization. When the book was published, it was widely accepted that this was their greatest achievement.

But, in fact, ESS/ST did many things besides promoting democracy. When considering the later development of Japanese S&T and its high economic growth, the ESS/ST's role in establishing technocratic management systems is crucial, although this was not mentioned in their official history. Kelly and his successors (Bowen Dees and others) played major roles in this particular 'achievement.'

From the Japanese perspective, Kelly has long been associated with the establishment of the JSC. The importance of his role here should not be underestimated. As previously mentioned, he had many friends among Japanese physicists and basic scientists, friends who were quite helpful in organizing the JSC. And as we have seen the Council was composed of university and academic, rather than industrial and governmental scientists.

However, Harry Kelly's doctorate was in physics and an average American PhD of physics is generally closer to applied scientists and engineers than are Japanese physicists. While Kelly had great sympathy for both physicists and academic scientists, being a son of a poor coal miner he, from the outset, approached science policy with a firm belief that science should be used for economic development. This position, as we have noted, was quite compatible with ESS policy.

Therefore, Kelly's interest in S&T was closer to MITI (created in May 1949 through a reorganization of the Ministry of Commerce), than to the Ministry of Education (Monbushō). Anticipating that the JSC would prove to be too academic to conduct science for economic recovery, Kelly established STAC (Scientific and Technical Administration Committee), as a coordinating committee for various administrative offices within the scientific community (cf. ch. 11). He also acted to ensure that the Ministry of Education did not achieve a monopoly on S&T research funding as a result of the demilitarization programs (cf. ch. 15), ensuring that funds were distributed to other ministries to prevent a possible imbalance towards basic research.

The ESS/ST's investigations at the beginning of the Occupation concluded that the shortcomings in Japanese science lay in the poor coordination of research laboratories, in spite of the rich pool of able scientific manpower. Based on this evaluation, the Office of Industrial Technology (Kōgyō Gijutsuchō) was established in August 1948, by integrating and coordinating various test laboratories, which formerly belonged to the Commerce Ministry (Shōkōshō). (After the Occupation was over, the status of the Kōgyō Gijutsuchō (as a quasi-ministry) was downgraded to Kōgyō Gijutsuin (as a quasi-agency) because of the opposition of other government departments. The English translation remains unchanged and, hence, the status change went unnoticed by the former GHQ. One of the ESS/ST's major achievements in the latter stages of the Occupation (now headed by Kelly's successor, Bowen Dees) was in the promotion of quality control.[9] In this sense, what is seen to have later developed into the 'Japanese style' of science and technology was already set in place during the Occupation.

Meanwhile Minka, who had initially regarded the Occupation Forces as liberators, were diametrically opposed to the industrialization of science. The primary concern of Minka's social science members was focused on opposing monopoly capitalism and appealing for political action against US Imperialism.

The JSC, requiring the democratic consensus of academic scientists, could not create a science policy for the people. Such planning cannot be achieved by the consensual deliberations of scientists. Hence, the democracy that the JSC enjoyed remained merely laboratory democracy. Since they could not assume the leadership role in science policy, the government had to create another organization to fill this role, namely the Science and Technology Council.

This did not stifle dissent from the JSC, though. Basic scientists who were influential in the JSC, had misgivings about science policy, afraid that it was too heavily inclined towards applied science. Their arguments, together with those of the social scientists, opposed the establishment of the Science and Technology Agency and the Science and Technology Council.

With the Nine-Point Economic Stabilization Program in 1948, the Occupation policy changed from ideologically instilling democratization to more concretely targeting economic recovery. Democracy slipped by the wayside as the Cold War deepened and the policies of the liberal New Dealers gave way to military conservatism. That is to say, the Occupation policy became politically conservative, but this very conservatism was economically beneficial for the Japanese. It was at this time that the ESS/ST turned its attention from the JSC to STAC. According to Dees, 'in the earlier half of the Occupation, GHQ was the policeman who watched and controlled the demilitarization process. In the latter half we changed to be the helper of economic recovery.[10] Throughout the Occupation period, and while enjoying laboratory democracy, Japanese scientists, remained isolated from and unaware of the changing Cold War environment in which academic science had lost its prewar eminence relative to the increasingly strengthened military-industrial complex. The shining road to high economic growth, underpinned by the US's own military and strategic motives, overshadowed the slogans of democracy.

Cold War Structure

The Establishment of the Cold War System

Conflict between the USA and USSR had already begun at the time of the Potsdam Conference. The USA had from the outset intended to exclude the USSR from participating in the Occupation of Japan.

We may mark the beginning of the Cold War as 26 July 1947 when the National Security Act was signed by President Truman, establishing a defense system.[11] In the same month, the Marshall Plan was implemented to begin the economic recovery of Western Europe.

The beginning of the Cold War, as we have noted, marks a great turning point in the Occupation policies for Japan. The motivation for lifting policies from exacting reparations includes the decision to make Japan a firewall against the aggression of international communism. Occupation polices turned to the issue of how to make Japan an independent industrial nation. In January 1948, Secretary of the Army K. C. Royal announced this change in direction. Accordingly in March, Undersecretary of the Army William Draper came to Japan to cut planned reparations by seventy-five percent in the first year and, in the following year, to totally abolish the severe demands of the Pauley Reparations Committee. In December, Washington ordered the Nine-Point Economic Stabilization Program for economic recovery. America wanted an Asian factory, and there were no other candidates for industrialization in Asia. As the promotion of export trade became one of the major policies of the Nine-Point Plan, this policy change undoubtedly accelerated Japanese economic recovery.

The Dodge Line

In early 1949 Joseph Dodge arrived in Japan as financial adviser with the rank of minister in SCAP. His experience included drafting the West German currency reform. His aim was to restructure Japanese industry for export. To this end he amalgamated the Ministry of Commerce and Industry (Shōkōshō) and the Office of Trade (Bōekichō) to create the Ministry of International Trade and Industry (MITI, Tsūsanshō).

Following the Nine-Point Program and Dodge's reconstruction policy (Dodge Line), Japanese industry began to increase its production, achieving sixty percent of the prewar mining and manufacturing index in the second quarter of fiscal 1949.

The Dodge Line ordered the reduction of government subsidies in an effort to make Japanese industry self-reliant. Various research funds were cut or reduced. ESS/ST and the Japanese Diet petitioned Dodge to exempt research funds from subsidy cuts but Dodge refused. The price-difference subsidy for the priority production system was abolished. A wide range of policies was applied to stop inflation. But these policies also severely damaged manufacturing industries such as the radio industry. In private, the Japanese accused Dodge of coming to Japan for the destruction rather than the reconstruction of Japanese industry.

The government embarked on the rationalization of industry by founding the Industry Rationalization Council (Sangyō Gōrika Shingikai) within MITI, resurrecting the title of a prewar government body. The new Council was different, though, in its emphasis on technological development. But this development would require funding. Accordingly, in 1950 and '51, research subsidies were introduced for the testing laboratories of various government ministries.

The Dodge Line of economic rationalization resulted in the mass dismissal of employees by both government bodies and corporations. Scientists and engineers in public and private research laboratories were not exempt. For instance, in April 1949 NEC (Nihon Denki) announced the retrenchment of 3,500 personnel and the closure of its research laboratory. In July, Toshiba cut 4,600 employees and its laboratory. Union activists and members of left-wing factions were special targets for dismissal. It became known as the 'Red Purge' (cf. ch. 26). Those public laboratories that had previously expanded by absorbing returnees from abroad were severely reduced in size. For example, the Railway Technical Research Institute of the Transportation Ministry was ordered to cut its employees in June 1949 and by April 1950 had been reduced to one quarter of its previous size.

According to Yoshirō Hoshino, the turning point in the modernization of Japanese industrial technology is not located in the Special Procurement (Tokuju) Programs of the Korean War in the 1950s, as economists commonly claim, but in Dodge's visit to Japan in 1949 and the ensuing rationalization policy.[12] Compared to the prewar rationalization policies, the promotion of efficiency and improved technology is a positive outcome of the postwar policy. Corporations took it seriously. For instance, in April 1954 a three-year plan for the rationalization of steel production facilities was implemented, with the aim of matching the advanced level of Western technology. Rolling mills were modernized as a direct result of this plan.

For S&T, the implementation of rationalization policy resulted in a priority allocation of scarce foreign exchange currencies and tariff protection of strategic industries. Allocated foreign exchange was spent for technology imports in 1950 and after. Toyo Rayon in June 1951 signed a technology agreement with American Dupont. Teijin and others were following suit to import foreign technology. Technology imports reached their peak in 1952. The strategic focus of these policies and their top-down implementation drove the nascent postwar democracy out of shops, factories, laboratories and government offices only to be replaced by the leadership of bureaucrats and managers. Technocracy was under way.[13]

The Korean War

Whether or not it was '*the*' turning point for the modernization of Japanese industry, the outbreak of the Korean War in June 1950, an important event in world history, had a direct and significant impact on Japan. Japan became the marshalling point for US military operations in Korea. The needs of the US military operations produced the Special Procurement (tokuju) boom. Japanese military facilities were salvaged from the reparations program and resumed their former activities. Some facilities were provided for the direct use of the Occupation Forces.

The Korean War did not radically promote Japanese S&T but it is noteworthy that quality control technology was urgently required to fulfill the specifications of military procurements. Formerly only a handful of engineers appreciated the value of quality control, but the US military requirement for total quality from the entire Japanese industrial sector generated a nation-wide movement that eventually became part of Japan's unique labor management system (cf. ch. 22).

The Korean War ended in 1953 producing some misgivings in the Japanese industrial sector that the Special Procurement would be terminated and a business slump would ensue. But, SCAP wanted to maintain Japan as the Asian supplier of military necessities for the Cold War. On this account, both parties agreed to the New Special Procurement Program (shin tokuju). The ban on the production of weapons and airplanes was thus lifted even before the Peace Treaty of San Francisco came into effect in April 1952. After the Peace Treaty was concluded, though, the Federation of Economic Organizations (Keidanren) seriously considered the resumption of Japan's weapons industries. Both the public and private sectors were more inclined to rearmament and weapons production at this time than at any other throughout Japanese postwar history.[14]

The Citizenry Sector's Relation to S&T

Grassroots Desires – Resolve Food Shortages

For the general public it was of primary importance to return to the prewar living standard of 1934. Japan entered the Pacific War with its economy and living standard in decline, and then lost everything upon surrendering.

The most serious shortages were of foodstuff. During the war some kind of order had been maintained through the absolute power of the military. Scientists and engineers were given the privilege of special rations and other benefits because the weapons production they engaged in was of prime importance.

After WWII Japan was in disorder and chaos. Black marketing was widespread. City dwellers suffered especially, being outside of the food (production) chain. Some scientists, freed from the wartime mobilization, tried to apply their specialist knowledge to solving food shortages, but nothing noticeable came of it.[15]

In 1947, with the support of President Nanbara of the University of Tokyo an Integrated Research Group was created at Tokyo University (Tōdai Sōgō Kenkyūkai). They announced a method to increase the yield of rice by X ray radiation that generated a great public response but was soon proven to be ineffective. Nothing else could be done until the conventional chemical fertilizer industry recovered.

Recovery in the chemical industry, particularly for chemical fertilizers, was urgently needed. With the help of the priority production system it rapidly achieved a radical increase to prewar levels of production.[16] By 1950, the heavily protected ammonium sulfate production industry was generating a large surplus, facilitating exports to South East Asia.

In the final analysis, the most important achievement of SCAP was land reform. The increase of ammonium sulfate production, the use of American pesticides and the application of the Soviet Michurin style agricultural production were all part of the land reform program. Although some specialists warned of the poisonous nature of the pesticides being used, their cautions were overshadowed by the pressing need for increased yields.

This was the only long-term way to overcome food shortages. The most pressing need for the postwar Japanese citizen was to overcome chronic malnutrition. In the shorter term, to address the shortage of rice, the source of most Japanese nutrition, SCAP provided relief aid in the form of surplus American wheat, with which bread was made for school meals. It was not until the middle of the 1950s that Japanese agricultural production was sufficient to feed its people. The mechanization and the large quantity of agricultural chemicals used, which changed the structure of Japanese agriculture, were yet to come. This did not occur until the structure of heavy and chemical industrialization in Japan was established. This is the theme of the post-Occupation period to be addressed in Volume 2 of this work.

Healthcare

American air raids had devastated the infrastructure of Japanese cities, including public sanitation facilities. When combined with the arrival of the overseas returnees infested with a wide variety of diseases, viruses, etc., immediately after the war, Japan's healthcare was in a parlous state. Efforts to improve hygiene and stop the spread of epidemics included what today must be seen as a most peculiar application of S&T, the application of DDT to the Japanese citizens who were deemed to be unsanitary. This practice, though, was merely an extension of the US policy of spraying their own troops with DDT in occupied territories. This practice was not limited to human sanitation either. Public health personnel, under the guidance of GHQ's PHW, sprayed DDT in the streets and in repatriation vessels. While many Japanese found the practice to be humiliating, it also must be seen to have been very effective in reducing the spread of disease (cf. ch. 20). It was not until 1949 that the translation of a US magazine article informed the Japanese public of the highly poisonous nature of DDT. Still, it was only much later that the Japanese appreciated the long-term environmental damage caused by this practice.

J-day also marked a change in population policy. Whereas the war effort had required a continual increase in population to ensure sufficient manpower for the military, the massive return of personnel from overseas, increased unemployment and extensive food shortages all combined to produce a need for population control policies. With the women's liberation movement in the background, the Eugenic Protection Act was enacted in July 1948 (cf. ch. 21). Its primary aim was the protection of the mother's body, but it directly contributed to birth control during the economic hardships of the postwar period.

When health insurance was introduced in the 1950s, the medical community was initially reluctant to treat insured patients. However, it was soon apparent that insurance enabled people to visit their doctors more often. Before long it was almost impossible to practice clinical medicine without insurance.

With the introduction of health insurance, great improvements in public sanitation and the resolution of food shortages in the early 1950s, the primary causes of death in Japan changed around 1958 from respiratory and digestive diseases to diseases of the brain and heart (cf. vol.2, ch. 36)

Censorship

In spite of the difficult economic situation, the publication of popular science magazines and books boomed. The publication of the magazine *Shizen* (Nature) in May 1946 was particularly significant for developing communication links between the scientific community and the general public. Freedom of expression recovered somewhat after the War, although the Occupation press code still controlled access to certain information.

For instance, when, in September 1945, the *Asahi Shimbun* (Asahi Newspaper) published an article which argued that the use of the atomic bomb was an infringement of the International Law that prohibited the use of poison gas, GHQ shut down their printing presses for three days in punishment.[17] Reports on the number of victims from the Hiroshima and Nagasaki bombs, as well as news of the intensifying nuclear R&D competition between the USA and USSR were kept away from the people. The Civil Intelligence Section (CIS) was empowered to open, translate into English and scrutinize private letters in an effort to censor any criticism of the Occupation Forces or information about the atomic bombs. It was only after the Occupation ended that the truth about Hiroshima and Nagasaki became widely available to the people. Their hunger for this information is evidenced by the fact that the edition of *Asahi Graph Magazine* featuring the first published photographs of atomic bomb casualties sold out immediately.

Conclusions

The progressive view of history in the postwar democracy period clearly distinguishes between the earlier half of the Occupation when the 'New Dealers,' sharing the ideology of Roosevelt and his associates, tried to conduct an experiment in Japan in an effort to accomplish what they could not do at home, and the latter half when, as the Cold War intensified, conservative reactionaries expelled the New Dealers from power. This view is shared by Edwin O. Reischauer, New Dealers and their sympathizers, who argued that with its absolute power, SCAP was able to implement radical reforms that no native government could ever accomplish. The original intention was for SCAP to take drastic action quickly, and then leave Japan to continue the policies on their own and as they saw fit. However, with the development of the Cold War, and especially the 'hot' spark – the Korean War – so close to Japan, SCAP stayed longer than anticipated. As the Occupation dragged out, anti-American sentiments grew stronger amongst the Japanese. This is a rather persuasive argument.

In the earliest phase of the Occupation, the greatest impact on S&T was the disbanding of military science and technology, which later induced the 'scrap and build' formula for the postwar development of Japanese S&T. Among more substantial New Deal reforms are various policies relating to S&T, e.g., the dissolution of Zaibatsu, educational reform and land reform. Kelly was a key figure in the formation of science policy in the New Deal period. He pursued democratization of the Japanese scientific community and was instrumental in the formation of the Science Council of Japan.

Postwar democracy was introduced as an initiative of SCAP, but from the Japanese perspective, it was not totally exogenous. For most Japanese intellectuals democracy was taken for granted as the liberation from wartime right wing, ultra-nationalism and militarism. Scientists and engineers also accepted it without much discord.

Kelly from very early in the Occupation maintained the ideal of S&T for economic recovery, but it was only after the political and institutional framework changed in the middle of the Occupation, that this ideal actually materialized as GHQ policy. Subsequently, GHQ's ESS/ST, the Japanese economic bureaucracy and most scientists and engineers had a common purpose. Then, science for economic recovery supplanted laboratory democracy as the mainstream of science policy. This turning point in Occupation policy has, so far, not been emphasized in the postwar history of science.

From another angle, the latter half of the Occupation is marked by the gradual retreat of SCAP's interventionist policy and the emergence of a re-empowered Japanese bureaucracy. Thus technocracy prevailed over democracy.

From the perspective of democracy, SCAP became reactionary in the context of the Cold War and the Korean War. However, the US policy decision to make Japan an Asian base for manufacturing stimulated S&T for economic recovery, particularly in the area of quality controls (cf. ch. 22). This change in policy direction did not go so far as to lift the ban on military S&T. With the Korean War and the deepening of the Cold War, the rearmament of Japan was discussed, weapons production recovered and a rumor began about the resumption of nuclear research, but SCAP never approved any resumption of military R&D. They set about to achieve the industrialization, not the militarization of Japan.

After the Occupation period, i.e., since 1952, the policy of S&T for economic recovery has been fully embraced by the Japanese as discussed in Volume 2. The transition between these two major actors in the formation of science policy, from the late Occupation GHQ to the post-Occupation Japanese, is not clearly documented, but they both valued and prioritized a substantive economic recovery over-and-above the ideology of democracy.

The reform-minded people moved from political and ideological democracy to efficient and economic technocracy. Thus, the paradigm for

Japanese science moved from the prewar pursuits of a military and political giant to a democratic nation and then to the technocratic pursuits of economic strength, all during the Occupation period.

Notes

1 Yoshirō Hoshino, *Gijutsu kakushin* (Innovation), Iwanami, 1958.
2 Nihon Kagakushi Gakkai (History of Science Society of Japan) (ed.), *Nihon kagaku gijutsu-shi taikei: Tsūshi 4 and 5* (A compendium of the history of Japanese science and technology: Historical overview, vols. 4 and 5), Daiichi Hōki Shuppan, 1964, 1966.
3 Takahide Nakamura, *Sengo Nihon keizai – Seichō to junkan* (Postwar Japanese economy: Growth and recurrence), 1968, pp. 243–4; and Yūjirō Hayashi (ed.), *Nihon no keizai keikaku – sengo no reki-shi to mondaiten* (Japanese economic planning: Postwar history and problems), 1957.
4 Shigeru Nakayama, 'Independence and Choice: Western Impacts on Japanese Higher Education,' in P. G. Altbach and V. Selvaratnam (eds.), *From Dependence to Autonomy: The Development of Asian Universities,* Kluwer, 1989, pp. 97–115.
5 Shigeru Nakayama, *Science, Technology and Society in Postwar Japan*, Kegan Paul International, 1991.
6 Nihon Kyōsantō Kagaku Gijutsubu (Japan Communist Party, Science and Technology Section), 'Nihon no kagakugijutsu no kekkan to kyōsanshugisha no ninmu (Defects of Japanese science and technology and the call to communists),' in *Zenei* vol. 2, no. 11, November 1946.
7 Yoshio Shiga, 'Shutsugoku zengo nokoto (What happened when we were released from prison),' in *Nihon Senryōgun* jō (Occupation Forces on Japan, vol.1), 1978.
8 Takahide Nakamura (ed.), *Senryōki Nihon no keizai to seiji* (Economy and politics of Occupied Japan), 1979, pp. 8–11.
9 Bowen C. Dees, *The Allied Occupation & Japan's Economic Miracle: Building the Foundations of Japanese Science and Technology 1945–52*, Japan Library, 1997.
10 Personal communication.
11 Seizaburō Shinobu, *Sengo Nihon seiji shi* (A history of politics in postwar Japan, vol.2), 1966, p. 621.
12 Yoshirō Hoshino, 'Sengo gijutsu-shi no jidai kubun (Periodization of the postwar history of technology),' in Shigeru Nakayama (ed.), *Nihon no gijutsuryoku* (Japan's technological power), Asahi Shimbunsha, 1986, pp. 75–86.
13 Kimihiro Masamura, *Sengo-shi jō* (Postwar history, vol.1), Chikuma Bunko, 1990, p. 266.
14 Takahide Nakamura, *Shōwa-shi 2* (History of the Shōwa Era, vol. 2), 1993.

15 Many proposals appeared in *Kagaku Shimbun* right after the War.
16 Katsumi Shimono, *Sengo Nihon kagaku kōgyō-shi* (History of the chemical industry in postwar Japan), 1987, pp. 24–5.
17 Gaikō kiroku (Diplomatic records), microfilm A'-0115, preserved in Gaikō Shiryōkan.

PART I
GHQ and Demilitarization Policy

1 The Scientific Intelligence Survey: The Compton Survey

Yukuo Sasamoto

The Occupation of Japan by the Allied Forces officially commenced on 2 September 1945 with the signing of the surrender documents agreed to in the Potsdam Declaration which had been accepted on 14 August 1945. Although officially it was only one of the Allied Powers, the United States proved to be, in reality, the sole Occupation Force. Accordingly, despite the proclaimed aim of the Occupation being to implement the Potsdam Declaration's objectives of demilitarizing and democratizing Japan, policies that were in the US national interest were given the highest priorities.

This chapter discusses the Scientific Intelligence Survey conducted in the earliest days of the Occupation. Often referred to as the Compton Survey, after Karl T. Compton, one of the initial leaders of the Survey group who attracted a lot of media attention, the objective of the study was to gather information about the scientific research developed by Japan's wartime military machinery. The US regarded this survey to be imperative. The survey commenced concurrent with the beginning of the Occupation – i.e., when General Douglas MacArthur, Supreme Commander for the Allied Powers (SCAP), landed at Atsugi Airport.

The US Occupation Forces conducted a range of investigations into Japan's science and technology (S&T), including the biological warfare research conducted by the 731 Unit of the Japanese Army (cf. ch. 6).[1] The investigation of the 731 Unit began as a part of the Scientific Intelligence Survey. A review of the Survey methods and Report reveals not only the priorities and procedures of the Occupation Forces, but also the general status of the Japanese military science research at the end of the war. This chapter thus aims to fill a lacuna in the historical studies of postwar Japanese science and technology.

Compton's Investigative Activities as Reported in the Print Media

Asahi Shimbun reported extensively on the Compton Survey on 16 September 1945, two weeks after the Occupation began. The article details the purpose of the survey, and includes descriptions of Compton and Tokyo

Imperial University, one of the targets of his study. The full text of the article is as follows:[2]

> **Marveling at the isotope separation equipment: Dr. Compton visits Tokyo Imperial University**
> Dr. K. T. Compton, an American scholar renowned for his involvement in the development of atomic bombs, suddenly visited the Hongō campus of Tokyo Imperial University by jeep at ten o'clock on the morning of 15 September. Ushered into the office of the University President, Dr. Compton had informal talks with President Uchida and professors of engineering, asking detailed questions about the role the University played in war activities and the methods of mobilizing scientists during the war. He then inspected the atomic laboratory on campus.
>
> Middle-sized, ruddy-faced and wearing spectacles with a thick amber-colored frame, Dr. Compton who wore plain khaki-colored military clothes, appeared to be more a serviceman than a scholar. High voltmeters and radiation detectors in the underground laboratory reflected the dim afternoon sunlight.
>
> Dr. Compton asked a number of scientific questions while inspecting the site with Chair Professor Sagane leading the way. The American scientist paid particular attention to an isotope separation instrument, requesting its graphs and nodding repeatedly. He looked back and smiled at the vegetable garden near a high voltmeter in the courtyard, joking: 'everything looks self-sufficient here.'
>
> Professor Sagane commented: "Dr. Compton, as the President of MIT as well as a member of the brain trust of President Roosevelt, played a central role in mobilizing American scientists and led the American scientific efforts which were crucial to achieving victory in the war".

This article reveals several points. The objective of the Compton Survey was threefold: (1) to study the role that universities played in the war; (2) to investigate the methods of mobilizing scientists; and (3) to examine the atomic laboratory of Tokyo Imperial University. President Shōzō Uchida and professors of the Faculty of Engineering received Dr. Compton. Professor Ryōkichi Sagane had information about the role Dr. Compton had played in the war.

While this study is widely known as the Compton Survey, it is important to recognize that Dr. Compton did not individually conduct the study. Rather, he was a only member of the Science Intelligence Survey group of the United States. Given that Dr Edward L. Moreland was the Chief of the Survey team, it would be more correct to refer to the survey as the Moreland Survey rather than the Compton Survey. However, the name 'the Compton Survey' was coined during the early publicity in the press about Compton's visits to key institutions and individuals, and it stuck.

Relationship between the Alsos Mission and the Scientific Intelligence Survey in Japan

During WWII, the Allied countries made continual and concerted efforts to obtain information about the Axis countries' wartime science research and developments, always with the aim of finding something that might turn the war to the Allies' advantage. The US considered it to be imperative to expeditiously gather accurate intelligence data on the development of nuclear weapons, especially in Germany. For this purpose, Brigadier General Leslie R. Groves, Commanding Officer of the Manhattan Project, organized the Scientific Intelligence Survey group. The group was dispatched to Italy in September 1943, when the US Fifth Army landed in southern Italy. The survey team was code-named the Alsos Mission (Alsos is Greek for 'Groves').

The Alsos Mission was headed by Lieutenant Colonel Boris T. Pash and comprised of fourteen members, of whom four were scientists; two were from the Office of Scientific Research and Development (OSRD), one from the Army and one from the Navy. Most members of the team returned to the US around March 1944 because their investigations were falling short of expectations.[3]

Meanwhile, Groves organized another Alsos Mission, larger than the Italian group, in April 1944 in order to collect more intelligence information about the development of atomic bombs in Western Europe. In April 1944, the newly organized Alsos group established an advisory committee made up of the directors of Navy Intelligence and OSRD, the commanding general of the Army Service Forces, and the Army General Staff 2 (G-2). The scientific director of the committee was Samuel A. Goutsmitt, a physicist on leave from the University of Michigan to work at the Radiation Laboratory of MIT. The members of the committee included both civilian and military scientists.

The new Alsos Mission belonged to the Army G-2 under the Supreme Headquarters Allied Expeditionary Forces (SHAEF), headed by Commanding General Dwight D. Eisenhower.[4] The Mission penetrated Occupied France (which was under the control of Germany) and, just before the German surrender, captured Dr. Werner K. Heisenberg, the leading German scientist in atomic bomb development. After some investigation, the Mission concluded that Germany had not reached the stage of manufacturing an atomic bomb.[5]

Major Robert R. Furman, a member of the Mission, was later dispatched to Japan as a Manhattan Project officer and investigated Japan's progress towards the development of an atomic bomb.[6]

While the Alsos Mission concentrated on atomic bomb intelligence, the United Kingdom and the United States established a special scientific and technical intelligence unit known as T-Forces whose objective was threefold:

> (1) 'to find out what the Germans knew about weapons, radar, synthetic fuel, synthetic rubber, torpedoes, rockets, jet engines, infrared, communications, and such other things as would help the Allies in the war,' (2) 'to gather information that could help to shorten the war against Japan after Germany's defeat – information on how much and what kind of scientific and technical know-how the Germans had passed on to the Japanese, and on what the Germans knew about Japan that might help in the Allied war effort,' and (3) 'to locate and detain – even intern – German scientists and technicians, interrogate them for the kind of information just identified, and prevent them from slipping away to seek wartime research and development projects.[7]

From Manila to Tokyo

The US Occupation authorities commenced the Scientific Intelligence Survey in Japan after conducting similar surveys in Germany and Italy. The document that Compton released in December 1945 explains the preparations that the Scientific Intelligence Survey group made for the Japanese survey.[8] At the time, Compton was President of MIT, a member of the U.S. National Defense Research Committee (NDRC), and Director of the Office of Field Service of the OSRD, whose aim was to maximize the scientific assistance available to the military in the theater of operations.

The duties of the OSRD included training soldiers to use and maintain new weapons and advising military staff officers on the strategic deployment of same. Compton arrived in Manila as head of the OSRD's Pacific Office, reporting directly to General Douglas MacArthur, Commander in Chief, Army Forces Pacific. As stated earlier, Compton was a member of the highest policy making group in the Manhattan Project and had explained the nature of the atomic bomb to MacArthur. The Pacific Office of OSRD strengthened the science and technology support system for the military's preparations for a major assault on Japan.

In the meantime, the US Army Forces in the Pacific established the Scientific and Technical Advisory Section (STAS) as a special intelligence section and appointed Edward L. Moreland as its head. Moreland was Dean of Engineering at the Massachusetts Institute of Technology, a member of the NDRS and a member of the OSRD.

At the end of the war in the Pacific, OSRD had approximately sixty scientists in its Pacific Office with specializations ranging from medicine to atomic bombs. Compton's group was keen to obtain answers to a number of questions such as: How well developed was Japan's high technology? What was Japan's capacity to develop and produce the means of warfare? What kind of scientific research should be prohibited until Japan became a peaceful nation? How was the S&T sector mobilized during the war? To answer these questions, Compton and Moreland organized the Scientific Intelligence Survey group, making use of scientists who were already based in the Pacific Region. The group was soon established with Moreland as its head.

The group made preparations for the forthcoming survey. The Military Intelligence Department in Washington sent Compton and others a list of the Japanese scientists and institutions to be targeted. The Military Intelligence Services in Manila updated and augmented the list on the basis of current (and local) information it had acquired. Washington provided a list of the names, occupations and addresses of all of the Japanese students who had graduated from the Massachusetts Institute of Technology, the California Institute of Technology, Harvard University, the University of Chicago and the University of California. Thus, an exhaustive list of Japanese scientists and researchers was compiled to enable the group to conduct a survey upon landing in Japan.

The Scientific Intelligence Survey group was amongst the earliest arrivals of the Occupation Forces in Japan. They immediately commenced their survey of Japanese scientists, scientific research institutes (both military and civilian) and delivered their findings to the relevant coordinating bodies in the US military. The Survey group was authorized to seize documents, put them under custody, summon relevant individuals and even place them under arrest if they refused to cooperate. After preparations in Manila, the Survey group landed in Yokohama in September 1945.

Objectives, organization and methods of the Scientific Intelligence Survey

Compton noted: 'Our first business with a Japanese scientist was to locate Hidetsugu Yagi, the "Vannevar Bush" of Japan, a former student at Harvard. This errand took us by jeep to his home in a far corner of Tokyo and off the main road.'[9] The 'Report on the Scientific Intelligence Survey' records that the interview with Yagi took place on 12 September 1945. Bush, a science administrator who had previously been the Dean of Engineering and Vice-President of MIT, headed the NDRC established in June 1940 to investigate and report on research and development aimed at improving combat methods. Later, he headed the OSRD and contributed to the military mobilization of science and technology.[10]

The Scientific and Technological Advisory Section (STAS), General Headquarters of the United States Army Forces, Pacific (GHQ,

USAFPAC) compiled a five volume report dated 1 November 1945, entitled *Report on Scientific Intelligence Survey in Japan, September and October 1945*. Volume 1 discusses the general objectives and organizations of the Survey and presents an overall summary of the findings.[11] Volume 2 focuses upon interview records of the former Japanese Army and Navy research scientists.[12] Volume 3 primarily consists of interview records with wartime R&D specialists from outside of the military institutions.[13] Volume 4 discusses Chemical Warfare R&D,[14] while Volume 5 addresses Biological Warfare.[15]

The Survey 'was undertaken...to take a quick preliminary investigation of scientific research and development in Japan during the war with two major objectives: (1) To discover the Japanese organization for carrying on war research and to appraise its effectiveness; and (2) To ferret out, as far as possible in a brief survey, any new discoveries or techniques of significance to war activities, and identify them for more detailed study by other appropriate agencies.'[16] It was noted that 'Aeronautics, the general field of Military Medicine, and Atomic Energy were not included within the scope of this Survey as they are covered by other agencies.'

'The initial technical staff of the Survey group, and their previous connections with war activities were:

- Dr. Karl T. Compton – Consultant (until 24 September 1945), Member of National Defense Research Committee, OSRD.
- Dr. David T. Griggs – Radiation Laboratory; Consultant Office of the Secretary of War , Advisory Specialist Group USSTAF and FEAF.
- Major William R. Hewlett, GSC – New Developments Division, War Department General Staff.
- Dr. Andrew Longacre – Radiation Laboratory; Office of Field Service, OSRD.
- Dr. Edward L. Moreland – Chief of Survey; Executive Officer, National Defense Research Committee, OSRD.
- Lt. Colonel Murray Sanders, CWS – Office of Chief CWS, Special Projects Division.
- Major Howard E. Skipper, CWS – Medical Division, Chemical Warfare Service.
- Dr. H. Kirk Stephenson – National Research Committee, OSRD; Office of Field Service, OSRD (Head of Research Section of Pacific Warfare Board).
- Lt. Harry H. Young, MAC (from October 1945) – Office of Chief, CWS, Special Projects Division.

In addition to these nine officially assigned staff, Lt. Colonel W. S. Moore, USMC, 'was designated by the Chief Surgeon, AFPAC, to maintain liaison with the Survey' and Lt. Gordon T. Wallis, Chemical Warfare Officer with

FEAF, ADVON, 'gave a great deal of help in the investigation of Chemical Warfare targets.'[17]

Methodologically, the survey was conducted through a series of interviews with key personnel from the relevant institutions to acquire information and collect the required data. There were more than one hundred and thirty-five interviews conducted, involving more than three hundred military officers, scientists and engineers.[18]

Interviewees

Compton's visit to Tokyo Imperial University, mentioned at the beginning of this chapter, was a part of the survey interviews, a record of which is available.[19] Compton himself returned to the United States at the end of September and did not participate in the entire survey project. His continuing involvement in the Manhattan Project and important role in the formulation of atomic bomb policies in the postwar United States made it difficult for him to continue a hands-on involvement in the survey of Japan.

The Survey group visited the Japanese Navy Technology Research Institute. This was documented in a Japanese report which described the sudden and unexpected nature of their visit, stating that 'the Group requested that the Institute submit its research organization structure and the detailed report and documentation about the present status and research outcomes of airwave weapon technology.'[20] Records of the interviews conducted during this visit (17 September 1945) have also been kept.[21]

As to the survey of the relevant individuals, Compton claimed that his group was able to secure cooperation from major Japanese scientists, commenting that 'Perhaps it was helped by the fact that I had known a number of their leaders during the past twenty-five years; and some of them had been guests in my home.'[22] His statement suggested that many of the scientists that his group investigated were those who had studied in the United States.

Documentation about how the Japanese scientists under investigation regarded the survey is very limited. This remains an area for future research.

The orders issued by Moreland are stored in the archive of the Central Liaison Office established by the Japanese Imperial Government to coordinate with SCAP. An order dated 22 September 1945 instructed the Japanese Imperial Government to advise Yoshio Tanaka, Ryōtarō Mitsuda, Kōtarō Honda and Daitarō Arakawa to report to interviews in a specified room in the Meiji Life building at a specified time on a specified date,[23] which indicates the Japanese Governments' involvement in the activities of the Compton Survey.

Overall Results of the Survey

Volume One of the STAS Report provides a summary of the Survey results, including the following:

1 The section entitled 'Cooperation from the Japanese' states: 'The Japanese interviewed were invariably courteous and, on the whole, reasonably cooperative.'[24]

2 The section entitled 'Japanese organization for research and development' makes the following observations:

> Prior to the war with the United States, Japan had shown an active interest in supporting research, and two organizations, the Japan Society for the Promotion of Scientific Research and the National Research Council, had been set up with considerable funds. During the war a third agency, the Board of Technology, patterned after the Office of Scientific Research and Development, in the United States, was set up, directly under the Prime Minister, for coordinating and supporting research, with an annual budget for research purposes of ¥20,000,000 (in addition to other funds for educational purposes).
>
> The total funds available for support of research through these three civilian organizations amounted to ¥44,000,000 in the last year of the war – roughly equivalent in purchasing power in Japan before the surrender to $44,000,000 in the United States...
>
> The Army and Navy had extensive organizations for research and development, and the larger manufacturers also had research laboratories, but nothing like so extensive, relatively, or so well organized and staffed as the corresponding laboratories in America.
>
> University laboratories, as such, did not take contracts for undertaking research on a large scale, which proved so successful in the United States. A comparatively few of the professors got into effective war work and consequently, a strong potential asset was wasted.
>
> There was no effective agency for coordinating research of the Army, Navy, and civilian scientists, and for bringing the full capacities of the civilian scientists to bear on military problems. One of the principal objectives of the Board of Technology had been to offset this coordination, but it failed completely to accomplish this purpose, due to the opposition of the Army and the Navy which were too powerful to be controlled.
>
> Another respect in which it failed was that it did not mobilize the civilian scientists for mass attack on major problems under contracts with universities or other large organizations, but allowed such efforts as the scientist made to be dissipated on small-scale projects under individual contracts.[25]

3 The report provides the following summary information on Japanese technical developments:[26]

> *RADAR*. Japanese research on radar at the close of the war had reached a stage roughly equivalent to that of the United States in early 1942. The operational radar showed strongly the influence of the antiquated British and American radar captured early in the war in Singapore and the Philippines. Research and development on new equipment was largely original and showed little influence from German or Allied developments except in the general trend toward shorter wave lengths and airborne navigation and bombing radar...
>
> Japanese research on radar failed to meet the needs of the war because of lack of proper direction, organization, and coordination. This arose in part from the fundamental failure of the Japanese High Command to realize the operational importance of radar. The original work of many individuals was first-class, and with a better organization, Japan could conceivably rise to the front rank of radar development.
>
> *COMMUNICATION*. No really significant work in the field of communication was found during the course of the investigation with the possible exception of the work in ionospheric measurements...
>
> *RADIO COUNTERMEASURES*. [T]he Japanese had developed and initiated procurement for 30 units of a 500 to 1,000-watt communication-type jammer covering the frequency range from 2 to 10 cm and were developing 100 and 1,000 jammers to be used against American 3-cm radar...
>
> *OTHER SIGNAL EQUIPMENT*. Under this grouping are considered the work on Meteorology, Flash and Sound Ranging, Measurements and Standards. Little really important work was found to have been done in any of these fields with the possible exception of Meteorology...
>
> *ROCKETS AND JET PROPULSION*. The Japanese Army started theoretical work in jet propulsion about 1931, and by the end of the war had in operation or development a line of spin stabilized rockets from 5 to 60 cm in diameter. The latter was scheduled for final testing in September, 1945. The Survey was unable to cover Navy rocket developments in detail but 20 cm and 45 cm spin stabilized rockets of good design had been seen in the Philippines by members of the Survey group...
>
> Research was under way on the development of long-range, liquid-fuel rockets using H_2O_2 and alcohol.
>
> The Navy was also working on the development of a jet-propelled fighter plane similar to the German Me-262 and had apparently had some help from Germans. Only a few test models had been built.

GUIDED MISSILES. The Japanese were experimenting with a heat-homing bomb for attack on ships. They had not actually made a successful model but thought they had arrived at a successful design (Mark 9) which was expected to be ready for test by September, 1945...

INFRARED. Some use was apparently made of near infrared for signaling purposes but no radical improvements had been made over well-known techniques. The Navy was also experimenting with far infrared for inspection of ships and airplanes at night by the heat from the boiler or engine exhaust.

UNDERWATER SOUND. In this field, more any other, the Japanese seem to have been helped by the Germans who had sent them plans, specifications, models, and technicians. Apparently good, but not outstanding, listening echo-ranging units had been made and standardized. The latter is a direct copy of the German prototype...

DEATH RAY. A real effort was under way to develop a "Death Ray" which it was hoped could be used to kill the crews of airplanes. Rabbits had been killed by high-frequency electro-magnetic waves at a distance of 30 meters with an exposure of 10 minutes...The earnestness of the Japanese effort is indicated by an appropriation of ¥1,000,000 for this work in 1945. There is, however, nothing in this development to indicate any serious threat to crews of airplanes, or in other military use...

ATOMIC BOMB. A special Atomic Bomb Survey group has inquired into Japanese activities in this field, and it was therefore not included in the scope of this Survey. Incidental questioning of scientists in the course of this Survey did not bring to light any information about atomic power or nuclear research in Japan not already known to the Atomic Bomb Survey group. It also seemed clear from this incidental inquiry that the Japanese had not made any real progress in the release of atomic energy.

MISCELLANEOUS DEVICES. The Japanese report they originally planned to make 20,000 of the paper bombing balloons for "attack" against the United States and that they had actually produced and released 9,000...

CHEMICAL WARFARE. Japanese laboratories had synthesized and examined more than 1,000 compounds as possible new poison gases, but they claim to have found none superior to the gases used in the United States.

CW munitions were, on the whole, poorly designed and inefficient.

The Japanese are evidently unaware of the poor protection of their gas mask canister against CW, and of the United States intention of taking advantage of this fact in case of chemical warfare.

4 The section on 'Aid from Germany and other sources' observes that '[w]ith very few exceptions, the Germans do not seem to have given the Japanese a great deal of technical help.'[27]
5 Under the heading 'Evaluation of Japanese research and development,' the report states:

> The Japanese made little progress in the development of new equipment requiring scientific skill even when measured in terms of their relatively small facilities.
>
> Probably the most important factor in this lack of progress was the failure of the Army and the Navy to make effective use of university scientists in helping to solve their technical problems...
>
> In general, the new technical developments were far behind corresponding American or German developments, and little ahead of American and German pre-war standards.
>
> Exceptions to the general rule of lack of progress were noted in the fields of Chemical Warfare, Meteorology, Ionospheric Measurements and Rocket Developments, in all of which some systematic work had been done, but nothing startling or in advance of Allied techniques had been developed.[28]

6 In Conclusion, the report states:

> Japanese progress was badly handicapped by lack of proper organization for research and development, and by the almost complete lack of cooperation between the Army and the Navy.
>
> They have a fairly large number of able scientists who could, unquestionably, have made significant contributions to the war effort, if they had been properly used.
>
> The Japanese are obviously now well aware of this situation, and will certainly be more effective if another emergency arises.[29]

As mentioned above, the data on biological warfare is not included in the microfilm collection of the National Diet Library of Japan that the author examined. Volume 5 of the Scientific Intelligence Report makes a few concluding remarks: First, the Japanese biological warfare plan would have perhaps been able to produce applicable weapons if the 1939 budget had allocated a sufficient amount of funds and the Japanese military and scientists had made efforts at mutual cooperation. Second, because Japan was afraid of the United States' reprisals, the Japanese would not have attacked the US military with biological weapons even if they had been able to do so. Finally, Japan is fully aware of reasons for their own failure in biological weaponry development and therefore would not repeat the same mistakes.

Notes

1 Tsuneishi, 1984 [5]
2 Asahi Shimbunsha (ed.), *Asahi Shimbun ni miru Nihon no ayumi: Shōdo ni kizuku minshushugi* (A history of Japan as found in the pages of *Asahi Shimbun*: Democracy built on burnt land), 1973, p. 85.
3 Jones, 1985 [6] pp. 280–2.
4 Jones, 1985 [6] pp. 285–6; Goutsmitt, 1977 [4]
5 Jones, 1985 [6] pp. 288–90; Goutsmitt, 1977 [4] pp. 69–70.
6 Jones, 1985 [6] p. 282; First Technical Service Detachment, *Memorandum To: General T.F. Farrell,* 14 August 1945. GHQ/SCAP Records, Box No. 7408, ESS(D)06311. Held by the National Diet Library of Japan. Major Furman headed the Tokyo group of the Manhattan atomic bomb survey team and was in charge of surveying Japan's research and development of atomic bombs.
7 Gimbel, 1990 [10] pp. 3–4.
8 Compton, 1945 [1]
9 Compton, 1945 [1] p. 102.
10 Sherwin, 1978 [11] pp. 53–4, 63.
11 GHQ, USAFPAC, STAS, 1945 [2] Volume I, Box No.9090 NRS-06247, 06248.
12 GHQ, USAFPAC, STAS, 1945 [2] Volume II, Box, No.7430 ESS(A)-09868, 09869, 09870, 09871.
13 GHQ, USAFPAC, STAS, 1945 [2] Volume III, Box No.8354 ESS(I)-00727, 00728, 00729, 00730.
14 GHQ, USAFPAC, STAS, 1945 [2] Volume IV, Box No.8354 ESS(I)-00730, 00731, 00732, 00733.
15 GHQ, USAFPAC, STAS, 1945 [2] Volume V.
This document is not included in the GHQ/SCAP files held at the National Diet Library. However, Tsuneishi, 1984 [5] contains a full Japanese translation of the original. Accordingly, Japanese researchers can make use of data from all five volumes. The reason why this volume is not included in the NDL files is unclear, but Tsuneishi, 1984 [5] p. 289, suggests that he discovered volume 5 in record groups 112 and 330 at the U. S. National Archive, not GHQ/SCAPIN record groups. Since the NDL files are based on the latter, volume 5 was not a part of the NDL possessions.
16 GHQ, USAFPAC, STAS; 1945 [2] Volume I, NRS-06247.
17 GHQ, USAFPAC, STAS, 1945 [2] Volume I, NRS-06247; Compton, 1945 [1] p. 100.
18 GHQ, USAFPAC, STAS, 1945 [2] Volume I, NRS-06247.
19 Compton, 1945 [1] p. 100.
20 Nakagawa,1990 [8] p. 298.
21 GHQ, USAFPAC, STAS, 1945 [2] Volume II, ESS(A)-09870.

22 GHQ, USAFPAC, STAS, 1945 [2] Volume III, ESS(I)-00728.
23 Ministry of Foreign Affairs, Document and Archives Division, 1976 [3] Reel number 0033.
24 GHQ, USAFPAC, STAS, 1945 [2] Volume I, NRS-06247.
25 GHQ, USAFPAC, STAS, 1945 [2] Volume I, NRS-06247.
26 GHQ, USAFPAC, STAS, 1945 [2] Volume I, NRS-06247, 06248.
27 GHQ, USAFPAC, STAS, 1945 [2] Volume IV, NRS-06248.
28 GHQ, USAFPAC, STAS, 1945 [2] Volume I, NRS-06248.
29 GHQ, USAFAPC, STAS, 1945 [2] Volume I, NRS-06248.
30 Tsuneishi, 1984 [5] p. 228.

Bibliographic Guide

[1] Karl T. Compton, 'Mission Tokyo,' in *The Technology Review*, vol. 48, no. 2, December 1945.

[2] General Headquarters, United States Army Forces, Pacific, Scientific and Technological Advisory Section, *Report on Scientific Intelligence Survey in Japan, September and October 1945*, GHQ/SCAP Records, National Diet Library Archives.
- Volume I, 'Brief Summary Report, Main Report (Objectives, Organizations and Results of Survey);
- Volume II, Appendix (Interviews and Other Materials Relating to Japanese Army and Navy Research and Development on Instruments and Equipment);
- Volume III, Appendix (Interviews and other Materials Relating to Wartime Research and Development Outside of Military Institutions);
- Volume IV, Chemical Warfare;
- Volume V, Biological Warfare.

[3] Ministry of Foreign Affairs, Document and Archives Division (ed.), *Gaimushō kiroku maikuro firumu* (Archival Microfilm of the Ministry of Foreign Affairs), First opening, 'Rengōgun shireibu zatsu raishin tsuzuri (Miscellaneous Communications with GHQ),' 1976.

[4] Goutsmitt, S.A., *Nachi to genbaku – Arusosu: Kagaku jōhō chōsa-dan no hōkoku* (The Nazis and the atomic bombs – Alsos: Report of the Scientific Information Survey Group), Kazuo Yamazaki and Tsūji Konuma (tr.), Kaimeisha, 1977.

[5] Keiichi Tsuneishi (ed. and tr.), *Hyōteki Ishii – 731 butai to beigun no chōhō katsudō* (Target Ishii – 731 Unit and American military intelligence activities), Ōtsuki Shoten, 1984.
This book is a translation of documents revealing the development of Japan's biological weaponry research de-classified at the U. S. National Archives. Out of some 8,000 pages covering the period from 1941 to 1948, Tsuneishi selected the most relevant pages and translated them into

Japanese. Chapter 5 of this book (pp. 226–91), the Sanders Report (15 August 1945 – December 1945), contains the full text of the Moreland Report, volume 5 of the Scientific Intelligence Survey. The author of the report was Murray Sanders who was a member of the Survey team. After compilation of the Survey reports dated 1 November 1945, he continued to survey the former Japanese military personnel. The book contains his post-Survey report (pp. 291–307).

[6] Vincent C. Jones, *Manhattan, the Army and the Atomic Bomb*, (United States Army in World War II, Special Studies), Center of Military History, United States Army, Washington, D. C., 1985.

[7] Jūkō Saitō, *Bōryaku-sen: Dokyumento rikugun Noborito kenkyūsho* (Plot warfare: Document – Noborito Research Institute of the Army), Jiji Tsūshinsha, 1987.

[8] Seizō Nakagawa, *Dokyumento: Kaigun gijutsu kenkyūsho – Erekutoronikusu ōkokuno senkusha-tachi* (Document: Navy Technology Institute – Pioneers in the kingdom of eletronics), Kōdansha Bunko, 1990.

[9] Kawasaki-shi Nakahara Heiwa Kyōiku Gakkyū (Nakahara Peace Education Classes in Kawasaki City) (ed.), *Watashi no machi kara sensō ga mieta – Bōryaku himitsu kichi Noborito kenkyūsho no nazo o ou* (We could see the war from our town – Investigating the mysteries of the Noborito Research Institute, the secret base of warfare plots), Kyōiku Shiryō Shuppankai, 1989.

[10] John Gimbel, *Science, Technology and Reparation: Exploitation and Plunder in Postwar Germany*, Stanford University Press, 1990.

[11] Martin J. Sherwin, *Hametsu e no dōtei – Genbaku to dai 2-ji sekai taisen* (The road to destruction – Atomic bombs and World War II), Mikio Katō (tr.), TBS Britannica, 1978.

2 Investigations of the Effects of the Atomic Bomb

Yukuo Sasamoto

Atomic Bomb Investigations in the History of Postwar Science and Technology

The US Military dropped a uranium atomic bomb on the city of Hiroshima at 8:15 am on 6 August 1945 and a plutonium atomic bomb on the city of Nagasaki at 11:02 am on 9 August. Let us consider just how the atomic bomb research carried out immediately after the dropping of the first ever atomic bomb used in real warfare has been treated in the history of postwar science and technology.

First, the 'Outline of the History of Japan's Science and Technology'[1] notes that, after the war, even though research activities had been halted, there were isolated incidents of research being conducted – notably investigations into atomic bomb damage in Hiroshima and Nagasaki. This work briefly touches upon the investigative research activities of Yoshio Nishina of the Institute of Physical and Chemical Research, Masao Tsuzuki of Tokyo Imperial University and Bunsaku Arakatsu of Kyoto Imperial University, while also making reference to the organization of the Atomic Bomb Casualty Research Special Commission by the Scientific Research Committee and the results of its investigations detailed in 'The Atomic Bomb Casualty Survey Report.'[2]

Secondly, Hiroshige, in 'A Social History of Science' [3] states that

> As a result of the decision to surrender, the mobilization arrangements which had been in place during hostilities started to be dismantled...On 5 September the Technology Institute was abolished as was the country's wartime research position. At that time scientists, with the exception of those involved in investigating atomic bomb damage and dealing with those people receiving medical treatment, spent their days in something of a state of limbo, quite unsure as to whether there was any hope of continued scientific research in the future under the US Occupation Forces.

While these writings are representative of the history of science and technology in postwar Japan, they do not sufficiently deal with the history of the atomic bomb investigations. In the following pages I would like to turn my attention to the aims, organization and outcomes of the US-Japanese investigations.

Initial Investigations

Giving-up on rendering assistance in Hiroshima and Nagasaki: The initial investigations

US forces had already landed on the shores of Okinawa when the US dropped the atomic bombs on Hiroshima and Nagasaki; the main islands of Japan were preparing themselves for the crucial battle yet to come. In Hiroshima, immediately after the atomic attack, General Shunroku Hata, Commander-in-Chief of the Second Army took overall control of the job of providing assistance to the afflicted and ordered the Army, Navy and the prefectural and municipal authorities to restore the provision of assistance to the needy. On 15 August, however, he relinquished command of this emergency damage management post.[4] The Second Army was the Home Islands Defence Corps which was directly affiliated with the Imperial Headquarters. The order for him to relinquish control was directly from Emperor Hirohito via the Imperial Headquarters.

The provision of assistance to the people of Hiroshima after 15 August was left to the prefecture and the city to carry out.[5] Assistance to Nagasaki was also abandoned. It was obvious, however, that such massive numbers of dead and injured could not be catered for by relying on prefectural and municipal authorities alone. It was unreasonable to have the prefecture and city take on the responsibilities of care abrogated by the nation and it was appropriate that the nation's abrogation of responsibility for the provision of assistance should be questioned.

The Army, Navy and Imperial Government conducted research into the new weapon known as the atomic bomb, with three key objectives: 1) developing an understanding of its nature; 2) studying its effect and the condition of those who had fallen victim to it; and 3) investigating the implications for defence and the establishment of future war strategies.[6] It is noteworthy that providing aid and treatment to the victims of the atomic bomb blast is not included amongst these research objectives.

To avoid wasting time in grasping an understanding of the nature of the new weapon the Imperial Headquarters dispatched a commission of inquiry to Hiroshima on 8 August. It was comprised of approximately 30 members, including the Head of the Commission HQ 2nd Division (Information) Chief Lieutenant General Seizō Arisue, Lieutenant Colonel Seiichi Nizuma of the War Department's Military Division and Yoshio Nishina of the Physics and Chemistry Research Institute. Nishina had participated in the Army's atomic bomb project and played an important role in the Commission's acknowledgement on 10 August that it was, indeed, an atomic bomb.[7] Regarding the effects of atomic bomb radiation, they reported that 'some people, even without obvious external injuries, died within 1–2 days.' It was evident that the human body suffered after-effects as a result of atomic bomb radiation.[8] This knowledge is at the base of the later 'information war' with the US.

At the same time as the Imperial HQ research team was dispatched, the War Department's Medical Bureau was dispatching their own group of military doctors whose purpose was to investigate the damaging effect of the atomic bomb on the human body and to establish a policy for treatment.[9] Their research included twenty-five pathological dissections being conducted by army doctors within two weeks of the bombing. Their results were reported,[10] but the ethical question regarding how, or from whom, permission was obtained to dissect these twenty-five bodies remains unanswered. There seems to have been a clear lack of concern shown for the victims' human rights.

Other initial mobilization investigations in Hiroshima include the Kyoto Imperial University Investigation Team,[11] which joined at the behest of the Kyoto Divisional Headquarters, the Osaka Imperial University Investigation Team and the Kure Naval District Investigation Group who joined at the behest of the Navy. In Nagasaki[12] the Nagasaki Military Police and the Kure Naval District Investigation Group joined in, as did the Kyūshū Imperial University Investigation Group[13] at the behest of the Western Division of the Army. In the Imperial Government's investigation, the Emergency Atomic Bomb Strategy Committee (also referred to as the Atomic Bomb Committee), which was established by the Cabinet immediately after the bomb was dropped, dispatched a research team from the Technical Institute to Hiroshima.[14] The findings of this team's investigations were published in the newspaper as part of a defence strategy against the new weapon.

Use of the Findings of the Initial Investigations: the Atomic Bomb Information War with the US

While the Army, Navy and Imperial Government pushed ahead with their initial investigations, the Information Bureau developed a publicity campaign to denounce the atomic attack. On 7 August it 'appealed to world public opinion and began a thorough propaganda campaign against the use of inhumane weapons.'[15] One of the aims of the propaganda campaign was to develop a defense to counterbalance against claims that atrocities were committed against Allied POWs.[16] On 10 August a Council in the Imperial presence decided to accept the Potsdam Declaration on the condition that the national polity be allowed to continue. At the same time as the government reported this to the Allied Powers it protested to the US government and the Committee of the International Red Cross against the atomic attack, saying the use of the 'new type of bomb' contravened international law. There was no reference to atomic bomb radiation casualties in the protest, which stated that 'combatants and non-combatants alike, men and women, old and young were all indiscriminately killed or wounded by the bomb blast and the heat it radiated.'[17] Nor was this protest based on arguments concerned with the human rights of the atomic bomb victims. In fact, the protest wears somewhat thin when one considers what the Imperial Government and Japanese Army had inflicted upon the people of China, Korea, Taiwan and various other Asian countries.

The Japanese Government, through Radio Tokyo and the Dōmei News Agency, commenced a continuing anti-US intelligence campaign immediately after the bombing of Hiroshima. This campaign took the form of continually updated public announcements of the numbers of dead and wounded caused by the atomic bombs and revelations of new information discovered in relation to sicknesses and disorders which resulted from atomic bomb radiation but had not been immediately apparent. The findings of the Imperial Headquarters Commission of Inquiry, the Army Investigation Group, and the Technical Institute's investigative team were utilized to the fullest.[18]

This sort of utilization of information about the damage caused by the atomic bomb was the thing which the Americans most feared. According to the *New York Times* (31 August 1945), Commanding Officer Major General Leslie Groves of the atomic bomb manufacturing project – the Manhattan Project – denied the findings, stating that 'Reports from Japan in relation to deaths attributable to the effects of atomic bomb radiation are pure propaganda.' The fact remains, however, that there were delayed disorders and sicknesses which resulted from atomic bomb radiation. When the US Occupation Forces landed in Japan, the Japanese broadcasts and propaganda about the atomic bomb casualties via the Dōmei News Agency were promptly banned by the Occupation Forces press code.[19]

From the Manhattan Project to the Atomic Bomb Consequences (Effect) Investigation

The Manhattan Project was a huge weapons development project which has had a significant impact on the history of science and, in fact, on the history of the world.[20] On 10 August, the day the Imperial Government announced its acceptance of the conditions of the Potsdam Declaration to the Allied Powers, Commanding Officer Major General Groves of the Manhattan Project, ordered the formation of a Group of Investigation into the consequences of the atomic bomb on Japan.[21] The Group of Investigation was headed by Brigadier General Thomas F. Farrell and Brigadier General James B. Newman Jr. The purposes of the investigation were to: 1) conduct medical research in relation to the investigation into the general effects felt by the atomic bomb victims in the cities of Hiroshima and Nagasaki, with particular attention being paid to investigating the consequences arising from radioactivity; 2) examine the physical effect on the buildings and other structures in Hiroshima and Nagasaki cities; and 3) report on Japan's atomic bomb production plans, the level of atomic power research and to conduct a mineralogical investigation which encompassed Korea.

The commission of investigation was comprised of: Group no 1 (the Hiroshima Group) with Colonel Stafford L. Warren, Head of the Manhattan Project Medical Department, as group leader; Group no 2 (the Nagasaki Group); and Group no 3 (the Tokyo Group) under Major Robert R. Furman. Furman had participated in the Alsos Research Team (cf. ch. 1)[22] which had investigated the development of nuclear bombs in Italy and Germany.

Under the Manhattan plan, 'the research into the atomic blast and the destructive capacity of radioactivity included the study of radiation weapons, the lethal amount of atomic blast radiation, the amount required to kill half the number of people and the zero casualty "threshold amount". As for research into the effects of radiation in relation to the management of the bombed area, biological and medical research was conducted into the 'accommodatable levels of radiation.'[23] There were also irradiation experiments carried out in relation to the human body.[24] This research was conducted under the direction of the Manhattan Project Medical Department, headed by Colonel Warren. Colonel Warren's investigation team had conducted research into the first atomic explosion test at Alamagordo in New Mexico on 16 July 1945.[25] Again, the absence of an objective to provide treatment to the victims is noteable.

Colonel Ashley W. Oughterson, advisor to the US Pacific Forces General Headquarters Medical Corps, submitted a memorandum to Brigadier General Guy Denit, Surgeon General of the Corps, entitled 'Research Report on the Injuries Resulting from the Atomic Blast' which Denit subsequently endorsed. As for reports on the Japanese dead and injured, it advised that this data would be obtained by questioning doctors and pathologists. The US Pacific Forces Medical Research Team was formed in response to this report.[26]

It is a simple fact that the Americans were not able to conduct research into the consequences of the bomb immediately after it was dropped. This was one reason why the US absolutely needed the co-operation of the Japanese researchers. The aim of the US investigators was not to establish a policy of relief aid and treatment for the atomic bomb victims. Rather, providing relief aid and treatment to the victims of the atomic bomb was left to the defeated country: Japan.

The Policy of Full Co-operation in Investigations with the Occupation Forces and Promoting Postwar Science

Strengthening of the Investigation Structures: Paving the Way for Co-operation with the Occupation Forces

The stationing of the Occupation Forces in Japan, strengthened the Japanese position in terms of atomic bomb investigations. Before the Supreme Commander of the Allied Powers, Douglas MacArthur, arrived in Atsugi, there were already research teams in operation. In Hiroshima, research was underway by the Army Medical College, Physics and Chemistry Research Institute and the Medical Faculty of Tokyo Imperial University Research Team[27] while in Nagasaki the Kyūshū Imperial University Medical Faculty Research Team had commenced their investigations. Emperor Hirohito expressed an 'Imperial Wish,'[28] dispatching Imperial Steward Nagazumi to Hiroshima and Imperial Steward Hisamatsu to Nagasaki on 1 September. On 2 September, the day of the signing of the surrender documents, the Imperial Headquarters, by Imperial Decree No. 1, commanded 'support for the Allied Forces' Occupation and their stationing in Japan.'[29] On 3 September, Foreign Minister Mamoru Shigemitsu negotiated military administrative questions directly with MacArthur at the Yokohama Headquarters. On the same day, at the request of the headquarters of the Chūgoku Military District, the Kyoto Imperial University's Investigation Team entered Hiroshima. Also in Hiroshima, Masao Tsuzuki of the Tokyo Imperial University's Medical Department announced the results of his research at a public lecture on the symptoms of the atomic bomb.[30] Meanwhile in Yokohama, a representative of the Imperial Government submitted an Atomic Bomb Victim Report to the US Military's Forward Headquarters.[32]

On 1 September the Army Medical College Military Doctors' Investigation Team – which had been dispatched by the War Department's Medical Bureau – revised the Disaster Investigation Report submitted to the Medical Bureau Chief, naming it the 'War Disaster Re-investigation Report' and continued with their research into atomic bombs. That report centred on the results of research into the late developing effects of radioactivity. One particularly noteworthy finding of this report (dated 10 September) was that, 'the possibility of the birth of abnormal children [as a result of atomic bomb radiation] cannot be discounted.'[32] Furthermore, on 10 September the Tokyo Imperial University Medical Department Obstetrics and Gynecology Research Team, which was commissioned by the Ministry of Health and

Welfare, commenced its research in Hiroshima.[33] The Americans' desire to deny the possibility of long-term effects is evidenced by the speed with which they shut down the information channels for these particular findings. Still, despite the fact that the Japanese research had scientific merit well beyond its propaganda value, we cannot overlook how the victims were used. From 30 August when MacArthur landed at Atsugi until 8 September, the Japanese had dissected twenty-five males and one female in addition to those twenty-five dissections referred to earlier.[34]

Why was it that, with the imminent commencement of the Occupation, the Japanese strengthened their atomic bomb investigations? It was because the Japanese Imperial Government changed tactics: instead of capitalizing on the damage caused by the bomb they decided to co-operate with the US in atomic bomb research. Having met defeat for the first time since the Meiji period and having had their home country occupied, Imperial Japan implemented a national policy of engaging with the US over the 'inhumane weapon.' This is to say they adopted a policy of researching the effects of the atomic blast and submitting all evidence to the US Forces. Although they stressed and publicized the aim of aiding and treating victims,[35] this was merely propaganda to conceal their policy of co-operation with the Occupation Forces.

Army Medical Lieutenant General Kenji Ibuka, the last Principal of the Military Medical School, aptly expressed the purpose of Imperial Japan. He said of the Military Medical School's research activities carried out after 15 August: 'It is our duty to diagnose and treat those who suffered these horrors. Researching the war injuries caused by specific weapons such as the atomic bomb is the greatest service we can provide as doctors and as Army surgeons. We considered it a memorial for colleagues who had died in the line of duty.' And he explained to the Manhattan Project Investigation Group that 'The results of our completed investigations of the Japanese medics, particularly the results which relate to medical science and biology, have already been tabled.'[36] In other words, they treated the victims and submitted research results to the Occupation Forces. The Military Medical School complied with the national policy of co-operating with the Occupation and thus was able to continue uninterrupted research into the injuries caused by the new weapon known as the atomic bomb.

However, the Military Medical School incorporates the 731 Unit.[37] We should not forget that the materials relating to experiments carried out on living patients were sent to the Tokyo Military Medical School and used for military research. In this connection, it behoves us to ask whether the Japanese Army would have forced the enemy Chinese troops to carry out research into injuries resulting from biological weapons – developed by the 731 Unit – in the battlefields in which such weapons had been used.

When you consider that amongst the atomic bomb victims there would have been many who were Japanese soldiers, it was 'immoral' – and thereby 'corrupt' – for the Japanese Government to co-operate with the Occupation Forces and to perform forensic dissections on the victims. This 'corrpution' appears not only in the Army, represented by Ibuka, but in Imperial Japan generally, headed by the Emperor.

If we look at the US movements, on 6 September MacArthur ordered the Imperial Government to give protection and assistance to the Manhattan Project Investigation Group's entry into Hiroshima.[38] This order is not recorded in the Foreign Ministry's 'Messages Received from the Allied Forces HQ,'[39] but can be found in the 'GHQ/SCAP Orders Collection' [40] edited by GHQ. Again, on 6 September, Brigadier General Farrell held a news conference with reporters from the Allied Forces at the Tokyo Imperial Hotel, and made the declaration that 'those expected to die have already died. As at the beginning of September there are no people suffering as a result of atomic bomb radioactivity.'[41] On 8 September the 1st Group under Farrell entered Hiroshima.[42] The 2nd Group entered Nagasaki on 9 September. In parallel with the Manhattan Project Investigation Group's entering Hiroshima, the War Office Medical Bureau issued a directive to the Military Medical School and a special relief team made up of 144 military doctors and nurses from the Military Medical School and the Tokyo No 1 Military Hospital was dispatched.[43] The real purpose here was to strengthen co-operation with the Manhattan Project Investigation Group. From 9 September, in places such as Tokyo, Kyoto, Osaka and Korea, the Manhattan Project 3rd Investigation Group commenced questioning both military personnel and civilians about atomic bomb production plans and interrogating people associated with atomic power research. At the same time they continued questioning people like Ryōkichi Sagane, Hideki Yukawa, Yoshio Nishina and Bunsaku Arakatsu.[44]

Establishment of the National Research Council

On 5 September the Ministry of Education abolished the Student Mobilization Bureau, and the Science Bureau was re-organized as the Science Education Bureau.[45] This was in line with the Government's science promotion policy. On 15 September the Science Education Bureau of the Ministry of Education issued the Science Research Council with a notification entitled 'Matters Relating to the Special Committee for the Investigation of the Effects of the Atomic Bomb' which ordered the establishment of the Special Committee for the Investigation of the Effects of the Atomic Bomb.[46] The formation of this Special Committee was announced in the newspaper but there is no record of either the course this notification took nor how it was decided to continue the atomic bomb investigations as a matter of national policy. Nor is there any record of the formation of the Special Committee or suggestion of a relationship with the US Forces. All the writer knows is that, at that time Kiyoshi Okano, who was the secretary of the Ministry of Education and who was involved in the Special Committee, made a pronouncement regarding the committee's formation. 'Yes, the Occupation Forces acknowledged it…I think Dr Nishina spoke about this quite sufficiently with the authorities and received their OK.'[47] Certainly, Yoshio Nishina had contact with the Occupation Forces. But the question is who manipulated him, and how was it that this huge investigation project became national policy?

For the reorganized Science Education Bureau the first jurisdictional concern was the atomic bomb investigation. In other words, making use of the atomic bomb damage was central to the Japanese Government's postwar science promotion policy. However, this national policy was contrary to the co-operation being given to the Occupation Forces. The notification states that the situation in Hiroshima and Nagasaki was investigated 'with all the scientific rigour our country could muster' and that 'this investigative research was a purely scientific study.' The following reference was made to the result of the investigative research: 'We count upon it to be thorough and most satisfactory as it is conducive to the stability of the lives of the local people.'

The atomic bomb was a product of science and technology but it was also a weapon of indiscriminate large scale slaughter. New weapons research could hardly be referred to as 'purely scientific.' In Hiroshima and Nagasaki there were many victims.[48] There was no way that a country in defeat should have conducted investigative research with an ill-defined goal such as promoting 'the 'stability of the lives of the local people.' And again we note the absence of any policy aimed at providing relief and treatment for the atomic bomb victims.

Furthermore, the notification announces the Committee Chairman (the National Research Council Chairman Haruo Hayashi), Vice-Chairmen (The National Research Council Vice-Chairman, Yoshio Tanaka and the Ministry of Education's Science Education Bureau Chief Kyōsuke Yamazaki), the heads of the nine divisional sub-committees and each of the divisional sub-committee members. The nine divisional sub-committees were Physical Chemistry/Physiography, Biology, Machinery/Metal/Raw Materials, Electric Communications, Civil Engineering/Constructions, Medicine, Farming/Fisheries, Forestry and Veterinary Animal Husbandry.[49] The specialist scientists who are referred to here are people who, until shortly before their appointments had supported the wartime science mobilization position. How did the large numbers of atomic bomb victims appear to the scientists?

While providing relief assistance and protection to atomic bomb victims was not referred to in the objective of the nation's research program, as far as victims were concerned, the nation's mobilization position itself represented a greater problem than the will of individual scientists. Either way, the scientists were once again mobilized to conduct investigative research in relation to the new set of circumstances without being given any opportunity to reflect on the war.[50]

The scope of the divisional sub-committees' areas of expertise corresponded with – and perhaps exceeded – the scope of the investigation being carried out by the Manhattan Project Investigation Group. This indicates that the Japanese researchers had extended beyond the sphere of simply co-operating with the Occupation Forces' investigation and that they had begun groping their way along the road to Japan's becoming an atomic power.

According to point 8 of MacArthur's Directive No 3 dated 22 September, research relating to atomic power was prohibited [51] but large scale Japanese atomic weapons research activities were not. It appears that investigations relating to the damage and effects of the weapon known as the atomic bomb were not included in the category of atomic power research. For the US, obtaining research data on the consequences of the atomic explosion was the primary aim, so they could not afford to ban the Japanese from atomic bomb research. On this point, the interests of both countries – the bomber and the bombed – coincided. Incidentally, on 27 September the Emperor paid a visit to MacArthur.

Co-operation with Medical Investigations

As for the investigations into the injuries inflicted on the human body, people from the Army Medical College and Tokyo Imperial University attended the Tokyo No 1 Temporary Military Hospital on 20 September, and discussed both the overall investigation activities to date and future strategies. On 22 September at the Medical Department of Tokyo Imperial University, the US research team (Manhattan Project Investigation Group, GHQ Military Medical Unit) met Professor Takeo Tamiya, Head of the Medical Department of Tokyo Imperial University, Professor Kanshi Sassa, Assistant Professor Jin Miyake and others.

At the meeting Colonel Oughterson demanded research co-operation on the medical front, saying 'The investigations are a joint Japanese-US venture, and in order to obtain research data it is necessary to have full co-operation from the Japanese.' As a result, two Japanese research teams were created, comprising personnel from Tokyo Imperial University Medical Faculty and the Institute of Physical and Chemical Research for Hiroshima and Nagasaki.[52]

It is noteworthy that neither the annals of Tokyo Imperial University Medical Faculty nor those of the University itself[53] make any reference to this atomic bomb research co-operation with the US forces. However, in a report edited by the US, a list of the members of the Tokyo Imperial University Investigation Team is cited together with words of appreciation.[54] This tells us that the ruling class of Japan consciously concealed the fact that Japan was co-operating with the atomic bomb investigations.

In the records covering that period there is no reference made to the medical research centering on the Japanese Military's Army and Navy Medical Schools. However, as we have already learned, co-operation between the Japanese military and the Occupation Forces continued. Disarmament and demobilization of the Japanese military were conditions of surrender and as the military could not publicly conduct research, it was necessary to work through the Medical Faculty of Tokyo Imperial University after the commencement of the Occupation.

On 12 October MacArthur ordered the formation of a Joint Commission to conduct research into the atomic bomb in Japan. Simultaneously he ordered the commanding officer of the US 6th Army, which had jurisdiction for the area west of Kyoto, to render assistance to the Joint Commission and issue them with special passes for access into prohibited areas in order to carry out appointed research duties.[55] The 'Joint Commission' referred to in this order was comprised of the Manhattan Project Group, the GHQ Military Medical Department and the Japanese Government group (represented by Masao Tsuzuki) and was chaired by Colonel Oughterson. The Japanese Government group was the investigation team formed as a result of the 22 September meeting at the Tokyo Imperial University Medical Faculty, referred to earlier.

On 11 November MacArthur ordered the Imperial Government to supply sufficient food provisions to the members of the Japanese investigation team, made up of sixty doctors and students, for the duration of the investigation, scheduled to conclude on approximately 1 December. This order is to be found amongst the Foreign Ministry's documents which are open to the public. It is recorded that after receipt of this order its execution was supervised and dealt with by the Central Liaison Office.[56] In Nagasaki the US Naval Technical Mission to Japan (headed by Naval Captain Shields Warren) joined the team and the investigation proceeded.[57]

The Medical Division of the Special Committee of the National Research Council was of the grandest scale, headed by Masao Tsuzuki who was also the Japanese Government representative on the Joint Commisson.[58] The Medical Division's research group consisted of 33 members of the medical chapter, 150 research officers and 1500 assistants.[59] The US Pacific Army General Headquarters Medical

Department ordered the Japanese investigation team to submit a written monthly report in Japanese, with an English translation, to the Joint Commission.[60] None of the many thousands of doctors who were engaged in these investigations requested either MacArthur, the Supreme Commander of the Allied Powers, nor the Japanese Government to establish a national relief and treatment agency for the atomic bomb victims. The doctors assumed the same posture as the nation; i.e., they had abandoned the idea of providing relief or treatment to the victims, utilizing them instead as mere research subjects.

At the Scientific National Council Special Committee on 24 October, the Ministry of Education abolished the Labor Management, Jet Propulsion, Aviation Fuels, Electromagnetic Wave Weapons, Magnetism Weapons, Noise Weapons and the People's Armament Weapon Special Research Committees (Directive No 9) and created anew the Transport Research Special Committee and the Atomic Bomb Casualty Research and Survey Special Committee (Directive No 11).[61] The government agencies that co-operated with the formally inaugurated Special Committee for the Investigation of the Effects of the Atomb Bomb in its activities included the War Department, the Navy Department, the Information Bureau, the Ministry of Home Affairs, the Ministry of Health and Welfare, the Agriculture and Forestry Ministry and the Ministry of Transport. Up until this point the special committees and the Japan-US Joint Commission had been pigeon-holed separately [62] but in fact together they comprised the investigation team in its entirety. In accordance with national policy, that represented the full level of research co-operation provided by the Japanese Imperial Government.

US Strategic Bombing Survey Team, the Atomic Bomb Documentary Film

The Special Committee established a film crew drawn from the Japan Movie Company and commissioned it, as a supporting body, to make a documentary film about the atomic bomb. It commenced filming on 25 September. On 29 October it completed its first shootings in Hiroshima and Nagasaki. On 12 December GHQ ordered a ban on shooting documentary films, but the film was nevertheless completed at the request of the US Strategic Bombing Survey. On 15 December GHQ's Military Medical Surgeon General ordered the Japanese Government to provide the Hiroshima/Nagasaki documentary film, held by the Japan Movie Co., to the Atomic Bomb Commission of the Military Medical Surgeon General's Department. Subsequently, on 17 December, the US Naval Technical Mission to Japan ordered the Japanese Government to provide every assistance to the members of the Japanese Investigation Group (physicists/

film makers) in relation to the research in the Nagasaki district, and on 18 December it commanded that the Military Medical Surgeon General ensure the Japan Movie Co. finish the positives of the Hiroshima and Nagasaki atomic bomb documentary films and submit them by 22 December.[63] This is the first time that the existence of a GHQ command to the Japanese Government relating to the atomic bomb documentary film in the official archives has been confirmed.

Thus it was that, at the end of April 1946, 'The Effect of the Atomic Bombs – Hiroshima and Nagasaki' (narration and subtitles in English) was completed. So far it was presumed that this film and the negatives were all seized by the Occupation Forces. But if it was part of the investigation conducted under the national policy of co-operation with the Occupation Forces, they should be regarded as having been 'provided to,' rather than 'seized by' the Occupation Forces.[64]

The goal of the US Strategic Bombing Survey (headed by Talakin D'Oliver) was to investigate all the effects of the US bombing and to report any information which would be useful in terms of national defence or war strategies to the Army and Navy Departments. These investigations were not limited to the atomic bomb attacks, but were concerned with the effects of the other air raids too.[65]

In November 1945 Britain also dispatched atomic bomb investigation teams to Hiroshima and Nagasaki to research the effects of the atomic bombs. In 1946 it published a report entitled 'The Effects of the Atomic Bombs Dropped on Hiroshima and Nagasaki.'[66]

The National Research Council's Special Committee for the Investigation of the Effects of the Atomic Bomb held its first briefing session on 30 November 1945, the second on 28 February 1946 and the third on 7 April 1947. The results of each divisional sub-committee were discussed. Through the organization of the Special Committee 'The work of each university, research facility and hospital, etc., was collated and codified. Research funds were distributed and research continued until 1947.' [67] It is not clear, however, what the pretext was for collaborating in the atomic bomb research with the Occupation Forces, nor what sums of money were involved.

The Army and Navy Ministries were abolished on 30 November 1945. The Army Medical College and other facilities were also disbanded. On this same date, the Army Medical College published a copy of the Atomic Bomb Survey Report, the 'Medical Report on Atomic Bomb Damage in Hiroshima' and submitted it to GHQ.[68] Incidentally the Navy's Kure Naval Station also submitted reports entitled 'Investigations into the Atomic Bomb in Hiroshima' and 'Research into the Biological Effects of the Atomic Bomb.'[69] The Japanese research results relating to the area of medicine were submitted to Masao Tsuzuki and the other results to Yoshio

Nishina. Later on these were submitted to the Economic and Scientific Section, Scientific and Technical Division (ESS/ST)[70] of GHQ/SCAP. The reports were translated either by the Japanese or GHQ.[71]

The Medical Section of the Special Committee held its first briefing session at the Tokyo Imperial University on 19 December 1945, with subsequent briefings on 23 January, 16 February and 25 February 1946. The head of this section, Masao Tsuzuki, submitted reports in English to the US members of the Joint Commission in April, May, June and September 1947.[72]

The following comment was made by the Committee Report in reference to the research activity conducted by the Special Committee in the investigations so far: 'realistically for the three years to the end of 1947, or virtually one and a half years from around the latter half of 1945, the work had stopped, never to be restarted. This was partly because of the radical changes seen in the educational and academic systems after the war, however, it was the Occupation that had the greatest effect.'[73] Reference is also made to the influence of the GHQ press code (cf. ch. 24).

However, the fundamental questions as to why the US, which had dropped the atomic bombs, allowed the victim nation to do research and why Japan co-operated in this research were never asked. Moreover, the fact that the Japanese research never included providing aid and treatment to the atomic bomb victims amongst its objectives was never acknowledged or criticized. Responsibility for this silence was side-stepped by the Japanese nation and attributed, instead, to the Occupation press code. This is consistent with Japan's failure to ask the US to take responsibility for its actions in dropping the atomic bombs.[74]

Meanwhile, at the Manhattan Project Commission, Major Furman, the Group 3 Leader, submitted a summary report of developments in relation to atomic and nuclear power in Japan to Brigadier General Newman.[75] On 30 November, Colonel Warren, the Group Leader of Groups 1 & 2, submitted to Brigadier General Groves a report on the preliminary surveys of the Hiroshima and Nagasaki atomic bombs.[76] The US – Japan Joint Commission compiled a report entitled, 'Medical Report on the Effect of Atomic Bombs in Japan by the US – Japan Joint Commission.'[77] According to the report entitled 'Casualties in Hiroshima and Nagasaki,' 'Most of the material was the result of survey research activities, conducted either voluntarily by Japanese doctors and researchers, or through the organization of the National Research Council Special Committee. It was submitted to the Americans ostensibly as a joint survey. Although it took the form of a submission, one cannot deny that there was intervention on behalf of the occupying nation.'[78] This, of course, is evidence of the two nations co-operating successfully in the investigations.

The Establishment of ABCC: the Atomic Bomb Effects Research Institutes in Hiroshima and Nagasaki

The US carried out two atomic bomb experiments at the Bikini atoll in July 1946. The Atomic Energy Law was effected on 1 August that year. The Manhattan Project facility was handed over to the Atomic Energy Commission on 31 December and the wartime atomic bomb manufacturing system became an atomic bomb mass production system in peace time.[79] The official history of the Atomic Energy Commission records that: 'Under normal circumstances, humans were not subjects for laboratory radiation experiments. However, the atomic bombs in Hiroshima and Nagasaki in August 1945 gave an exceptional (unusual and promising) opportunity for measuring the effects of radiation on mass groups of humans.'[80] 'Measuring the effects of radiation on mass groups of humans' was what the US atomic bomb manufacturing interests needed more than anything. It was for this purpose that the early stage research results of Hiroshima and Nagasaki were compiled.

It was under these circumstances that the Secretary of the Navy, James Forrestall, advised President Harry S. Truman to continue research into the sicknesses which came about as a result of the after-effects of atomic bomb radiation. Having received this advice, Truman, on 26 November 1946, ordered the US Academy of Science – National Research Council to establish the Atomic Bomb Casualty Commission 'to conduct continuous and long-term biological/medical research into the effects of the atomic bomb on humans.'[81]

The US Academy of Science – National Research Council sent an ABCC Preliminary Research Group (led by Dr Austin M. Brues, Dr Paul Henshaw and three others) to Japan. They arrived in Tokyo on 25 December 1946 and conducted a preliminary study until 30 December with a view to long-term continuous research.[82] We cannot list all of the Japanese people interviewed for this research because they are too numerous, but most of them were medical professionals who had been engaged in radiation research. It is worth mentioning, however, that they interviewed Ryūtarō Azuma at the Medical Bureau of the Ministry of Health and Welfare.[83] This indicates that the focus of radiation research had shifted towards medical research with co-operation from the Japanese Ministry of Health and Welfare.

The National Research Council published the 'Atomic Bomb Casualty Commission, General Report,' which included the results of the preliminary research, on 25 March 1947.[84] The report recommended the ABCC conduct research into cancer, leukemia, shortened life-span, reduced vigor, altered development, sterility, changes in vision, abnormal pigmentation, epilation, changed immunity levels and a type of genetic research called 'genetic system change.' In particular, in relation to genetic research, the Council recommended the organization of a system of registering pregnancies in Hiroshima, Nagasaki and control areas (Kure City was chosen for this), to conduct follow-up research on abnormal childbirth. It recommended that co-operation from gynecologists and midwives be sought for the purpose of establishing a comprehensive register of births and deaths in order to establish a system of periodical checks on the offspring of atomic bomb victims. The Council recommended that this genetic program be handed over to the Japanese Government.

A member of ABCC Preliminary Research Group, First Lieutenant James V. Neel, started haematology research on atomic bomb victims at the Hiroshima Red Cross Hospital in April 1947. Meanwhile, the Japanese Ministry of Health and Welfare established the Institute of Health on 21 May 1947 (re-named the National Institute of Health on 31 May 1949). On 23 May, the Director of the GHQ Public Health and Welfare Section, Captain Crawford F. Sams, met Neel from the ABCC to suggest the possibility of utilizing this newly established Institute of Health to assist with the execution of the ABCC's genetic program.[85] On 6 June, Marine Colonel Warren from the ABCC, Marine Colonel Carl Tessmer, Rokuzō Kobayashi, the Director of the Institute of Health, Kikuo Hamano, the Director of the Preventive Hygiene Bureau of the Ministry of Health and Welfare, and Masao Tsuzuki from the Science Council of Japan and others had a meeting to discuss the relationship between the ABCC's genetic program and the Institute of Health. At the meeting, the Japanese representatives explained the 'Outline of the Genetic Research Plan on Atomic Bomb Casualties'[86] and outlined the co-operation sought from the Japan Genetic Science Council.[87]

The 'Outline of the Genetic Plan' states that 'as the effectiveness of atomic energy and the methods involved in its use advance in the future, research into its effect on the human body will not only be a concern for Japan but for the welfare and public health of the human race as a whole. It is of the utmost importance that the Institute of Health, which should be the contact point of all international medical fields, leads the way in this research in Japan, the country which has suffered the effects of atomic blasts first hand.' Again, there was no mention of a plan for treating the atomic bomb victims in this paper.

The fact that the interests of the victims, suffering as they did, were discarded and the purpose of the research established as being the possible benefits of the use of nuclear power in the future and the effect it has on genetics, implies that the Japanese themselves had altered the purpose behind their atomic bomb research co-operation.

Subsequently the ABCC and the Institute of Health held many discussion sessions with a view to studying problems relating to the genetic program.[88] In the US, Neel presented an outline of the atomic-bomb victim genetic plan in the *Science* magazine of 10 October 1947. The Institute of Health appointed Fumio Kida (Professor of Kumamoto Medical College) in charge of the genetic plan.[89] On 24 December, Hamada, Director of the Prevention Bureau of the Ministry of Health and Welfare, compiled a program for an agreement to provide for the Asano Library to be used as ABCC's research rooms for twenty years. In return, ABCC would pay two million yen to Hiroshima City.[90]

On 14 February 1948, the Allied Army Headquarters announced that they would establish research institutes for the ABCC in Hiroshima and Kure, with Marine Colonel Tessmer taking office in March. Later the Japanese Ministry of Health and Welfare established the Atomic Bomb Effect Research Institute (Cabinet Order No 273), to operate branches of the Institute of Health in Hiroshima City, Nagasaki City and Kure City. Hiroshi Maki was the first Director, serving concurrently as the Deputy Director of the ABCC.[91] In Nagasaki, R. Brewer and others from the ABCC commenced research at the Nagasaki Public Health Centre in July.

The ABCC in Kure was completed in October 1948. On 17 December the Ministry of Health and Welfare presented Brigadier General Sams with a report (in English) entitled 'The Atomic Bomb Casualty Investigation Project.'[92] This planning paper is an important document, providing a summary of Japan's research co-operation up to that point and a clear description of the Japanese Government's policy from then on.

This paper reports that the atomic bombs in Hiroshima and Nagasaki may have had various biological effects on the victims, in particular, 'the genetic effects, alteration in the normal patterns of child growth, development of hematopeutic disorders and malignant changes of keloid are among those which should be studied thoroughly medically.' It is noteworthy that the 'nuclear age' was already predicted: 'Since the results of this survey are expected to contribute a great deal to the protection of the welfare of mankind in wartime as well as in the coming peaceful "atomic energy era", we should not lose this golden opportunity to solve this problem once and for all for the benefit of mankind.' Since the treatment of victims was not an objective of the Special Committee for the Investigation of the Effects of the Atomic Bomb of the National Research Council, it was not mentioned in the report.

Moreover, this research project was a joint enterprise by the ABCC and the National Institute of Health. The support received from GHQ, the National Research Council and the Japanese Ministry of Health and Welfare was stated in the plan. It was reported that, as a matter of policy, the National Institute of Health would establish Atomic Bomb Effect Research Institutes in Hiroshima and Nagasaki and that the joint research project would run for twenty years. In addition it also mentioned proposals for setting rules by which the Board of the Atomic Bomb Biological Committee would operate, staff numbers and budgetary considerations. The budget was about 3.83 million yen for 1948 and 14.99 million yen for 1949. Surely budgets of this size would have been sufficient to build a national hospital for the victims of the atomic blasts.

The ABCC opened a temporary research centre at Gaisen-Kan in the town of Usjina, Hiroshima City in July 1949. This was followed by the construction of another research centre in Hijiyama Park in November 1950. It was in July 1950 that the ABCC's Nagasaki quarters were completed and ready for use.

The ABCC conducted a survey of the casualty population in 1949. In addition, a national survey of surviving victims was conducted as an adjunct to the national census of 1 October 1950 following a request from the ABCC who tallied the survey results.[93]

Special attention should be paid to the survey concerning the effects of radiation in relation to genetics. In this survey, over 72,000 cases of pregnant women, both atomic bomb casualties and non-casualties, were examined in Hiroshima City, Nagasaki City and Kure City (the control site) as part of a study of the effect of radiation on pregnant women, their spouses and the children who were born (including dead foetuses). While the 'deformities' of children were checked, the survey revealed that the effect of the radiation had been underestimated. A summary of the results of the survey (conducted from 1948 to 1952) was published in *Science* magazine in the US in 1953,[94] and a complete report paper was published in 1956.[95] The ABCC donated this report to the Japanese National Diet Library. The receipt stamp, dated 20 June 1960, is on the back page of the report. Although this report was not kept secret in Japan, as far as I am aware the purpose of the survey, the method by which it was conducted and its results essentially have not been critically appraised until today.

The San Francisco Peace Treaty was signed on 8 September 1951. This coincided with the 'The Atomic Bomb Casualty Survey Report Summary' which was published in August 1951. GHQ ordered the Japanese Government to submit this report to them on 4 September.[96] In December, with the co-operation of the ABCC and the Hiroshima Medical Society, the effects of the atomic blasts were made public with the staging of the Atomic Bomb Effects Presentation. In January 1952, people involved in the ABCC, the National Institute of Health and the Science Council of Japan held a briefing session and discussion conference. These were followed by a special presentation on illnesses caused by atomic bomb radiation, given by members of the Hiroshima Medical Society at its fourth general meeting in February.[97] In April, a 'Symposium on Radiation and Atomic Bomb Sicknesses' was held at the general meeting of the Japan Haematology Society.[98]

On 28 April 1952, the Japan Peace Treaty and the Japan-United States Security Treaty came into effect. Japan, excluding Okinawa and some other areas, became an independent country.

First Steps Towards a 'Nuclear Nation': A Historical Irony

Having witnessed the US-Japan investigations carried out in Hiroshima and Nagasaki immediately after the atomic bombs were dropped, I feel a strong need to question the path that the Japanese Government followed and the role that it played in the research efforts. It is certainly surprising that the interests of the US, the country which first developed nuclear power and which first utilized it in the creation of atomic bombs, and Japan, the first country ever to suffer an atomic attack, should have matching interests in the ensuing research.

I have not focussed on any individual researcher, doctor or scientist because Japan, as a nation, was mobilized by a national policy regarding atomic investigative research. It was not that I wished to ignore the question of individuals' will or freedom to study and research, but rather that it can be presumed such freedom hardly existed. However, surely there must have been an inquiring mind present amongst the scientists who wanted to survey the actual scene of the atomic bombings irrespective of the nation's policy direction. Therein lies the magic of science.

In conclusion I would like to mention that I once heard from an atomic bomb victim the expression 'the ten dark years,' in reference to the period between the beginning of the Occupation and 1957 when the A-Bomb Victims Medical Care Law ('a law relating to the medical treatment of the atomic bomb victims') was enacted. These words are etched deeply on my memory. As discussed, the Japanese Government did not act in the interests of the atomic bomb victims. It did not do what they wished. On the contrary, it thoroughly utilized both the victims of the atomic bombs and non-victims as subjects for its research. In this sense, the existence of Hiroshima and Nagasaki undoubtedly represented a 'golden opportunity' in terms of realizing the 'atomic energy era' which was referred to in the research planning paper submitted to the Occupation Forces by the Ministry of Health and Welfare. Thus we see that Japan, itself a victim of the atomic bomb, has taken the first step towards becoming 'a nuclear nation.' This is a historical irony and one of the aspects of life that was changed as a result of the liberation of nuclear power.

Notes

1 Nihon Kagakushi Gakkai (History of Science Society of Japan), 1964 [1] p. 44.

2 Nihon Gakujutsu Kaigi Genshi Bakudan Saigai Chōsa Hōkokusho Kankōkai Henshū Iinkai (Science Council of Japan, Editorial Committee on Atomic Bomb Damages), 1951 [4] & 1953 [5]
3 Hiroshige, 1973 [2] p. 215.
4 Hiroshima Genbaku Iryō-shi Henshū Iinkai (Editorial Committee on a Hiroshima Atomic Bomb Medical Treatment History), 1961 [8] p. 93.
5 Hiroshima Genbaku Iryō-shi Henshū Iinkai (Editorial Committee on a Hiroshima Atomic Bomb Medical Treatment History), 1961 [8] p. 93.
6 'Dai 3-bu Genbaku Shiryō-dan no katsudō (Part 3, activities of the Atomic Bomb Investigation Group),' in Hiroshima-ken (ed.), 1972 [12]; Hiroshima-shi Nagasaki-shi Genbaku Saigai-shi Henshū Iinkai (Hiroshima and Nagasaki Municipal Editorial Committee for Magazines on Atomic Bomb Damages), 1979 [13] pp. 387–95; Nishina Kinen Zaidan (The Nishina Commemorative Foundation), 1979 [14].
7 Nishina Kinen Zaidan (The Nishina Commemorative Foundation), 1979 [14] pp. 47–52, 55–8; Yomiuri Shimbun, 1968 [10]; Tsuneishi, 1990 [72] pp. 820–5.
8 Nishina Kinen Zaidan (The Nishina Commemorative Foundation), 1979 [14] p. 51.
9 'Dai 3-bu Genbaku Shiryō-dan no katsudō (Part 3, activities of Atomic Bomb Investigation Group)', in Hiroshima-ken (ed.), 1972 [12]
10 Hiroshima-shi Nagasaki-shi Genbaku Saigai-shi Henshū Iinkai (The Hiroshima and Nagasaki Municipal Editorial Committee for Magazines on Atomic Bomb Damages), 1979 [13] p. 388.
11 Bunsaku Arakatsu, 'Hiroshima-shi ni okeru genshi bakudan-teki chōsa' (Atomic bomb research in Hiroshima City),' in *Osaka Asahi Shimbun* (Osaka Asahi Newspaper), 14–17 September, 1945, at Nishina Kinen Zaidan, 1979 [14] pp. 204–8; Shigeki Sugiyama, 'Hiroshima-shi ni okeru genbaku chōsa (Atomic bomb research in Hiroshima City),' in *Osaka Asahi Shimbun* (Osaka Asahi Newspaper), 18 September, 1945, at Nishina Kinen Zaidan, 1979 [14] pp. 209–10.
12 'Shūshō, dai 3-setsu, genshi bakudan saigai gakujutsu chōsa (Last Chapter, paragraph 3, scientific research on atomic bomb casualty),' in Nagasaki Shiyakusho (ed.), 1979 [15]
13 Kenichi Shinohara, 'Butsuri gakusha no taiken – Nagasaki no hōshanō chōsa (A personal experience of a physicist – radiation research at Nagasaki),' in Nishina Kinen Zaidan, 1979 [14] pp. 181–4.
14 Matsumae, 1987 [17] pp. 53–58; Ministry of Foreign Affairs, 1952 [6] p. 541.
15 Shimomura, 1949 [3] p. 199.
16 Braw, 1988 [44] pp. 49–50, 191; Hiroshima-ken Ishikai (Hiroshima Prefectural Medical Association), 1970 [11] pp. 50–1.
17 *Asahi Shimbunsha* (Asahi Newspaper), 11 August , 1945.

18 Nakajima, 1984 [18] pp. 54–7; Sasamoto, 1989 [19] pp. 115–18; Kitayama, 1988 [20] pp. 289–327; Braw, 1988 [44] pp. 49–50, 190–2.
There is hardly any Japanese record regarding this matter. The above records are based on records in the US and Australia. This suggests a cover-up by Japan.
19 Braw, 1988 [44] p. 50.
20 Smythe, 1951 [22]; Groves, 1964 [25]; Gruef, 1970 [27]
21 Kesaris, 1977 [29] book 1, vol. 4, ch. 6.
22 Goutsmitt, 1977 [28]
23 Nakagawa, 1985 [33] pp. 56–7.
24 *Mainichi Shimbun* (Mainichi Newspaper), 25 October 1986.
25 Hacker, 1987 [38] ch. 4.
26 Nishina Kinen Zaidan (The Nishina Commemorative Foundation), 1979 [14] p. 121–2; A. A. Liebow, 'Saigai tono sōgū – Hiroshima igaku nikki (Encounter with disaster: a medical diary of Hiroshima 1945),' in Hiroshima-ken Iishikai [74] nos. 2–3, vol. 20, 1967, pp. 81–180.
27 Hiroshima-shi Nagasaki-shi Genbaku Saigai-shi Henshū Iinkai (The Hiroshima and Nagasaki Municipal Editorial Committee for Magazines on Atomic Bomb Damages), 1979 [13] p. 293.
28 Hiroshima-shi (Hiroshima City Office), 'Shōwa 20-nen Hiroshima-shi jimu hōkoku-sho narabini zaisan-hyō (Office report and financial tables of Hiroshima City in Shōwa Yr 20).'
29 Toshio Matsumura, *Daihon'ei Rikugun-bu 'tairikumei tairikushi sōshūsei'* (The Army of the Imperial Headquarters; the orders of the Army of the Imperial Headquarters; the instructions of the Army of the Imperial Headquarters: Complete edition) 10 vols., MT Shuppan, 1994
30 Imabori, 1959 [7] pp. 97–8.
31 Oughterson and Warren, 1955 [54] p. 7.
32 Hiroshima-ken (Hiroshima Prefecture), 1972 [12] p. 419.
33 Yasushi Mitsuya and others, Tōkyō Teikoku Daigaku Igaku-bu Sanfujin-ka (Dept of Obstetrics and Gynecology, Faculty of Medicine, Tokyo Imperial University), 'Genshi bakudan no josei sei kinō ni oyoboshitaru eikyō (Effects of the atomic bomb on women's sexual functions),' in Nihon Gakujutsu Kaigi…, 1953 [5] vol. 2, pp. 735–9.
34 Sasamoto, 1989 [19] pp. 119–20.
35 *Asahi Shimbun* (Asahi Newspaper), 29 August 1945.
36 Riku-gun Gun'i Gakkō, Rinji Tōkyō Daiichi Riku-gun Byōin (Army Medical College, Provisional Tokyo First Army Hospital), 'Genshi bakudan ni yoru Hiroshima sensai igakuteki chōsa hōkoku (Medical survey report on war casualty in Hiroshima by an atomic bomb),' in Nihon Gakujutsu Kaigi…, 1953 [5] vol. 1, p. 286.
37 Tsuneishi, 1984 [43]
38 C. F. Sams, *DDT kakumei: Senryō-ki no iryō fukushi seisaku o kaisō suru*

(The DDT revolution: Reflections on medical welfare policy during the Occupation), Eiji Takemae (tr.), Iwanami Shoten, 1986, pp. 32–5.
39 Ministry of Foreign Affairs, Archives and Documents Section, 1976 [21]
40 GHQ/SCAP, 1945–1952 [58] vol.1.
41 Imabori, 1959 [7] pp. 148–9.
42 Nishina Kinen Zaidan (The Nishina Commemorative Foundation), 1979 [14] p. 121; Marcel Junod, 'Hiroshima no sanka (Disaster in Hiroshima),' in *Sekijūji Shimbun*, 1 May, 1 June and 1 July 1983.
43 Imabori, 1959 [7] pp. 100–1.
44 GHQ/SCAP Records, [73] Documents regarding Manhattan Project Investigation Group-Third Group, Box No.7408, ESS(D)06309-06311; Yukuo Sasamoto, 'Genbaku chōsa' (Atomic bomb research), in Nakayama and Yoshioka (ed.), 1994 [77] pp. 19–26.
45 Kindai Nihon Kyōiku Seido-shi Shiryō Henshū Iinkai (Editorial Committee for the Historical Data of Modern Japan's Educational System), 1964 [41] vol. 19, p. 395; Abe, 1988 [45] p. 11.
46 Kindai Nihon Kyōiku Seido-shi Shiryō Henshū Iinkai (Editorial Committee for the Historical Data of Modern Japan's Educational System), 1964 [41] vol. 28, pp. 117–19.
47 Chūgoku Hōsō (China Broadcasting), *Genbaku kiroku eiga no sugao* (Real face of atomic bomb documentary film), broadcast on 6 July, 1984.
48 Hiroshima-shi Nagasaki-shi Genbaku Saigai-shi Henshū Iinkai (Hiroshima and Nagasaki Municipal Editorial Committee for Magazines on Atomic Bomb Damages), 1979 [13] Part III, pp. 59–383.
49 Committee members (as of 14 September 1945) published on the notice are as follows:
- Masao Kinoshita (Science Advisor, Ministry of Education);
- Physics/Chemistry/Geology Section: Masaharu Nishikawa (Section Chairman, Prof of Tokyo University), Yoshio Nishina (Institute of Physical and Chemical Research), Sakuhei Fujiwara (Scientist of Meteorological Institute), Bunsaku Arakatsu (Professor of Kyoto University), Seishi Kikuchi (Professor at Osaka University), Ryōkichi Sagane (Professor of Tokyo University), Kenjirō Kimura (ditto), San'ichirō Mizushima (ditto), Sakae Shinoda (Professor of Kyūshū University), Takeo Watanabe (Professor of Tokyo University);
- Biology Section: Kaname Okada (Section Chairman, Professor of Tokyo University), Yuzuru Ogura (Professor of Tokyo University), Teizō Ezaki (ditto);
- Machinery/Metal/Material Section: Masaichi Majima (Section Chairman, Professor of Tokyo University), Shōichi Noguchi (Professor of Tokyo University), Tokushichi Mishima (ditto);
- Electricity/Power/Communication Section: Shōji Sedō (Section Chairman, Professor of Tokyo University), Kan'ichi Ōhashi (Director of

Electric Laboratory), Shigeru Yonezawa (Director of Investigation Section, Engineering Bureau, Communication Institute);

- Engineering/Architecture Section: Yutaka Tanaka (Section Chairman, Professor of Tokyo University), Kiyoshi Mutō (Professor of Tokyo University), Korokurō Hirose (ditto);
- Medicine Section: Masao Tsuzuki (Section Chairman, Professor of Tokyo University), Masanori Nakaizumi (Professor of Tokyo University), Takehiko Kikuchi (Professor of Kyoto University), Shōzō Ōno (Professor of Kyūshū University);
- Agriculture/Marine Products Section: Ikusaku Amamiya (Section Chairman, Professor of Tokyo University), Yoshichi Asami (ditto), Issui Kawamura (Professor of Kyūshū University);
- Forestry Section: Ihachirō Miura (Section Chairman, Professor of Tokyo University), Kentarō Nakamura (Professor of Tokyo University);
- Veterinary Animal Husbandry Section: Kiyoshi Masui (Section Chairman, Professor of Tokyo University), Kiyotsuna Sasaki (Professor of Tokyo University).

50 Hitoshi Yoshioka, *Kagakusha wa kawaru ka – Kagaku to shakai no shisō-shi* (Will scientists change? A history of the ideology of science and society), Shakai Shisōsha, 1984.

51 Ministry of Foreign Affairs, Special Material Section, (ed.), *Nihon senryō jūyō bunsho* (Important documents of the Occupation of Japan), vol. 1, 1949 (re-published in 1989), pp. 79–86.

52 Hiroshima-shi Nagasaki-shi Genbaku Saigai-shi Henshū Iinkai (Hiroshima and Nagasaki Municipal Editorial Committee for Magazines on Atomic Bomb Damages), 1979 [13] p. 393; Nishina Kinen Zaidan (The Nishina Commemorative Foundation), 1979 [14] pp. 121–4.

53 Tokyo Daigaku Hyakunenshi Henshu Iinkai (Editorial Committee for a One-hundred Year History of the University of Tokyo) (ed.), *Tokyo Daigaku hyakunen-shi,* (The one-hundred year history of the University of Tokyo), 10 vols., University of Tokyo Press, 1985–1987.

54 Oughterson and Warren, 1955 [54] p. 432.

55 Nishina Kinen Zaidan (The Nishina Commemorative Foundation), 1979 [14] p. 121.

56 'Hiroshima Genshi Bakudan Chōsa-kai Nihon-gawa iin eno haikyū' (Regarding rations to the Japanese Committee members of Hiroshima Atomic Bomb Investigation Committee) (11 November, 1945, an order from GHQ/SCAP to the Japanese Government, CLO No.110, supervised by the Secretariat), Reel No.A'-0028, in Ministry of Foreign Affairs, Document and Archives Division (ed.), 1976 [21].

57 Reports of the US Naval Technical Mission to Japan, 1945–1946, 1983 [75] Reel No.10.

58 Hiroshima-shi (Hiroshima City Office), 1981 [16].

59 In addition to the previously mentioned committee members, the research group included Takeo Tamiya (Professor of Tokyo University), Kanshi Sassa (ditto), Hitoshi Miyake (Assoc. Prof. of Tokyo University), Ren Kimura (Prof. of Kyoto University), Shōgo Funaoka (ditto), Shun'ichi Mashita (ditto), Shigeteru Sugiyama (ditto), Shigeki Mori (ditto), Kōichi Takagi, Ryōjun Kinoshita (ditto), Nobuyoshi Fuse (ditto), Hiroyuki Fukushima (ditto), Shōichi Kaminaka (Prof. at Kyūshū University), Yoshisada Nakajima (ditto), Kōsaku Ono (ditto), Tōichirō Sawada (ditto), Michinori Hayashi (Prof. of Okayama Medical University), Kōhei Furuya (Prof. of Nagasaki Medical University), Kanryō Sera (Prof. of Kumamoto Medical University), Kenji Ibuka (Director of Army Medical College), Masatami Hirai (Army Medical Major General of Army Medical College), Seijirō Yokokura (Navy Medical Major General of Navy Medical College), Izumi Kanai (ditto), Minoru Katsumata, (Technical Officer at Ministry of Health and Welfare), Yoshio Furuya (ditto), Shigeru Takaori (Railway Officer). Nihon Gakujutsu Kaigi…, 1951 [4] p. 2.

60 Hiroshima-shi (Hiroshima City), 1981 [16] pp. 53–7.

61 Kindai Nihon Kyōiku Seido-shi Shiryō Henshū Iinkai (Editorial Committee for the Historical Data of Modern Japan's Educational System), 1964 [41] p. 119.

62 Hiroshima-shi Nagasaki-shi Genbaku Saigai-shi Henshū Iinkai (The Hiroshima and Nagasaki Municipal Editorial Committee for Magazines on Atomic Bomb Damages), 1979 [13] pp. 390–4.

63 'Hiroshima Nagasaki no kiroku eiga yōkyū no ken' (Regarding the request of Hiroshima/Nagasaki documentary film) (15 December 1945, an order from Chief Surgeon, G-2 to the Japanese Government), 'Nagasaki chiku Chōsa-dan ni bengi kyōyo kata meirei no ken (Regarding an order to supply conveniences to the Nagasaki area investigation group),' 17 December 1945, from the person in charge of the Technical Dept, US Naval Technical Detachment to the Japanese Government, supervised by Dept. 1; 'Hiroshima Nagasaki genshi bakudan eiga poji sakusei kata meirei no ken (Regarding an order to produce positive film of Hiroshima Nagasaki atomic bomb),' 18 December 1945, an order from Chief Surgeon, G-2 to the Japanese Government, supervised by Dept. 1; 'Nihon-sha no Nagasaki genshi bakudan higai eiga sakusei keii kata meirei no ken (Regarding an order concerning the details to produce a film by a Japanese company on damage caused by the Nagasaki atomic bomb),' 18 December, 1945, an order from the Bureau of War Department Intelligence, G-2 to the Japanese Government, supervised by Dept. 1; Reel No. A'-0033, in Ministry of Foreign Affairs, Document and Archives Division (ed.), [21] 1976.

64 Kano and Mizuno, 1965 [55]; Hiroshima-shi Nagasaki-shi Genbaku Saigai-shi Henshū Iinkai (The Hiroshima and Nagasaki Municipal Editorial Committee for Magazines on Atomic Bomb Damages), 1979 [13]

pp. 394–5; Nishina Kinen Zaidan (The Nishina Commemorative Foundation), 1979 [14] pp. 131–94; Chūgoku Hōsō (Chūgoku Broadcasting), *Genbaku kiroku eiga no sugao* (Real face of atomic bomb documentary film), 1984.

65 Hiroshima-shi Nagasaki-shi Genbaku Saigai-shi Henshū Iinkai (The Hiroshima and Nagasaki Municipal Editorial Committee for Magazines on Atomic Bomb Damages), 1979 [13] p. 395; The US Strategic Bombing Survey, 1946 [50]

66 *Eikoku chōsadan hōkokusho…*(The British Investigation Team Report…), 1946 [51].

67 Hiroshima-shi Nagasaki-shi Genbaku Saigai-shi Henshū Iinkai (The Hiroshima and Nagasaki Municipal Editorial Committee for Magazines on Atomic Bomb Damages), 1979 [13] pp. 391–2.

68 Imabori, 1959 [7] pp. 101–2; Sasamoto, 1989 [19] pp. 135–6; GHQ/SCAP Records [73]. The report translated into English, 'Medical Report on Atomic Bomb Damages in Hiroshima, Army Medical College, Tokyo Provisional First Army Hospital' (Box No.7407, ESS(D)06277-06279) is in the GHQ/SCAP Archives.

69 Nihon Gakujutsu Kaigi Genshi Bakudan Saigai Chōsa Hōkokusho Kankō Iinkai (The Science Council of Japan, Publication Group on Investigation Reports on Atomic Bomb Damages, Editorial Committee), 1953 [5]

70 GHQ/SCAP Records [73] ESS/STD data.

71 GHQ/SCAP Records [73]. All the reports of the first and second meetings by the Japanese Special Committee for the Investigation of the Effects of the Atomic Bomb were translated into English (TS-00072-00074 in Box AG-7 (7)). There is also a report entitled, 'Atomic Bomb Survey of Hiroshima City (General Survey), Kure Naval District,' which was translated into English by Kure Naval District, Box No. 7408, ESS(D)06317.

72 Hiroshima-shi (Hiroshima City Office), 1981 [16] pp. 53–7.

73 Hiroshima-shi Nagasaki-shi Genbaku Saigai-shi Henshū Iinkai (The Hiroshima and Nagasaki Municipal Editorial Committee for Magazines on Atomic Bomb Damages), 1979 [13] p. 392.

74 Sasamoto, 1989 [19] pp. 136–7.

75 GHQ/SCAP Records [73] To: Brig. Gen. J. B. Newman, From: R. R. Furman, Major, CE., Subject: 'Summary Report, Atomic Bomb Mission, Investigation into Japanese Activity to Develop Atomic Power,' 30 September 1945, Box No. STD-1, TS-00207.

76 Warren, 1947 [46]

77 Oughterson and Warren, 1955 [54]

78 Hiroshima-shi Nagasaki-shi Genbaku Saigai-shi Henshū Iinkai (The Hiroshima and Nagasaki Municipal Editorial Committee for Magazines on Atomic Bomb Damages), 1979 [13] p. 394.

79 Kawakami, 1974 [69] pp. 96–126.
80 Hewlett and Duncan, 1969 [68] p. 244.
81 Hiroshima-shi Nagasaki-shi Genbaku Saigai-shi Henshū Iinkai (The Hiroshima and Nagasaki Municipal Editorial Committee for Magazines on Atomic Bomb Damages), 1979 [13] p. 395.
82 GHQ/SCAP Records [73], documents of Atomic Bomb Casualty Committee preliminary investigation, Box No. 7408, ESS(D)06313. Hiroshima-shi (Hiroshima City Office), 1981 [16] pp. 58–103.
83 GHQ/SCAP Records [73] Atomic Bomb Casualty Commission, Report No.1, For the Period November 22–30, 1946. Box No. 7408, ESS(D)06313.
84 Atomic Bomb Casualty Commission, 1947 [59]
85 GHQ/SCAP Records [73] Public Health and Welfare Section, To Col. C. F. Sams, Chief, Public Health and Welfare Section, GHQ, SCAP, From J. V. Neel, 1st Lt, M. C., Subject: Memorandum on a Conference Concerned with the Genetic Program of the Atomic Bomb Casualty Commission, 23 May 1947, Box No. 9354, PHW01727.
86 Hiroshima-shi (Hiroshima City Office), 1981 [16] pp. 348–9.
87 GHQ/SCAP Records [73] Public Health and Welfare Section, Memorandum For Records: Subject: Conference with Dr. Paul E. Henshaw of ESS Section concerning the Genetic Program of the Atomic Bomb Casualty Commission, 9 June 1947, Box. No. 9354, PHW01727.
88 GHQ/SCAP Records [73] ABCC documents, Box. No. 9354, PHW01727.
89 Kokuritsu Yobō Eisei Kenkyūsho (National Institute of Health), 1948 [60] no. 2, p. 1.
90 Kokuritsu Yobō Eisei Kenkyūsho (National Institute of Health), 1948 [60] no. 2, p. 1.
91 Kokuritsu Yobō Eisei Kenkyūsho (National Institute of Health), 1948 [60] no. 2, p. 1.
92 GHQ/SCAP Records, [73] Ministry of Welfare, Japanese Government, To: Chief, Public Health & Welfare Section, from Chief, Liaison Section, Minister's Secretariat, Ministry of Welfare, Subject: The Atomic Bomb Casualty Investigation Project. Through: Central Liaison and Coordinating Office, (16 December 1948), Box. No. 9354, PHW01706; Ministry of Foreign Affairs, Archives and Documents Section, 1975 [21]. Submittal of Proposed Plan for Atomic Bomb, (17 December, 1948, a memorandum from the Japanese Government to GHQ/SCAP, CLCO4156, LSAW271), Reel No. A'-0024; Takemae, 1994 [76] vol. 18, p. 15052.
The main text is excluded from this report by Ministry of Foreign Affairs; the full report is only available in the Archives of GHQ/SCAP mentioned above [73]. I hope the report will be released by the Japanese Government.
93 Chūgoku Shimbun-sha (Chūgoku Newspaper) (ed.), *Nenpyō Hiroshima yonjū-nen no kiroku* (A chronological table, a record of forty years of Hiroshima), Chūgoku Shimbunsha, 1986, p. 68.

94 J. V. Neel, W. J. Shull, D. J. MacDonald, N. E. Morton, M. Kodani, K. Takeshima, R. C. Anderson, J. Wood, R. Brewer, S. Wright, J. Yamazaki, M. Suzuki and S. Kitamura, 'The Effect of Exposure to the Atomic Bombs on Pregnancy Termination in Hiroshima and Nagasaki: Preliminary Report,' in *Science*, Vol.118, November 6, 1953.

95 Neel and Shull, 1956 [53]

96 'Genshi bakudan "saigai chōsa hōkoku" yōkyū no ken' (Regarding the request of Atomic Bomb 'Casualty Investigation Report') (Order from GHQ/SCAP to the Japanese Government on 4 September, 1949), Reel No. A' – 0031, in Ministry of Foreign Affairs, Documents and Archives Division (ed.), 1976 [21]

97 Hiroshima-shi Nagasaki-shi Genbaku Saigai-shi Henshū Iinkai (The Hiroshima and Nagasaki Municipal Editorial Committee for Magazines on Atomic Bomb Damages), 1979 [13] p. 398; Hiroshima-ken Ishikai (Hiroshima Prefectural Medical Association), 1952 [74] vol. 5, p. 29.

98 Hiroshima-shi Nagasaki-shi Genbaku Saigai-shi Henshū Iinkai (The Hiroshima and Nagasaki Municipal Editorial Committee for Magazines on Atomic Bomb Damages), 1979 [13] p. 399; Nihon Ketsueki Gakkai (Japan Haematology Association) (ed.), *Hōshanō narabini genbaku shōgai* (Radiation and atomic bomb damage), (Ketsuekigaku Tōgi-kai hōkoku dai5-gō) (Haematology Conference report no. 5), Nagai Shoten, 1953.

Research Notes – Submission of the Atomic Bomb Casualty Report by the Japanese Imperial Government

As stated in the main text, a representative from the Japanese Imperial Government submitted the Atomic Bomb Casualty Report to the Forward Headquarters on 3 September 1945. The fact that this report was submitted is noted in the Medical Study Report released by the Americans. After finishing this manuscript, I checked some GHQ/SCAP materials in the possession of the National Diet Library and found the report that was presumed to have been submitted. It was in the GHQ/SCAP Records (Top Secret), full Translation of 'Reports on Effect of Atomic Bomb at Hiroshima,' 6 September 1945, Translation by G-2, GHQ, AFPAC Advisory Committee, Box No. 450, AG(B)00528, owned by the National Diet Library.

It is in two parts. The first part is entitled 'Report of Investigation of Hiroshima Air Raid Damage, 20.8.13' (dated 13 August 1945, 6 pages) and is an English translation of a report by an 'Army Medical Committee' of 30 members. It contains information on the causes of casualties. The leader of the study group was Colonel Kōichirō Toda MD. The second part is entitled 'Report on the Hiroshima catastrophe (regarding radioactivity)' (dated 15 August 1945, 4 pages) and is a report by an army medical group of four members. It contains research on radioactive materials caused by the atomic bomb. This research was by Lieutenant Keisuke Mizonō, MD of the Army Medical College, Lieutenant Kōichi Muraji MD, Lieutenant Kazuharu Kimura MD and Lieutenant Hidehiko Tamaki MD, of the Institute of Physical and Chemical Research. The group that compiled the first part was a study group based on the Department of the Army's Hiroshima Casualty Study Group, and that of the second part was a radiation study group sent by the Army Medical College.

I have more or less come to the conclusion that this report was submitted by the Imperial Government on 3 September due to the following fact found in the records of the Manhattan Project Investigation Group: One of the leaders of the group, Brigadier-General Farrell, sent a top secret telegram to Major General Groves, Commander of the Manhattan Project, dated 16 September and reported on the English translation of the report which had just been completed. Because Farrell had not yet had time to type a copy of the translated manuscript he was unable to send Groves the report. Farrell himself summed up the contents of the manuscript and presented it. (GHQ/SCAP Records, CINCAFPAC ADV, WARCOS, TOP SECRET, URGENT, Approved by T. F. Farrel, 6 Sept. 1945, untitled, Box No.AG-8, TS-00081, in the possession of the National Diet Library). The report put out by the Japanese on their investigations into radioactivity concluded that there wasn't enough radioactivity at the epicenter to cause harm to humans and that they were confident that it would be possible to cultivate vegetables at the Army training barracks. Farrell reports this also, but as for the conclusion that denies the existence of residual radioactivity, when considered in terms of the Japanese atomic damage exposure line, they appear to be compromising towards the US position. This is evidence of the Japanese and Americans' co-operation regarding the direction of research. Incidentally, the report mentions that 6 September was the day on which MacArthur ordered the Imperial government to co-operate with the Manhattan Project Investigation Group.

Bibliographic Guide

Atomic Bomb Investigations in Postwar Science and Technology History

[1] Nihon Kagakushi Gakkai (History of Science Society of Japan) (ed.), *Nihon kagaku gijutsu-shi taikei, Tsūshi 5* (A compendium of the Japanese history of science and technology, vol. 5), Daiichi Hōki Shuppan, 1964.

[2] Tetsu Hiroshige, *Kagaku no shakaishi – Kindai Nihon no kagaku taisei* (A social history of science: The structure of modern Japanese science), Chūō Kōronsha, 1973.

The Initial Investigation

[3] Kainan Shimomura, *Shūsen-ki* (Notes on the end of the war), Kamakura Shoten, 1949, republished by Kodansha Bunko, 1985.
[4] Nihon Gakujutsu Kaigi Genshi Bakudan Saigai Chōsa Hōkokusho Kankō Iinkai (Science Council of Japan, Editorial Committee on Atomic Bomb Damages) (ed.), *Genshi bakudan saigai hōkokushū: Sōkatsu-hen* (A report on atomic bomb damages: Summary volume), Nihon Gakujutsu Shinkōkai (Japan Society for the Promotion of Science), 1951.
[5] Nihon Gakujutsu Kaigi Genshi Bakudan Saigai Chōsa Hōkokusho Kankō Iinkai (Science Council of Japan, Editorial Committee on Atomic Bomb Damages) (ed.), *Genshi bakudan saigai hōkokushū: Dai-1 dai-2 bunsatsu* (A report on atomic bomb damages), 2 volumes, Nihon Gakujutsu Shinkōkai (Japan Society for the Promotion of Science), 1953.
[6] Ministry of Foreign Affairs (ed.), *Shūsen shiroku* (A historical record of the end of the war), Shimbun Gekkan-sha, 1952.
[7] Seiji Imabori, *Gen-suibaku jidai – Gendai-shi no shōgen* (The age of atomic and hydrogen bombs – A testimony of modern history), vol. 1, San'ichi Shobō, 1959.
[8] Hiroshima Genbaku Iryo-shi Henshū Iinkai (Editorial Committee on a History of Medical Treatment after the Hiroshima Atomic Bomb), *Hiroshima genbaku iryō-shi* (A history of the medical treatment after the atomic bomb), Zaidan Hōjin Hiroshima Genbaku Shōgai Taisaku Kyōgikai (Counter-Strategy Council on Hiroshima Atomic Bomb Damages, Inc.), 1961.
[9] Nihon Gensuikyō Senmon Iinkai (The Japan Council against Atomic and Hydrogen Bombs, Specialist Committee) (ed.), *Gensuibaku Hakusho: Kakusareta shinjitsu* (The white paper on atomic and hydrogen bombs – Hidden facts), Nihon Hyōronsha, 1961.
[10] Yomiuri Shimbun, *Shōwa-shi no tennō* (Shōwa Emperor in history), 4 volumes, Yomiuri Shimbunsha 1968.
[11] Hiroshima-ken Ishikai (Hiroshima Prefectural Medical Association) (ed.), *Genbaku no ki* (A record of the atomic bomb), vol. 1, 1970.
[12] Hiroshima-ken (Hiroshima Prefectural Government) (ed.), *Hiroshima kenshi: Genbaku shiryō-hen* (A history of Hiroshima Prefecture: Volume on the atomic bomb), 1972.
[13] Hiroshima-shi Nagasaki-shi Genbaku Saigai-shi Henshū Iinkai (The Hiroshima and Nagasaki Municipal Editorial Committee for Magazines on Atomic Bomb Damages), *Hiroshima Nagasaki no genbaku saigai*

(Damages caused by the Hiroshima and Nagasaki atomic bombs), Iwanami Shoten, 1979.

[14] Nishina Kinen Zaidan (The Nishina Commemorative Foundation) (ed.), *Genshi bakudan – Hiroshima Nagasaki no shashin to kiroku* (The atomic bombs: Photographs and records of Hiroshima and Nagasaki), Kofūsha Shoten, 1979.

[15] Nagasaki Shiyakusho (Nagasaki City Office) (ed.), *Nagasaki genbaku higaishi* (A compilation of damages caused by the atomic bomb dropped on Nagasaki), vol. 1, Nagasaki Kokusai Bunka Kaikan, 1979.

[16] Hiroshima-shi (Hiroshima City Office) (ed.), *Hiroshima shin-shi: Shiryō-hen (Tsuzuki shiryō)* (The new Hiroshima City: volume on data, vol. 1 [Tsuzuki Data]),' 1981.

[17] Shigeyoshi Matsumae, *Waga Shōwa-shi* (My history of the Shōwa period), compiled by Hisaya Shirai, Asahi Shimbunsha, 1987.

[18] Tatsumi Nakajima, 'Hiroshima: Sono kageri wa fukaku (Hiroshima: Its deep and dark clouds),' in *Gekkan Shakai-tō* (The Socialist Party monthly), July 1984.

[19] Yukuo Sasamoto, 'Genbaku shōdō chōsa ni okeru Nihon-gun no yakuwari (The role of the Japanese Military in the initial atomic bomb investigation),' in *Rekishi to shakai* (History and society), no. 9, 1989.

[20] Setsurō Kitayama, *Rajio Tokyo: Senji taiseika Nippon no taigai hōsō – III Haiboku e no michi* (Radio Tokyo: Japanese wartime overseas broadcasts, vol. III: The road to defeat), Tabata Shoten, 1988.

[21] Ministry of Foreign Affairs, Document and Archives Division (ed.), *Gaimushō kiroku: Rengō-gun no hondo shinchū narabini gunsei kankei ikken, Rengō-gun tono aida no raiōshin tsuzuri – Rengō-gun shirei-bu ōshin tsuzuri – Rengō-gun shirei-bu ōshin tsuzuri* (Foreign Ministry records – The two-way communications between the Allied Powers and the Ministry over the stationing of Allied troops in mainland Japan and over the military administration – A pad containing communications from the Allied Powers) A' 1. 0. 0. 2-2, 1976, held by the Foreign Ministry Foreign Diplomacy Historical Data Section at the National Diet Library.

From the Manhattan Project to the Atomic Bomb Influence (Effect) Investigation

[22] Smythe, H.D., *Genshi bakudan no kansei: Smythe hōkoku* (The Completion of the atomic bomb: The Smythe Report), Yoshio Nishina, (editorial supervision), Asao Sugimoto, Eizō Tajima, and Eiichi Kawasaki (tr.), Iwanami Shoten, 1951.

[23] F. Neibel, and C. Bailey, *Mohaya kōchi nashi – Hiroshima genbaku tōka no himitsu* (Not even any high ground – Secrets concerning the dropping of the atomic bomb on Hiroshima), Masahiro Sasagawa and Reiko

Sugibuchi (tr.), Kōbunsha, 1960.

[24] G. Hewlett and Oscar E. Anderson Jr., *The New World, 1939/1946: A History of the United States Atomic Energy Commission*, University Park, 1962.

[25] Leslie R. Groves, *Watashi ga genbaku keikaku o shiki shita* (I directed the atomic bomb plan: The inside story of the Manhattan Project), Kengo Tominaga and Yuzuru Sanematsu (tr.), Kōbunsha, 1964.

[26] A. K. Smith, *Kiki to kibō – Amerika no kagakusha undō* (A Peril and a Hope: The Scientists' Movement in America 1945–1947), Tetsu Hiroshige (tr.), Misuzu Shobō, 1968.

[27] S. Gruef, *Manhattan keikaku – genbaku kaihatsu gurūpu no kiroku* (The Manhattan Project: Records of the group that developed the atomic bomb), Seitarō Nakamura (tr.), Hayakawa Shobō, 1970.

[28] S. A. Goutsmitt, *Nachi to genbaku – Arusosu: Kagaku jōhō chōsa-dan no hōkoku* (The Nazis and the atomic bomb – Alsos: Report of the Scientific Information Survey Group), Kazuo Yamazaki and Tsūji Konuma (tr.), Kaimeisha, 1977.

[29] Paul Kesaris (ed.), *A Guide to the Manhattan Project: Official History and Document*, A Microfilm Project of University Publications of America, Inc., 1977, held by the National Diet Library.

[30] Martin Sherwin, *Hametsu e no dōtei – Genbaku to dai 2-ji sekai taisen* (The road to destruction – Atomic bombs and World War II), Mikio Katō (tr.), TBS Britannica, 1978.

[31] Gordon Thomas and Max Morgan-Witts, *Enola Gay – Dokyumento genbaku toka* (Enola Gay – Documenting the dropping of the atomic bomb), Sen Matsuda (tr.), TBS Britannica, 1980.

[32] Yasuo Nakagawa, 'Kokusai hōshasen bōgo kijun no hensen to genshiryoku kaihatsu (Changes to the international radiation protection criteria and the development of atomic energy),' in *Kobe Daigaku kyōyo-bu kiyō ronshū* (Journal of the College of Liberal Arts at Kobe University), no. 33, March 1984.

[33] Yasuo Nakagawa, 'Manhattan keikaku no hōshasen hibaku kanri to hōshasen eikyō kenkyū (The research of the Manhattan Project on radiation exposure management and radiation effects),' in *Kobe Daigaku kyōyō-bu kiyō ronshū* (Journal of the College of Liberal Arts at Kobe University), no. 36, October 1985.

[34] Vincent C. Jones, *Manhattan, the Army and the Atomic Bomb*, (United States Army in World War II, Special Studies), Center of Military History, United States Army, 1985, held by the National Diet Library.

[35] Yasuo Nakagawa, 'Hiroshima Nagasaki no genbaku hōshasen eikyō kenkyū: Kyūsei-shi kyūsei shōgai no kashō hyōka (Research into the effects of the Hiroshima and Nagasaki atomic bombs' radiation: The underrated sudden deaths, sudden injuries),' in *Kagakushi kenkyū* (Studies of science history), Period II, vol. 25, no. 157, 1986.

[36] Yasuo Nakagawa, 'Hōshasen ni yoru banhatsuteki eikyō no kashō hyōka (I) (The underrated late-developing effects of radiation (I)),' in *Kagakushi kenkyū* (Studies of science history), Period II, vol. 26, no. 163, 1987.
[37] Yasuo Nakagawa, 'Hōshasen ni yoru banhatsuteki eikyō no kashō hyōka (I) (The underrated late-developing effects of radiation (I)),' in *Kagakushi kenkyū* (Studies of science history), Period II, vol. 26, no. 164, 1987.
[38] Barton C. Hacker, *The Dragon's Tail: Radiation Safety in the Manhattan Project, 1942–1946*, University of California Press, 1987.
[39] Yasuo Nakagawa, *Hōshasen hibaku no rekishi* (A history of radiation exposure), Gijutsu to Ningensha, 1991.
[40] Akira Yamagiwa and Seiitsu Tachibana (eds.), *Shiryō Manhattan keikaku* (Reference material – The Manhattan Project), Seiitsu Tachibana (tr.), Ōtsuki Shoten, 1993.

Total Co-operation with the Investigations of the Occupation Forces and the Postwar Science Promotion Policy

[41] Kindai Nihon Kyōiku Seido-shi Shiryō Henshū Iinkai (Editorial Committee for the Historical Data on Modern Japan's Education System) (ed.), *Kindai Nihon kyōiku seido shiryō* (Historical data on modern Japan's education system), vols. 19 & 28, Dai Nippon Yūbenkai Kōdansha, 1964.
[42] Keiichi Tsuneishi, *Kieta saikin-sen butai: Kantō-gun dai 731-butai* (The lost germ warfare corps: The Kwantung Army No. 731 Unit), Kaimeisha, 1981 (Chikuma Bunko, 1993).
[43] Keiichi Tsuneishi, *Hyōteki Ishii – 731 butai to Amerika chōhō katsudō* (The Target: Ishii – 731 Unit and American military intelligence activities), Ōtsuki Shoten, 1984.
[44] Monica Braw, *Ken'etsu 1945–1949: Kinjirareta ganbaku hōdō* (Censorship 1945–1949: Forbidden reporting on atomic bombings), Seiitsu Tachibana (tr.), Jiji Tsūshinsha, 1988.
[45] Akira Abe, *Kyōiku kankei hōrei mokuroku narabini sakuin: Shōwa-hen* (Educational ordinances: An inventory and index – Shōwa period, III), Kazama Shobō, 1988.

American Strategic Bombing Investigation Team, The Atomic Bomb Documentary Films, Survey Summaries

[46] S. L. Warren, *Preliminary Medical Report, Hiroshima-Nagasaki Bombings*, 27 Nov. 1947. Appears in Kesaris [29].
[47] The Manhattan Engineering Project, *Final Report of Findings of the Manhattan Project Atomic Bomb Investigating Groups at Hiroshima and Nagasaki*, 19 April 1946. Appears in Kesaris [29].

[48] The Manhattan Engineering Project, *The Atomic Bombings of Hiroshima and Nagasaki*, 30 June 1946. Appears in the Department of Foreign Affairs Document Section.

[49] The Manhattan Engineering Project, *Photographs of the Atomic Bombings of Hiroshima and Nagasaki*, 30 June 1946. Appears in Kesaris [29].

[50] The United States Strategic Bombing Survey, *The Effects of Atomic Bombs on Hiroshima and Nagasaki*, US Government Printing Office, 1946. Microfilm held by the National Diet Library. Japanese translation: 'Amerika gasshūkoku senryaku bakugeki chōsadan hōkoku' in Yoshito Ōsumi and Sadao Kamata (tr.), *Nagasaki Zosen Daigaku kenkyū hōkoku* (The Nagasaki Shipbuilding University research report), vol. 18, no. 1, 1977.

[51] *Eikoku chōsadan hōkokusho: Hiroshima oyobi Nagasaki ni tōkasareta genshi bakudan no kōka* (The British Investigaton Team Report: The effects of the atomic bombs dropped on Hiroshima and Nagasaki), 1946. Appears in Yamagiwa and Tachibana, 1993 [40].

[52] Reports of the U.S. Naval Technical Mission to Japan, 1945–46.

[53] J. V. Neel and W. J. Shull, *The Effect of Exposure to the Atomic Bombs on Pregnancy Termination in Hiroshima and Nagasaki*, National Academy of Sciences, National Research Council, USA, no. 491, 1956, held by the National Diet Library.

[54] Ashley W. Oughterson and Shield Warren (ed.), *Medical Effects of the Atomic Bombs in Japan*, National Nuclear Energy Series: Manhattan Project Technical Section, Division VIII – vol. 8, McGraw-Hill, 1955. Held by the National Diet Library.

[55] Ryūichi Kanō and Hajime Mizuno, *Hiroshima nijūnen – Genbaku kiroku eiga seisakusha no shōgen* (Twenty years after Hiroshima: Testimonies of the atomic bomb documentary filmmakers), Kōbundō, 1965.

[56] Akira Iwasaki, *Senryō sareta sukuriin – Watashi no sengo-shi* (The occupied screen – My postwar history), Tōyō Keizai Shinpōsha, 1975.

[57] Jun Furukawa, 'Senryō to chōhō: "Genbaku eiga" fairu to kiroku eiga no yukue (The Occupation and intelligence – The files of "Atomic Bomb Movies" and the future of documentary films), *Senshū hogaku ronshū* (Law journal of Senshū University), joint 55 and 56 editions, February 1992.

Structure of Atomic Bomb Casualty Research: ABCC and National Institute of Health

[58] GHQ/SCAP, SCAPIN-A, 1945–1952. Held by the National Diet Library.

[59] Atomic Bomb Casualty Commission, *General Report*, National Research Council, January 1947. Appears in Kesaris [29].

[60] Kokuritsu Yobō Eisei Kenkyūsho (National Institute of Health), *Kokuritsu yobō eisei kenkyūsho nenpō* (Annals of the National Institute of Health), no. 1 (1947), no. 2 (1948), no. 3 (1949), no. 4 (1950), no. 5 (1951), no. 6 (1952).

[61] 'ABCC to wa nani ka: Genshi Bakudan Shōgai Chōsa Iinkai o tazunete (What is the ABCC – We visited the Atomic Bomb Casualty Commission),' in *Kagaku Asahi* (Science Asahi), February 1950.

[62] 'Genbaku Shōgai Chōsa Iinkai (ABCC) wa nani o shite kitaka (What has the Atomic Bomb Casualty Commission (ABCC) been doing?),' in *Kagaku* (Science), vol. 22, no. 6, June 1952.

[63] National Research Council, Division of Medical Sciences, Committee on Atomic Casualities, Field Agency, *Semi-Annual report, 1 July 1952 thru 31 December 1952*, Headquarters, Hiroshima, Japan.

[64] Kokuritsu Yobō Eisei Kenkyūsho/Genbaku Shōgai Chōsa Iinkai (National Institute of Health/Atomic Bomb Casualty Commission) (eds.), *Kokuritsu Yobō Kenkyūsho/ABCC nijūnen no ayumi* (Twenty years of the National Institute of Health and the ABCC), Kokuritsu Yobō Eisei Kenkyūsho, 1966.

[65] Kokuritsu Yobō Eisei Kenkyūsho (National Institute of Health) (ed.), *Yoken nijūnen no ayumi* (Twenty years of the National Institute of Health), 1967.

[66] Yoshio Sugihara, 'ABCC mondai ni tsuite (On the ABCC issue),' in *Nihon no kagakusha* (Japanese scientists), vol. 2, no. 3, December 1967.

[67] ABCC Rōdō Kumiai Honbu (The ABCC Labor Union Headquarters) (ed.), *Tōron shiryō: Amerika no genbaku shōgai chōsa kenkyū kikan ABCC saihensei kōsō – Sengo o shorishi hibakusha kyūen zenjinrui-teki kikan to suru tameni* (Discussion data: the American Atomic Bomb Casualty Commission's 'restructuring concept – Proposal to address postwar issues and to make it an instrument for the aid of the bomb victims and for all human kind),' ABCC Rōdō Kumiai Honbu (ABCC Labor Union Headquarters), not-for-sale item, 1972.

[68] Richard G. Hewlett and Francis Duncan, *Atomic Shield, 1947–1952* (A History of the United States Atomic Energy Commission, Vol. 2), University Park, Pennsylvania University Press, 1969.

[69] Kōichi Kawakami, *Genshiryoku no seiji keizai-gaku* (The political economy of atomic power), Heibonsha, 1974.

[70] Genbaku Shōgai Chōsa Iinkai (ABCC) 1947–1975 (Atomic Bomb Casualty Commission 1947–1975), *ABCC-Yoken kyōdō kenkyū hōkokusho* (Joint research report of ABCC and the National Institute of Health), ABCC, 1975.

[71] M. Susan Lindee, 'What is a Mutation? Identifying Heritable Change in the Offspring of Survivors at Hiroshima and Nagasaki,' in *Journal of the History of Biology*, vol. 25, no. 2, Summer 1992.

[72] Keiichi Tsuneishi, 'Riken ni okeru uran bunri no kokoromi (The uranium splitting tests at the Institute of Physical and Chemical Research),' in

Nihon butsuri gakkai-shi (Journal of the Japan Physics Association), vol. 45, no. 11, November 1990, pp820–5.
[73] General Headquarters, Supreme Commander for the Allied Powers (GHQ/SCAP) Records. Held by the National Diet Library.
[74] Hiroshima-ken Ishikai (Hiroshima Prefectural Medical Association), *Hiroshima igaku* (Hiroshima medicine), vol. 5, nos. 5 and 6, 1952, vol. 20, nos. 2 and 3, 1967.
[75] Reports of the U.S. Naval Technical Mission to Japan, 1945–1946, Scholarly Resources Inc., 1983. Held by the National Diet Library.
[76] Eiji Takemae (editorial supervisor), Yukuo Sasamoto, Kazumoto Takano, Yukuo Wakamatsu (eds.), *GHQ e no Nihon seifu taiō bunsho sō shūsei* (The Japanese Government's policy towards the GHQ; Total collection of documents), MT Shuppan, 1994.
[77] Shigeru Nakayama and Hitoshi Yoshioka (eds.), *Sengo kagaku gijutsu no shakai-shi* (A social history of postwar science and technology), Asahi Shimbunsha (Asahi Sensho), 1994.

3 Destruction Of Cyclotrons

Shigeru Nakayama

From its invention by E. O. Lawrence in the 1930s, the cyclotron type of particle accelerator was central to nuclear research. The beam it produced could be applied to every aspect of science and technology, including biology and medicine. Thus, its installation was symbolic of the approaching nuclear age. Before WWII, Japan had been second only to the United States in the number of cyclotron installations.[1] Although the destruction of cyclotrons by the Occupation Army was 'accidental,' it sparked a conflict between the military Occupation Forces and the international scientific community. The military considered the cyclotrons to be weapons – or potential weapons – to be destroyed whereas the scientists considered them as instruments of basic research to be preserved. Hence, their destruction marked a turning point in the Occupation science policy with a powerful symbolic meaning and strong resonance for the scientific community.

The Occupation Forces' Science Policy

The Occupation Forces issued a memorandum on science and technology policy on 2 September 1945 which: ordered the Japanese to keep all research and development facilities intact until further notice; ordered a basic report on all scientific activity being conducted by laboratories and other scientific and technological organizations, and; established guidelines defining what was permissible research and what was prohibited.

Following the precedent established during the occupation of Nazi Germany only months earlier, they confiscated useful research facilities and resources as war reparations.

On 22 September 1945 SCAP (the Supreme Commander of the Allied Powers) appointed an investigation team whose responsibilities included reporting on Japanese scientific activity and the status of (prohibited) nuclear research.[2] The Scientific Intelligence Survey Team headed by MIT President Karl T. Compton and Dean Edward L. Moreland arrived in Japan in late September 1945. The team included Russell Fischer, a scientist who had prior experience with the Survey team sent to Germany and was thus able to compare the two occupations. The team visited Rikagaku Kenkyūsho (the Institute of Physical and Chemical Research, abbreviated hereafter as Riken) and other sites, reporting to US President Harry S. Truman on 4

October 1945. They reported that nuclear research and development had been discontinued because of the bombing of Tokyo by the US Air Force.[3]

International Solidarity of Scientists

Yoshio Nishina, the Director of Riken, wanted to utilize the Japanese cyclotrons for peaceful purposes. Before the war Nishina Laboratories at Riken had used the cyclotron for biological research; they simply wanted to continue this research in the postwar world. To this end Nishina formally requested permission from Compton to use his cyclotrons. Nishina and Compton, although situated on opposite sides in the War were both internationally recognized nuclear scientists. Nishina had faith that Compton, as a scientist, would understand and believe his claim that the research was for peaceful and non-threatening purposes, even though the military did not. Nishina proudly represented Japanese scientists to the international community on this issue.

Compton responded that he had come to Japan as a scientist-advisor without any power but had advised MacArthur to grant the permission sought. Following the advice of Moreland, Nishina wrote to MacArthur, seeking permission to utilize the cyclotron for biology, medicine, chemistry and metallurgy.[4] He avoided mentioning physics, fearing that it might be construed as relating to nuclear weapons and therefore jeopardize the entire proposal. On the following day, 16 October, Moreland's memo to MacArthur said there was no problem in resuming cyclotron use. On 17 October a memorandum was issued and on 19 October GHQ Public Relations announced that 'Nishina's proposal is permitted; based on the report of Dr. Compton and others that cyclotrons cannot be used for research against the interest of Allied Forces.'[5]

In summary: the science advisors to the Occupation Forces supported Nishina's petition to use the cyclotrons, and; GHQ military officers, trusting in their scientists' advice, granted Nishina permission without reference to Washington.

Lack of Information in Washington

The world of physicists is very cosmopolitan. For the officers in daily contact with Japanese scientists, it was utterly unthinkable that they might secretly intend to resume military research. Washington, however, was different. They were not so quick to shift out of a wartime mindset or reconsider Japan's status as an enemy. Washington did not comprehend the rapid changes that occurred in the Japanese world-view after the surrender. Hence their opinions and judgments were dominated by abstract general principles and were much more conservative than GHQ's. They issued directives derived from these general principles which did not fit the Occupation reality.

One might even say that in a very non-scientific manner, the Washington bureaucrats responsible for Occupation science policy dismissed any evidence from the 'field' (GHQ, SCAP, special investigators, etc.) that conflicted with their already formed views. But speculation aside, we know that the formulation of Washington's science policy for the occupied area did not include any consultation with Compton or Moreland – their appointed 'frontline' advisory team – on the question of what to do about the cyclotrons. Washington did not accept the special exemption given to Nishina, relying on their general principles rather than their scientists' advice. Needless to say, the relationship between the military bureaucrats and scientists – both American and Japanese – was less than friendly.

This tension and distrust was not limited to the Washington bureaucrats: Nishina's permission was first challenged by Major J. A. O'Hearn of GHQ's Economic and Scientific Section Industry Division (ESS/ID), who was not satisfied that this matter should be decided solely on the scientists' advise. After examining directives from Washington, O'Hearn concluded that the use of the cyclotrons should be more restricted. On 27 October a memorandum was issued prohibiting its use for chemistry and metallurgy, although it was still permitted for biology and medicine.[6] The justification seems to have been that biological and medical applications were for the treatment of cancer and had nothing to do with military affairs, while chemistry and metallurgical research must have some relation to nuclear weapons. After issuing his memorandum, O'Hearn requested feedback from Washington about his action.

Presumably in response to O'Hearn's query, a telegram was sent from the Joint Chiefs of Staff to the Commander-in-Chief of the Pacific and Chinese Area on 30 October with a directive to confiscate all research facilities with atomic potential, put those in charge of the facilities in custody, report on the activities of each facility and prohibit further research activity. The Joint Chief of Staff also requested that the Commander-in-Chief report on all actions that SCAP took in Japan.[7]

On 7 November a message addressed to MacArthur was drafted in the name of the Secretary of the Army, Robert P. Patterson, ordering the destruction of all cyclotrons as soon as all of the technical information about them had been collected.[8] The precise process of the decision-making in Washington on this issue remains unknown to this day. Unconfirmed hearsay (by Fischer) has it that Brigadier General Leslie Groves appointed two young staff officers to the Manhattan Project, who then drafted this message while Groves was away and transmitted it through Secretary of the Army Patterson to MacArthur.[9]

On 19 November, in accordance with the directive from Washington, SCAP cancelled all earlier memoranda to Nishina, implemented a total ban on cyclotron use by the Japanese and ordered the destruction of all existing cyclotrons.[10] The officer in charge of this operation was cautious enough to have widely distributed copies of the decision to various related offices in Washington requesting their comments, but nobody paid any attention. On 20 November all four cyclotrons in Japan (two in Riken, one each in Kyoto and Osaka Imperial Universities) were confiscated by SCAP officers, together with related documents. On 24 November all of the cyclotrons were destroyed.[11] Thus SCAP personnel under the command of Major O'Hearn faithfully executed the official directive from the Secretary of the Army.

The Response in the Scientific Community

The scientific community could not understand what the military had done. Everybody in the community thought that the mission of the military was only to confiscate and destroy Japanese military facilities. The cyclotron was such a precious resource for research that it should not have been destroyed. Shock and outrage at the destruction was certainly not limited to Japanese and American scientists. Chinese physicists were angry because they had wanted one for reparation[12] as did the Australians.[13]

The US scientific community had no intention of utilizing the confiscated Japanese cyclotrons. Nevertheless, the prevailing feeling amongst them was that it was shameful that the US had let their military destroy scientific facilities. Scientists at the Oak Ridge National Laboratory complained to the Department of the Army. Karl Compton wrote a letter of protest to Patterson (*The New York Times*, December 15, 1945) in which he mentioned that his examination of the two cyclotrons at Riken revealed that they were not powerful enough for military use but adequate for biological and medical research.[14] Numerous journalists blamed and ridiculed the military for its stupidity.

Nishina trusted that the scientific community shared – or at least understood – his point of view, even though the military might not. Nishina had a wide range of acquaintances in the international physics community, having studied in Europe and participated in the beginnings of quantum mechanics. They shared a scientistic value system and group ethic that transcended national barriers.

When he was informed that the cyclotrons had been destroyed, he immediately thought the news was mistaken. When he questioned SCAP about this, Major O'Hearn said 'of course' the US scientists, including Compton, had agreed with the destruction. This news robbed Nishina of all hope. A little later, though, he read articles in foreign newspapers about scientists' protests against the misconduct of SCAP and realized that O'Hearn had lied to him. Infuriated, Nishina sent a letter of protest to SCAP and demanded permission for the importation of radioactive isotopes from the USA in compensation for the destruction of the cyclotrons.[15]

Nishina firmly believed in the classical rational view of science 'for the sake of the development of science' and trusted that the world scientific community shared his belief. However, his experience showed that neither the military nor the people shared this view. The US scientific community initiated some fund-raising to compensate Nishina but it never materialized, having been abandoned in fear that such a gesture might conflict with popular sentiment – a popular sentiment that saw no reason to help Japan, their enemy. In spite of their cosmopolitan world-view, the scientific community was clearly involved in a political contest. In the background of the scientists' protest was growing political strife between scientists and the military about who should control nuclear energy. Scientists began to use the cyclotron incident when lobbying governments to demonstrate why the control of nuclear power could not be entrusted to the military.[16]

Responsibility of the Military

It is still puzzling that on 28 November – after the cyclotrons were destroyed – Dwight D. Eisenhower, the Chief of General Staff of the Department of the Army, sent a top-secret telegram to MacArthur querying the possibility of delivering the cyclotrons to the US before destruction.[17] MacArthur and his staff were surprised to receive this message that directly contradicted the earlier directive from the same Department. Later MacArthur was to write in his autobiographical memoir that in spite of the fact that he had opposed the directive from Washington to destroy the cyclotrons, he was held responsible for its execution.

GHQ in Tokyo was not happy about being blamed for the destruction of the cyclotrons. They found themselves between the confused directives from Washington and media accusations fueled by the scientific community. Accordingly, MacArthur sent a telegram requesting an investigation into the directive to destroy the cyclotrons. Washington replied that they had ordered confiscation but not destruction. GHQ responded that they had neither advised nor instigated the destruction of the cyclotrons, but had simply obeyed an order from the Secretary of the Army. The Secretary of State, Cordell Hull acknowledged that MacArthur was right.

In the beginning of the controversy, Secretary of the Army Patterson had considered the cyclotrons as machines for making atomic bombs and could see no reason not to destroy them. After the controversy, Patterson finally admitted (in a press interview on 10 January 1946) that MacArthur was right; Patterson confessed that he had not had the time to carefully read the directive prepared by his staff for the destruction of cyclotrons. It appears that besides Groves, there was nobody amongst the Army's upper echelons who was well versed in the operation and application of a cyclotron. Only after the incident was disclosed did they realize the gravity of their decisions.

The policy making process that resulted in the destruction of cyclotrons was a matter of concern not only to the US. They had a worldwide effect. Science policy makers from the other Allied Powers – especially Britain – who had complaints about the *de facto* US dominance of the Occupation Forces, sent a recommendation to GHQ that the destruction of Japanese scientific facilities and the prohibition of research should not be enforced in a hurry as they could have no harmful effect on the Occupation Forces.[18] Scientists did not consider the application of demilitarization policy to be urgent since they regarded the possibility of the Japanese continuing or resuming the war effort to be very low.

The Arrival of Kelly

The directive of 24 November was for the destruction of five cyclotrons in Japan, including (mistakenly) two cyclotrons at Osaka Imperial University (where only one existed). Major O'Hearn, who was responsible for the operation, discovered similar instruments for nuclear research, such as a Cockroft and a Van-de-Graaf high-voltage generating device for accelerating particles at the Kikuchi Laboratory in Osaka University. He wondered whether one of them might be another cyclotron, but was unable to decide this himself. He badly needed the advice of expert scientists. He requested Washington to immediately send two scientists who had participated in the Manhattan Project.[19]

In response to his request, two scientists engaged at MIT in radar research were appointed: Gerald W. Fox and Harry C. Kelly. According to Kelly, the day after the cyclotron incident an Army Colonel, who commissioned Kelly to go to Tokyo as a science advisor, visited him.[20] The Colonel never told him what was required of a science advisor. The Colonel mentioned the destruction of the cyclotrons, but not the ongoing trouble the Army had vis-à-vis the scientific community. Kelly, however, could guess the situation behind the proposal.

Kelly asked why he had been chosen. The response was 'Because you had nothing to do with the Manhattan Project or making atomic bombs. In other words, nobody can learn the secret of atomic bombs from you. On the other hand, we want you to collect whatever information the Japanese have about atomic bombs.' Officially, Kelly was assigned to Intelligence, a job that military people were familiar with and understood.

Despite the fact that the Colonel ordered Kelly to go to Japan immediately, he was late in arriving. When Kelly came to Japan in early 1946, SCAP did not know how to deal with a scientist and remained indifferent to him. In the meantime, the issue of responsibility for the cyclotron incident had been resolved.

Kelly's colleague, Fox, could not get along with his military superior, Brigadier General John W. O'Brien and soon resigned. Kelly was left alone. Rather than gathering intelligence about Japanese military facilities, Kelly preferred to work as a liaison between SCAP and the US scientific community, as well as liaising between SCAP and the Japanese scientific community. As early as 1946, he recommended that the ESS/ST science policy should develop S&T for economic recovery, rather than focusing on the demilitarization and surveillance of Japanese science as the military perspective would have it.

Aftermath

After the cyclotron incident, Nishina appealed to the world scientific community in an effort to take advantage of a widespread perception that the destruction of the cyclotrons was a disaster of the Occupation policy. Fueling the guilty consciences of the US scientific community, he hoped to win better conditions for Japanese researchers. Kelly and his associates at ESS/ST – people who had day-to-day contact with Japanese researchers – were sympathetic. In fact it was through the efforts and cooperation of Kelly that Nishina was able to send a proposal to SCAP to get permission for Japanese researchers to use the cyclotron at Lawrence-Berkeley Laboratory.

Although the cyclotrons were destroyed and nuclear research was prohibited, it took repeated petitions to SCAP before, finally on 2 May 1946, special permission was granted to Nishina for 'making and using stable isotopes of carbon, nitrogen and sulphur, for basic research into the improvement of agriculture, fishery, forestry and medical care.'[21]

However, as mentioned in section 3 above, general military staff that had no contact with Nishina and other Japanese scientists had no sympathy for them or the development of Japanese science. Furthermore, the Washington military cadres were primarily concerned with the developing Cold War. Having no direct contact with the people of the occupied area, they had no interest whatsoever in the pleas of Japanese scientists for freedom of research. Thus, when Dr. Paul Henshaw, responsible for Japanese basic science at ESS/ST, suggested that Nishina send a note to *Nature* magazine to denounce the US Army for their misconduct in destroying cyclotrons, it was Kelly who was concerned about possible negative repercussions and advised against it.[22] It is possible to view the entire cyclotron incident as an unintended consequence of Nishina's original request for special exemption from the prohibition of nuclear research. It raised a question that the Washington bureaucrats could only respond to by the application of a general principle that led to the destruction, not only of Nishina's cyclotron, but of the other cyclotrons at Osaka and Kyoto as well.

Nishina and others, such as Seishi Kikuchi of Osaka University, repeatedly petitioned for the reconstruction of cyclotrons and permission to conduct nuclear research but the military were not interested. The principle of prohibition was maintained throughout the Occupation period. Japanese scientists had, perhaps, overestimated the influence of the scientific community on military decision-makers.

With the constant effort and support of Kelly, the reconstruction of the Riken cyclotron was eventually permitted to commence in 1951 under a new name: Kagaku Kenkyūsho (Scientific Laboratory). In contrast, the US military had given the Germans permission for the peaceful use of cyclotrons in March 1949.

In comparison to the Western physics community – who had been almost totally involved in the war mobilization, the Japanese physicists had maintained relatively more freedom even during World War II. Certainly this was partially due to the fact that the Japanese military did not expect much from academic researchers, but the effect was still relatively greater scientific freedom. Thus it is quite understandable how Nishina could maintain his scientist belief in 'science for science's sake,' a belief that science was somehow outside of or above national and international politics. It seems that Japanese scientists had expected that the world scientific community would soon 'return' to the idyllic prewar state of affairs, as soon as the second war was over and they were liberated from wartime mobilization. Yet, as soon as the war ended there was conflict between the scientific community and the military regarding the control of nuclear weapons – and therefore all nuclear research. Then, as the Cold War deepened, scientists were again involved in the maneuvering of the military-industrial complex. The classical ideal of freedom of research was greatly curtailed. Nishina stands out for his efforts to maintain the ideal of prewar academic science, but he was perhaps naïve in his approach to the postwar reality that the world scientific community faced.

Notes

1 Weiner, 1974 [4]

2 SCAPIN 47; No.3, issued 22 September 1945. On the prohibition of nuclear research, the Japanese Imperial government 'will prohibit all

research or development of work which has as its object affecting mass separation of Uranium 235 from Uranium, or affecting mass separation of any other radio-actively unstable elements.'

3 *Report of the Scientific Intelligence Survey in Japan*, dated 1 November.
4 Letter to General MacArthur from Dr. Yoshio Nishina, dated 15 October 1945, TS-00213 (TS is an abbreviation for Top Secret).
5 SCAPIN 154, 17 October 1945, Memo for record RCK/JAO'H/rhw 16 October 1945, Permission for the Use of Cyclotrons at Riken.
6 SCAPIN 196, 27 October 1945, ESS-IN Memorandum for The Imperial Japanese Government, 27 Oct. 1945 TS-00213 Memo for record, J A O'H.
7 JCS serial #17, or TS-00213.
8 Special Security Message GSB-183, dated 9 November 1945, TS-00213.
9 Personal communication from Dr. Samuel Coleman.
10 SCAPIN 307, 17 November 1945.
11 K. T. Compton [2]
12 Joseph and Dorothy Needham, *Science Outpost*, 1948.
13 Morris S. Low, 'Mosquito Coils, Bearings and Batteries; The Australian Scientific Mission in Occupied Japan, 1945-47,' paper presented at Japanese Studies Association of Australia, Sixth Biennial Conference, 6-9 July 1980, p. 8.
14 *New York Times*, 15 December 1945. Cf. From SCAP To JCS – URGENT: Memo for record, JAO'H. TS-00213.
15 Yoshio Nishina, 'An Account of the Destruction of the Cyclotrons in the Nishina Laboratory,' 20 December 1945, TS-0037.
16 Alice Kimball Smith, *A Peril and a Hope: The Scientists' Movement in America*, 1945-47.
17 Groves 1964 [7] chapter 31.
18 *Policy Towards Pure and Applied Scientific Research in Japan.*
19 GHQ/SCAP, OUTGOING MESSAGE, ESS/IN RCK/JAO'H/LOM, date 24 November.
20 'Charles Weiner's Interview with Dr. Harry C. Kelly at Raleigh, North Carolina, October 27, 1975,' preserved in MIT Archives.
21 SCAPIN 1128-A, 2 May 1946 (ESS(A)-09777).
22 Personal communication (April 1989) of Robert Multhauf, a historian of science, who worked for ESS/ST at the time.

Bibliographic Guide

[1] Occupation Forces Sources:

- The most standard secondary source is David L. Putnam, 'Re-organization of Science and Technology in Japan (1945 to September 1950),' in *History of the Non-military activities of the Occupation of Japan* (Kokkai Toshokan (National Diet Library) microfilm number CHS-04304).

- Sources for the cancellation of the use of cyclotron to Nishina and its destruction is classified in TS (top secret) 00213. It includes a report by Yamazaki (Kelly's interpreter) on the investigation into the confusion surrounding the directives for the destruction of the cyclotrons.
- The Occupation of Japan was not executed through direct military control but rather by indirect governance through the Japanese bureaucracy. Hence, SCAP sent directives to the Japanese Government in the form of its memoranda (SCAPIN).

[2] Directives of SCAP for the destruction of cyclotrons:
A dispatch of the Public Relation Section of GHQ to those concerned in SCAP beforehand, dated 19:30, 23 November 1945. It conveys the directive of MacArthur to begin the destruction of cyclotrons at 10:00 on the following day but keep it secret until it was officially announced at 12:00. It also briefly describes what the Occupation Forces had so far discovered in their investigations of the Japanese cyclotrons and nuclear research, and includes short biographies of Yoshio Nishina, Bunsaku Arakatsu and Seishi Kikuchi. We can judge from their investigation report that Japanese nuclear research had not yet reached a 'problematic' level. It is unclear why the cyclotrons were destroyed. The dispatch consists of three typewritten pages.

[3] Japanese accounts:
On Riken
- Shin'ichirō Tomonaga and Hidehiko Tamaki (ed.), *Nishina Yoshio, Denki to kaisō* (Yoshio Nishina: biography and recollections), 1952.
- Hidehiko Tamaki, 'Aru kenkyūsho no unmei – Eikō to kunan no Rikagaku Kenkyūsho (A fate of a research institute: Glory and hardship for Riken),' in *Kagaku Asahi*, November 1961.

On Osaka University
- Junkichi Itō, 'Handai no kyū saikurotoron shimatsuki (Disposal of Osaka University cyclotron),' in *Iwanami kōza gendai butsurigaku geppō* (Monthly bulletin of the Iwanami series on contemporary physics), no.9, 1965.

On Kyoto University
- Shigeru Nakayama, 'Harii Kerii (Harry Kelly),' in Kagaku Asahi (ed.), *Sukyandaru no kagakushi* (History of scandalous science), Asahi Shimbunsha, 1988.
- Tsūji Konuma and Yoshio Takata, 'Nihon no genshiryoku kenkyū ni tuiteno dainiji taisengo no senryō seisaku (Occupation policy on Japanese nuclear research after World War II),' in *Kagakushi kenkyū* (Studies of the history of science), pp. 138–46, Autumn 1992.

[4] American accounts:
- Charles Weiner, 'Cyclotrons and Internationalism: Japan, Denmark and

the United States, 1935–1945,' in *Proceedings of XIVth International Congress of the History of Science*, 1974 (Tokyo, 1975).

Later, the author wrote 'Japan's Nuclear Bomb Project,' in *Science*, no. 199, 17 February 1978, which was widely recognized as evidence of Japanese wartime nuclear bomb research. In spite of the author's accurate evaluation and precaution not to be misunderstood, it had the effect of alleviating the guilt amongst Americans for having dropped the atomic bombs.

[5] Repercussion of cyclotron destruction:
- 'Scientists at Oak Ridge Decry Wrecking of Japan's Cyclotrons,' New York *Herald Tribune*, 26 November 1945.
- K. T. Compton, quoted in the *New York Times*, 6 December 1945.

[6] Kelly's arrival:
- Harry C. Kelly Collection, North Carolina State University Library, Raleigh, North Carolina.
- Kelly's last position was the Vice Chancellor of North Carolina State University, Raleigh. His widow donated his archives to the Library.

[7] Others (secondary):
- Leslie R. Groves, *Now It Can Be Told: The Story of the Manhattan Project*, 1964.

4 Demilitarization and the Peaceful Remobilization of Manpower

Yukuo Sasamoto

Following its defeat in the Second World War the Japanese military was disbanded. At the same time the structure for mobilizing science for the war effort was dismantled. This chapter looks at the following questions: What was the aim of the Allied Forces in disbanding the Japanese military and dismantling the system of wartime science mobilization? How did the Japanese implement the disbanding? And what problems accompanied the demilitarization? Finally, how was the peaceful conversion of the manpower that had been engaged in the wartime science mobilization managed after the war?

Overview of the War-time Science Mobilization Structure

Tetsu Hiroshige has already provided an overview of the wartime science mobilization system in *Kagaku no shakaishi – Kindai Nippon no kagaku taisei* (A social history of science – The structure of modern Japanese science). On 14 October 1943 controls were announced on the Research Mobilization Council and the Provisional Wartime Researchers' Establishment System and on 24 March 1944 the first sixty-six wartime researchers were appointed. In the 1944 budget, brought down by the Cabinet on 26 November 1943, the promotion of science and technology research was the top priority for consideration amongst the important policies set out. This prioritization was 'deemed to mean "strengthening the mobilization of science and technology and delivering scientific research with the aim of strengthening our physical fighting capacity – air power in particular".' On 6 November 1943 the Institute of Physical and Chemical Research 'established six "research departments to finalize the war" and, with the participation of researchers from outside the Institute, closely occupied themselves with war-time research.'[1]

But little progress was seen in the research itself due to a lack of research data. Hence, on 10 July 1944, as the war entered its final phase, the Cabinet decided on a general plan for the mobilization of science and technology. This 'involved reorganizing the mobilization of technicians, chiefly with the aim of increasing aircraft production, and re-allocating those with technical expertise away from areas where the need was low and deploying them in aviation. Training centers were established in factories, and office workers who had been trained in the humanities were retrained as technicians. The number of science students was further increased and the disciplines of wartime technical guidance personnel and wartime technical personnel were established. It was arranged that mobilization activities could be carried out beyond the boundaries of the undertaking.'

Furthermore, on 10 August the Cabinet decided that employees of government agencies (technicians) should be mobilized and sent to the aircraft factories. On 24 August the Ministry of Education decided on a means of mobilizing student intelligence: 1000 post-second year university and vocational training schools were to be excused from physical labor mobilization and engaged in war-time scientific research. On 29 August the Cabinet made a decision on the issue of the militarization of science and technology. The Committee for the Application of Military and Naval Technology was established consisting of members from the army and navy as well as others from outside the forces. The committee's aim was to centrally coordinate the research and production of new weapons. Further, since 'seven of the ten external members were key members of the Science Research Council, university researchers were forced to conduct research as directed by the military. It was also decided to expand and strengthen the Scientific Research Council.'[2] On 5 December Hidetsugu Yagi was appointed to the position of President of the Technical Institute with hopes for the development of a new scientific weapon.

Japan's Defeat in the War and the Disbanding of the Japanese Military

On 10 August 1945 the Japanese Government announced to the Allied Forces that it would accept the Potsdam Declaration with conditions. On 11 August the Allied Forces, through James Francis Byrnes, the US Secretary of State, responded to the Japanese Government's conditional acceptance of the declaration. His response included the following statement:

> The Emperor will be required to authorize and ensure the signatüre by the Government of Japan and the Japanese Imperial General Headquarters of the surrender terms necessary to carry out the provisions of the Potsdam Declaration and shall issue his commands to all the Japanese military, naval and air authorities and to all the forces under their control wherever located to cease active operations, and to surrender their arms, and to issue such other orders as the Supreme Commander may require to give effect to the surrender terms.

At the Imperial Conference on 14 August Emperor Hirohito accepted the Potsdam Declaration and organized representations to the American,

British, Chinese and Soviet Governments to this affect. Article 9 of the Potsdam Declaration clearly states that 'The Japanese military forces, after being completely disarmed, shall be permitted to return to their homes with the opportunity to lead peaceful and productive lives.' On 15 August the Japanese Government expressed its wishes to the four governments. In reference to affecting disarmament it said '...it is suggested that the best and most effective method would be that under the command of His Majesty the Emperor, the Japanese forces are allowed to disarm themselves and surrender arms of their own accord.'[3] At noon on 15 August the Emperor announced the end of the war to the populace by Imperial Edict via a recorded message on the radio.

At the time of the surrender settlement, the strength of the Japanese army and navy was

> 169 general divisions of the army, 4 tank divisions and 15 air divisions amounting to some 5.5 million personnel of which approximately 2.4 million were in Japan and approximately 3.1 million were overseas. The navy had lost most of its warships. Most naval servicemen were in army units and could hardly be thought of as being navy personnel, but even so the numbers are the greatest recorded to date, with approximately 1.3 million in Japan and approximately 400 thousand overseas; a total of 1.7 million.[4]

The government and military leaders feared the troops might rebel in response to the surrender and, hand-in-hand with the decision to surrender, they made use of the authority of the Emperor. On 17 August the Emperor delivered an Imperial Rescript to the army and navy servicemen, ordering them not to rebel. Furthermore, members of the Imperial household were dispatched to important military posts overseas to represent the Emperor and persuade the people stationed there to comply with the surrender. In the end only a small number of military personnel vowed to continue the fight and no large-scale military rebellion eventuated.[5]

On 2 September 1945, aboard the US battleship Missouri, the surrender documents were signed and Japan formally accepted surrender. The wording of the documents ordered an unconditional military surrender: 'We hereby proclaim the unconditional surrender to the Allied Powers of the Japanese Imperial General Headquarters and of all Japanese armed forces and all armed forces under Japanese control wherever situated.'[6] Simultaneously all authorities in Japan accepted General Order No 1 (Army and Navy) of the Supreme Commander of the Allied Powers (SCAP). General Order No 1 confirmed the directive of the Emperor and made the surrender of the Japanese military clear. The Order also spelled out what information was to be reported to SCAP: e.g., the disposal of weapons and military facilities and the surrender of Japanese forces to the four Allied Powers, both within Japan and elsewhere. Furthermore, the Order specified which parties the Japanese troops in different places were to surrender to:

a) The senior Japanese Commanders and all ground, sea, air and auxiliary forces within China (excluding Manchuria), Formosa and French Indo-China North of 16 degrees North latitude, shall surrender to Generalissimo Chiang Kai-Shek.
b) The senior Japanese Commanders and all ground, sea, air and auxiliary forces within Manchuria, Korea North of 38 degrees North latitude, Karafuto, and the Kurile Islands, shall surrender to the Commander-in-Chief of Soviet Forces in the Far East.
c) (1) The senior Japanese Commanders and all ground, sea, air and auxiliary forces within the Andamans, Nicobars, Burma, Thailand, French Indo-China South of 16 degrees North latitude, Malaya, Sumatra, Java, Lesser Sundas (including Bali, Lombok, and Timor), Boeroe, Ceram, Ambon, Kai, Aroe, Tanimbar and islands in the Arafura Sea, Celebes, Halmahera and Dutch New Guinea shall surrender to the Supreme Allied Commander, South East Asia Command.
(2) The senior Japanese Commanders and all ground, sea, air and auxiliary forces within Borneo, British New Guinea, the Bismarcks and the Solomons shall surrender to the Commander-in-Chief, Australian Military Forces.
d) The senior Japanese Commanders and all ground, sea, air and auxiliary forces in the Japanese mandated Islands, Bonins, and other Pacific Islands shall surrender to the Commander-in-Chief, US Pacific Fleet.
e) The Imperial General Headquarters, its Senior Commanders, and all ground, sea, air and auxiliary forces in the main islands of Japan, minor islands adjacent thereto, Korea South of 38 degrees North latitude, Ryukyus, and the Philippines shall surrender to the Commander-in-Chief, US Army Forces, Pacific.[7]

From the perspective of the Japanese armed forces overseas, this order, 'insofar as it sealed the road to disarmament, the effect, as it were, was that of an "Anti-Insurgency Declaration", both as far as the liberation of people other than those of the Big Four powers was concerned and in terms of the independence of Asian people.'[8]

The Japanese army and navy issued an order to surrender – based on the surrender document and General Order No 1 – to all Japanese forces on 2 September 1945. The Special Continental No 1 Order of the Imperial Headquarters Military Division commanded that:

1. Hostilities towards the enemy will cease and all troops will disarm; 2. Weapons and equipment are to be maintained in their present condition and

delivered to the Allied Forces Military Commander; 3. Damaging or burning the following items is prohibited – maintain in their present state: a) munitions, b) transportation facilities and telecommunication facilities, c) military facilities and buildings, and all plans thereof, d) factories, workshops, engineering plants, research centers, testing centers, examination centers, data relating to technology, licenses, plans, drawings, discoveries, weapons, ammunition; 5. The manufacturing or division of war apparatus is prohibited; 6. Assistance for occupation and stationing of Allied Forces.[9]

Disbanding of the Japanese Military

In principle the disbanding of the Japanese military was to take place after the signing of the surrender documents on 2 September 1945, but the military had already begun to disband independently and irrespective of the Allied Forces plan, following an order issued on 18 August to that effect. On 23 August the Minister for the Army issued detailed instructions for the military units in Japan to disband. However, with up to 3.5 million members of the armed forces and other personnel overseas at the time, disbanding and repatriating required an enormous shipping capacity. The number of ships capable of being pressed into service was only 280,000 tons on the high seas and 140,000 tons in local waters, and half of that was required to meet the transportation needs at home. Hence the number of ships that could be pressed into service to transport disbanded personnel was very limited.

The naval warships employed to transport disbanded personnel included 'an all up total of 101 large and small vessels, from a single aircraft carrier down to the special subchasers. The total capacity to carry personnel was approximately 22,000 per voyage. Under these circumstances it could be expected to take over four years to repatriate all the personnel who were overseas. Consequently on 23 August the Ministry for the Army telegraphed its military commanders overseas, instructing them to act in accordance with the execution of the repatriation process: '1. While the standing rule is to the effect that the Japanese Army will be disbanded after repatriation, be attentive to convening and disbanding as many troops as possible where they are presently stationed. 2. Ask for co-operation in the autonomous execution of disarmament in accordance with the situation on the ground locally.'[10]

Certain military units were identified as special cases for immediate demobilization. For example, an order was given on 22 August for the disbanding of the railway shipping units in Japan to be completed as soon as possible. The demobilization and disbanding of units whose personnel were 'involved in the farming industry, particularly those for whom there is a need, such as those involved in transport or communications, those engaged in mining (especially coal) and those from other authorities who are presently serving in a military capacity' was declared a high priority.

The repatriation of special units such as chemical warfare and secret intelligence, regardless of whether they were in Japan or overseas, was given an even higher priority.[11]

An example of this sort of special repatriation can be seen in the case of the biological warfare units. On 8 August 1945 the Soviet Union had declared war on Japan and commenced attacks. On the day that they entered the war General Otozu Yamada, Commander of the Kwantung Army, issued an order to the Ishii and Wakamatsu Biological Warfare Units for the destruction of any records they held. He was afraid that biological weapons would be handed over to the Soviets and might subsequently be used against Japan. A strategy was developed which involved the destruction of records concerning 'the circumstances surrounding the bomb, the destruction of wooden structures and the "log" of massacres that had occurred in jails.' On 11 August the members of the Pingfan Ishii Unit Headquarters Corps, having destroyed their records, departed by train. None was captured by the Soviets. On 21 August the train pulled into Pusan and 'the members of the unit were gathered into groups of about 100 and put aboard landing craft. They were taken to the coast at Hagi in Yamaguchi Prefecture, where they were dismissed, each to make his own way home. Meanwhile, their leaders had returned to Japan by plane from Pyongyang.'[12] Here we can see that the Japanese military units had destroyed their records well before SCAP issued General Order No 1 directing them not to destroy anything, and to maintain records in their present state.

After the signing of the surrender documents on 2 September 1945, SCAP proceeded to abolish the Japanese military structures one after another. The Imperial Headquarters went on 13 September, the Military Command Section on 15 October, the General Staff Headquarters on 30 November and the Army and Navy Ministries on 1 December. The Japanese military reported to General Douglas MacArthur (SCAP) on the progress of troop demobilization (the report [in English] is dated 1 October and entitled 'The Process of Japanese Demilitarization').[13] On 16 October General MacArthur broadcast a radio announcement to the United States outlining the Occupation policies and the status of their implementation. He emphasized that 'the process of Japanese disarmament and military demobilization has taken place quickly and...exceedingly smoothly, the likes of which has no precedent in history. There has not been one bullet fired nor one drop of blood shed.' In other words, the demobilization of the Japanese military in Japan went very much according to plan.[14]

For the most part, the disbanding of the military units within Japan was completed by mid-October. The Ministry of Foreign Affairs records that in Japan there were approximately 2.19 million troops and that '2.14 million soldiers had been disbanded [97 percent] and 42,000 were still in service.' The Ministry's records show that, as of 27 October, there were approximately

3.68 million returning soldiers and 'the number demobilized was 100,000, or 3 percent of the total. On 15 April 1946 the number was 750 thousand, or 20 percent and on 15 June it was 1.91 million, or 52 percent.'[15]

Dismantling the Japanese Military and Cooperation with the US Occupation Forces [16]

As discussed in Chapter 2 ('Investigations of the Effects of the Atomic Bomb'), there was substantial cooperation from the Japanese in dismantling the Japanese military and related matters. The cooperation afforded to the atomic bomb research conducted by the US Occupation Forces is especially noteworthy. The Army and Navy Medical Divisions, the Army's Military Medical College, the Naval Military Medical College, the Army Hospital and the Navy Hospital all made tremendous contributions to these investigations through their research into the effects of the atomic bomb on the human body. Before the Japanese military establishments such as the Army and Navy Ministries were abolished, the Army Medical College and the Provisional Tokyo First Army Hospital compiled a document, dated 30 November, entitled 'Medical Report on Atomic Bomb Damage in Hiroshima.'[17] This report was translated into English and provided to the US Occupation Forces.[18] At the beginning of this report Lieutenant General Kenji Ibuka, the last Head of the Army Medical College says:

> When, on 15 August 1945, the Imperial Rescript terminating the war was announced, general conditions changed suddenly and various disturbances and influences began to affect this area. Consequently the policy on survey and research was changed and emphasis was placed on allowing a wider latitude for medical examinations and/or obtaining a complete medical survey of the victims.
>
> We were of the opinion that since the war was no longer to be continued it was our duty to attend to the wounds of the victims; that the study of wounds caused by a special weapon such as the atomic bomb was of the greatest public service, both as a physician and as an Army medical officer; and also that this would be a means of continuing the spirit of the comrades who had died whilst pursuing their duties.

The report also refers to the relationship between the US Forces and the atomic bomb research teams:

> In early September, a party led by Brigadier General Farrell and Brigadier General Newman of the US Army arrived in Japan to conduct a survey of the results and effects of the atomic bomb in Hiroshima. Professor Tsuzuki and Major Motohashi accompanied this party as guides and provided up-to-

> date information from the survey conducted by our medical scientists, especially on the results from the medical and biological standpoints.[19]

As discussed in Chapter 2, this statement reflects the Japanese Government's policy of co-operating in the research that the US Occupation Forces were conducting into the effects of the atomic bomb.

On 4 January 1946 the Headquarters of the Supreme Commander of the Allied Powers (GHQ/SCAP) issued the Purge of Public Officials Order to the Japanese Government (SCAPIN550). It ordered a purge of all public officials who had been involved in the execution of the war in a political, financial or educational capacity and the leaders of nationalist groups. This purge was not limited to those who could be referred to as militarists but also included many 'army and navy career employees, special police employees and officials from the Army and Navy Ministries.'[20]

With the abolition of the Army and Navy Departments on 1 December 1945 the Ministry of the Army changed its name to the No 1 Demobilization Ministry and the Ministry of the Navy changed its name to the No 2 Demobilization Ministry. Some of their functions and personnel were retained to oversee the demobilization. The Historical Fact Section was established by the No 1 Demobilization Ministry Secretariat with the aim of recording and preserving the details of the former military organization and personnel.[21] To this end the commissioners of the Demobilization Ministry employed many ex-military officers and bureaucrats, all with civil servant status. The No 1 and No 2 Demobilization Ministries later became the Demobilization Office with the No 1 and No 2 Demobilization Bureaus established within it. In 1948 its business was taken over by the Demobilization Bureau of the Returned Soldiers' Assistance Office of the Ministry of Health and Welfare.

The G-2 (Staff Section No 2 – Information) of GHQ gathered together former army and navy commissioned officers in the Historical Section of GHQ to compile and edit MacArthur's War History. They were set to work mainly under Takushirō Hattori (ex-Colonel, former General Staff Office Strategy Section Chief), the Head of the No 1 Demobilization Bureau's Historical Fact Section. Hattori served in the Historical Section until 1952. Similarly a group of former military intelligence officers established an Agency for Cooperation with the US Occupation Forces to assist in intelligence activities regarding the growing Soviet threat.[22]

The Peaceful Remobilization of Manpower

There seems to have been virtually no research conducted into the post-war movements of soldiers or civilians who were engaged in wartime scientific research. Hence, there is little data available. Pending further research into this question I can offer no more than an outline of these movements.

One instance where we do have a relatively good picture of the movements of former army and navy technicians is in the Japanese National Railways Technical Research Institute. In August 1945, facing defeat in the war, many considered 'the revitalization of the Japanese railways to represent a significant step in the rebuilding of Japan.' So the National Railways set out to gather 'the best scientific and technical brains from all over the country' to the Railways Technical Research Institute. It attracted many 'excellent technicians from the former army, navy and air force.' When the Central Aviation Research Institute facilities were added to the existing facilities, it resulted in the largest composite research facility in the nation (cf. ch. 9).[23]

The Central Aviation Research Institute was abolished in December 1945 by Imperial Edict No 734. 'This piece of land is 85,146 m^2 in area, with buildings accounting for 9,967 m^2. In accordance with the national policy of appropriating all military research facilities including land, buildings and all other equipment and appointments existing at the research institute, it was given to the railways under the Railways Technical Institute Expansion Plan. From January 1946, the Ministry of Transportation, began to employ the former institution's researchers, too.'[24] The chronological table provided by *Jūnen no ayumi* includes the following description: 'On 5 September 1946, the Railways Technical Institute incorporated the facilities and personnel of the Central Aviation Research Institute and those of some military establishments into its organization, reorganizing itself and beginning to expand.[25]

Mr Shōji Ikeda, an employee of the Railway Technical Research Institute at the time, captures the importance of this expanded research capacity saying:

> Demilitarization threatened to paralyze our research activities across the country. Nobody could convincingly disagree with the conviction of the Institute Head, Mr Nakahara, who maintained that technical research was the only way to pick ourselves up from the defeat suffered in the war.
>
> First, the Occupation Forces ordered that we ensure the maintenance and safe custody of the facilities at the former Aviation Research Institute. Second was the important question of attracting the top researchers from the former army and naval forces. Competent and capable people are essential for technical research so, following a decision by the then Vice-Minister for Transportation, Mr Hirayama, many researchers were appointed from the former military and naval forces. In fact the total number of researchers of university graduate standard reached as many as 270. In order to provide a place for so many researchers to work a rapid fit-out of the research facility was required.

> In the meantime, temporary use was made of other facilities such as the former Naval Technology Research Institute in Meguro, the former Army Munitions Factory in Sagamihara and the former Army Telecommunications Research Institute in Kokubunji, but this was not without problems.'[26]

We have no details about the 270 former army and navy technicians referred to here, but the figure seems to be somewhat exaggerated. My calculations, based on the 1950 'National Railways Employees Name List (technical graduates)' indicate that there were approximately thirty-five who listed army service on their career records and approximately another fifty-eight who listed naval service, making a combined total of ninety-three.[27] These are the technical graduate figures for the National Railways as a whole, not just the Railways Technical Research Institute. There are, however, records showing in excess of one hundred researchers added to the Railways employee register at this time, including thirteen former employees of the Central Aviation Research Institute.[28]

In addition to the former army and navy technicians who made their way to the National Railways, we are able to identify former (non-military) naval technicians as a relatively cohesive group. This was the case in the Extra Ministerial Bureau of the Ministry of the Navy's pre-war Naval Institute Headquarters, which was reorganized in 1941. On 16 August 1945 a Vice-Ministerial Conference decided that the jurisdiction for the Naval Facilities Division should be transferred to the Transportation Ministry and assigned to a new organization within the Ministry to be called the Transportation and Construction Administrative Office. 'The naval facilities personnel and the Transportation Ministry's Underground Construction Headquarters were brought together. Things were arranged in such a way that each Facility Division and Facility Corps was deployed as a regional construction unit to the main cities within the nine railway office jurisdictions which existed at that time.'[29] An official announcement dated 28 August 1945 stated 'All personnel who were employed at former naval establishments are to be part-timers (unpaid) at the Transportation and Construction Administrative Office.'[30] There were as many as ten or twenty thousand 'technical officers, naval engineers and assistant engineers' employed by the Naval Facilities Division at the close of the war.[31]

During the war Technical Captain Yōji Itō, of the Electricity Research Division of the Naval Technology Research Institute, had mobilized Yoshio Nishina, Tsunesaburō Asada and others who started up 'informal talk-fests for those interested in physics' with a view to conducting atomic bomb developmental research. After the war, however, Ito compiled and edited the technical history of the navy as Chairman of the Technical History Section of the Data Research Group incorporated in March 1946 as the Ministry of Health and Welfare No 2 Demobilization Bureau's

Historical Fact Research Section [The Data Research Group] (authorized by the Ministry of Education).[32]

Navy technical officers contributed significantly to the development that occurred in the electronics industries in post-war Japan. Their contribution can be attributed to 'the existence of what was referred to as the Navy's cutting edge research in electronic weapons technology, research relating to radio waves, telecommunications, magnetism and sound, as well as the availability of technical officers who had been involved in the development of these' both before and during the war.[33]

For example, the founder of the Tokyo Telecommunications Industries (later Sony), Mr Akio Morita, had been a Naval Technical Lieutenant attached to the No 1 Technology Workshop Aviation Optics Division in Yokosuka. There were many other Naval Technical Captains and Lieutenants amongst the development researchers. For example, Kenjirō Takayanagi, Head of the Radio Wave Weapons Section of the No 2 Naval Technology Workshop (treated as a Second Lieutenant), sought employment for fifty young, willing technical officers from the Workshop at the NHK Broadcasting Technology Research Institute. He himself, however, was a victim of the public officer purge and was unable to find employment as a technical officer at that time.[34] He later joined the Victor Company with about twenty young technical officers in July 1946.[35] Other examples could be cited, such as Technical Captain Atsuyoshi Ōuchi of the Yokosuka Industrial Workshops Navigational Experimentation Division, and many other naval technology officers and technicians who served in a supervisory or leadership capacity in Nihon Denki.[36]

In summary, the demilitarization of the Japanese armed forces and industrial infrastructure was an unprecedented success, no doubt in large part because the Government and the people accepted their defeat and cooperated with the Occupation Forces. This, in turn, was assisted by the fact that the demilitarization did not mean the total destruction of facilities and equipment. The redeployment of these facilities and of the personnel who could best utilize them was a major contributor to the development of various industries, and thus to the reconstruction of the Japanese nation.

Notes

1 Hiroshige, 1973 [1] p. 212.
2 Hiroshige, 1973 [1] p. 214.
3 Etō, 1986 [2] p. 501.
4 Fujiwara, 1987 [6] p. 1.
5 Fujiwara, 1987 [6] p. 7.
6 Hattori, 1965 [7] p. 962.
7 Ministry of Foreign Affairs, 1952 [3] pp. 800–1.

8 Ara and Awaya, 1979 [4] p. 194.
9 Hattori, 1965 [7] p. 963.
10 Hattori, 1965 [7] p. 957.
11 Hattori, 1965 [7] p. 957.
12 Tsuneishi, 1981 [10] p. 229.
13 GHQ/SCAP, 1945 [11] Box No.7541, ESS (D) 07038.
14 Ara, 1994 [5] pp. 16, 22.
15 Ara and Awaya, 1979 [4] p. 195.
16 Officially the Allied Forces occupied Japan, but for all intents and purposes it was actually an occupation by the United States alone. The US Army maintained almost total control over Occupation policy, especially of everything related to atomic weapons – and therefore all nuclear research. Hence, while the officially correct term of reference is the 'Allied Occupation Forces' it is often more accurate to refer to the 'US Occupation Forces.'
17 Nihon Gakujutsu Kaigi Genshi Bakudan Saigai Chōsa Hōkokusho Kankō Iinkai, 1953 [13]
18 GHQ/SCAP [14] Box 7407, ESS (D) 06277-06279.
19 Nihon Gakujutsu Kaigi Genshi Bakudan Saigai Chōsa Hōkokusho Kankō Iinkai, 1953 [13] p. 286.
20 Baerwald, 1970 [15] pp. 181–2.
21 Sasaki, et. al., 1991 [16] p. 208.
22 Sasaki, et. al., 1991 [16] pp. 208–9.
23 Tetsudō Gijutsu Kenkyūsho Gojūnenshi Kankō Iinkai, 1957 [17] p. 2.
24 Tetsudō Gijutsu Kenkyūsho Gojūnenshi Kankō Iinkai, 1957 [17] p. 107; Tetsudō Gijutsu Kenkyūsho, 1967 [18] p. 3.
25 Tetsudō Gijutsu Kenkyūsho (Railway Technology Research Institute of the Japan National Railway), *Jūnen no ayumi – sōritsu rokujusshūnen* (History of the last ten years at the sixtieth anniversary), 1967.
26 *Kokutetsu shokuin meibo (Gijutsu gakushi)* (A name list of Japan National Railways employees [engineering graduates]) (as at 10 August 1950), [19]
27 Tetsudō Gijutsu Kenkyūsho Gojūnenshi Kankō Iinkai, 1957 [17] p. 881.
28 *Kokutetsu shokuin meibo (Gijutsu gakushi)* (A name list of Japan National Railways employees [engineering graduates]) (as at 10 August 1950), [19]
29 Kaigun Shisetsu-kei Gijutsukan no Kiroku Kankō Iinkai, 1972 [20] p. 92.
30 Kaigun Shisetsu-kei Gijutsukan no Kiroku Kankō Iinkai, 1972 [20] pp. 92–3.
31 Kaigun Shisetsu-kei Gijutsukan no Kiroku Kankō Iinkai, 1972 [20] p. 91.
32 Nakagawa, 1990 [21] p. 313.
33 Nakagawa, 1990 [21] p. 12.
34 Nakagawa, 1990 [21] p. 306–7.
35 Nakagawa, 1990 [21] p. 309.
36 Nakagawa 1990 [21] 13 p. 310.

Bibliographic Guide

[1] Tetsu Hiroshige, *Kagaku no shakaishi – Kindai Nippon no kagaku taisei* (A social history of science – The structure of modern Japanese science), Chūō Kōronsha, 1973.

[2] Jun Etō, (editorial supervision), Ken Kurihara and Sumio Hatano (ed.), *Shūsen kōsaku no kiroku* (A record of moves to end the war), vol. 2, Kōdansha Bunko, 1986.

[3] Ministry of Foreign Affairs, *Shūsen shiroku* (A historical record of the end of the war), Shimbun Gekkan-sha, 1952.

[4] Kei Ara and Kentarō Awaya, 'Nippon haisen to kyū-shihaitaisei no kaitai (Japan's defeat and the disbanding of the old system of domination),' in *Rekishigaku kenkyū* (Historical research), special edition, October 1979.

[5] Kei Ara, *Nihon senryō-shi josetsu* (An introduction to research into the history of the Occupation of Japan), Kashiwa Shobō, 1994.

[6] Akira Fujiwara, *Nihon gunji-shi* (A Japanese military history), vol. 2, Nihon Hyōronsha, 1987.

[7] Takushirō Hattori, *Daitōa sensō zenshi* (A complete history of the Great East Asia War), Hara Shobō, 1965.

[8] Seizō Arisue, *Arisue kikan no shuki* (Notes on the Arisue Secret Service), Fuyō Shobō, 1983.

[9] General Headquarters, Supreme Commander for the Allied Powers, *History of the Non-military Activities of the Occupation of Japan, Introduction*, 1945 through 1951. *Nihon senryō – GHQ seishi* (The Occupation of Japan: GHQ's official history), vol. 1, Nihon Tosho Sentaa 1990.

[10] Keiichi Tsuneishi, *Kieta saikin-sen butai: Kantō-gun dai 731-butai* (The lost germ warfare corps: The Kwantung Army No. 731 Unit), Kaimeisha, 1981 (Chikuma Bunko, 1993).

[11] General Headquarters, Supreme Commander for the Allied Powers (GHQ/SCAP) Records, *Japanese Army Empire Demobilization, Empire Demobilization Trends*, 1 October 1945, Box No. 7541, ESS (D) 07038. The National Diet Library Archives.

[12] Jun Endō (ed.), *Senryō shiroku 1 – Kōfuku bunsho chōin keii* (A record of the history of Occupation 1: Circumstances surrounding the signing of the surrender documents), Kōdansha Gakujutsu Bunko, 1989.

[13] Nihon Gakujutsu Kaigi Genshi Bakudan Saigai Chōsa Hōkokusho Kankō Iinkai (The Science Council of Japan, Editorial Committee on Reports on Atomic Bomb Damages) (ed.), *Genshi bakudan saigai hōkokushū* (A report on atomic bomb damages), 2 volumes, Nihon Gakujutsu Shinkōkai (Japan Society for the Promotion of Science), 1953.

[14] General Headquarters, Supreme Commander for the Allied Powers (GHQ/SCAP) Records, Box No. 7407, ESS (D) 06277-06279. The National Diet Library Archives.

[15] H. Baerwald, *Shidōsha tsuihō* (Purging leaders), Rinjirō Sodei (tr.), Keisō Shobō, 1970.
[16] Atsushi Sasaki, Shunsuke Tsurumi, Ken'ichi Tominaga, Masanori Nakamura, Kimihiro Masamura, and Yōichirō Murakami (eds.), *Sengo-shi daijiten* (An encyclopedia on postwar history), Sanseidō, 1991.
[17] Tetsudō Gijutsu Kenkyūsho Gojūnenshi Kankō Iinkai (Committee for Publication of the Fifty Year History of the Railway Technology Research Institute of the Japan National Railways) (ed.), *Nihon kokuyū tetsudō gojūnenshi* (The fifty year history of the Railway Technology Research Institute of the Japan National Railways), Zaidan Hōjin Ken'yūkai, 1957.
[18] Tetsudō Gijutsu Kenkyūsho (Railway Technology Research Institute of the Japan National Railways) (ed.), *Tetsudō gijutsu kenkyūsho sōritsu rokujū shūnen – Jūnen no ayumi* (Sixtieth anniversary of the Railway Technology Research Institute – Ten years progress), 1967.
[19] *Kokutetsu shokuin meibo (Gijutsu gakushi)* (A name list of Japan National Railways employees [engineering graduates]) (as at 10 August 1950), editor and year of publication unknown.
[20] Kaigun Shisetsu-kei Gijutsukan no Kiroku Kankō Iinkai (Committee for Publication of 'A Record of Naval Facility Engineers and Technicians') (ed.), *Kaigun shisetsu-kei gijutsukan no kiroku* (A record of naval facility engineers and technicians), 1972.
[21] Seizō Nakagawa, *Dokyumento: Kaigun gijutsu kenkyūsho – Erekutoronikusu ōkoku no senkusha-tachi* (Document: Navy Technical Research Institute – Pioneers in the kingdom of electronics), Kōdansha Bunko, 1990.
[22] Takenori Maema, *Jetto enjin ni toritukareta otoko* (Men possessed by jet engines), Kōdansha Bunko, 1992.

5 Military Research and Its Conversion: Naval Radar Development

Miwao Matsumoto

Between Prewar and Postwar

It is often claimed that the seeds of Japan's rapid postwar economic growth were sown during the years of military conflict. It is suggested that part of the war's legacy was to provide Japan with a skilled workforce, pioneering research and a strong and systematic research culture. While many anecdotal accounts support this assertion – particularly in areas such as aeronautical engineering and chemistry – no empirical data remains which could confirm the existence of an overall trend and thereby prove (or disprove) the general theory. In contributing to this debate, this article investigates the socio-historical connection between military research and the science and technology of the postwar period, with particular reference to the developments in radar made by the Imperial Japanese Navy. While acknowledging that Tetsu Hiroshige[1] has already outlined the relationship between the promotion of research and the training of skilled personnel during the war and the developments in postwar science and technology, the objective of this chapter is to elucidate the dynamism of this relationship.

The Establishment of a Military Research Organization: The 'Invisible Spin On'

The development of radar in Japan was divided between the military, industrial and academic sectors. The military's radar research was located in places like the Naval Technology Research Institute and the Army's Tama Technology Research Institute. Companies such as Nihon Radio, Sumitomo Communications and Tokyo Shibaura Denki were at the forefront of industrial developments in radar, while the Imperial universities were at the centre of academic research. To what extent was the research in these different sectors coordinated and connected? What were the lines of communication and command?

In reality both industrial and academic research into radar technology was unregulated by the unified military objective that accompanied the wartime mobilization and dominated the military research organizations.

In reply to the inquiry of the US Scientific Intelligence Survey (hereinafter referred to as: the 'Compton Commission of Inquiry'; and its report: the 'Compton Report') conducted immediately after the war (11 September 1945),[2] Dr Hidetsugu Yagi, renowned for his invention of the Yagi Antenna, spoke about the internal structure of wartime military research: 'The army and navy certainly never co-operated with each other. Each considered that it would be preferable to lose the war than to co-operate'.[3] Concerning the military's attitude to academic research, he said: 'The army and navy treated scientists exactly as if they were "foreigners" (for example it seems that they attempted to keep information relating to even the enemy's weapons secret from scientists).'[4] This was the same military that 'strictly controlled all research, issuing orders concerning the priority of projects in non-military research organizations'.[5]

Such evidence suggests that rather than being seen as a template for postwar research and development and the rebuilding of the nation,[6] the experience of wartime mobilization, with its lack of inter-organizational co-ordination and integration, instead provided a negative example from which lessons of what *not* to do could be drawn.

In the absence of any systematic co-ordination of the organizations conducting wartime research into radar, it was left to individuals, through a network of informal personal contacts, to create and maintain inter-organizational links that harnessed each branch of research to the military objective. While this network is difficult to discern and trace, it in fact represents the socio-historical key to understanding how the peaceful transition of manpower was made possible. We refer to this informal contact between military and non-military personnel during weapons research and development as the 'invisible spin on'.

Responding to questions from the Compton Commission of Inquiry, Kinjirō Okabe, a student of Hidetsugu Yagi and a pioneer in advocating the split-anode magnetron,[7] identified seven people as the 'main' Japanese researchers interested in the magnetron, the key component technology of radar.[8] They were: K. Morita (Tokyo Institute of Technology); K. Oai (Kawanishi Machinery Manufacturing Co.); T. Okada (Nihon Radio Co.); S. Uda (Tōhoku Imperial University); S. Mizuma (Navy); T. Itō (Navy); and F. Baba (Mitsubishi Denki Co.). As indicated, these key researchers were evenly spread across the three sectors involved in radar research, with two being employed by the military, three others working within industry and the other two holding academic positions. One of the two military-based researchers, Tsuneo Itō, had been a student of Yasushi Watanabe at Tōhoku Imperial University before being selected by the Naval Technology Research Institute. While still working at the university, Tsuneo Itō, along with Kanjirō Takahashi and others, succeeded in developing the 'Tachibana-type' magnetron (the 8 split-anode magnetron – Okabe's was a 2 split-anode magnetron).[9] The 'Tachibana-type' magnetron was able to emit a succession of waves in the order of 10W with a wavelength of 10 cm and as such represented a breakthrough in radar technology. Before then (circa 1936–37) the world's shortest wavelength microwave had been the 16 cm wave, utilizing the BK tube.[10] The 'Tachibana-type' magnetron could thus be seen as an epoch-making proto-type, one which now attests to the fairly high level of basic radar research achieved in Japan prior to the war.

Within the network of informal contacts between the three sectors involved in wartime radar research, one man, Yōji Itō,[11] performed a nodal function on behalf of the military. Among many other things, it was Yōji Itō who recruited Tsuneo Itō to the Naval Technology Research Institute. Takeshi Nawa, then Head of the Naval Institute's Radiowave Electricity Research Department, went so far as to nominate Yōji Itō as the one person 'on whom the success or failure of the development of radar in Japan depended.'

After graduating from the Electrical Engineering Department of the Engineering Faculty of Tokyo Imperial University, Yōji Itō studied overseas at the Dresden University of Engineering, which was, at the time, the Mecca for students of light electrical engineering. He took a German doctorate there studying under G.H. Barkhausen, who had also taught Hidetsugu Yagi. Upon returning to Japan in 1929 Yōji Itō joined the Naval Technology Research Institute. In 1936 he received his engineering doctorate from Tokyo Imperial University for his analytical research into the 'Tachibana-type' magnetron (the back-coupling theory in the area of low frequency waves).

Becoming increasingly involved in the development of radar through his role as a technical officer within the Naval Technology Research Institute, Yōji Itō began expanding the network of informal contacts across the military, academic and industrial sectors that was to become pivotal to radar research during the war. This network was to become so extensive that Takeshi Nawa described it as, 'consisting of universities, various major companies and broadcasting corporations together with people from public research facilities. It was just huge…[and] impossible to fit into one bag.'[12]

How then did Yōji Itō manage to create this network of researchers, and what was the exact nature of the radar development which was thus facilitated? It is to these questions and the more general circumstances surrounding the development of radar in Japan that we must now turn our attention.

The Circumstances Surrounding the Development of Radar

Much of the success of Yōji Itō's ability to co-ordinate the various sectors involved in wartime radar research should be attributed to the smooth informality of intra-sector contact. This is well illustrated by an anecdote regarding his remarkably frank attitude in promoting scientists' transfer from academia to the military. At the beginning of 1942 Yōji Itō and Seishi Kikuchi, a physicist of Osaka Imperial University, paid a visit to the home of Takeshi Nawa, the Head of the Naval Institute's Radiowave Electricity Research Department. Itō is reported to have said, 'Kikuchi thinks he should quit the university and come over to the Navy, so I am hoping that the Department Head will make the same request'.[13] Naturally, Nawa obliged. As this encounter suggests, the links between the military and academic sectors forged by Itō were more personal and nuanced than those described in the broad terms of the Compton Report (see Table 5-1).

With regard to the military and industrial sectors, Nihon Radio Co. operated as the nexus between the two. In 1939, in order to develop the practical application of the 'Tachibana-type' magnetron (what became known as the radio range-finder), the military disclosed some of its

Table 5-1: The Naval Technology Research Institute Electric Wave Electricity Research Department Research Division and Laboratory Organization.

Division/Laboratory Name	Research Laboratory	Objectives
Tokyo University Division	Science Department Research Laboratory	Electric Wave Application Weapons Research
	Medical Department Research Laboratory	as above
Tōhoku University Division	No 1 Research Laboratory	Electric Wave Basic Research
		Electric Wave Application Weapons Research
Osaka University Division	No 1 Research Laboratory	Basic Material Research
	No 2 Research Laboratory	Exceedingly Shortwave Vacuum Tube Research
	No 3 Research Laboratory	Electricity Material Parts Research
Kyoto University Division	Dean of Engineering	Electric Wave Basic Research/Material Research
Nagoya University Division	Kinbara Research Laboratory	Research related to Statics and the Dellinger Phenomenon
Waseda University Laboratory Annex	No 1 Research Laboratory	Electric Wave Tube Research
	No 2 Research Laboratory	Electric Wave Diffusion Circuit Research
		Anti-shaking, Anti-reverberation Methods Research

NB: This table was compiled from information contained in *Nawa Takeshi tsuisōroku* (Takeshi Nawa's memoirs), 1973, pp. 244–5.

analyses of the 'Tachibana-type' magnetron to Nihon Radio and thereby created a network of personal contacts between the two parties. Nihon Radio also enjoyed a relationship of technical co-operation with the German Telefunken Company. At the time Nihon Radio was the third largest manufacturer of vacuum tubes in Japan, behind only Tokyo Shibaura Denki and Sumitomo Tsushin.

Those connected within this intricate network – with Itō at the nerve centre in the Naval Technology Research Institute – included: Shigeru Nakajima of Nihon Radio (and Itō's younger brother), Shōzaburō Yamazaki, Hirokazu Satō and Takashi Okada (the T. Okada mentioned in the Compton Report).[14] As with the military and academic network, the relationship between the Naval Technology Research Institute and Nihon Radio was ostensibly a relationship of equals. For the most part, however, it was in fact uni-directional, with Nihon Radio tacitly conforming to the Naval Institute's agenda. Nihon Radio's main role was to test and manufacture radar equipment, but in at least one instance the company, working on information obtained by Yōji Itō, had the opportunity to select the technologies. At that time the patent for the high capacity vacuum tube – requisite technology for radar manufacturing – was held by GE and the rights to this patent in Japan were held by Tokyo Shibaura Denki. Suspecting there had been subrogate manufacturing, Nihon Radio took it upon itself (albeit in response to Itō's prompting) to turn their attention to the BK tube developed by Barkhausen, under whom Itō had studied.[15]

What were the lasting achievements of Japan's wartime development of naval radar technology? Viewed in terms of technological advancement, the discovery of the so-called Giken oscillation, based on the analytical research into the 'Tachibana-type' magnetron is quite significant. So, too, was the development of the cavity magnetron (capable of emitting 10 cm waves of 300W successively), which was born out of efforts to suppress the excessive heat phenomenon of the 'Tachibana-type' magnetron.[16] Beyond a narrow definition of radar technology one can identify further accomplishments associated with radar detection, radio navigation, radio apparatus allowing a differentiation between the enemy and allied aircrafts and radio altimeters.[17] Despite these 'successes,' from the perspective of wartime mobilization, this research can be depicted as having 'run out of time' due to the under-developed status of associated technology – notably high sensitivity receiving apparatus.[18]

When considered in socio-historical terms, the most striking achievement of the development of naval radar was the informal yet very close human networks that it created between the military, academic and industrial sectors. This led to a large-scale organization of the networks which precipitated the establishment of the Second Naval Technology Research Institute. This was created by merging the research and development activities of the Electric

Waves, Magnetism and Acoustics Departments which were established in April 1944 and which were, in turn, created from the July 1943 establishment of the Electric Wave Research Department headed by the Commander of the Third Division of the Technical Headquarters of the Navy.

The expansion of these operations led to a corresponding expansion in the informal network amongst the technological and scientific community which was developing on a multitude of fronts (see Table 5-2). As a direct result of this organization Yasushi Watanabe and Seishi Kikuchi were joined by Kenjirō Takayanagi, known for his role in the development of television (NHK Broadcasting Technology Research Institute), Michio Hatoyama (the Institute of Physical and Chemical Research, later the Electro-Technical Laboratory), Hiroshi Kiyomiya (the Electro-Technical Laboratory), Ikumatsu Takeda (Ryojun Engineering University), Kōichi Shimoda (Tokyo Imperial University Graduate School) and others. Thus, a smorgasbord of people destined to become key postwar players in the field of microelectronics were brought into contact through this informal wartime mobilization network. In hindsight it is relatively easy to detect these connections and identify the fruit they bore in the postwar period.[19]

To conclude, I would now like to explore the significance of the delayed effect that military research had when converted to peace time pursuits.

Table 5-2: Second Naval Technology Research Institute divisional research themes structure

1 Radar Department
- a) Radar for small patrol planes (FK-3)
- b) Radar for small patrol planes (N-6 Type 1)
- c) Radar for night fighter use (FD-2)
- d) Radar for night fighter use (Cyoku-3)
- e) Radar for night fighter use (FK-4)
- f) Radar for large plane use (Kasumi-51)
- g) IFF (M-13)
- h) Radar detector for plane use (FT-13, FT-C)
- i) Wireless homing device (FP)
- j) Radar to guide fighters (Hama-61, Hama-62, Hama-63)
- k) long range radar (Hama-14)
- l) Radar for submarine use (Model 13)
- m) Radar detector for submarine use (E-27)
- n) Simplified short-wave direction finder
- o) Radar to guide boats (TH)
- p) Radio altimeter for plane use (FH-1)

2 Communications Department (details omitted)

3 Acoustic Department (details omitted)

4 Light and Heat Department (details omitted)

5 Magnetic Detection Department (details omitted)

NB (1-f) is the Rotterdam type. See the Compton Report Vol. II (GHQ/SCAP Records Box No 7430 ESS(A)-09868), App. B.

The Conversion of Military Research to Peacetime: 'Spin Offs'

The wartime mobilization process created some 5,940 naval technical officers, 714 of whom were reportedly in the electricity and telecommunications fields.[20] After the war, living in a society without the Navy, the survivors of these ranks – many of whom had been part of the informal inter-personal network earlier discussed – provided the personnel base for Japan's budding postwar microelectronics industry. Many of this number found work in the burgeoning industrial sector, with corporations like Sony, Mitsubishi Denki, Nihon Victor, Sharp and Matsushita Denki, Tokyo Shibaura Denki, Nihon Denki and Nihon Radio amongst others. Others filled the ranks of the government bureaucracy (i.e. NHK, Electro-Technical Laboratory and Telecommunications Laboratory), which rapidly replaced the military in the role of providing informal leadership to the various arms of technological research. Still others became academics in the laboratories of Tokyo University, Tōhoku University, Osaka University, Tokyo Institute of Technology, Tokyo University of Education, Waseda University and Keio University.[21] This pool of ready-trained electronics and telecommunications personnel was a key 'spin off' benefit in the conversion of military research to peacetime endeavors (see, for example, Table 5-3) and represented a continuation and expansion – minus the military sector – of the informal networks established during the wartime mobilization period.

It is often suggested that this phenomenon was the key to the so-called 'economic success of Japan'. Here I wish to cite one typical example of the conversion of military research to peaceful pursuits, one which neither repeats the usual 'success story' account, nor the concomitant tales of overcoming difficulties, but rather reveals something of the external conditions and the frame of reference of those who advocated such conversions. I must point out that until the establishment of the three radio wave laws in 1950 (the Electric Wave Law, the Broadcasting Law and the Electric Wave Supervisory Committee Establishment Law) all radio wave R&D was under the administration of GHQ (General Headquarters, Supreme Commander Allied Powers) and any research that might lead to the development of weapons was prohibited.

The conversion of wartime research into peacetime pursuits was obligatory and heavily promoted by the postwar administration. It was in this environment, in 1947, that Yōji Itō established his own wireless equipment company. Before seeking approval from GHQ, Itō turned for advice to his immediate superior from his Navy days, Takeshi Nawa. Fumio Tazaki, a consultant with the company at the time, recalls: 'When it came to choosing a name for the company, Nawa said that because there was a great deal of interference from GHQ, "You wouldn't want any interference coming from outside given that people who held important posts in relation to naval electric wave weapons during the war are establishing a company as a manufacturer of wireless equipment". He advised that, due to the current climate, they should adopt a name which would not be immediately associated with a radio and electric equipment company...He asked if there was not a gentle name that would be appropriate for the company, the name of a flower, perhaps – "freesia" or "tulip", for example. After much discussion, Itō's company was called "Kōden Sangyō" Ltd (Light Electricity Industrial Company).'[22]

In September 1947 Nawa, in his role as one of Kōden Sangyō's sponsors, attended a meeting held to promote the establishment of the company. Once Kōden Sangyō had received GHQ approval, Nawa was instrumental in securing its finance by introducing Itō to the Teikoku Bank.[23] So, in effect, Itō's position within the old Naval Institute's network was pivotal to the establishment of his peacetime company, instrumental both in obtaining GHQ approval and procuring funds.

A document relating to Itō's overseas travels in 1954 provides us with a privileged insight into his own 'opinion in relation to Japan's application of science and technology.'[24] In chapter five of this document Itō focuses his attention on the relationship between war, military research and the development of technosciences. His thesis is neatly summarized in the chapter's opening sentence: 'The vanguard of scientific technology is military technology. This is one of the saddest aspects of the fate of humanity'.[25] He reiterates this central assertion when he concludes: 'Consequently it was natural that after the war the technology that had been developed for the purposes of war would be immediately utilized, with some remodeling, in the wider culture and for commerce...From the point of view of the happiness of the human race this is very contradictory but it can't be helped'.[26]

What if any were the lessons that Japan could learn from Itō's understanding of the transitions from military to civilian research? Itō believed that: 'In relation to science and technology...one option is to go with the American strength, another is to work towards becoming a proper member of the international community, not pointlessly relying on the goodness of other countries but finding the ability to be independent and self-reliant. Unless we do this we will come to rely on the US, initially feeling gratitude and later regret, with this subsequently leading to chaos. The US will offer guidance and assistance, but come to grow tired of Japan's waywardness, and eventually will cut her off'.[27]

If in philosophy there exists both idealism and pragmatism, we see that in the philosophy of Itō the two converged around the issue of transforming Japan's military research into peacetime pursuits. Contrary to what one might have expected, the highly regarded Itō – who from the prewar period

Table 5-3: 'Spin off' products from the wartime naval radar development: A case of Nihon Radio in 1991

	Radar for ships	Loran	Fish-finder	Supersonic tester
a) Total number	49,654	49,438	105,000	
b) Gross sales	200	49	101	818
c) Exported number	40,695	29,000	70,667	
d) Sales by export	157	20	70	458
d/b(%)	78	41	69	56

NB The unit of sales is 100 million yen. See S. Nakajima, 'Hi senryō-ka ni okeru seisan hinmoku sentei to sono go no seisan jōkyō (Nihon Musen no baai), (Product selection under the Occupation and the subsequent production situation [The case of Nihon Radio]),' (dated 19 February 1992), Record of the address given by the Diffusion of Technology Working Group of the 149th committee meeting of the Japan Society for the Promotion of Science.

through to the postwar period was seen as a pragmatist with an ability to get things done[28] – was at heart an idealist who deeply lamented the contradictory role war played in the research and development of technosciences and the happiness of mankind. While pragmatism has dominated the history of Japan's scientific and technological research, Itō's buried idealism forces us to reflect to what degree others involved in wartime military research were, at some level beyond success or failure of the research, motivated by the possible future peacetime applications of the technologies they were developing.

Notes

1 See Tetsu Hiroshige, *Kagaku no Shakai-shi: Kindai Nihon no kagaku taisei* (A social history of science: the structure of modern Japanese science), Chūō Kōronsha, 1973, pp. 84–276.

2 The report issued from this inquiry was called the Compton Report, named after Karl T. Compton, then President of MIT and a key advisor to the Commission.

3 The Compton Report [1], 1945, vol. 3, GHQ/SCAP Records Box No. 8354 ESS(I)-00727, App. 3-A-1.

4 The Compton Report [1] 1945, vol. 3, GHQ/SCAP Records Box No. 8354 ESS(I)-00727, App. 3-A-1.

5 The Compton Report [1] 1945, vol.3, GHQ/SCAP Records Box No. 8354 ESS(I)-00727, App. 3-A-1.

6 See Hiroshige, 1973, pp. 252–76.

7 The magnetron is a vacuum tube with the pivotal direction of a bi-polar tube with a magnetic field. It is also referred to as the Magnetic Electric Tube (so named by Yōji Itō). It was invented by the American A. W. Hull

in 1927. Kinjirō Okabe of Tōhoku Imperial University focussed on the emittance of electric oscillations ('A' type and 'B' type oscillation) and in 1927 succeeded in splitting the anodes and increasing the rate of oscillation and further decreasing the wavelength through joining the same circuit with the split anode. This is the split-anode magnetron (the 3 split, the 4 split and the 8 split magnetrons were later developed). Generally the frequency of a 'n' split magnetron is from 1-200GHz. This is utilized as an oscillator for microwave radar.

8 The Compton Report [1] 1945 Vol III (GHQ/SCAP Records Box No. 8354 ESS(I)-00727), App. 19-1.

9 According to Nakajima [14] 1988, pp. 19–34, the wavelength of this magnetron is 20 cm but this is the value of the 'Tachibana-type' M2. The 'Tachibana-type' M312 made 10 cm waves possible. For more details or the classifications of the split-anode magnetron developed by Japan at the time see Nakajima [10] 1947.

10 'BK' refers to Barkhausen and Kurz, the researchers who invented the oscillation tube.

11 Ko Itō Yōji-kun Kinen Ronbun-shū Kankō-kai [7] 1956, p. 31.

12 Ko Itō Yōji-kun Kinen Ronbun-shū Kankō-kai [7] 1956, p. 32.

13 Ko Itō Yōji-kun Kinen Ronbun-shū Kankō-kai [7] 1956, p. 31.

14 Ko Itō Yōji-kun Kinen Ronbun-shū Kankō-kai [7] 1956, p. 45.

15 Ko Itō Yōji-kun Kinen Ronbun-shū Kankō-kai [7] 1956, pp. 43–5.

16 See Ko Itō Yōji-kun Kinen Ronbun-shū Kankō-kai [7] 1956, pp. 44–5 for more details regarding Giken oscillation. For details about the cavity magnetron, see the Compton Report [1] 1945, vol. 3, GHQ/SCAP Records Box No. 8354 ESS(I)-00727, App. 7-3; 7-4.

17 For details on this and an overview of the electric wave weapons researched and developed by the Navy and a perusal of their use and capacity, see Nihon Musenshi Henshū Iinkai [9] 1951, vol. 10, pp. 380–401.

18 See the Compton Report [1] 1945, vol. 1, GHQ/SCAP Records Box No. 9090 NRS-06247, App. 3-A-1.

19 For example, see Y. Nakagawa, *Kaigun Gijutsu Kenkyujo* (Naval Technology Research Institute), Nihon Keizai Shimbun Co. 1987.

20 Nakagawa, *Kaigun Gijutsu Kenkyujo* (Naval Technology Research Institute), Nihon Keizai Shimbun Co. 1987, p.17.

21 These included prominent academics such as Shinichirō Tomonaga (then at Tokyo University of Humanities & Sciences, a Nobel laureate in physics in 1965), and T. Miyajima (also then Tokyo University of Humanities & Sciences) and Masao Kotani (then Tokyo Imperial University) who were involved in magnetron theory analysis using the 'dynamics of particles' method, and Sōgo Okamura (then Naval Technology Commissioned Officer) who was involved with ship detecting experimentation using microwave radar.

22 See Takeshi Nawa, *Nawa Takeshi tsuisōroku* (Takeshi Nawa's memoirs) [6] 1973, p. 312. Kōden Sangyō Ltd currently trades as Kōden Manufacturing Ltd.
23 See *Nawa Takeshi tsuisōroku* (Takeshi Nawa's memoirs) [6] 1973, pp. 313–14.
24 See Ko Itō Yōji-kun Kinen Ronbun-shū Kankō-kai [7] 1956, pp. 326–36.
25 Ko Itō Yōji-kun Kinen Ronbun-shū Kankō-kai [7] 1956, p. 313.
26 Ko Itō Yōji-kun Kinen Ronbun-shū Kankō-kai [7] 1956, p. 333.
27 Ko Itō Yōji-kun Kinen Ronbun-shū Kankō-kai [7] 1956, p. 333.
28 See Ko Itō Yōji-kun Kinen Ronbun-shū Kankō-kai [7] 1956, pp. 347–557.

Bibliographic Guide

[1] Report on Scientific Intelligence Survey in Japan (known as the Compton Report), 1945, GHQ/SCAP Records Box No. 9090 NRS-06247 & No. 8354 ESS(1)-00727.
[2] K.T. Compton, 'Mission to Tokyo,' in *The Technology Review*, vol. 48, no. 2, 1945.
[3] *M. J. Peck and M. J. Scherer, 'The Weapons Acquisition Process,' Manuscript submitted to Graduate School of Business Administration, Harvard University, 1962.*
[4] R. W. Home, 'Science on Service,' Manuscript, March 1987.
[5] H. E. Guerlac, *Radar in World War II*, Tomash Publishers, 1987.
[6] Takeshi Nawa, *Nawa Takeshi tsuisōroku* (Takeshi Nawa's memoirs), Nawa Takeshi Tsuisōroku Kankōkai (Takeshi Nawa Reminiscences Publication Committee), 1973.
[7] Ko Itō Yōji-kun Kinen Ronbun-shū Kankō-kai (The late Yōji Itō Commemorative Works Publication Committee), *Itō-san no omokage* (The memory of Mr Itō), 1956.
[8] Moto Gunrei-bu Tsūshin Kachō no Kaisō Kankō-kai (Publications Committee on the Recollections of the Communication Chief of the Former Naval General Staff Division), *Moto gunrei-bu tsūshin kachō no kaisō* (Recollections of the Communications Chief of the former Naval General Staff Division), 1981.
[9] Nihon Musenshi Henshū Iinkai (The Editorial Committee on the History of Japan Wireless), *Nihon Musenshi* (A history of the wireless in Japan), 13 vols., Denpa Kanri Iinkai (Supervisory Committee of Electrical Waves), 1950, 1951.
[10] S. Nakajima, *The Study of the Microwave Magnetron*, Japan Radio Co., 1947.
[11] S. Nakajima, *Hi senryō-ka ni okeru seisan hinmoku sentei to sono go no seisan jōkyō (Nihon Musen no baai)*, (Product selection under the

Occupation and the subsequent production situation [The case of Japan Wireless]), 19 February 1992, Record of an address given by the Diffusion of Technology Working Group to the 149th committee meeting of the Japan Society for the Promotion of Science.
[12] S. Nakajima, 'Mitō gijutsu tono tatakai (The fight with unexplored technology)', Record of the address delivered to the Tōhoku Branch of the Electronic Communications Society, at Akita University, 3 June 1983.
[13] S. Nakajima, 'Gijutsu to keiei – Tokuni benchăbijinesu no igi ni tsuite (On technology and management: the significance of nurturing venture businesses),' Record of address delivered to the Nittsū Shōsangyō Training Institute, 27 November 1979.
[14] S. Nakajima, 'Reedaa kaihatsu no rekishi (The history of the development of radar),' in *Kōden Gihō* (Technological reports of Koden Co. Ltd.), no. 4, 1988, pp.19–34.
[15] S. Nakajima, 'Japanese radar development prior to 1945,' in *IEEE Antennas and Propagation Magazine*, vol. 34, no. 6, 1992, pp.17–22.
[16] S. Nakajima, 'Japanese magnetrons,' in *Electronics*, May, 1946, pp.114–15.
[17] S. Nakajima, 'Magunetoron to reedaa no kenkyū: Taiheiyō Sensō shūketsu made (Magnetron and radar research: Until the conclusion of the Pacific War),' in *Denki gakkai kenkyū-kai shiryō: Denki gijutsushi kenkyū-kai, HEE-94-1–9* (Research Group materials of the Electricity Society: Research group on electrical technology history/HEE-94-1–9), 1994, pp. 49–58.
[18] Shigeru Ōno Materials.
[19] M. Matsumoto, 'A hidden pitfall in the path of prewar Japanese military technology,' in *Transactions of the Newcomen Society for the Study of the History of Engineering and Technology*, vol. 71, no. 2, 2000, pp. 305–25.

Datum [1] is the most important of the materials written in English. Data [6], [7] and [8] were distributed to those concerned as not-for-sale items and include valuable data in terms of key person conversion to peacetime pursuits. In particular, in Takeshi Nawa's memoirs more than 10 pages are devoted to records of the interaction with the Compton Commission of Inquiry which visited the Naval Technology Research Institute. Datum [7] contains a record of Dr Itō's thesis written under Dr Barkhausen and his report which was written on the split-anode magnetron and submitted to the Naval Technology Research Institute. This is valuable in terms of checking the potential of the research base.

Datum [9] was compiled with the aim of avoiding losing materials. It was compiled with co-operation from Electric Communications Co-operative Inc with impetus for this being received from the enforcement of the three electric wave laws on 1 June 1950. It is one of the most standard collections of data relating to the history of radio technology, covering a period from the mid 19th century through 1941 (with some articles extending into the war or later). The discussion of radar is primarily in vol. 3, *Musen Kenkyū-shi* (The history of wireless research), vol. 9, *Rikugun Musen-shi* (The history of Army wireless) and vol. 10, *Kaigun Musen-shi* (The history of Navy wireless). These are valuable in terms of gleaning information relating to the start of radar development and the usage and capability of radar developed during the war.

The thesis and materials of S. Nakajima are on a par with datum [7] in terms of being basic materials from which we can glean information about the development of radar at that time and the circumstances surrounding the prewar and wartime development of radar seen from the most intimate vantage point of Yōji Itō's younger brother.

Datum [18] is newly discovered material on developments in naval radar during the wartime mobilization, which have been kept privately by Shigeru Ōno. Although it was unavailable for the present writer when preparing this article and most of its contents are confined to technical matters, it deserves serious attention for further details.

Datum [19] sheds new light on the long-neglected aspect of 'Spin Off' from a serious failure of military technology in the Navy just before the outbreak of World War II that has been kept secret.

6 The Ishii Unit

Keiichi Tsuneishi

The Ishii Unit existed in the former Japanese colony of Manchuria (in the North Eastern part of China) from the early 1930s until 1945.[1] Shirō Ishii, a Lieutenant-General Army Surgeon, was the founder of the Ishii Unit and its commanding officer for a long period of time. Amongst his achievements was establishing the 'Prevention of Epidemics Laboratory' in the Army Medical College in Tokyo and the 'Tama Unit' in Nanking. Both were part of a network he was creating – known as the 'Ishii Agency' – for the research and development of biological weapons.

The Ishii Agency was a network of medical practitioners across the length and breadth of Japan engaged in collecting a wide variety of pathogenic organisms. The organisms were required to manufacture vaccines, the first step in the development of biological weapons. Research funds and resources were provided to medical practitioners who co-operated with the unit in its research and development. Teachers of medicine were induced with facilities and resources to cooperate in directing their medical students and technicians to collect specimens and data for this research.

In some cases the data was obtained from experiments conducted on living people. The vast network of the Ishii Agency enabled some scientists to access experimental research data without doing the experiments themselves. They were thus able to achieve scientific results dependent upon human experiments, and achieve fame and recognition in their fields, without ever being associated with the human experimentation. Thus, after the war they escaped the condemnation and ostracism of their more directly complicit colleagues.

Perhaps if these people had been subject to public scrutiny and criticism the Japanese medical world might have re-structured itself. The aim of this chapter is to demonstrate that it did not do so. An examination of the Ishii Unit reveals that certain aspects of the Japanese medical world, particularly in relation to human experiments, still remain unchanged from the way they were before and during the war.

A Brief History of the Ishii Unit

Shirō Ishii was assigned to a secret research unit in China as early as 1931 under the Kwantung Army Strategy Military Operation Chief of Staff,

Colonel Kanji Ishihara. There was no formal name for the unit, but they called it 'the Tōgō Unit.' Its mission was to conduct experiments involving humans.[2] In August 1932, as a second-class army surgeon (equivalent to Lieutenant-Colonel), Ishii succeeded in establishing the Prevention of Epidemics Laboratory in the Army Medical School (PELAMS) and got himself appointed chief of the laboratory. In 1936, The Tōgō Unit became the Ishii Unit; officially called Manchuria Unit 731, formally inaugurated in the Harbin suburb of Heibō (Pingfan). During the war with China in 1937, the Ishii Unit equipped and maintained the Epidemic Prevention Water Supply Departments' (EPWSD) network. Ishii himself recounts this as follows:

> The Kwantung Army Military Epidemics Prevention Department was established as a fixed military institution in Manchuria with Colonel Ishii in the northern office at the time of the outbreak of the "China incident" in 1937. It oversaw the Chinese virus operations and occupied the Beijing Tiantan Central Institute for the Prevention of Epidemics. The Northern China EPWSD was established under the then Army Medical College instructor Colonel Kikuchi. In May 1939, the Central China EPWSD was established in the Central China Nanking General Hospital, and in the same month the Southern China EPWSD, headed by Colonel Iwao Tanaka, was established in the Zhong Shan University in Canton in Southern China...[3]

The Ishii Agency's facilities were systematically completed by February 1942, with the establishment of the Nanpō Military EPWSD in Singapore.[4]

The five different EPWSDs from Harbin to Singapore each belonged directly to the military. Besides these five departments, a number of other Water Supply Epidemic Prevention Sections (WSEPS), derived from divisions of the army and other bodies, existed in a supporting capacity. These were all smaller than the military departments. The number of regular staff members of this network totaled more than 20,000. Thus, Ishii was in command of the equivalent of approximately two army divisions. Fewer than ten percent of them were researchers or doctors.

The military departments controlled and coordinated all research and development, deciding the roles allocated to each WSEPS. At the lower levels of the network, at the army division level, research activities included things such as inoculation with vaccines, the supply of purified water, and the diagnosis of illnesses.

It has been confirmed that experiments were conducted on humans by the Kwantung Army Military EPWSD as well as the Southern China Army EPWSD.[5] Many medical researchers were assigned to these two military squads. Researchers from the Medical Department of Kyoto Imperial University were sent to the former and researchers from the Medical Department of Tokyo Imperial University were dispatched to the latter. After the war, some of these researchers moved on to prominent positions in the Japanese medical community. For example someone from the Kwantung Army EPWSD was appointed Dean of the Medical Department of Kyoto University while another became President of Kyoto Prefectural Medical University. Someone from the Southern China EPWSD was appointed the Head of the National Preventative Health Laboratory, established after the war on the model provided by the Epidemics Laboratory at Tokyo University.

The data from the EPWSD human experiments was supplied to the professors of the Medical Departments of the Imperial Universities, which were affiliated with Army Medical College, as needs dictated. With this data, the professors were able to improve vaccines and carry out other research and experimentation.

The 'Dōjinkai' played an important role in the relation between the medical world and the Japanese military in China, both before and during the war. The Dōjinkai was established in 1902 with the objective of 'spreading the skills of medical pharmacy to various Asian countries and improving medical hygiene conditions.'[6] However, after the so-called 'China incident' the group was clearly incorporated into the military's substructure. In 1938, the Dōjinkai organized 'anti-epidemic facilities' in Beijing, Shanghai, Nanking and other places, as part of the military installations at those locations. It was stated that 'The work of the Central China Epidemics Prevention Head Office is the responsibility of the Central China Dispatch Force in the area under military occupation.'[7] Yoneji Miyagawa, the Head of the Epidemics Laboratory at Tokyo University, became the vice-president of the Dōjinkai in 1938. Fumimaro Konoe became its president in 1939.

Investigations by the Occupation Forces and Demilitarization of the Medical World

Soon after the Occupation began, the US Forces began to investigate what they called the Biological Tactics Unit, the Epidemic Prevention Water Supply Department (EPWSD) discussed above. The investigation by the Occupation Forces was in two parts: the sixteen months leading up to January 1947 and the ensuing twelve months. The turning point in January 1947 was when the US investigators learnt that the Ishii Unit and others had conducted experiments on human beings. They were alerted to this fact by the Soviet Union's call for Ishii and others to be charged and tried for conducting such experiments.[8]

The US investigators had completed their initial investigations into human experiments in March 1946. These investigations were conducted by Army Surgeon Lieutenant Colonel Murray Sanders and Veterinary Surgeon Lieutenant Colonel Arvo Thompson. Both of these men belonged to Camp Detrick (known as Fort Detrick after 1955) which had been engaged in the research and development of biological weapons since 1944. In the first stage of this investigation, the SCAP (The Supreme Commander for the Allied Powers) used his personal authority to award an exemption for war crimes to those involved with the former Unit.[9] In 1946 the Dōjinkai was disbanded along with right-wing associations by Notification No 19, based on Imperial Order No 101.[10] However, other than this, the investigations conducted by the Occupation Forces into Japan's medical world did not go much beyond the level of the Kyūshū University human dissection incidents.[11]

Sanders, who came to Japan at the end of August 1945, was a member of the Compton Investigation Group (cf. ch. 1, 5), whose purpose was to conduct research into the effects of the atomic bombs in Japan, chemical weapons (poison gas) and biological weapons (viruses). He carried out his investigations from September to November. In total he carried out thirty interrogations, involving more than twenty people, and including former Army Cabinet Minister Yoshijirō Umezu.[12] He was, however, unable to interrogate the two unit leaders, Ishii and former Army Surgeon Lieutenant General Seiji Kitano, because he was unable to confirm their identities. After Sanders left, Thompson came to Japan and continued investigations until March. He interrogated more than 15 former unit members, mainly doctors, including Ishii and Kitano.[13]

By 1942, the US military had begun to suspect that Japan was using biological weapons in the war against China and the Information staff of No 2 Section (G-2) commenced investigations. Before the war ended, G-2 knew that the central person behind the biological research strategy was Army Surgeon Lieutenant General Shirō Ishii, that he pursued his interests at the 'Epidemic Prevention Water Supply Department,' and that the operation was divided between the Kwantung Army unit in the suburbs of Harbin and the unit in Nanking. They also knew that biological weapons had been utilized several times by the Japanese against China.[14]

The initial investigations focused on the institutions supporting the research and development of biological weapons, that is: the Army Medical College, the medical departments of Japanese universities and medical universities, and especially the Epidemics Laboratory of Tokyo Imperial University. Sanders' and Thompson's investigations clearly indicated that 'the institution central to Japan's biological strategy was located in Heibō (Pingfan), near Harbin in Manchuria. This place was a huge self-contained institution with some 3000 members from 1939 to 1940.' In Harbin, the 'Ujigata' and others bombs were researched and developed as germ bombs. More than 6000 of them were tested in the outdoor laboratory.[15] However, Sanders and Thompson had found no evidence of experiments being conducted on humans.

The 'Research Results' of the Ishii Unit; Strife between the United States and the Soviet Union

In January 1947, a public prosecutor from the Far East Military Court in the Soviet Union demanded that Major General Charles Willoughby, the head of G-2, hand over Ishii and others for conducting experiments involving human beings. When G-2 re-examined Ishii and others 'they *tacitly* admitted the fact that there had been human experiments.'[16]

These revelations created a problem for the US Government: how to deal with the war crimes exemption previously granted to the members of the Ishii Unit. In the Nuremberg trials conducted after Germany's surrender, Nazi doctors had been tried for conducting human experiments. The US Government was concerned that putting one country on trial and not the other might sow the seeds of trouble. GHQ/SCAP spent more than eight months addressing the Government's concerns and apprehensions before it decided to uphold the war crimes exemption previously granted.[17]

As a friendly gesture, the US allowed the Soviets to question Ishii and the others but refused to share any information with them, including information already obtained from and about Ishii. After learning that human experiments had been conducted, Doctor Norbert H. Fell was dispatched in April 1947 to re-interrogate Ishii and others in positions similar to his. Doctors Edwin Hill and Joseph Victor joined him in these investigations in October.

While they primarily questioned the same people that Thompson had already interrogated, some new names came up in this investigation. Two of them were professors from the Medical Faculty of Tokyo University – Professor Saburō Kojima and Professor Shōgo Hosoya. They had both been researchers in Ishii's laboratory at the Army Medical College.

The Hill and Victor report, though, was not made public until the early 1980s. It was thus some thirty years after the investigations before the Japanese public could find out who knew what about human experimentation during the war. According to the report, Kojima knew that the Ishii Unit was dealing with experiments involving humans. He admitted that he had knowingly accepted data derived from human experiments for his work in the development of vaccines.[18] Similarly, the report revealed that Hosoya's trial vaccine had been continually administered to humans in experiments conducted by the unit.[19] The report does not clearly establish the relationship between Hosoya and the experiments, but, in the context of the situation in Japan at the time, it is extremely unlikely that he refused data derived from human experiments by Ishii and others.

By appointing professors from Tokyo University and Kyoto University as the exclusive researchers in the Army Medical College's Prevention of

Epidemics Laboratory (PELAMS), Ishii was able to ensure the research cooperation of these institutions and could directly oversee researchers connected across the country. Essentially, PELAMS was the central controlling hub of the Ishii Agency, coordinating the collection and distribution of research data from all research activities, including human experiments. Researchers co-operating in Ishii's activities benefited from the research data and other materials produced through the network.

The suppression of the Hill and Victor report suggests that the objective of the US investigations was to obtain valuable data about the experiments conducted on humans rather than to punish or administer justice. Hence, until this report was released in the early 1980s, nobody knew that two professors from Tokyo University had been under investigation by the US Forces. These professors were thus able to pursue their careers untainted by their involvement in human experimentation. So, when the National Institute of Infectious Diseases (NIID) was established, Kojima moved there from Tokyo University, assuming the position of Vice-President. He later became the President. Similarly, Kitano and other former Ishii Unit members were later appointed to conduct scientific observations in Antarctica as special members of the Science Council of Japan. If the role of the Prevention of Epidemics Laboratory had not been finally revealed, the involvement of these prominent Japanese researchers in human experimentation would never have been disclosed, and their reputations would have remained unsullied.

The Korean War and the Science Council's Rejection of a Counter Resolution

A resolution seeking ratification of the 1925 Geneva Agreement, which banned biochemical weapons, was presented to the Thirteenth General Assembly of the Science Council of Japan (JSC) in the autumn of 1952.[20] The resolution was presented at this time because there was suspicion that the US military might have used biological weapons in the Korean War.

On 8 May 1951, Heong-young Park, the North Korean Minister of Foreign Affairs, sent a telegram to the chairman of the United Nations Security Council, protesting against the use of biological weapons by the US Army. On 22 February 1952, Minister Park issued a formal statement to the Chairman of the United Nations General Assembly, condemning the US for its use of biological weapons. On 8 March, the Chinese Foreign Affairs Minister Chou En Lai expressed support for the North Korean claims. The International Committee of Scientists was organised to investigate North Korea's accusations against the United States. The committee conducted on-the-ground investigations and presented a report confirming the North Korean claims on 31 August.[21] The committee membership included Doctor Joseph Needham, a biochemical scholar who had been assigned to the British Embassy in China before the war and later became a science historian.

The committee arrived at the conclusion that the tactical employment of biological weapons by the US in Korea was a reproduction of the tactics the Japanese had used in China. The committee reported that the US military attacked China and Korea with weapons developed from the knowledge created by the Ishii Unit's research into biological weapons technology.

This was the background for the presentation of the resolution to ban biological weapons. The Seventh Sectional Chapter (the Medical Section) of the JSC rejected this resolution, providing the following reason:

> When we say that biological weapons are useless, it should be understood that such weapons are useless when compared to the weapons we presently have. Perhaps we should say they are ineffective – as a landscape gardener would be in trying to stop the continental drift…If some idiot wants to put time and effort into coming up with such an impractical thing, I would most certainly and strongly advise him to abandon such a stupid pursuit. You may rest assured on that point.[22]

At the time that this statement was made, the membership of this sectional chapter consisted of many of Ishii's teachers, bacteriologists and researchers from Ishii's Prevention of Epidemics Laboratory, and others who were well informed about Ishii's activities. It is inconceivable that these eminent scientists considered themselves to have been engaged in a stupid pursuit. Therefore, we must consider that the statement above was offered merely to quash the resolution and deflect attention away from the practices of the Japanese medical community.

For a long time, it has been believed that the Seventh Sectional Chapter quashed the resolution because they wanted to stand behind their medical colleagues, former Unit members such as Ishii and Kitano. However, as seen in the document 'Research Results of the Ishii Unit,' Ishii's organization was providing a venue for experiments on humans for the Japanese medical world. When the resolution was presented, the leaders of the medical world, who were also members of the JSC, were still utilizing the results of Ishii's human experiments for their own research purposes and thereby achieving recognition for their research programs. In order to protect themselves, they needed to keep a lid on the fact that Japan had had a program of biological weapons research and development. In order to do this they had to strenuously reject the resolution that would have ratified the Geneva Agreement.

It is not an overstatement to say that the Japanese medical world had no intention of reviewing its position on the Ishii Unit and the many associations of the Ishii Agency. As we have seen, the Ishii Unit bestowed a lasting, but tainted legacy on the Japanese medical science community – the implications of which are as yet not fully known, but which manifest throughout the medical research institutions. Many of Japan's most eminent medical researchers are tainted by this legacy, but have largely escaped scrutiny and repercussions for their actions, and this in large part due to the US Occupation Forces complicit and self-serving suppression of the report about their investigations into the matter.

Notes

1 *Kantō-gun bōeki kyūsui-bu ryakureki* (A brief history of the Department of Water Supply for the Prevention of Epidemics, Kwantung Army) [6] 1982; and Endō, 1974 [14] pp. 162–3.

2 Endō, 1974 [14] pp. 162–3.

3 *Gun igakkō boueki kenkyū hōkoku* (Report on prevention of epidemics by the Army Medical College), Rikugun Gun'igakkō Gunjin Bōekigaku Kyōshitsu (Epidemics Prevention Department of the Army Medical College) [1] 1940, p. 41.

4 *Report on Japanese Biological Warfare (BW) Activities,* Army Service Forces, Camp Detrick, Frederik, Maryland [21] 31 May 1946, p. 5.

5 Keiichi Tsuneishi, *Kieta saikin-sen butai* (The germ strategy unit that disappeared), Kaichōsha, 1981, pp. 141–76; Keiichi Tsuneishi and Tomizō Asano, *Saikin-sen butai to jiketsu shita futari no igakusha* (The germ strategy unit and two doctors who committed suicide), Shinchōsha, 1982, pp. 147–275.

6 Dojinkai, 1943 [5] p. 5.

7 Dōjinkai, 1943 [5] p. 462.

8 A document dated 9 January 1947 in Memorandum for Record, Subject: Request of Russians to Arrest and Interrogate Japanese Bacteriological Warfare Experts in Whom U.S. Has Prior Interest. Presented by AC of S G-2, GHQ, FEC, 27 March 1947 [23]

9 'Interview with Doctor Sanders,' in *Asahi Shimbun*, 15 August 1983.

10 Aoki, 1975 [13] p. 192.

11 Similar investigations are recorded in *Report of Japanese Activities in Medical Sciences*, The Final Report of the Committee for the Technical and Scientific Investigation of Japanese Activities in Medical Sciences, [20] December 1945 and in Report of Investigation Division, Legal Section, GHQ, SCAP, Inv. Dev. No. 330, [22] 4 April 1947. Information on artificial blood and related things was collected. However, besides those prosecuted for the 'Kyūshū University Incidents' or the abuse of prisoners in concentration camps in various areas, doctors were not committed for trial.

12 General Headquarters, United States Army Forces, Pacific, Science and Technology Advisory Section, *Report of Scientific Intelligence Survey in Japan, vol. 5, Biological Warfare*, September, October and 1 November 1945 [17] and *Supplementary Biological Warfare Information*, 9 November 1945 [18]

13 *Report on Japanese Biological Warfare (BW) Activities,* Army Service Forces, Camp Detrick, Frederik, Maryland, 31 May 1946 [21]

14 Refer to the documents of the Record Group 319 (R.G)'s 1000 pages of approximately 2000 pages of documents called the Gray File in the Suitland Branch Office in the American National Archives Library. About 200 pages of the most important parts of those 1000 pages are translated in 'Target Ishii' (translated by Keiichi Tsuneishi, Ōtsuki Shoten 1984).

15 General Headquarters, United States Army Forces, Pacific, Science and Technology Advisory Section, *Report of Scientific Intelligence Survey in Japan* Volume V, Biological Warfare, September and October 1945, 1 November 1945 [17]; *Supplementary Biological Warfare Information*, 9 November 1945 [18]; and *Report on Japanese Biological Warfare (BW) Activities,* Army Service Forces, Camp Detrick, Frederik, Maryland, 31 May 1946 [21]. *Memorandum for Major General Clayton Bissell*, AC/S, G-2: Subject: Japanese Biological Warfare Activities, 26 October 1945 [19] terminated the existence of Ishii Unit, but the Unit's human experimentation was not yet known.

16 Telegram to Major General Alden Waitt from C-52423, CINCFE, 6 May 1947 (R.G.153).

17 SWNCC Series of documents of 351, 1947–1948 (R.G. 153), and a telegram to SCAP from W-97605, the Joint Chiefs of Staff, 13 March 1948 (R.G.153).

18 Hill and Victor, 1947 [26]; includes the record of an interrogation of Kojima on 18 November 1947.

19 Hill and Victor, 1947 [26]; includes the record of an interrogation with the engineer, Kaoru Ishimitsu on 25 November 1947.

20 Rō Wake, *Bio-chemical Weapons*, Chūkō Shinsho, 1966, p. 179.

21 *Report of the International Scientific Commission for the Investigation of the Facts concerning Bacterial Warfare in Korea and China*, Peking, 1952 [29]

22 The minutes of the proceedings of the science council. Rō Wake, *Bio-chemical weapons*, Chūkō Shinsho, 1966.

Bibliographic Guide

Data on the people concerned before the war

[1] *Gun igakkō bōeki kenkyū hōkoku* (Report on prevention of epidemics by the Army Medical College), Rikugun Gun'igakkō Gunjin Bōekigaku Kyōshitsu

(Epidemics Prevention Department of the Army Medical College), typed printing.

[2] Army Surgeon Major Naeo Ikeda, and Army Engineer Saburō Araki, *Hashōfū dokuso narabini gahō sesshu ji ni okeru kin 'kuronakishii' ni tsuite (Shidō Nagayama chūsa)*, (About muscle 'chronaxie' upon the injection for tetanus toxin [Supervisor, Lieutenant Colonel Nagayama]). Date of publication unknown (around 1940).

[3] *Kiidan shageki ni yoru hifu shōgai narabini ippan rinshō-teki shōjō kansatsu* (Observations on skin injury and general clinical symptom caused by *kii* bombs), prepared by the Kamo unit, (Notes suggest that Major Ikeda was in charge), (The time unknown, but it is about human experiments conducted in September 1940). Typed printing.

[4] Rikugun Gun'i Gakkō (Army Medical College), *Rikugun Gun'i Gakkō gojūnen-shi* (Fifty-year history of the Army Medical College), 1936.

[5] Dōjinkai (Dōjinkai Foundation), *Dōjinkai gojūnen-shi* (Fifty year history of Dōjinkai), 1943.

[1] contains 61 reports from among reports number 10 to number 947 that the College received between 29 July 1939 and 8 May 1944. All of them are stamped to be either 'secret' or 'top secret.' According to the former officers of the laboratory for the prevention of epidemics, 100 to 200 copies of every volume were printed and provided to only the people concerned. The people who received all the copies were a small number of army surgeons including Ishii. The 61 reports were mostly concerned with virus reinforcement and the development and effects of vaccines. [2] and [3] are the reports of human experiments by the Ishii unit. The Kamo unit in [3] is another term for the Ishii unit. 'Kii bomb' refers to yperite (mustard gas), a kind of vesicant. [4] and [5] are as per the titles. Neither the Medical College nor Dōjinkai had opportunities to compile their history after this. In reference to the Dōjinkai after this time, see [13].

Data bearing official opinions of the Japanese government.

[6] *Kantō-gun bōeki kyūsui-bu ryakureki* (A brief history of the Department of Water Supply for the Prevention of Epidemics, Kwantung Army), 1982.

[7] 'Lab report describes Japanese atrocities on POWS,' in *Los Angeles Times*, 11 August 1988.

[6] consists of only three pages of data handed in to Social Labor Committee of the House of Representatives from the Ministry of Health and Welfare on 6 April 1982. The data reveal the fact that the Kwantung Army in China had the Water Supply Department for the Prevention of Epidemics and had five branches. The reporter Linda R.

Monroe, a public relations officer for the Japanese ambassador in Washington, acknowledged the existence of the Ishii Unit [7], but denied the conduct of human experiments and the use of biological weapons against China.

Recollections of the People Involved

[8] Seiji Kitano, 'Bōeki hiwa' and 'Zoku bōeki hiwa (The secret story of the prevention of epidemics),' in *Nihon iji shinpō* (Japan medical affairs bulletin), nos. 1947 (1961) and 2516 (1972).

[9] Kiyoshi Kawashima, *Kushū Ryōka* (Haiku collection: Lonely flowers), self published, 1974.

[10] Toshio Yoshimura, *Kiju kaiko* (Reminiscing on my seventy-seventh birthday celebration), Yoshimura Sensei Kiju Kinen Gyōji-ka (Celebration Committee of Mr Yoshimura's Seventy-seventh Birthday), 1984.

[11] Sueo Akimoto, *I no ronri o tou: Dai 731-butai deno taiken kara* (Questioning the ethics of medicine: From the experiences of the 731 Unit), Keisō Shobō, 1983.

[12] 'Daughter's eye-view of Lt. Gen. Ishii,' in *Japan Times*, 5 September and 29 August 1982.

[13] Giyū Aoki, *Dōjinkai shinryō bōeki-han* (Clinical group for the prevention of epidemics of Dōjinkai), Igaku-bu Saikingaku Kyōshitsu Suiyō-kai (Wednesday Meeting Group of the Department of Bacteriology in the Faculty of Medicine), 1975.

[14] Saburō Endō, *Nitchū jūgonen sensō to watashi* (The fifteen-year Sino-Japanese War and me),' Nitchū Shorin, 1974.

[15] Tsuyoshi Miyatake, *Shōgun no yuigon: Endō Saburō nikki* (The will of a general: The diary of Saburō Endō), Mainichi Shimbunsha, 1984.

[8] to [11] were all written by former members of the Ishii unit. Kiyoshi Kawashima in [9] was among a few of those members, who stood trial in the court proceedings recorded in [27]. [12] is an article based on an interview given by Ishii's oldest daughter. Saburō Endō in [14] and [15] was a former army Lieutenant General and looked after the Ishii unit during his periods as the Kwantung Army Strategy Chief of Staff (1932–34) and Deputy Chief of Staff (1939–40). [15] was based on the diary Endō kept for over 60 years. The relevant parts of the diary (about 6 pages) have already been obtained.

Materials to Understand Japan's Medical World after the War

[16] *Gakujutsu Kaigi ketsugi-an: Dai 13-kai sōkai* (The resolution submitted to the thirteenth general meeting of the Science Council of Japan Resolution). As mentioned in the text, it was rejected at the meeting.

American Data : Before the Exposure of Human Experiments

[17] General Headquarters, United States Army Forces, Pacific, Science and Technology Advisory Section, *Report of Scientific Intelligence Survey in Japan, vol. 5, Biological Warfare*, September, October and 1 November 1945.
[18] *Supplementary Biological Warfare Information*, 9 November 1945.
[19] *Memorandum for Major General Clayton Bissell*, AC/S, G-2: Subject: Japanese Biological Warfare Activities, 26 October, 1945.
[20] *Report of Japanese Activities in Medical Sciences*, The Final Report of the Committee for the Technical and Scientific Investigation of Japanese Activities in Medical Sciences, December 1945.
[21] *Report on Japanese Biological Warfare (BW) Activities,* Army Service Forccs, Camp Detrick, Frederik, Maryland, 31 May 1946.

[17] and [8] are reports compiled by Sanders. [17] corresponds to volume 5 of the Compton Report. Together with [19] one can find 'Summary of Information Extracted from a Report by a Member of the Staff of the Army Medical College, Tokyo,' a document submitted by Ryōichi Naitō, deputy to Ishii. This report turned the attention of the Occupation authorities to the Ishii unit in Harbin. [20] is a report amounting to about 100 pages which suggests that the Occupation Forces scrutinized the Army Medical College as well as the Medical Faculties of the Imperial Universities. [21] is often referred to as the Thompson Report.

American Military Data: After the Exposure of Human Experimentation

[22] Report of Investigation Division, Legal Section, GHQ, SCAP, Inv. Dev. No. 330, 4 April 1947.
[23] Memorandum for Record, Subject: Request of Russians to Arrest and Interrogate Japanese Bacteriological Warfare Experts in Whom U.S. Has Prior Interest. Presented by AC of S G-2, GHQ, FEC, 27 March 1947.
[24] Brief Summary of New Information about Japanese B. W. Activities, 20 June 1947 (005).
[25] Letter from Norvert H. Fell to A C of S, G-2, GHQ, Far East Commission, 24 June 1947 (006).
[26] Hill and Victor, Summary Report on B. W. Investigations, 12 December 1947.

According to [22] GHQ's Legal Division acquired information about the Ishii Unit's human experimentation prior to G-2. However, G-2 was surprised at the report and prohibited the Legal Division from conducting any further investigations.

[24] and [25] are the so-called Fort Detrick Reports. The numbers 005 and 006 are the custody numbers at Fort Detrick.

[26] is known as the Hill and Victor report.

Materials around the Time of the Korean War

[27] *Saikinsen-yō heiki no junbi oyobi shiyō no kado de kiso sareta moto Nippon gunjin no jiken ni kansuru kōhan shorui* (Court proceedings of former Japanese soldiers who were charged over preparations and use of bacteriological weaponry), Gaikokugo Tosho Shuppan (Foreign Language Book Publishers), Moscow, 1950.

[28] 'Naichi ni ikiteiru saikin butai (Bacteriological warfare corps still operating within Japan),' in *Shinsō* (Truth), no. 40, April 1950, pp. 6–13.

[29] *Report of the International Scientific Commission for the Investigation of the Facts concerning Bacterial Warfare in Korea and China*, Peking, 1952.

[27] represents the records of the trial held in Khabarovsk from 25 to 30 December 1949. The accused were twelve former Japanese soldiers. The trial took up not only the Ishii Corps but also the biological weapons corps of the Kwantung Army's Department of Epidemics Prevention for Military Horses, with the latter having run experiments on domestic animals. The records contain some 600 pages and were widely read within Japan upon publication. Accordingly, many observers suspected that they were simply the propaganda documents of the Soviet Union. However, upon close reading, one can conclude that the Soviet Union, which must have acquired much more materials than documented here, published the only confirmed information in these court proceedings. The main body of [29] contains about 60 pages. Its appendix amounts to some 600 pages. The main text was translated into Japanese upon publication.

7 Military Science and Technology in Peacetime

Shigeru Nakayama

Militarily Biased Development of Science

Japanese Marxist historians of science characterize the development of Japanese science and technology (S&T) from the late nineteenth century to the end of World War II as 'militarily biased.' This characterization has been widely accepted during the postwar period. However, what counts as 'regular development' and what 'biased' is not easily defined unless a sufficient international comparison is made. If we look closely at history, we find that when Imperial University was founded in 1885, the military-oriented departments such as Zōhei (Military Technology), Kayaku (Ammunition) and Zōsen (Shipbuilding) were created as permanent institutions at the request of the Army and Navy. These disciplines do not follow Western precedents, as do disciplines such as chemistry and civil engineering. In this respect, at least, the Japanese development of S&T was unique.

Western universities established before the creation of nation-states were reluctant to incorporate the sciences into their institutions. When they finally did, they continued to resist teaching applied or technical science for a long time, focusing instead on 'pure' scientific research. Hence, militarily oriented research was at best a peripheral activity for a university. Thus in France, for instance, the Ecole Polytechnique was created in the military-government-industrial complex to develop and teach military technology more efficiently than could be achieved in the university system. In contrast, the modern Japanese universities were created by a nation-state and were thus, from their inception, an integral subsystem of its bureaucracy for whom, under the imperialistic competition of the late nineteenth century, 'catching up' with Western military S&T was of primary importance. The very existence of a new nation-state depends upon military strength, which in modern terms, means that it depends upon its applied S&T. But neither the public nor the private sectors in Japan had adequate facilities established for research and the education of a modern scientific workforce. In response to the military's needs, then, the Japanese university became a cornerstone of modern scientific research and education. Thus the development of university science – and S&T in general – in Japan was indeed militarily biased in comparison with early Western institutions.

The first 'university-affiliated' research laboratory (ARL) was the Aeronautical Laboratory of Tokyo Imperial University, created soon after World War I. This laboratory was created as a direct response to – and *to respond to* – military demand.[1] This marks the beginning of a tendency towards a military focus that accelerated during WWII, culminating in the creation of Daini Kōgakubu (the Second School of Technology) of Tokyo Imperial University in 1942 by the military government. Thus, the evidence clearly supports the Marxist characterization of Japanese S&T developing with a 'military-bias.'[2]

Rearrangement of Military Biased Disciplines

Preparations before the Occupation Forces Arrive: the Case of Tokyo Imperial University

The basic objective underlying the Allied Powers' Occupation policy was the demilitarization of Japan. Far from being simply a process of confiscating weapons and dismissing soldiers and sailors, the objective entailed a radical reconfiguration of Japanese society. Militarily oriented research institutes, university departments and professorial chairs were all subject to possible abolition or restructuring. And since all of the national universities had war-related institutions, Occupation policy affected almost the entire Japanese university sector. To illustrate the nature and extent of its impact we will look at Tokyo Imperial University as an example.

The Japanese at the time of surrender in August 1945 had no concept of what being occupied would entail until the Occupation Forces actually arrived. While some were optimistic about the situation, most were pessimistic. There were rumors everywhere. The School of Technology was so militarily oriented that they were afraid of some kind of reprimand. All of the teachers who had been employed by Military Departments resigned.

In the meantime, a university-wide faculty meeting was held to discuss reforming the university structure. On 25 August the University President submitted a memorandum to the Education Minister. The memorandum summarized proposed interim changes to the school-department structure while acknowledging that they had to wait until the Occupation Forces arrived and issued orders before implementing the changes. The chairs of Aeronautical Medicine and Weaponry Medicine that had been created in the School of Medicine during the war, and Aeronautical Physics in the Department of Physics, School of Science had to change their titles. The most drastic changes were expected in the School of Technology, where the following changes to department titles were proposed.

Table 7-1 Proposed titles for University Departments

Existing titles	Proposed titles
The First School of Technology:	
Department of Aeronautical Engine	Combustion Engineering
Aeronautical Fuselage	Applied Mathematics
Military Technology	Precision Manufacturing Technology
Ammunition	absorbed into Applied Chemistry
The Second School of Technology:	
Aeronautical Fuselage	Fluid Technology
Military Technology	Precision Technology

On 31 August the President sent a second proposal to the Education Minister, this time proposing to change the titles of professorial chairs in war-related departments. With a few exceptions, where the department title changes, most of the chair titles are also subject to change. One of the exceptions was in the Department of Shipbuilding, where the department title remained unchanged, while the chair title Warship was changed to Special Ship. It did not occur to the Japanese that the Occupation Forces would also demand such changes in the field of Radio Technology.[3]

SCAP Prohibitions

In early September, various directives were issued from the Supreme Commander Allied Powers (SCAP) in rapid succession. It was only then that the Japanese began to understand what the 'Occupation' really meant. On 2 September SCAP Directive No 1 ordered the preservation of all facilities and information related to R&D. On 22 September Directive No 3 (SCAPIN 47) prohibited research on atomic energy and 'The Basic Direction Concerning the Occupation and Control of Japan Immediately after the Surrender' (Joint Chiefs of Staff JCS#10) was issued. The latter directive, as part of a policy to destroy the foundations of Japanese military power, prohibited research and education directed specifically towards military development.[4] It also implied that the war-related departments and chairs must be restructured or abolished.

On 18 November SCAPIN 301 prohibited all 'teaching, research or experiments in aeronautical science, aerodynamics or other subjects related to aircraft or balloons.' In accordance with this order, on 24 November the director of the Department of School Education, the Ministry of Education, sent a memorandum entitled 'On the Order Based on the Directive concerning Aeronautics' to the presidents of all relevant universities and polytechnics. It ordered that all of the departments of aeronautical science had to change their titles. There were eight universities with aeronautical science departments: the Imperial Universities of Tokyo, Kyoto, Kyūshū,

Tōhoku, Osaka and Nagoya, and the Institutes of Technology in Tokyo and Ryojun. The other three Imperial Universities – Hokkaidō, Keijō (now Seoul) and Taihoku (now Taipei) – did not have such departments.

At Tokyo Imperial University the demand to restructure generated much inter-departmental strife. The Humanities and Social Sciences Departments, had both suffered during the militaristic government for their ideological differences. They had been stridently opposed to the creation of the Second School of Technology, which was established to satisfy military demands. During the debates about the restructure of the universities the Humanities and Social Science Departments demanded that the chairs from the Second School be re-allocated to themselves. But the First School of Technology also had designs on absorbing the Second School *in toto*.[5] Beginning in November, professors of Schools and Departments met a number of times to negotiate but no conclusion was reached. Thus, to ensure compliance with SCAP directives, on 9 January 1946, Imperial Edict No 7 was issued, abolishing five chairs in Aeronautical Engineering and five chairs in Aeronautical Fuselage. This was immediately followed by Imperial Edict No 8 which established five new chairs in Physical Technology and five in Combustion Engineering. Throughout the process, besides the titles of the chairs, nothing changed in the departments themselves; neither the professors nor the department personnel changed.

This re-labeling extended into private schools and polytechnics as well. The Department of Aeronautics at Kansei Gakuin Polytechnic School was abolished in September 1945 and on October 20 Kantō Gakuin School of Aeronautical Technology was renamed Technological Polytechnic School. Schools whose names included 'Kōkū' (Aeronautics), 'Kahei' (Gun and Ammunition), 'Tōa' (East Asia) and 'Kōa' (Rising Asia) were all renamed.[6]

As mentioned above, the Japanese had not foreseen the prohibition of radio-related research. On 24 December SCAP ordered that research in electronic technology was permitted only with specific clearance from the General Headquarters (GHQ) of SCAP. Six research fields were specifically and categorically excluded from the permission granted to the Electro-technical Laboratory of the Communications Ministry: television, radio and radar countermeasures, pulse modulated multi-channel electronic proximity devices, voice secrecy system (speech scramblers), electronic code and cipher systems (SCAPIN 494). On 16 January 1946 GHQ granted permission for research activities to resume at Nippon Electric Company Laboratory with exceptions similar to those listed above (SCAPIN 609) and on 19 January extended the prohibition to include 'radio and radar aids to navigation at Electro-Technical Laboratory' (SCAPIN 626). When the Japanese realized that such research conflicted with Occupation policy, they changed the title of laboratories and departments from 'electronics' and 'radio wave' to 'electric.' For example, the Ultra Short-wave Laboratory of the Hokkaidō University became the Applied Electric Laboratory. The Electronic Technology Laboratory of the Tokyo Institute of Technology was renamed the Electric Science Laboratory. With the exception of the Laboratory of Military S&T, these were changes in name only.

From the spring of 1946, Harry C. Kelly of GHQ's Economic and Scientific Section, Science and Technical Division (ESS/ST) began to redefine Occupation science policy away from demilitarization (and its accompanying prohibition of research) and towards encouraging the research deemed to be crucial for economic reconstruction. On 25 May he relaxed the restrictions on research fields, announcing that SCAP prohibitions applied only to military research (SCAPIN 984). In July the restrictions on research were further relaxed (SCAPIN 1067, 1084). The prohibitions on radio wave research were withdrawn. Permission was granted to Kenjirō Takayanagi at the Technology Laboratory of NHK for television research and to Kinjirō Okabe at Osaka Imperial University for a magnetron.[7]

Dissolution and Restructuring of a Laboratory and Department

The affiliated research laboratories (ARLs) of universities had greater structural flexibility than did the departments and schools of universities. The ARLs had been established and organized to respond to the changing needs of the military and could thus easily adapt to a changing situation. Again, many of them changed in name only, but the more overtly militaristic facilities had to employ a variety of measures in order to survive.

The Aeronautical Laboratory of Tokyo Imperial University, the Aeronautical Medicine Laboratory of Nagoya Imperial, and Katsuzai Kōgaku (the Reviving Materials Technology) Laboratory of Kyūshū Imperial Universities were all disbanded in January 1946. On 30 March the first two of these were renamed Science and Technology Laboratory and Environmental Medicine Laboratory respectively, and were officially revived. When the Aeronautical Laboratory was disbanded and renamed, half of the staff resigned while the remaining half redirected their research projects away from aeronautics and other prohibited fields. The Reviving Materials Technology Laboratory, though, had been so recently established and so short lived that it was never revived.

ARLs were directly affiliated to the university headquarters but were independent of the university's school-department structure. Their faculties were largely self-governing, including autonomy in personnel management; thus, they were not easily disbanded. In contrast, laboratories belonging to the schools were much smaller in scale and much less autonomous. Most of them were simply abolished. By April 1946 the restructuring of the affiliated laboratories at the Imperial Universities was completed.

University departments, though, are not as flexible as research laboratories, partly because they consist of students as well as faculty members. Restructuring these departments requires students to change disciplines, and in some cases the department of enrolment. Thus, unlike the ARLs, the restructuring of university departments means real change for the people involved.

But there were other factors to consider, too. For example, Kelly visited the Second Technology School of Tokyo Imperial University on 10 February 1947 with the intention of restructuring the school as a laboratory. For Kelly, the objective in restructuring the laboratories was not simply to create jobs for unemployed scientists and technicians, but to rationalize the research system on the principle that fewer laboratories would be bigger laboratories with better quality equipment and should therefore function better.[8] He thus deemed it appropriate to merge the Second Technology School with the Science and Technology Laboratory. However, the faculty members' desire for survival was so strong that, in 1949, they managed to establish an independent Production Technology Laboratory. This was the last of the war-related laboratories to be restructured.

Amongst the other laboratories affiliated with Tokyo Imperial University, the South Asian Natural Science Laboratory was renamed the Regional Natural Science Laboratory but its restructuring was not easy, producing no significant achievement. It was finally abolished in 1952.[9]

National Laboratories

Immediately after the War, all of the military laboratories were closed and their equipment confiscated. Only a few of these were restructured and reopened with SCAP's permission. For instance, in December 1945, the Shimoda Laboratory of the Second Navy Engineering Research was moved to the Ministry of the Interior and The Electrotechnical Laboratory of the Communications Ministry was reopened. The Ministry of Military Procurement was abolished, and instead the Ministry of Commerce and Industry was revived. All of the laboratories of the former survived as part of the latter. Generally speaking, right after the war the researchers who worked for the military laboratories were absorbed into other non-military national laboratories. However, many were later retrenched as part of the rationalization of administrations.

The Office of Engineering had been the center of war mobilization research. It was dissolved and its three primary functions were allocated to other government offices: general investigation went to the Bureau of

Table 7-2 The change of war-related chairs, departments and laboratories, contrasting old and new

The number at the end of chair title is the number of chairs; '+' means number of new chairs added to old chairs. Prepared by Shigeru Nakayama.

	Old title	**New Title**
Tokyo Imperial University		
Affiliated Laboratory	Aeronautical Laboratory	Science and Technology
	Southern Natural Science	Regional Natural Science
	Second Technology School	Production Technology Laboratory
Dept: First Technology	Military Technology	Precision Technology
	Ammunition	absorbed by Applied Chemistry
	Aeronautics	Physical Technology
Second Technology	Aeronautical Fuselage	Physical Technology
Chairs: Medical School	Aeronautical Medicine	Physiology
	Weaponry Medicine 2	Hygiene 2
1st Technology School	Aeronautics 5	Applied Mathematics 5
	Aeronautical Engine 3	Internal Combustion 5
	Military Technology 5	Precision Manufacturing Technology 4
	Ammunition 4	Gas Thermodynamics 1 Blasting Operations 1
	Chemical Weapon 1	Applied Chemistry +2
2nd Technology School	Aeronautical Fuselage 5	Applied Mathematics 4
	Aeronautical Engine 5	Combustion Engine 3
	Military Technology 4	Precision Technology4
Science School	Aeronautical Physics 1	Physics 1
Economics School	Colonial Policy 1	National Economy 1
Kyoto Imperial University		
Dept: Technology School	Aeronautical Technology	Applied Physics
Chairs: Medical School	Aeronautical Medicine 1	Environmental Medicine 1
Science School	Aeronautical Physics 1	Physics +1
Technology School	Aeronautics 6	Applied Physics 6
	Aeronautical Electricity 1	Electrical Technology 1
Economics School	East Asia Economic Policy 1	Economic Policy 1
Humanities School	Japan Spiritual History 1	Abolished
Kyūshū Imperial University		
School-Attached Laboratory	Reviving Materials	abolished
	Tropical Agriculture and Forestry	abolished
Chairs: Technology School	Aeronautics 5	Engineering Dynamics 5
Medical School	Volks-Colonial Hygiene 1	Volks Hygiene 1
Tōhoku Imperial University		
School-Attached Laboratory	Aeronautical Medicine	abolished
	Metal Laboratory, Aeronautical Division	abolished
Chairs: Technology School	Aeronautics 5	Engineering Dynamics 5
Medical School	Aeronautical Medicine 1	Physiology 1
Hokkaidō Imperial University		
School-Attached Laboratory	Ultra Short Wave	Applied Electricity
Chair: Technology School	Fuel 4	Applied Chemistry 4

...

Table 7-2 continued

		Old title	New Title
Osaka Imperial University			
School-Attached Laboratory		Aeronautical Science	Applied Dynamics
		Ultra Short Wave	Applied Electricity
Chair:	Technology School	Fuel 4	Applied Chemistry 4
Osaka Imperial University			
School-Attached Laboratory		Aeronautical Science	Applied Dynamics
			Disaster Prevention Science
Dept:	Technology School	Aeronautics	Engineering Dynamics
			Structural Engineering
Chairs:	Medical School	Aeronautical Medicine 1	Hygiene 1
	Technology School	Aeronautics 5	Engineering Dynamics 5
Nagoya Imperial University			
School-Attached Laboratory		Aeronautical Medicine	Environmental Medicine
Dept:	Technology School	Aeronautics	Physical Technology
Chairs:	Medical School	Aeronautical Medicine 2	Environmental Medicine 2
	Technology School	Aeronautics 6	Physical Technology 6
Tokyo Institute of Technology			
Laboratory		Electronic Aeronautics	Electric Science
Nagasaki Institute of Medicine			
Laboratory		East Asian Local Disease	Local Disease
Tokyo University of Industry			*Hitotsubashi University*
Laboratory		East Asian Economy	Economics

Cabinet Investigation, patents and standards to the Patents and Standards Office, and scientific research to the Science Education Bureau of the Education Ministry.

Agency of Industrial Science and Technology

Amongst SCAP's first tasks in Occupation was an investigation of the research activities of the Japanese wartime mobilization. ESS/ST's evaluation of the various research sites visited and other reports concluded that, although the Japanese public laboratories had many able researchers, the coordination between laboratories and among researchers was poor or simply non-existent. Thus, they were ineffective at producing results.

ESS/ST developed policies for restructuring laboratory management based on these findings. Kelly advised that a new Agency of Industrial Science and Technology be created to facilitate the coordination of the various public laboratories. It was established in January 1948. From the beginning, Kelly had intended that the Metal Institute of Tōhoku University would be included,[10] but soon discovered that trans-Ministry coordination was hardly realizable. Only those laboratories belonging directly to the various bureaus of the Ministry of Commerce and Industry were incorporated into the affiliated organization of the Agency of Industrial Science and Technology. The organization consisted of the Coordination Department, the Standard Division, the Central Inspection Institute of Weight and Measures, the Mechanical Engineering Laboratory, the Tokyo Technology Laboratory, the Osaka Technology Laboratory, the Brewing Laboratory, the Ceramics Laboratory, the Textile Technology Laboratory, Geological Survey of Japan, the Electrotechnical Laboratory, the Industrial Art Institute, the Fuel Laboratory and the Mining Engineering Laboratory. The mission of the Coordination Department was to coordinate between government administration and private industry. The department was managed by the Working Council of Industrial Technology, which decided its highest level of policy through the vote of its members and whose main objective was the integration of R&D laboratories with the administration of industry and commerce. Furthermore, in order to reflect the diverse interests of the business and scientific communities, the director of the Council of Industrial Technology installed an advisory board consisting of members from various communities.[11]

This externally initiated reorganization was not easily affected. The Ministry's administrative clique was afraid of the precedent set by the Ministry of Construction, where the engineering clique had achieved power equal to the administrative clique; they were afraid that this situation might be duplicated in Commerce and Industry. When the Occupation ended and the Japanese Government was rationalized, the status of the Agency of Industrial Science and Technology was downgraded from Chō (Agency) to In (Institute) but this title change never appeared in English.[12]

Railway Technical Research Institute

All of the major industry sectors in Japan were radically redirected after the war, with one exception: the National Railways. It was a stronghold of S&T research and development that continued to work without interruption,[13] even though it was war-damaged. Immediately after the war the structure of the National Railways began to expand, initially through the absorption of returnees from Mantetsu (Manchuria Railway). The Railway Technical Research Institute experienced the greatest expansion of all the Railway institutions.

On 31 December 1945 the Railway Technical Research Institute adopted a plan that would eventually result in it becoming the largest general research laboratory in Japan. The plan was to acquire and incorporate the facilities of the Central Aeronautical Laboratory of the former Ministry of Transportation, which would have to be closed because of the prohibitions on aeronautical research, and to employ scientists and engineers from the old military arsenals and from heavy industry in the private sector.

Before the war there had been 385 personnel in the Test and Design Department. By the day of surrender they numbered 1,162 (including 610 transferred from the Central Aeronautical Laboratory). The later addition of personnel took it to a peak of 1,567 at the end of October 1947. By then they were no longer restricting their research to projects directly related to running a railway, but rather had absorbed the research foci of other organizations into the new Basic Research Department which, at the end of 1946, created three science departments for basic research.[14]

The Occupation Forces confiscated the facilities of the Central Aeronautical Laboratory for reparations. Thus, although the Railway Technical Research Institute was responsible for the management and maintenance of these facilities, they could not use them freely.[15] The wind tunnel of the old Aeronautical Laboratory of Tokyo Imperial University was also listed as an item for reparation. In September 1946, the Railway Institute requested the release of it for use in basic research for a new bullet train,[16] but their petition was refused by SCAP. [17] It has been said that this basic research led to the production of a successful bullet train but it is impossible to attribute credit for this success to any single element; the bullet train is the product of an integrated research and development system.

The Railway Technical Research Institute provided a temporary shelter for the scientific and engineering personnel who became unemployed at the time of surrender. There was a hidden agenda here – to protect and preserve prohibited aeronautical technology. Among those researchers who moved from the military laboratories were seventeen scientists holding the rank of major or above and two others who were forced to resign in June 1948 during the purge of public sector officials. Famous among these men are Kitarō Matsumoto, who designed the warships Musashi and Yamato, and Toshio Kondō, who was a key person in aeronautical engineering as well as a professor of Navy College.[18]

In May 1949 Shannon, from GHQ's CTS (Civil Transportation Section), suggested that the Railway Technical Research Institute was useless and should be abolished in accordance with a plan to curtail the manpower at the National Railways Corporation; a plan based on the recognition that the Railways industry was generally in decline.[19] The members of the Laboratory enlisted Kelly to argue for their survival,[20] but because of the section rivalry within GHQ, the matter was left undecided. The laboratory cadres then lobbied the Cabinet. On 22 June 1949 the Prime Minister consulted the newly created Science Council of Japan (JSC) 'on the matter of the Railway Technical Research Institute.'[21] The Council decided in favor of the Laboratory's survival on 29 June.

Zaidan Hōjin Rikagaku Kenkyūsho (Riken)

Riken was an exceptionally productive research institution in prewar Japan, training excellent young researchers for the Japanese scientific community. It is organizationally a holding company consisting of thirty-six companies that were financed by Riken's research outcomes. Hence, in spite of the pending grief for the researchers who worked there, it was destined to be disbanded because of GHQ's anti-monopoly policy.

The director, Masatoshi Ōkōchi was identified as a war criminal and had to step down. Yoshio Nishina succeeded to the directorship. He and his colleagues attempted to save Riken by pointing out its special significance in appeals to Kelly. Kelly pushed the Anti Trust and Cartels Division of ESS to preserve the research function of Riken and discussed ways to rescue it financially. The Education Ministry and personnel of Riken hoped to make it a zaidan hōjin (juridical foundation under governmental control) but the proposed structural rearrangement did not satisfy SCAP. [22] Kelly's idea was that, instead of making Riken a zaidan hōjin focused on basic science research, it should perform industrial research that would contribute to an economic recovery for Japan. After pursuing a number of avenues and 'pulling some strings,' Kelly finally set the stage for General Douglas MacArthur to send a letter to Prime Minister Shigeru Yoshida permitting the re-establishment of Riken as the Kagaku Kenkyūsho (Kaken; the Scientific Research Institute) Corporation in March 1948.

Nishina became the President of the new organization. He tried to retain its research funding through the sale of penicillin but its prewar glory was never recovered.[23] In the later course of development, the main stream of industrial research moved from dedicated research corporations like Riken to the central research institutes of big corporations.

Conclusion

Science and technology was developed in Japan with a military bias from the late nineteenth century until the end of World War II. The Allied Powers' demilitarization program affected radical change to the scientific community, including a reorientation of research programs, the abolition and amalgamation of research laboratories and a restructuring of university departments. Of the research laboratories extant before the war, only the Railway Technical Research Institute at the National Railways emerged with its original research programs and facilities relatively intact. In fact, through consolidations and amalgamations, the Railway Technical Laboratory played a significant role as the largest basic science research organization in postwar Japan.

Notes

1 Tokyo Daigaku Hyakunenshi Henshū Iinkai (ed.), *Tokyo Daigaku Hyakunen-shi, Bukyoku-shi* (Hundred year history of Tokyo University, Section history), vol. 3, 1987 [5] p. 43.

2 Nihon Kyōsantō Kagaku Gijutsubu (Japan Communist Party, Science and Technology Section), 'Nihon no kagaku gijutsu no kekkan to kyōsan shugisha no ninmu (The defect of Japanese science and technology and the call for communists),' in *Zen'ei* (Vanguard), nos. 10-11, 1946.

3 'Hisho' [3]. See especially the President's proposal addressed to the Minister of Education dated 31 August 1945.

4 According to Edwin Reischauer, the early Occupation policy was based on a Marxist interpretation that the Japanese military aggression was spurred by the monopolistic capitalism of the Zaibatsu. This prohibition must have been modeled on the precedent of the Occupation policy for Germany. However, while the US Occupation science policy for Germany provided detailed specifications prohibiting nuclear physics and rocket technology, the policy for Occupied Japan was not so specific. Cf. Office of Military Government for Germany (US), *Military Government Regulations Title 27. Control of Scientific Research* (effective date 30 July 1946) Part 2, Prohibited Research, ESS(E)06455-7.

5 Tokyo Daigaku Hyakunenshi Henshū Iinkai (ed.), vol. 3, 1987 [5]

6 For the table outlining the renaming of private schools and universities, see Takashi Hada, 'Sengo kyōiku kaikaku to kagaku gijutsu no shomondai (Problems of postwar education reform and S&T),' in Sōsuke Watanabe (ed.), *Kōwa dokuritsu gono wagakuni kyōiku kaikaku ni kansuru chōsa kenkyū* (Investigation on the Japanese educational reform after the peace treaty and independence), March 1992, p. 9.

7 *Kagaku Bunka Shimbun*, 13 June 1946.

8 Imaoka, 1987 [20] p. 258.

9 Tokyo Daigaku Hyakunenshi Henshū Iinkai (Editorial Committee for a One-Hundred Year History of the University of Tokyo) (ed.), *Tokyo Daigaku hyakunen-shi* (The one hundred-year history of the University of Tokyo, Departmental histories), vol. 4, 1987.

10 *Naagoya Kōgyō Gijutsu Shikensho nijūgonen-shi* (Nagoya Technology Laboratory; Twenty-five year history), 1978, p. 339.

11 Kōgyō Gijutsu-chō (Agency of Industrial Science and Technology), *Kenkyū hakusho* (Research white paper), 1951, p. 49.

12 B. Dees, the successor of Harry Kelly, was not informed of this downgrade (perscom with Shigeru Nakayama).

13 Tetsudō Gijutsu Kenkyūsho (Railway Technical Research Institute), 1957, [31] p. 2.

14 Tetsudō Gijutsu Kenkyūsho, 1957 [31] pp. 40-4. Nihon Kokuyū Tetsudō (National Railways), *Nihon Kokuyū Tetsudō hyakunen-shi* (One hundred year history of Japan National Railways), vol. 10, 1973, pp. 214-15. Tetsudō Gijutsu Kenkyūsho, *Jūnen no ayumi – sōritsu shichijūshūnen* (History of the last ten years of Railway Technical Research Institute – at the seventieth anniversary), 1977, pp. 1, 3, 191-3, 280-2.

15 Tetsudō Gijutsu Kenkyūsho, 1957, [31] p. 81.

16 Un'yushō Tetsudō Gijutsu Kenkyūshochō (Director of the Railway Technical Research Institute, Transportation Ministry), 'Baishō wariate yori kikai o jogaisuru ken (On the release of machineries from the reparations list) 6 September 1946,' ESS(B)11723-34.

17 'Memo for record' JWO'B, 5 February 1947, ESS(B)11723-34.

18 Tetsudō Gijutsu Kenkyūsho, 1957 [31] pp. 51-2.

19 Tetsudō Gijutsu Kenkyūsho, *Jūnen no ayumi – sōritsu rokujūshūnen* (History of the last ten years at the sixtieth anniversary), 1967, pp. 210-11.

20 Tetsudō Gijutsu Kenkyūsho, 1957 [31] p. 785.

21 Nihon Gakujutsu Kaigi (Science Council of Japan), *Nihon Gakujutsu Kaigi nijūgonen-shi* (Twenty five year history of the Science Council of Japan), 1977, p. 551.

22 Shōkōshō Sōmukyoku (Commerce and Industry Ministry, General Management Section), *Zaidan Hōjin Rikagaku Kenkyūsho Shinkaisha setsuritsu ni tomonau shiryō* (Sources of Riken for establishing a new company) [23]. B. Dees' testimony in Tamaki, 1978 [21] p. 172.

23 Uchida, 1978 [21] p. 172.

Bibliographic Guide

SCAP Research Prohibitions and their Repeals

[1] SCAPIN (preserved in Kokkai Toshokan (Diet Library) Kensei shiryōshitsu).

1.1 SCAPIN 47: Directive No. 3, 22 September 1945.
Important directive prohibiting research on nuclear energy, etc. All amendments to research prohibitions refer to this as the most basic directive.

1.2 SCAPIN 301: Commercial and Civil Aviation, 18 November 1945.
Notice to abolish all government institutions related to aeronautics by the end of the year. It prohibited all education, research and experimental activity in aeronautical science, aeronautical mechanics and anything related to aeroplanes and balloons, even in private enterprise.

1.3 SCAPIN 494: Operation of Electrotechnical Laboratory, 24 December 1945.

Permission for resuming research activity at the Electrotechnical Laboratory of the Communications Ministry. Its activities were generally permitted by SCAP with the exception of six research fields (e.g., television) which remained prohibited.

1.4 SCAPIN 609: Operation of Nippon Electric Company Laboratory, 16 January 1946.
Permission for reopening the laboratory of Nippon Electric Company, although seven research fields (e.g., television) remained prohibited.

1.5 SCAPIN 626: Operation of Electrotechnical Laboratory, 19 January 1946.
Added wireless and radar research to the prohibited list for the Electrotechnical Laboratory.

This is a minor addition, which was mistakenly omitted from the list in SCAPIN 494 (1.3 above). In 1949, when the Japanese shipbuilders were attempting to win contracts to build ships for overseas buyers, it became apparent that it was virtually impossible to sell a ship without navigational radar equipment pre-installed. The prohibition was then relaxed by MacArthur on the pretext that radar is little more than a compass.

1.6 SCAPIN 984; Amendment of SCAPIN 47 (Directive No 3), 25 May 1946.
This amendment was part of the trend towards relaxing research restrictions, changing the requirement for monthly research reports to bi-annual reports.

1.7 SCAPIN 1067: Operation of the Nippon Electric Company Laboratory, 16 July 1946.
Relaxation of research prohibitions for the Nippon Electric Company Laboratory.

1.8 SCAPIN 1084: Operation of Electrotechnical Laboratory, 25 July 1946.
Relaxation of research prohibitions for the Electrotechnical Laboratory of the Communications Ministry. It permitted electronic research, including television, which was not directly connected to military affairs.

[2] Kanpō (Government Gazette).
Responses from the Japanese Government to the SPAPIN.

Universities' Responses

[3] 'Hisho'.
Preserved in Tokyo University Archive. Administrative; documents from the President's office arranged annually. Documents concerning the renaming of war-related departments appeared in 1946. The process between proposals and Imperial Edicts can be found in handwritten copies or casual printings. Provides reasons for renaming faculty chairs.

[4] Tokyo Daigaku Seisan Gijutsu Kenkyūsho (ed.), *Tokyo Daigaku Daini Kōgakubushi* (History of the Second School of Engineering, University of Tokyo), November 1968.

Silver Jubilee celebration of the Opening of the School. Edited by Teijirō Muramatsu.

[5] Tokyo Daigaku Hyakunenshi Henshū Iinkai (Editorial Committee for a One-Hundred Year History of the University of Tokyo) (ed.), *Tokyo Daigaku hyakunen-shi, Bukyoku-shi* (The one hundred-year history of the University of Tokyo, Section history), vol. 3, University of Tokyo Press, 1987.
The section concerning the School of Engineering was written by Sozai Koyano. The description of the Second School of Engineering is not much different from [4] but interviews with secretariats provide insights into the renaming of faculty chairs.

[6] Tokyo Daigaku Seisan Kenkyūsho (Institute of Industrial Science, University of Tokyo), 'Nijūshūnenshi (Twenty year history),' in *Seisan Kenkyū* (Industrial Science), vol. 21, no. 5, May 1965, pp. 167–367.

[7] Kyoto Daigaku Shichijūnen-shi Henshū Iinkai (ed.), *Kyoto Daigaku shichijūnenshi* (Seventy year history of Kyoto University), 1967.
On the renaming of the aeronautical department, see p. 643.

[8] *Kyoto Daigaku Kōgakubu hachijūnenshi* (Eighty year history of the School of Engineering, Kyoto University), 1980.
Description is similar to [7]

[9] *Tōhoku Daigaku gojūnenshi* (Fifty year history of Tōhoku University).
This is an excellent account which became the model for writing the history of Japanese universities in the postwar period. The change of chairs in the aeronautical department is provided in jōkan (vol. 1) pp. 468ff.

[10] *Kyūshū Daigaku gojūnenshi, Tsūshi* (Fifty year history of Kyūshū University, Outline history), 1967.

[11] *Kyūshū Daigaku gojūnenshi, Gakujutsushi (Jō)* (Fifty year history of Kyūshū University, History of academic matters, vol. 1), 1967.

[12] *Kyūshū Daigaku shichijūgonenshi, Shiryōhen (Ge)* (Seventy-five year history of Kyūshū University, Source materials), 1989, p. 6.

[13] *Hokudai Hyakunen-shi (History of Hokkaidō University*), 1980, pp. 1211–12, 1218.

[14] *Osaka Daigaku gojūnenshi, Tsūshi* (Fifty year history of Osaka University; Outline history), 1985, p. 567.
On the abolition and change of wartime facilities, see pp. 237–40.

[15] 'Osaka Daigaku Kōgakubu no Sengo-shi (Postwar history of School of Engineering, Osaka University),' in *Osaka Daigaku-shi kiyō* (Journal of the history of Osaka University), May 1981, pp. 93–110.

[16] Minoru Imoto, 'Haisen zengo no nikki (A diary before and after the end of the war),' in *Osaka Daigaku-shi Kiyō*, vol. 1, May 1981, pp. 111–16.

[17] *Nagoya Daigaku gojūnen-shi, Bukyoku-shi* (Fifty year history of Nagoya University, Section history), vol. 2, November 1989, p. 351.

[18] *Tokyo Kōgyō Daigaku kyūjūnenshi* (Ninety year history of the Tokyo Institute of Technology), 1975, p. 625.

[19] *Tokyo Kōgyō Daigaku hyakunenshi, Tsūshi* (Hundred year history of Tokyo Institute of Technology, Outline history), 1985, p. 592.
[20] Kazuhiko Imaoka, *Tokyo Daigaku Daini Kōgakubu* (Second School of Engineering, Tokyo University), 1987.

Postwar Reorganization of Riken

[21] 'Tokushū, Rikagaku Kenkyūsho rokujūnen no ayumi (Sixty year history of Riken),' in *Shizen*, vol. 33, no. 13 (rinji sōkan), December 1978.
Primarily a panel discussion by former Riken members
Includes: Hidehiko Tamaki, 'Kaken no umareizuru kurushimi (Painful birth of Kaken)', Saburō Uchida, 'Konkyū to doryoku no nakano Kaken (Kaken in despair and effort)', and 'Shiryō Riken no ayumi (Sources for the history of Riken)' with a chronological table added.
[22] Shin'ichirō Tomonaga and Hidehiko Tamaki (eds.), *Nishina Yoshio: Denki to kaisō* (Yoshio Nishina, his biography and recollections), 1952.
[23] Shōkōshō Sōmukyoku (Commerce and Industry Ministry, General Management Section), *Zaidan Hōjin Rikagaku Kenkyūsho Shinkaisha setsuritsu ni tomonau shiryō* (Sources of Riken for establishing a new company).
On 8 August 1947 the Katayama Cabinet decided to submit 'Zaidan Hōjin Rikagaku Kenkyūsho ni kansuru sochi ni kansuru hōritsu (Law concerning Riken)' to the Diet. This text was prepared by an official of Shōkōshō Sōmukyoku (Commerce and Industry Ministry, General Management Section) for the preparation of the Bill. The final chapter is '10. Kaiso ni kansuru rengōgun shireibu tono kōshō keika no taiyō (Outline of the negotiation process with SCAP on the reorganization).'
[24] SCAP ordered every university and research institute to report their activity every three months; later this was relaxed to annual reports. These reports are in the Occupation documents preserved at the Recording Center, Maryland, US, and in the Gaimushō Shiryōkan (Diplomatic Archive of the Ministry of Foreign Affairs). This collection constitutes a valuable resource for information about the state of affairs in the days immediately after WWII.

Miscellaneous

[25] Akira Abe, *Kyōiku kankei hōrei narabini sakuin Shōwa-hen* (Educational ordinances: An inventory and index, Shōwa Period III), 1988.
[26] 'Gunkankei gakuto (Military-affiliated Students),' in *Kokumin no kagaku* (Science for the national people), vol. 1, no. 5, 1946, p.18.
[27] Nihon Kagakushi Gakkai (ed.), *Nihon kagaku gijutsu-shi taikei 5; Tsūshi 5* (A compendium of the Japanese history of science and technology: vol. 5, Historical overview), 1964.

Includes a chronological table of wartime decrees concerning the restructuring of wartime research organizations originally listed in *Kanpō* (Government Gazette) [2]

[28] Nihon Kagakushi Gakkai (ed.), *Nihon kagaku gijutsu-shi taikei vol.18, Kikai gijutsu* (A compendium of the Japanese history of science and technology vol.18, Mechanical engineering), 1966.
The Departments of Military and Aeronautical Engineering were discussed in the mechanical engineering volume. It is regrettable under the pacifist historiography of science that military engineering, most characteristic of prewar S&T, was not treated independently – as is its due – in this series.

[29] Keiichi Yamada and Shūichi Tsukahara (eds.), *Kagaku Kenkyū no raifu saikuru* (Life cycles of scientific research), 1986.
Chapter 6 'Kotonatta senmon bun'ya eno tenkan – kōkūkōgaku no keesu sutadii (Turn-over to different discipline – case study of aeronautical engineering)' provides a statistical study of career changes amongst the graduates of Aeronautical Engineering Departments.

[30] Kiyoshi Tomizuka, *Hachijūnen no shōgai no kiroku* (Records of my eighty years), 1975.
Includes his diary from immediately after the surrender. The author was a professor at the Aeronautical Research Institute of Tokyo Imperial University. During the War he served as a board member of Genron Hōkokukai (The Society for Nationalist Opinion). Because of this role, he was purged from his official post after the War. He was a supporter of the Japanese military during the War, then afterwards he claimed to have warned the Japanese Government about their lack of rational judgment which cost Japan the War. His opportunism invited criticism after the War. He was, however, an optimistic scientist. In the postwar period, he claimed that Japan should develop precision industries like the clock and watch industries of Switzerland rather than depending on heavy industry.

[31] Tetsudō Gijutsu Kenkyūsho Gojūnenshi Kanko Iinkai, (Commitee for publication of the Fifty Year History of the Railway Technology Research Institute), *Nihon kokuyū tetsudō gojūnen-shi* (Fifty year history of the Railway Technology Research Institute), 1957.

Part II

Academic Research and its System Under the Occupation Power

8 The Role of Advisory Missions

Shigeru Nakayama

The military staff at General Headquarters, Supreme Commander for the Allied Powers (GHQ/SCAP) did not have the expertise required to formulate science policy. GHQ's Economic and Scientific Section, Science and Technical Division (ESS/ST) was staffed by scientists and engineers, but they did not have sufficient authority to direct the Japanese scientific community. GHQ therefore requested expert advisors to be sent from the USA. Their opinions influenced not only the Occupation policy itself, but also had enduring implications for the later development of Japanese science and technology (S&T). This chapter aims at measuring the extent of their influence through an examination of their advice and its implementation.

The Process of Inviting Major Missions and their Role

Strictly speaking, there were only two Science Advisory Missions sent to Japan during the Occupation. Both came at the invitation of the ESS/ST. Officially, the first mission was called the Science Advisory Group and the second the Science Advisory Mission, but we shall refer to them as the First and Second Science Advisory Missions.

There were also a number of specialized S&T advisory missions that visited Japan during the Occupation. They can be divided into two groups.[1] The first group arrived in Japan during the early stages of the Occupation to collect information about Japanese wartime science and to advise SCAP on which facilities and equipment to red tag for reparation (cf. ch. 1 and 2). With the exception of an Australian Science Mission that arrived in October 1945 for the purpose of extracting reparations, this activity was conducted solely by the US Forces. The second group of advisory missions was invited from the US to provide advice to the Occupation Forces on science and technology. In this respect, there was some overlap between the first and second groups. Brigadier General John W. O'Brien and some of his associates from the first group remained in Japan as advisors to the ESS/ST on science and technology matters in general. The second group, however, was specifically tasked with influencing the Japanese scientific community to implement research and develop practices already proven to be effective in the US.

Each of these advisory groups was recruited and dispatched at the request of a specific section of GHQ, who authorized the request to the

Department of the Army in Washington DC. The Department of the Army then appointed mission members from scientific societies and universities. They arrived in Japan within a year. Each GHQ section that had invited a mission played host to that mission, briefing them on the situation and requirements, accompanying their observation tours and arranging meetings with Japanese scientists. Each mission provided recommendations to GHQ based on their assessments and left for home (with a few exceptions as noted above).

Missions Invited as Advisors on Occupation Policy

While ESS/ST was responsible for S&T related missions, the Civil Information and Education Section (CIE) was responsible for educational matters. Where their responsibilities overlapped, tasks and responsibilities were allocated to one of each: for example 'research in higher education' was assigned to CIE. S&T policy responsibilities were further fragmented with the Natural Resource Section (NRS) – in charge of resource-related matters – and Public Health and Welfare (PHW) – responsible for medical care and hygiene – each requesting their own advisory missions.

Education Missions

The First Educational Mission was dispatched at the request of CIE to reform the educational system along the lines of the American model (called the 6-3 system: 6 years of primary school, 3 years of junior high school, 3 years of high school). It arrived in Japan in March 1946 with George D. Stoddard as its head. Since S&T was properly the domain of the ESS/ST, the Education Mission avoided duplication and limited its advice to the CIE to science and general education.[2]

At the university level the Mission's advice centered on the educational rather than the research function. In the higher education system, they criticized the prewar college education curriculum for emphasizing specialized vocational education too early. They advised that the system should be oriented towards general education aimed at producing good citizens. The content of the science curriculum itself was left for the Science Advisory Group which arrived a little later and was attached to ESS/ST.[3]

Postwar educational reform was of a magnitude comparable to land reform and the dissolution of the Zaibatsu. Reform began at the lowest level of primary education, gradually moving upward. It took several years for the reform wave to reach higher education, which involved more specialized scientific disciplines. By the time the reforms reached the graduate education level, the Peace Treaty had been signed in San

Francisco, the Occupation was over, and GHQ had gone home. Thus educational reform was not as thoroughly implemented in the Japanese higher education system as it was at the primary and secondary levels. An example of where the reforms were not completed can be seen in the Education Mission's advise that Japanese universities should give their faculties sabbatical leave – a change that was never realized.

CIE has been criticized for implementing the Education Mission's advice to adopt an American-type system of higher education too rigidly. However, the university accreditation standard instituted under the aegis of the University Accreditation Association (an adaptation of the US mutual assessment system) was still quite flexible. It was only after the responsibility for the university accreditation process was assumed by the University Accreditation Council of the Japanese Education Ministry that bureaucratic formality prevailed and the autonomy and flexibility of the university was curtailed. The rigid formalization of graduate courses that dictate two years for a Master's degree and a further three years for a PhD is but one example of this rigidity.[4]

The Second Educational Mission came to Japan to assess how the advice of the First Mission had been implemented. They arrived in August 1950 and submitted their report on 22 September. When compared to the First Mission's idealism, this report was more realistic. The Second Mission advised that the number of new universities should be considerably reduced in order to maintain the quality of universities in the hard economic circumstances of postwar Japan. But the foundations of many new universities were already underway and it was virtually impossible to terminate their birth.

The First Science Advisory Mission

The First and Second Science Missions were conceived by Dr. Harry C. Kelly. His aim was to build a brand new organization for scientific research. When he returned to the US for a summer vacation in 1946, he visited the National Academy of Sciences and requested they send a science advisory mission. He wanted their imprimatur on the re-organization that he was planning for Japanese science and technology. Consequently the First Science Advisory Mission came to Japan in the summer of 1947, headed by Dr. Roger Adams. After a nationwide observation tour, they held discussions with Japanese representatives in Tokyo on 24 July. The Mission's final report provided a model for the Japanese Scientific Research Organization Renewal Committee, which was at the time engaged in renewal planning. The report was not released, though, until March 1948, supposedly 'in order to not cause useless confusion by leakage to the public.'

This explanation, though, is dubious: the Science Advisory Mission was formed to advise the Renewal Committee on the creation of a new research organization. But as Dr. Bowen Dees made clear in an interview, ESS/ST had requested the mission in an effort to overcome CIE's conflicting views on this issue (cf. ch. 10).

The First Science Advisory Mission's recommendations were based on Kelly's briefing and their own observations over six weeks. Apparently influenced by the Education Mission that had come earlier, they were highly critical of the Education Ministry and tried to curtail its power. The Science Mission advised that higher education and research should be separate from the Education Ministry. They recommended, instead, that democratically elected Advisory Council on Higher Education and Research should be established as an equivalent to the American Academy of Science. A central role of the Advisory Council would be to appoint a Commission on Higher Education and Research, a government organization that would be tasked with implementing the Council's advice and overseeing the practices of the higher education and research systems. A significant part of the Commission's role, then, would be to allocate scientific research funds.

This recommendation to establish an Advisory Council was at the heart of the conflict between the CIE and ESS/ST. While the Science Advisory Group understood higher education, especially graduate school, to be integrally related to research, and therefore included in the scope of their responsibilities, Colonel Nugent, the head of CIE considered this to be his territory and fervently complained. Mediating this 'turf war' took some time and caused a delay in the publication of the Science Mission's report. The conflict was resolved through a compromise: higher education would not be addressed in the report.

ESS/ST was not alone in their support of the Science Mission's recommendations: the NRS and PHW[5] also saw that scientific research was crucial to higher education. In fact, even the CIE had no real objection to the contents of the recommendation. Nevertheless, resolving this interdepartmental strife took so long that the Science Council of Japan (JSC: Nihon Gakujutsu Kaigi) was established prior to the release of the Science Mission's report. Thus the Mission's influence on the research community was less effective than might have been expected.

CIE independently proceeded with reforms to the Education Ministry, vesting in it the authority and responsibility for the administration of higher education, including science. Accordingly, the Ministry of Education changed their English title to the Ministry of Education, Science and Culture and assumed control of research in educational institutions. Seiji Kaya, a technocratic physicist, was appointed Director of the Science and Education Bureau of the Ministry. At this point, the higher education sector became the exclusive territory of the Education Ministry; MITI, the

Health Ministry, Agricultural Ministry and still later the Science and Technology Agency had little influence and no authority to interfere.

Beyond the issues of education, it is noteworthy that the First Science Advisory Mission also recommended restructuring the three existing Japanese science organizations. In 1946, ESS/ST initiated changes that resulted in the Imperial Academy becoming a purely honorary organization. At the same time the functions and responsibilities of the National Research Council (Gakujutsu Kenkyū Kaigi) and the Japan Society for the Promotion of Science (JSPS: Nihon Gakujutsu Shinkōkai) were to be taken over by the newly created Science Council of Japan (JSC). But the JSC did not exactly conform to the Science Mission's recommendation (cf. ch. 9). One significant difference was the inclusion of the humanities and social sciences in the JSC. The recommended dual structure of an Advisory Council and a Commission was not realized. However, the First Science Advisory Mission's intent to create an organization outside of the power of the Education Ministry was fulfilled when the Secretariat of the JSC was located in the Prime Minister's Office (Sōrifu). But later, the Education Mission created its own Science Council (Gakujutsu Shingikai) and restructured the JSC (cf. vol. 3, ch. 6).

The First Science Advisory Mission had also suggested restricting the Education Ministry's authority to interfere in the administration of private universities but nothing ever came of it.

Recommendation of the Second Science Advisory Mission

On 27 November 1948, more than a year after the First Mission produced its recommendations, the Second Science Advisory Mission arrived in Japan to assess the implementation of the First Mission's final proposal for establishing the Japanese Scientific Research Organization Renewal Committee (Gakujutsu Sasshin Iinkai; cf. ch. 9). Again the US National Academy of Science sent five members, with Dr. Detlev W. Bronk as its head. The objectives of the Second Mission were not as clear as the First Mission's had been. They arrived shortly before the JSC commenced, thus there was little chance that their advice might influence the new organization. They conducted a twenty-day observation tour from 29 November to 19 December and then wrote a report of recommendations that simply ratified ESS/ST science policy. However, they did recommend promotion of further international exchange, an institutionalization of the exchange program for scientists and students, and differential funding so that limited funds could be used effectively to create a few science centers equipped for quality research and education. They also emphasized the importance of applied research directed at addressing the economic difficulties of the day.[6]

With only one exception, the Second Mission consisted of basic scientists. Their initial intention had been to strengthen basic science research in Japan but upon their arrival they observed the economic difficulty of postwar Japan. Ishidore Rabi, a member of the mission, plotted to divert the US Army's research funds to rescue impoverished Japanese scientists. He proposed that GHQ should turn Japan into a research base at the front.[7] Kelly argued, instead, that science facilities should promote applied research that could contribute to an economic recovery.[8] Apparently Kelly was more convincing, because the Second Mission came around to this view.

Kelly then spent several months investigating Japanese research institutes, nominated several candidate laboratories and proposed Rabi's idea to the Department of the Army in Washington. We have, to date, found no evidence of any response from the Army. Nor do we have even hearsay evidence that Rabi's plot was implemented. What we do have is evidence that imported machinery and literature was allocated to the eight most powerful laboratories as Kelly had recommended.[9]

Professional Education Mission

During the final stages of the CIE-initiated education reforms – when the reforms reached higher education – they required advice on a broad range of professional and vocational matters. CIE requested this advice from the other GHQ sections – including ESS/ST and PHW, whom CIE had previously marginalized on the issue of higher education policy – who, in turn, invited relevant advisory missions from the US. In August 1950, PHW organized for the Medical Education Mission to come to Japan, with Cyril N. H. Long, Dean of the Medical School at Yale, as its head. All twelve members of the Mission were medical scientists. The Mission proposed that the Research Institute of Infectious Diseases (Densenbyō Kenkyōsho) – affiliated with Tokyo Imperial University – be transferred to the Ministry of Health and Welfare and restructured on the model of the US National Institute of Health (NIH). Of course, the people at Tokyo Imperial University opposed this suggestion. The Mission did, however, successfully organize a few six-week training seminars in Tokyo and Kansai which were welcomed by the members of the Japanese medical community who had been suffering from a shortage of overseas research information ever since the war began. This Mission was so successful that a Second Mission was invited in the summer of 1951.[10]

Similarly, a Dental Education Mission was invited by PHW in 1950 with the same objectives as the Medical Mission. It was not as effective, though, since Lieutenant Colonel Ridgeway, the officer in charge of the mission, was not very enthusiastic about the reform of dental education.[11]

Report of the Engineering Education Mission to Japan

In October 1950 Dr. Bowen Dees succeeded Kelly at ESS/ST and promptly requested an advisory mission from the American Society for Engineering Education. In response, an advisory mission headed by Professor Harold L. Hazen of MIT and fifteen associates arrived in Japan in July 1951. To avoid any conflict, Dees had consulted with CIE beforehand. The Japanese Education Ministry organized a committee headed by Koroku Wada, President of the Tokyo Institute of Technology, to receive the Mission.

The Mission's advice was simply a model for reproducing the US system of engineering education. The Mission was critical of the fact that public servants were responsible for engineering education in Japan. They recommended a more flexible system. Their visit, however, was in the final days of the Occupation and soon after their report was completed GHQ was disbanded. Thus, the Mission's recommendations were not as influential as others had been on primary and secondary education. Nevertheless, they provided the Japanese with a model of professional education to aim for, i.e. a US-style education complementary to the US dominated postwar world of science and technology.

The Japanese Institute for Engineering Education was organized and first met on the occasion of the Mission's visit in July 1951. In July 1952, the Japan Association of Engineering Education (Nihon Kōgyō Kyōiku Kyōkai) was founded. The members of the Mission continued to offer assistance such as supplying a variety of materials and hosting engineers' visits to the United States even after the mission itself was officially over.[12]

Innovative members of the Education Council often advocated the recommendations of the Mission but their impact on the system was negligible. Some of their key points are as follows:

1 The practice of officially employing teachers and staff members as civil servants and maintaining the chair system, as they do at the national universities, is oppressive and should be abolished.

2 Educational standards should be monitored by an independent professional organization similar to the American Society for Engineering Education to ensure quality.

(Implementing either of these two recommendations was utterly impossible in a system where the Education Ministry dominated engineering education.)

3 It should be easier for educational institutions to accept donations from private enterprises.

4 A Board of Trustees should be established to ensure close relationships are developed and maintained with the industrial sector.

(These two points were perceived to potentially create legal problems for educational institutions in the public sector as well as inviting opposition from the left-wing student movement who objected to the increasing capitalist domination of the higher education system.)

5 To encourage/foster closer student-teacher relationships.

6 To reject nepotism in faculty appointments.

(These final two suggestions were considered to be a matter of campus culture, which was dependent upon the actions of the people on a particular university campus, and not a concern for the Education Ministry or curriculum reform. No institutional change was brought up on this matter.)

Higher degrees in Japanese universities had been awarded along European lines, where doctoral dissertations are researched and examined without any course-work. In contrast, American graduate schools are primarily educational institutions rather than organizations for authorizing and examining the research of candidates, especially in professional schools such as medicine and engineering. Thus the Mission recommended the adoption of a system for conferring "course-work doctorate" degrees as in US universities. This advice was formally accepted but higher degree course-work was never as developed as in the US universities. Even during the high economic growth of the 1960s, when the number of science and technology students rapidly increased, the implementation of the US model was not greatly advanced.

The Mission also pointed out that the Japanese university system was weak in the disciplines of chemical engineering and environmental engineering. Chemical engineering departments expanded during the high economic growth of the 1960s, but environmental engineering chairs were not established until the 1970s when pollution and other environmental issues became impossible to ignore any longer.

Missions on Industrial Issues

Let us turn now to examine a selection of the US advisory missions that were closely related to industry.

1 The American Statistics Mission. Headed by Mr. Rice, the Deputy Director of the Office of Budget in the Department of Treasury, the Mission arrived in Japan in 1947 and warned that GHQ would face insurmountable problems in a few years unless they improved the domestic production of consumer goods, particularly textiles.

2 Strike Mission. In August 1947, The US Department of the Army commissioned a Mission headed by Clifford Strike of the Overseas Investment Consultants in New York. With thirty associates, all with engineering expertise, they advised SCAP on the reconstruction of the mining and manufacturing industries necessary for the economic recovery of Japan.

3 The Pope Report. In April 1949, the Department of the Army comm-

issioned Frederic Pope to review the Japanese chemical industry. He toured 42 factories and produced a report entitled 'Japanese Chemical Engineering' which outlined the weaknesses in the Japanese chemical industry and recommended how it could be reformed and developed. He also outlined the causes of high costs and provided concrete advice for the production of salt, the generation and dissemination of thermo-electric power and how to adopt floating ore washing.

4 The Machine Tools Mission. In the final phase of the Occupation, GHQ was promoting Japan as 'the factory in East Asia.' To this end, GHQ requested a Mission be sent to investigate the possibility of procuring machine tools from Japan that were in short supply in the US at the time. The Mission arrived in Japan on 5 February 1951 and made observations until 21 February. To their disappointment, the quality of Japanese machine tools did not satisfy the US standards for defense production. They wrote a report before leaving Japan on 26 February.

Refusal of Missions by Other Nations

The Allied Powers were not entirely happy with the *de facto* American domination of the Occupation. They demanded permission to send other science missions to Japan to gather S&T information and exact reparations from the Japanese. All of these requests were under Kelly's jurisdiction. ESS/ST's general policy was to refuse all Soviet proposals and to deal with British, Chinese and Indian demands on a case-by-case basis. The US took precautions to not lose the hegemony they had achieved.

For instance, the UK repeatedly demanded that a British Technology Mission be given access to collect information about Japanese progress in S&T. GHQ consistently rejected this demand on the grounds that, in accordance with the Far Eastern Commission, such a mission must be fully international rather than from a particular nation. Hence, British frustrations at the news of various science missions consisting only of Americans generated a litany of complaints to the US State Department. In order to pacify them, ESS/ST obtained permission from GHQ to provide the British Embassy in Japan with the recommendations of the First Science Mission, which had not yet been provided to the Japanese, on the condition that the British would never publicize it.[13] British interference in GHQ's science and technology policy was never allowed.

Conclusion

US missions came to Occupied Japan, made observations, consulted with the Japanese and made recommendations for reforming Japanese institutions. Believing they had something like a 'calling' to the task, they tried to transplant US systems of science, technology and education to Japan, truly believing these to be the best in the world. An institutional system is by its very nature closely inter-related to a regional culture. The system they aimed to reproduce was, of course, the system that had developed in and maintained US culture.

For the Japanese during the Occupation, however, the advice was generally considered to be an absolutely imperative order. At first the Japanese tried to assimilate the US system into their institutions, beginning with the formal aspects rather than the substance.

The recommendations from each Mission were addressed to GHQ, rather than the Japanese. As much as possible, GHQ governed Japan indirectly through the Japanese bureaucracy, trying to avoid any coercive action. Hence the final responsibility for implementing recommendations lay with the Japanese. Unlike Meiji Japan, and unlike developing countries, Japan had already established modern systems appropriate to their culture before the war. Thus, the elements of recommended reforms that were readily compatible with the prewar Japanese system were readily implemented without any backlash. GHQ's indirect governance approach, though, provided a gap through which the entrenched Japanese bureaucrats – like those in the Education Ministry – intervened to resist reform. But also, partly because of the Japanese' lack of understanding of the US systems, many of the reform plans remained half-baked.

Before the war the Japanese scientific community had admired European science and regarded US science as B-grade. During the Occupation the US model became the only choice. In this transition from the European model to the US model the US Science Missions contributed a great deal.

Notes

1 The term 'groups' here is used loosely, referring to the different 'waves' or 'stages' of the advisory missions. In other words, each mission was independent of the others. We have grouped them chronologically according to the broad orientations or focus of their advice in terms of changes to Occupation Forces' policy.

2 There were negotiations between Harry C. Kelly of ESS/ST and Mark T. Orr of CIE regarding this issue. At this point the demarcation of ESS/ST and CIE was settled. (Orr's testimony to S. Nakayama) ESS(A)09787 C13, 19 December 1946.

3 Hata, 1992 [15] p. 27.

4 American graduate school has no firm time limit. In 1990, Japanese graduate schools became more flexible on this point.

5 PHW proposed its own special education system for doctors, dentists and nurses on the grounds that medical education is quite different from

ordinary education. CIE, though, insisted on a unified education system. After intense conflict, CIE eventually prevailed. *Daigaku Kijun kyōkai jūnen-shi* [17] 1957, pp. 61–9.
6 ESS(E)14225/D10-D14, p. 5, 18 December 1948. Memorandum to SCAP from The Second U.S. Scientific Mission to Japan.
7 Shigeru Nakayama, 'Senryōka Nihon kenkyū kichika jiken – Ishidore Rabi (Studies of Japan under the Occupation – the so-called military base incident: Ishidore Rabi),' in *Sukyandaru no Kagakushi* (History of science scandals), Asahi Shimbunsha, 1989, pp. 212ff.
8 Bowen Dee's telephone discussion with Samuel Coleman, 19 March 1986.
9 Testimony of Bowen Dees.
10 *Daigaku Kijun Kyōkai jūnen-shi* [17] 1957, p. 63.
11 *Daigaku Kijun Kyōkai jūnen-shi* [17] 1957, p. 66.
12 *Daigaku Kijun Kyōkai jūnen-shi* [17] 1957, p. 78.
13 Incoming Message, from CSCARD Ro. To CINCFE. NR:WAR92530 [13] 1947.

Bibliographic Guide

Reports

[1] US Department of Education, 'Development in Japanese Education in terms of the Report of the United States Education Mission to Japan,' Wanamaker Papers. Available in printed form together with Japanese translation *Beikoku kyōiku shisetsudan hōkoku*, Kokusai Tokushinsha, 1946.
US Department of Education, 'Report of the Second United States Education Mission to Japan,' in Akio Igasaki and Kōichirō Yoshihara (eds.), *Sengo kyōiku no genten 2* (The origin of postwar educational reform), Gendaishi Shuppankai, 1975.
[2] Kokuritsu Kyōiku Kenkyūsho (National Institute of Educational Research, Japan), *Kaigai gakujutsu kenkyū: Hōkokusho, senryōki Nihon kyōiku ni kansuru zaibei shiryō no chōsa kenkyū* (Survey of sources preserved in the US on Japanese education during the Occupation period), March 1988.
[3] Natural Resources Section, 'Report of Mission and Organization of the Natural Resources Section, 1 April 1947,' Trainor Collection (preserved at Kokuritsu Kyōiku Kenkyūsho), microfilm roll no. 56.
[4] *Report of the Scientific Advisory Group to the National Academy of Sciences, USA*. This was written in August 1947 during the Mission's stay in Japan but because of the internal strife in GHQ, it was not released until March 1948 when the Japanese Renewal Committee published their reform plan.
Cf. 'Reorganization of Science and Technology in Japan,' in *History of the Nonmilitary Activities of the Occupation of Japan, 1945–1951*, available at Diet Library, Tokyo, Japan.

[5] 'Report of the United States Cultural Science Mission to Japan,' Trainor Collection (preserved at Kokuritsu Kyōiku Kenkyusho), microfilm roll no. 65. When the First Science Advisory Mission came to Japan on 29 July 1947, the Humanities Advisory Mission was requested. With the good office of the Education Department of CIE, five members including Edwin O. Reischauer, Harvard Associate Professor, arrived in Japan in March 1948. They visited universities nation-wide to conduct seminars and meet with Japanese scholars before submitting their report. They found that Japanese universities were weak in promoting cultural sciences, the libraries were poor, and the professional chair system was inhibiting. They advised that the scholars' community needed to be more internationally oriented.

[6] GHQ/SCAP, Memorandum to SCAP from the Second U.S. Scientific Mission to Japan, ESS (E) 14225/D10-D14, 18 December 1948, p. 5.
This report is less significant than the First Mission's but in addition to the report, it includes a list of the Japanese laboratories they investigated and their evaluation in 'Leading Research Laboratories in Japan.' This list is informative about the state of affairs in Japanese laboratories at the time. The Mission observed that the Japanese laboratories had excellent researchers but there was no coordination between them.

[7] Report of the Engineering Education Mission to Japan, 1951, p. 79. Included in ESS (D)-12396 (ESS-49196)(1951.9), EDD (D)-12397 (ESS-49197) (1951.9).
This report addresses the purpose of engineering education for undergraduates, the general education curriculum, the methods of teaching, organization and finance, appointment of teachers and staff, research and consulting activities, academic degrees in engineering, libraries and textbooks, and disciplines not yet developed in Japan such as chemical and environmental engineering. It discusses American cases and suggests reform measures.

[8] Kōgyō Kyōiku Kenkyū Shūkai Chūō Un'ei Iinkai, *Kōgyō Kyōiku Kenkyū Shūkai Hōkoku* (Report of the Institute for Engineering Education), 1951, p. 149.
The American Engineering Education Mission to Japan together with Japanese (university people and some businessmen) gathered for seven seminars. The proceedings were edited and published with the US advisors' contributions in English and the Japanese responses in Japanese.

[9] GHQ/SCAP, Correspondences between ESS and the Engineering Education Mission to Japan, ESS (E)-06482 (ESS-29873)(1951.6–1952.1).

[10] GHQ/SCAP, ESS (E)-06483 (ESS-29874)(1951.6–1952.1). Book catalogue donated by the Engineering Education Mission to Japan, including English translation of President (of Tokyo Imperial University) Nanbara's report 'Japanese Engineering Education.'

[11] Unitarian Service Committee, Inc., 'Report of Medical Mission to Japan, 27 November 1951,' Trainor Collection (preserved at Kokuritsu Kyōiku

Kenkyusho), microfilm roll no. 65.

[12] American Dental Association Mission to Japan, 'Report, July 1951,' Trainor Collection (preserved at Kokuritsu Kyōiku Kenkyūsho), microfilm roll no. 60.

[13] Incoming Message, from Cscard Ro. To Cincfe. NR:WAR92530. Through the Far Eastern Commission the UK criticized the destruction of cyclotrons by the US Army (cf. ch. 3) and expressed dissatisfaction at being alienated from Occupation science policy. ESS/ST's response to the UK is also recorded.

[14] CAS(C)-5624 (CAS-12343). On general education at the higher education level (1949.2–1949.7). The proceedings of the Kansai branch of the General Education Commission (including natural science) set up within the Association of University Accreditation (Daigaku Kijun Kyokai) (English translation).

[15] Takashi Hata, 'Sengo kyōiku kaikaku to kagaku gijutsu no shomondai (Postwar educational reform and various issues of science and technology)', in Sōsuke Watanabe (ed.), *Kōwa dokuritsugo no wagakuni kyōiku kaikaku ni kansuru chōsa kenkyū* (Research on Japan's educational reforms after the conclusion of the Peace Treaty), Kokuritsu Kyōiku Kenkyūsho, March 1992, pp. 1–50.

[16] Tsuchimochi, Hōichi Gary, 'Beikoku Gakujutsu Komondan hōkokusho to sengo Nihon no kōtō kyōiku kaikakuan – Adams danchō bunsho o chūshin ni (The report of the United States Scientific Advisory Group and Japan's postwar reform of higher education with special reference to the Adams papers)', Kokuritsu Kyōiku Kenkyūsho (National Institute for Educational Research), Sōsuke Watanabe (ed.) *Kōwa dokuritsu go no wagakuni kyōiku kaikaku ni kansuru chōsa kenkyū* (Research on Japan's educational reform after the conclusion of the Peace Treaty), March 1992, pp.51–64. Also in *Daigaku ronshū* (Research in higher education), no. 20, 1991.

Other

[17] Daigaku Kijun Kyōkai (University Accreditation Association), *Daigaku Kijun Kyōkai jūnen-shi* (Ten year history of the University Accreditation Association), 1957.

Research Note

The source-materials of Japanese wartime research were largely burned and scattered at the time of surrender in order to conceal their wartime activity. For that reason, the reports of the US investigation teams who came to Japan at the beginning of the Occupation are the most valuable source materials for Japanese military research. During the confusing state of affairs in the period immediately after the war it was rather difficult to make a general account of the activity of Japanese laboratories, both public and private. Hence the investigation of the Second Science Advisory Mission is valuable as a record of the state of laboratory activity during the Occupation.

All of the Missions submitted official reports. In addition to these there are administrative materials recording the exchanges between the Missions and GHQ sections which reveal the insider's voice in a manner that is absent from the official reports. Most of these sources are available at Kokuritsu Kyōiku Kenkyūsho (National Institute of Educational Research) or Kokkai Toshokan (Diet Library) both in Tokyo, Japan.

9 The Reorganization of Research Structures

Shigeru Nakayama

The Reform of Academic Structures

The New Deal faction in the General Headquarters (GHQ) of the Occupation Forces wielded almost absolute power as they pushed to implement systemic changes in Japan; to reconstruct the society in ways they could only dream of at home. The Natural Resource Section's (NRS) land reform, the Civil Information and Education Section's (CIE) 6-3 system of educational reform and the Economic and Scientific Section's (ESS) disbanding of the Zaibatsu are all examples of New Deal programs.

Likewise, Harry C. Kelly at the ESS Science and Technical Division (ESS/ST) had a desire to reform the old science research organizations and create a brand new academic structure. The outcome of his effort was the Science Council of Japan (JSC). The ESS/ST's self-evaluation is straightforwardly reflected in their official history, 'The Reorganization of Japanese Science and Technology,' which presents the creation of the Science Council of Japan as their most significant achievement.[1]

However, while Kelly is widely credited with the inception and foundation of the JSC, the JSC that was established and the system it oversees are certainly not as Kelly conceived them, remaining too closely tied to regional milieu and the government bureaucracy. Although SCAP had absolute power to command, it was also guided by a principle of indirect governance. Hence the detailed process of implementing policy was left to the Japanese authorities. This chapter examines the process by which the JSC was established and tries to clarify the discrepancies between the American conception and the Japanese implementation.

Liaison Group

Kelly was the head of basic science at the ESS and later became the deputy director of ESS/ST. While most of the military officers in other sections of the Occupation Forces treated the Japanese as enemies, Kelly had realized as early as April 1946 that the underlying objective for Occupation policy should be the economic recovery of Japan (the phrase 'science and technology for economic recovery' appears in a memo dated 23 April 1946). He was irritated by the inefficiency and slowness of the old system of scientific research and intended a radical reform. He therefore personally assumed the task of restructuring scientific research organizations as required to achieve this goal.

To facilitate the reform process he created the Japan Association of Science Liaison (JASL) in June 1946, beginning with his personal contacts (Jurō Horiuchi of Hokkaidō Imperial University and Hiroshi Tamiya of Tokyo Imperial University). JASL was Kelly's private consulting body as well as the liaison between GHQ and the Japanese scientific community. This was an encouraging start, leading him to believe that perhaps he could reform the academic organizations through JASL. His early success with JASL encouraged him to create further liaison groups: for technology, the Engineering Liaison (JAEL); agriculture, Agricultural Liaison (JAAL); and medicine, Medical Liaison (JAMG).

Soon after the surrender, though, most Japanese scientists were in such an apathetic mood that they could hardly decide anything by themselves, and what they did decide, they could not implement. The liaison groups demonstrated a preference for following GHQ's orders, rather than initiating their own reform plans. Basically they just wanted to know GHQ's intentions. Although the members of the liaison associations were close to GHQ through their personal contacts with Kelly, they were not elected and had no authority to represent the views of the Japanese scientific community. Furthermore, they were afraid of the power of the old Imperial Academy and were reluctant to stand at the front of the reform movement in conflict with the Academy. As an individual example, Ryōkichi Sagane, the most fluent English speaking member of the JASL, was apprehensive of being isolated from the Japanese scientific community for personally being too close to GHQ. Yoshio Nishina was also anxious about this point and visited ESS/ST on 26 October 1946 to express his concerns. But since it was Occupation policy to restructure via indirect governance, the reform had to be primarily an endogenous process, which left Kelly frequently irritated by the slowness of Japanese reform.

Reorganization of Three Old Academic Organizations

There were three major academic organizations in prewar Japan, the Imperial Academy (Teikoku Gakushiin), the National Research Council (Gakujutsu Kenkyū Kaigi) and the Japan Society for the Promotion of Science (JSPS: Nihon Gakujutsu Shinkōkai). In the autumn of 1945, the Education Ministry began planning reforms for these organizations to fit them in to the postwar environment. On 28 September 1946, the Education Ministry submitted its reform plan to ESS/ST to strengthen the Imperial

Academy and to retain the management of all research funding within the Education Ministry itself.[2] On 28 November, a special session of the National Research Council agreed to the Education Ministry's reform plans.

However, Kelly, who had gone home for a vacation in the summer of 1946, took advantage of the occasion to contact the American National Academy of Sciences. He asked them to send a mission to provide advice on reforming the Japanese scientific organizations.

Upon his return to Japan in late September, he immediately convened a meeting of the Liaison Group to demand a proposal for reforming Japan's scientific and technical infrastructure. The Science Liaison Group responded with a reform plan that was quite critical of the three existing organizations, particularly the conservative and gerontocratic Imperial Academy.

Through his meetings with the liaison group scientists Kelly realized that the existing academic and scientific structure (the three scientific organizations and the Ministry of Education) was an impediment to scientific development and concluded that a wholly new organization had to be created.[3] Anticipating the defensive moves to conserve the old organizations and protect their power, on 15 November 1946 Kelly called upon the Liaison Associations and initiated countermeasures. Kelly met with the head of the Imperial Academy, Hantarō Nagaoka on 18 November and attempted to impress on him – some times in harsh language – the urgency of the problems facing Japan and Japanese science and the need to quickly and effectively reorganize scientific and research institutions, and *de facto* declared the abolition of three existing institutions.[4] At the same time, he officially sanctioned JASL to organize an institution based on the consensus of the entire scientific community.

Among the three existing scientific institutions, the Imperial Academy was the most authoritative institution – and the most gerontocratic in character – and dominated the Japanese scientific community. The other two were more heterogeneous and included the generation of JASL members. For the Imperial Academicians, it was utterly unthinkable that they might lose their established power to JASL, which consisted of the generation of their pupils. But JASL was so apprehensive of the Imperial Academy that they were not able to convey Kelly's message to them. Kelly was hesitant to directly order the creation of a new institution out of respect for the Occupation policy of indirect governance, but he was less reluctant about using his power to abolish the old system in order to clear the way for the Japanese reformers.

Sewaninkai (Preparatory Committee)

Even with Kelly's authority, the Imperial Academy's resistance was so entrenched that JASL at one stage abandoned its role as a leading reformer.

This prompted the Ministry of Education's Scientific Division Chief, Kinji Shimizu, to intervene. Since JASL did not have the authority of representing the Japanese scientific community, Shimizu called for a Sewaninkai (intermediary preparatory committee) in January 1947 from a wide range of disciplines. Consisting of 44 scholars from a variety of disciplines, including the humanities and social sciences, the Sewaninkai's role was simply to be the board of electors to select the members of the Academic System Renewal Committee (Gakujutsu Taisei Sasshin Iinkai).

In order to understand what happened with the Sewaninkai, let us turn briefly to what was 'normal' in this type of situation in Japan at the time. It was customary for the Education Ministry to consult and choose the senior professors of Tokyo Imperial University. This university consists of seven faculties: law, economics, humanities, science, engineering, medicine and agriculture. Coordination and decision making about academic matters is determined by the balance of power within the old classification of science and learning. The classification of science disciplines is generally very conservative. When the Sewaninkai was created, new interdisciplinary disciplines that did not belong to any of the 'traditional' seven faculties were excluded from consideration.

Selection of the members of the Renewal Committee began under the supervision of the Sewaninkai in April 1947. The different disciplines maintained different academic customs so there was no unified election system that would be universally applicable, but a compromise was reached between various powers. The selection process was by no means equitable; it did not produce a fair representation of age, locality or academic sectarianism. This cronyism was severely criticized by the newspapers. The old establishment style of the Imperial University prevailed, where everything was decided in hidden negotiations between bossy senior professors. It became apparent that open elections and equal representation for the younger generation was impossible under the Education Ministry's ministrations. Instead the top-down hierarchy typical of government office was introduced to the academic world. In the process the formality of the selection procedures had been maintained but any prior passion for reform was quashed.

Kelly's original idea of creating a copy of the US Academy model had, by then, been expanded to the applied sciences (such as engineering, medicine and agriculture) that he could control. He had no control over the social sciences and humanities because they did not fit into his program of 'science for economic recovery.' He was not content with the direction that the Sewaninkai was heading but at the time could see no other way. Later, after the Renewal Committee began to function, he appointed himself an advisor on election procedures.

Left-Wing

Of the various organizations of left-wing scientists the best known is Minshushugi Kagakusha Kyōkai (Minka: the Association of Democratic Scientists), an umbrella group for leftist scientists and science organizations. Minka was launched as early as January 1946 but does not appear in ESS/ST documents until much later.

The Preparation Committee (Junbi Iinkai) for the creation of the Research Restoration Council (Kenkyū Hukkō Kaigi) first met on 9 April 1947 at the National Science Museum in Ueno to draft its prospectus and organizational structure. It was intended as a Council of the unions representing scientists and technicians at public, private and university laboratories and to complement the Economic Restoration Council (Keizai Hukkō Kaigi). It also included some humanities researchers.[5]

The new socialist government of Tetsu Katayama was formed on 1 June 1947. The chairman of the Imperial Inventions Association, Shun'ichi Ono, was the spokesman of the Preparation Committee. Ono was a Socialist Party member and had a number of friends in the new cabinet. The inauguration of the Katayama cabinet was a welcome event for him and seemed to represent a propitious time for his group. Thus, on 10 June, when members of the Preparation Committee met with Kelly, O'Brien and their associates in ESS/ST, Ono apprised Kelly of his potential influence.

This was about the time that the Sewaninkai was finalizing its selection of the members of the Renewal Committee. The young left-wing members of the researchers' union were alienated from the backroom maneuverings of the establishment, and put pressure on ESS/ST to ensure that their opinions were also reflected in the forthcoming scientific organization. GHQ had not listened to the Economic Restoration Council so far, but with the change of government Ono believed that the Council's views would be given their proper weight.

Kelly and his associates were certainly sympathetic to the young scientists' desire to be represented on the Science Council, but from their very first meeting they were greatly concerned about the activities of the leftist scientists, and believed that Ono was a Communist Party member. At the time, the ESS/ST was not well informed about the leftist scientific organizations, and began to thoroughly investigate Minka and the Japanese Communist Party's position. In fact, Kelly himself visited the offices of Minka.[6]

There were conflicting assessments of the Research Restoration Council within ESS/ST. Lieutenant Colonel E. C. Allan, the Technical Division chief of ESS/ST, suggested that if the Sewaninkai were likened to an employers' group, GHQ should ensure that the Ono faction was appropriately (proportionally) represented on the Renewal Committee.

Kelly disagreed, arguing that there was no point in confusing the issue by comparing academic reorganization with a labor-management contract. He declared that GHQ's goal was to create an organization of scientists to promote research, not to establish a political organization.

On 7 July an Expanded Preparation Committee meeting of the Research Restoration Council was held at the National Science Museum at Ueno. Some fifty scientific societies, research organizations and university laboratories (both public and private) were represented at this daylong discussion.

Kelly and Allan both addressed the meeting. Kelly stressed the importance of pursuing scientific reorganization through the Sewaninkai, warning that any other approach at that stage risked disrupting the reorganization process. Allan, again, was more sympathetic to the leftist Restoration Council. The Research Restoration Council discussed their relationship to the Sewaninkai and concluded that they should endeavor to ensure their members were democratically represented in the new Academic System Renewal Committee. As GHQ understood it the Restoration Committee was intent upon changing the feudalistic nature of the Japanese scientific community.[7]

On 29 July, Ono and Hideomi Tsuge – representing the Restoration Committee and Minka, respectively – and other leftist organization representatives met with representatives of the mainstream scientific community: Naoto Kameyama of the National Research Council and Seiji Kaya of the Sewaninkai. The leftists criticized the process for electing the Renewal Committee members while those representing the *status quo* defended themselves. This debate was reported in the newspapers and attracted the attention of the entire Japanese scientific community.[8]

The leftists eventually succeeded in getting their members represented on the Renewal Committee. Kinnosuke Ogura, the Chairman of Minka, was appointed to the Committee's Science Section. Tsuge of Minka, Mizuho Inanuma of the History of Science Society of Japan, and Gensaku Fujimori of the Democratic Association of Engineers were appointed to its General Section[9] and members from the Government Test Laboratory and the Japan Medical Association joined its Agriculture and Medicine Sections.

Scientific Research Organization Renewal Committee

The task of the Sewaninkai was completed on 10 August with the selection of 108 members of the Renewal Committee. Fifty-five members were Professors from Tokyo Imperial University, more than half of the Committee's total. Eighty percent (91 members) were graduates of Tokyo Imperial University. Kyoto Imperial University was the second most represented organization with eight graduates, which indicates how Tokyo-centered the Committee was. The average age of members was 52 years.

Kelly was not happy with the Tokyo University dominance. He expressed his dissatisfaction in a handwritten memo to the visiting US Scientific Advisory Group that arrived in Japan on 19 July for a 40-day observation tour. He informed them that the Sewaninkai did not submit the proposed election procedure to ESS/ST until it was too late to alter. He also said that the procedure sounded good on paper but was wrong in practice because the members of the Executive of the Sewaninkai were mainly old school and nominated old school people as electors, thus continuing the process of old school ties. If ESS/ST had applied pressure to change this they would have violated GHQ's indirect governance policy. Consequently ESS/ST could do little more than watch the undemocratic process of the Sewaninkai be reproduced in the Renewal Committee, quite contrary to Kelly's ideas for reformation.

A number of Japanese were also critical of the absence of democratic process in Sewaninkai's activities. Various newspaper editorials commented on the dominating influence of the Tokyo Imperial University clique in the Renewal Committee.

Thus, the Renewal Committee started to work under the scrutiny of all of the sectors of Japanese society. The attentive Renewal Committee paid respect to the three established scientific organizations, consulted with GHQ, invited membership from others – including Minka – and finally aimed at formulating the policy recommendation of the Renewal Committee.

On 25 August the first Renewal Committee meeting was held at the Prime Minister's Official Residence, joined by Prime Minister Katayama and the First Science Advisory Mission. Subsequently, a general assembly of the Renewal Committee was held monthly. The General Assembly selected Kankurō Kaneshige as its head and hosted a welcome party for the First Advisory Mission. Three days later, the Mission submitted a report to GHQ entitled 'The Reorganization of Science and Technology in Japan,' then left Japan. The report was occasionally consulted by ESS/ST but its publication was delayed so that it would not influence the deliberations of the Renewal Committee.

From the outset of Kelly's reform efforts, a central role was assumed by JASL. All of its active members, such as Hiroshi Tamiya and Ryōkichi Sagane, were purely academic scientists. Kelly was a physicist and was close to these internationally recognized scientists but basic scientists were not the crucial element in his attempts to develop 'science for economic recovery.' During the later deliberations of the Renewal Committee, however, technocratic scientists, such as Seiji Kaya, Naoto

Kameyama, and Kankurō Kaneshige arose. They were to become the spokesmen for the rapid growth of Japanese science and technology. Seiji Kaya was particularly influential. He was the vice-chairman of the Renewal Committee for the reorganization of scientific organizations and, in the autumn of 1948 became the Head of the Bureau of Science and Education at the Education Ministry.[10] He was well suited for liaison between the scientific community and the Japanese bureaucracy. His technocratic way of thinking was in tune with the ESS/ST perspective.[11]

At the second general assembly in September the various liaison groups, learned societies and universities submitted their own reform proposals. At the third general assembly in October, a conflict of opinion appeared between the natural and social scientists. The natural scientists and engineers believed they should attempt to influence government policy through their technocratic opinions with the overarching aim of rationalizing politics. The social scientists wanted to stay out of politics in the interests of academic freedom.

Minka also proposed a reform plan that would give strong political power to a new scientific organization modeled after the Soviet Academy of Science.[12] Their proposed Supreme Council of Scientists (Saikō Kagakusha Kaigi) would have the authority to 'command the government' but the proposal was dismissed for being too 'highbrow.' However, their proposal to extend the right to be nominated and to vote to include non-academic researchers – i.e., researchers from private laboratories – was accepted. Accordingly, voter qualification was extended to include everyone who had written scientific papers and was endorsed by a learned society.

At the fourth general assembly in November, the conflict between the natural sciences and social sciences continued and new conflicts emerged between public and private universities, between university and private laboratories and between Tokyo and other areas. The power struggle was so intense that at one time it jeopardized the survival of the Renewal Committee.[13] GHQ became irritated by the slowness of the Renewal Committee's reforms and exerted pressure for progress.

At the fifth assembly in December, budget restrictions were applied by the government to force them to finish their work quickly (by March 1948). Only after this pressure was exerted did the debate and discussion become more concrete. It was generally agreed that the forthcoming new scientific organization must be a 'Council.' Then they argued about its relationship to the government and the proposal to limit the right to vote to scientists.

In January 1949 the sixth general assembly erupted with disputes about the Council's relationship with the three old school scientific organizations. At this stage, the general direction was decided vis-à-vis the old organizations but the relationship of the forthcoming organization with the government ministries, especially with the Education Ministry, was not yet certain. Then, suddenly the Katayama cabinet was dissolved. The political community and bureaucracy were so confused that it was difficult to conduct any negotiations with government offices. However, in February the seventh general assembly agreed on the title of the new organization: the Science Council of Japan (JSC). The members of the Council were to be chosen by a nation-wide election of scientists.

On 23 March 1948, a meeting of all the GHQ sections was held for the formal approval of these general principles, particularly regarding the relationship between the government and the JSC. At the eighth Renewal Committee assembly from 25–27 March, the governing principles of the JSC were to be laid out. The original draft of the Renewal Committee said 'The government shall seek the opinion of the Science Council of Japan.' The word 'shall,' however, was later crossed out and replaced with the word 'may.' This correction appeared in the final draft and was later adopted in the approved Science Council Act, resulting in the Council's recommendations having no binding authority, and therefore being increasingly ignored by the government, a development that defines the history of the JSC.

Both ESS/ST and the mainstream of the Renewal Committee had misgivings that the Education Ministry would eventually dominate the new organization. In order to limit their influence, an Interim Report of the Renewal Committee presented a plan to reform the Education Ministry. For the same purpose the Secretariat of the JSC was located within the Prime Minister's Office rather than the Education Ministry.[14] Furthermore ESS/ST suggested the establishment of the Scientific and Technical Administration Committee (STAC) to facilitate liaison between the Science Council and the government's administrative offices.[15]

Kankurō Kaneshige, Chairman of the Renewal Committee, after considering the opinions of various groups and obtaining the approval of GHQ, drafted: the Science Council of Japan Act (Nihon Gakujutsu Kaigi Hō), the Science Council of Japan Statute (Nihon Gakujutsu Kaigi Kaisoku), Rules for Electing Members of the Science Council of Japan (Nihon Gakujutsu Kaigi Kaiin Senkyohō) and the General Rules for the Science, Technology and Administration Council (Kagaku Gijutsu Gyosei Kyōgikai Yōkō). The Imperial Academy tried to prevent the Diet from passing the Science Council of Japan Act, but Kelly, afraid the Act which had taken two years to get together might fail to get through the Diet, put pressure on the Chairman of the Educational Committee of the House of Representatives. Finally, on the basis that the Act was the product of sufficient democratic discussion, the Diet passed it without debate or even discussion[16] on the last day of the parliamentary session. The Act was passed into law on 10 July 1948.

Election Result

The Japanese scientific community had no experience of holding an election for their own governance. Nevertheless, after much confusion in the process, 944 candidates were finally listed for the 210-member Council. On 20 December 35,354 scholars and scientists voted. Although most of the senior academics expected to be elected, the election results often proved otherwise. Many of the members of the Renewal Committee who had prepared the Act and Election Rules were not selected. In the Medical section not one of the Tokyo University professors was elected. Neither were Kankurō Kaneshige and many Sewaninkai members. ESS/ST considered the election results to be, on the whole, much more satisfactory than the selection of the Renewal Committee members.[17]

Out of the 210 elected members, 112 were from the Tokyo area (including 32 Tokyo University faculty members), 91 were old Imperial University faculty members, 25 were from other National University faculties, eight were from Prefectural University faculties, 18 were from government organizations and public laboratories, and 64 were from the private sector including 40 from private university faculties. Thus the concentration of memberships to the academic establishment circle was avoided. According to a GHQ investigation, 61 candidates were members of the Japan Communist Party, out of which 26 were elected. There were also 40 Party sympathizers elected.[18] More than ten percent of the candidates supported by Minka were successful. Thus constituted, the Science Council of Japan was inaugurated in January 1949.

The Aftermath of the Three Old Scientific Organizations

The three old scientific organizations had been suppressed by GHQ but had continued their functions until the inauguration of the JSC. Then, they were obliged to reorganize. Imperial Academy President Hantarō Nagaoka and his successor Sanryō Yamada were repulsed by the idea of subjecting the Academy to the control of the new Science Council of Japan. Nagaoka sent an official letter of complaint to John W. O'Brien but GHQ took no action, commenting that they had no responsibility in the matter. It was, they said, the responsibility of the Renewal Committee. Kelly had been harshly critical of Nagaoka's preference for basic science and accused the Imperial Academy of being incapable of dealing with the problems confronting postwar Japan. He had strongly suggested that Nagaoka should dissolve the 'useless' Academy. Eventually the Renewal Committee reduced the Academy to a purely honorary institution whose members would be elected by the Science Council. This meant the subordination of the Academy to the Council. Later, when Yamada became President of the Academy, he tried – and eventually succeeded – in re-establishing the Academy as a body independent of the Council.[19]

The JSC and its National Research Council (NRC) was a national affiliate of the international academic organization, International Council of Scientific Unions (ICSU) established in 1931. Its function was transferred to the Research Liaison Council (Kenkyū Renraku Iinkai), a sub-committee of the JSC. All of the NRC personnel were absorbed into the Secretariat of the JSC.

The Japan Society for the Promotion of Science was nominally a private sector organization and thus could not be abolished by official sanction, but it was considerably weakened when its government subsidies were withheld. It was preserved as an affiliated body of the JSC with the number of personnel reduced from 48 to 16.

The Significance of the Science Council of Japan

In spite of all the obstacles encountered during the process of its formation, the newly established JSC was an organization with no precedent anywhere in the world. It was almost everywhere the norm that members of an organization like the National Academy would be selected by the existing members. Thus, it was usual that the new members were scientists whose research conformed to the established paradigm. In order to break down this 'old school' dominance, the British Association for the Advancement of Science and others were established in such a way that its officers were elected by a membership of free association. In contrast, the JSC members were directly elected by all of the members of the scientific community and had an official role in formulating science policy and advising the government. Such an organization did not exist even in the 'socialist' countries. Thus it was a remarkable organization, which became internationally recognized as the first parliament of scientists in history.

Notes

1 Putnam [7]
2 Putnam [7] pp. 61–2.
3 Putnam [7] p. 52.
4 Nakayama [9] p. 359. Includes ESS/ST sources.
5 *Kagaku Bunka Shimbun*, 20 April 1947.
6 Tsuge, 1980 [14]
7 *Kagaku Bunka Shimbun*, 15 July 1947.
8 *Kagaku Bunka Shimbun*, 5 August 1947.
9 *Kagaku Bunka Shimbun*, 10 August 1947.

10 *Kagaku Bunka Shimbun*, 15 September 1948.
11 Bowen Dees' telephone interview by Samuel Coleman.
12 Masao Gorai, 'Nihon Gakujutsu Kaigi sōritsu no koro (Reflections on the circumstances in which the JSC was established), in *Nihon no kagakusha* (Japanese scientists), vol. 5, no. 12, Dec. 1970, pp. 34–5.
13 *Kagaku Bunka Shimbun*, November 10, 1947.
14 'Kagaku Gijutsu Kyōgikai narabini Kagaku Sinkō o Tantō suru Shumushō nikansuru An oyobi gian (Draft plans and bills concerning the principal ministry to be in charge of the Science and Technology Committee and the promotion of science,' in *Gakujutsu Taisei Sasshin Iinkai Sōkai haihu siryō, tsuzuri (dai8kai)* (Papers for distribution at the Plenary Sessions of the Renewal Committee of the Science and Technology System), 1948.
15 Nihon Gakujutsu Kaigi, 1974 [6] p. 270.
16 Putnam [7] p. 94.
17 Putnam [7] p. 101.
18 Putnam [7] p. 102.
19 Okano [15]

Bibliographic Guide

Primary Sources

[1] GHQ/SCAP records (edited at the Japanese Diet Library and available on microfilm), Box No. 7399 from Folder title/number: (1) Science Reorganization – Research Restoration Council, to Folder/title number: (7) Renewal Committee. It includes an attached index of the vast collection of ESS/ST documentation from the beginning of JASL to the formation of the Japan Science Council.
[2] 'Dai-1kai Nihon Gakujutsu Kaigi kaiin senkyo kisoku (Membership election rules of the first meeting of the Science Council of Japan),' in *Kanpō* (Gazette); 'Gakujutsu Taisei Sasshin Iinkai Senkyo Kanrikai kisoku (Election administration rules of the Renewal Committee of the JSC)' in *Kanpō* (Gazette).
[3] Renewal Committee, 'Gakujutsu Taisei Sasshin Iinkai sokkiroku (Stenographic record of the Renewal Committee of the JSC)', in the library of the Science Council of Japan.
[4] 'Gakujutsu Taisei Sasshin Iinkai sōkai haihu shiryō tsuzuri (Papers for distribution at the general meetings of the JSC's Renewal Committee),' in Nihon Kagakushi Gakkai (ed.), *Nihon kagaku gijutsu-shi taikei 5; Tsūshi 5* (An Outline of the history of Japanese science and technology: vol. 5, Historical overview), Daiichi Hōki, 1964.

[5] *Nihon kagaku gijutsu-shi taikei, Tsūshi 5*. Includes comments from *Asahi Shimbun* and *Nihon Keizai Shimbun*, plus the member lists of Gakujutsu Sasshin Iinkai, Nihon Gakujutsu Kaigihō, etc.

Secondary Sources

[6] Nihon Gakujutsu Kaigi Nijūgonen-shi Henshū Iinkai (Science Council of Japan Editorial Committee of the Twenty-five Year History) (ed.), *Nihon Gakujutsu Kaigi nijūgonen-shi* (Twenty-five year history of the Science Council of Japan), Nihon Gakujutsu Kaigi (Science Council of Japan), 1974.

[7] Putnam, David L., 'Reorganization of Science and Technology in Japan (1945 to September 1950)', in *History of the Non-military activities of the Occupation of Japan*, Kokkai Toshokan (National Diet Library), microfilm number CHS-04304.

[8] Comments on the Science Council of Japan in *Shizen* and *Kagaku Shimbun*. *Shizen's* perspective was left wing and liberal, while *Kagaku Shimbun* presents a technocratic view.

[9] Shigeru Nakayama, 'The American Occupation and the Science Council of Japan,' in Everett Mendelsohn (ed.), *Transformation and Tradition in Sciences*, Cambridge University Press, 1984.

[10] Takashi Hata, 'Sengo kyōiku kaikaku to kagaku gijutsu no shomondai (Postwar educational reform and various issues of S&T),' in Sōsuke Watanabe (ed.), *Kowa dokuritsu go no wagakuni kyōiku kaikaku ni kansuru chōsa kenkyū* (Research on educational reforms in Japan after the conclusion of the Peace Treaty), Kokuritsu Kyōiku Kenkyūsho, 1992.

[11] *Nihon Gakujutsu Kaigi geppō* (JSC monthly), vol. 15, no. 1, 1974, Nihon Gakujutsu Kaigi Nijūgoshūnen Kinen Tokushū (Special edition to commemorate the twenty-fifth anniversary of the JSC).

[12] 'Nijūgoshūnen o mukaeru Nihon Gakujutsu Kaigi (The JSC reaches its twenty-fifth anniversary),' in *Nihon no kagakusha* (Japanese scientists), vol. 9, no. 9, 1974, pp. 434–55.

[13] 'Kagaku Renraku Shōgaikai no shōkai (Introducing the Science Liaison Association),' in Nihon Butsuri Gakkai (ed.), *Nihon butsuri gakkai-shi* (Japanese journal of physics), vol. 1, no. 1, appendix 1946. The statues of JASL were published in July 1946.

[14] Hideomi Tsuge, *Minka to Watashi* (Minka and me), Keisō Shobō, 1980. The active participant and one-time secretary of Minka wrote his memoirs in his final days.

[15] Kiyoshi Okano, 'Gakujutsu gyōsei (Science administration),' in *Gakujutsu geppō* (Science monthly), vol. 29, no. 1, April 1976. An Education Ministry bureaucrat described the moves of the Renewal Committee.

10 Science Engineering Education in Japanese Universities after World War II

Takashi Hata

Science and technology and university education in modern Japan

The faculties of engineering formed an important part of the Japanese Imperial Universities from the start. National policy makers promoted natural science education in general, and engineering education in particular. This was in sharp contrast to universities in Europe that were built around the studies of classics and humanities long before the Industrial Revolution and had resisted the introduction of modern sciences into their curriculum. It was only through a complex process that the Revolution inevitably brought about their introduction.

With the growth of the heavy industry sector after World War I, Japan witnessed an increased demand from industry to apply science and develop technology. Japan's national mobilization program intensified such calls. One can see the strong connection between science and technology in the fact that a particular term, *kagaku gijutsu* (science and technology), was coined during World War II and took firm root thereafter. The technocrats in the Division of Science at the Cabinet Planning Agency initially formulated and disseminated the phrase[1] during the establishment of the so-called new science and technology system. The term was defined to mean 'scientific technology' as well as 'science related to technology,' which includes nearly all of the natural sciences.

In contrast, the Ministry of Education classified science into three research categories – 'basic, applied and practical' – and defined technology as the area which aims at practical and productive use, thereby arguing against the concept of *kagaku gijutsu*.[2] The Ministry's move reflected its jurisdictional sectionalism against the Cabinet Planning Agency which appears to have endeavored to exercise comprehensive control over the entire domain of science and technology. Chiefly prompted by military imperatives, the Japanese government tended towards uniting science and engineering education.

Education Policy and Science and Technology in the Early Postwar Years

After the Japanese acceptance of the Potsdam Declaration – i.e., after the end of World War II – Japan's universities were required to either remove or reorganize old practices. During the war the institutions of higher education were subjected to military requirements. Their first move after the war was an attempt to return to the situation that had existed before the outbreak of the war. On the afternoon of 15 August 1945, the Imperial University of Tokyo held a meeting of Deans at which it was agreed that war-related research should come to a full stop. The President of the University met with the Minister of Education on 24 August. On 4 September he reported to the University Senate that departmental names within the Faculties of Medicine and Engineering were to be altered while both the name and research objectives of the Aeronautical Research Institute were to be changed.[3] At the Tokyo Institute of Technology in late September, President Koroku Wada initiated an advisory committee made up of full and associate professors and instructed them to review wartime research activities and formulate a reform plan.[4] Wada, Deputy Chief of the Cabinet Technology Agency during the war, played a significant role in the reorganization as the Chairman of the University Accreditation Association.

In the meantime, at Osaka Imperial University, Dean of Science Seiichi Ueno, advocated the expansion of universities. He claimed: 'The closure of the university is out of the question. But there might be some problems with the Aeronautical Department and the Shipbuilding Department. So, before we are subjected to accusations, we should take steps to change their names to peaceful ones. The age of chemistry is just around the corner. We should make every effort to expand our University. The growth of Japan depends upon the growth of universities.'[5] In Ueno's approach, one can detect both the crafty strategy of adapting to the postwar situation and the optimism of developing science. These moves on the part of universities precede those of the Ministry of Education which drew up policy documents for higher education immediately after the war. They include the 'Outline of postwar school arrangements' discussed at the Ministry meeting on 18 and 20 September[6] and the classified 'Outline of special vocational school arrangements' written around the same time.[7]

This period marks the beginning of a trend to prioritize the promotion of science as the central feature of postwar education policy. On 15 August, Prime Minister Kantarō Suzuki delivered a speech, which attributed the loss of the war to the backward state of Japanese science. On 3 September, Education Minister Tamon Maeda called a ministerial meeting in which high-ranking officers pointed out 'rote learning and abstract education' (Vice-Minister, Seiichi Ōmura), 'the poverty of humanities and exclusively utilitarian treatment of science' (Chief of Bureau of Science, Jirō Arimitsu) and 'the loss of rational education and training aimed at the development of individuality since the introduction of the national school (kokumin gakkō) system in 1941' (Hideo Nakane, Head of the Document Section and the Personnel Evaluation Section, Minister's Secretariat). They all agreed that Japan was defeated because Japanese science had failed to catch up with the more advanced nations and to strike root extensively throughout the country.[8] In his speech on 5 September at the 88th extraordinary session of the Imperial Diet, Hitoshi Ashida, Member of the Diet, listed the scarcity of scientific technology as the third reason for the war defeat. On the same day, the Ministry of Education implemented the first organizational reform that abolished the Bureau of Student Mobilization and reorganized the Bureau of Science into the Bureau of Science Education. On 15 September a document was produced (Direction of education for the construction of a new Japan) which outlined the postwar education policy in a systematic fashion for the first time. The policy emphasized the promotion of science and the maintenance of the national polity.

Science education was understood to be for personality formation rather than the application of science to industry and technology. In a radio address on the evening of 9 September, Education Minister Maeda recognized 'the importance of science as a part of high human cultural learning'.[9] The report of the Civic Education Reform Committee submitted around this time stressed the significance of 'cultivating the spirit of free observation and rational empirical thinking liberated from authority and tradition' and removing 'irrational and emotional attitudes'. The uplifting of the scientific spirit was directly connected to the desired images of society and citizenry immediately after the war.[10]

Accordingly, an emphasis was placed on science education at the primary and middle school levels as well as the adult education level, and not on the higher education and university-level education or the training of researchers. Teaching staff at universities and tertiary-level vocational colleges were expected to assist in science education at lower levels of education.[11]

However, the specific programs tended to be an extension of the wartime policy of the promotion of science education. The Science Education Bureau of the Ministry of Education drew up the 'Items for lasting planning'[12] in which individual plans for the promotion of science education resembled the items covered by the 'Outline for the establishment of the new framework of science and technology' which the Cabinet authorized on 27 May 1941. During the war, the Ministry was already concerned about the declining science standard arising from student mobilization and the general decline of the levels of education. Amongst the efforts to redress this was a special science education program for the gifted in the autumn of 1944.[13]

However, the implementation of the Occupation's educational reform revealed the limitations of these strategies to promote science education. The special science education program for the gifted was abandoned in 1947 and the Study Committee for Special Science Education was dissolved. In October 1946, the Ministry 'could not help finding it extremely difficult to sustain the program as a nationally supported system in the light of drastic changes in the circumstances surrounding domestic education.'[14] This reflected the growing gulf separating the idea of special education for the gifted and the principle governing the newly introduced educational system. Furthermore, a nationwide survey which investigated the conditions of science education revealed a shortage of teachers and pointed out that 'they find it difficult to teach pupils who require considerably specialized knowledge.'[15] Thus, science education reached an impasse with scarce teacher numbers and underdeveloped facilities and equipment.

The Start of Postwar Higher Education Reform and Science/Technology

The American Education Mission to Japan, which was more influential than Japanese policy makers in shaping science education reform, made comprehensive suggestions as to how Japanese universities ought to be organized. The Mission's report emphasized that they should contribute to the formation of the civic spirit among students, a philosophy that facilitated the introduction of liberal arts education to institutions of higher education. The report touched only in a very limited way on the role that universities should play in their relationship with industry and the economy of the nation. The Mission considered natural science education primarily as a means for the formation of worldviews. This line of thinking was in part attributable to the fact that the Mission was made up mainly of educationists and educational psychologists, not natural scientists. The American bias in favor of a humanities orientation squarely matched the science education policy which the Ministry of Education advocated in 1945. In the process of reorganizing universities under the new system, both the Ministry and the CIE (Civil Information and Education Section) defined their programs within the broad framework of the Occupation policy. Thus, neither saw science and engineering education in relation to industry. At this point, the repatriation policy of the Occupation tended to be punitive, going so far as to suggest that Japan's surplus industrial capacity be trimmed. This orientation is most conspicuous in the report of the Special Committee on Japanese Repatriation, the so-called Strike First Report.[16]

In close connection with the 'Outline of the plan for national land reconstruction' prepared by the National Land Bureau of the Ministry of Home Affairs on 1 September 1946,[17] the School Education Bureau of the Ministry of Education drew up the classified 'Plan for school arrangements.'[18] The former document regarded 'the reconstruction of agriculture' as the prime task of the nation and made no reference to the promotion of science and engineering. The latter document proposed a scheme to establish a hierarchical system of institutions of higher education. Within this framework, the paper focused on: the regional distribution of institutions of higher education; the balance of disciplines among them; detailed plans to transform the imperial universities into general universities; establishing imperial universities for women in the Tokyo and Kinki areas; and elevating national medical colleges to medical university status. Regarding specialist education for industry, the plan referred only to special vocational schools and middle schools where it was regarded as 'desirable to strengthen the basic components in the curriculum and to add 'scientific and technological' elements to it.' The prime objective was to expand and reinforce the studies of agriculture and fisheries.

With the reforms that put special vocational colleges at risk, some groups, of course, debated the long-term provisions for science and technology education. In accordance with instructions from the CIE, the Japanese Education Reform Committee started examining the reform of the school system at its seventh plenary meeting held on 18 October 1946.[19] In this meeting, Teizō Toda, a member of the Japanese Educators Committee, criticized the three-tier stratified system of expert technical education composed of universities, special vocational colleges and vocational high schools. He went on to propose that the special vocational college (senmon gakkō) sector should be abolished and the university sector should be expanded. In opposition to this view Kyōsuke Yamazaki, who was both Vice-Minister of Education and Professor of Engineering at the Imperial University of Tokyo, pointed out the importance of maintaining the three tiers of engineers, foremen and craftsmen and argued the necessity of technological education at the special vocational school level. Over time, the argument in favor of the removal of the prewar stratified system gained ground. The plenary session held on 27 December decided that university education should be expanded to a total of four years. With the Japanese intellectuals aiming primarily at the dislodgement of the stratification of higher education structure, a majority of policy makers at this time showed little interest in altering it to adapt to the changing industrial structure of postwar Japan.

Formulation of University Establishment Standards and Science/Technology Education

The abovementioned trend was discernible in the process of formulating the standards for establishing universities. The process involved debates over the concrete substance of science and technology education in universities under the new system. At its plenary session held on 12 May 1947, the University Accreditation Association arrived at a broad agreement over the 'Outline of standards for establishing universities.' On 7 July, the Association changed the title of the document to the 'Standard for the establishment of universities' and approved it. At the same time, the Association proceeded to study the standards for specialized education and stated in the above document that 'separate criteria will be formulated.'[20]

The Science Subcommittee of the Association that assumed the chief responsibility for studying the standards for science and engineering education consisted of Koroku Wada and others. They submitted a 'Proposal' in about May or early June. They also produced the 'Outline of criteria for the establishment of universities (in the area of engineering),'[21] but the exact date of its submission is unknown. The latter is an interesting document that reveals how the above-mentioned academics attempted to design postwar Japanese engineering education. Defining their objective, they declared that 'engineering education should aim at contributing to the advancement of human culture and knowledge and the development of the nation and its scholarship and industry.' In the eighth section, the submission stated that the 'contents of university education must be built upon a proper balance between academic and basic science to promote both cultural literacy and engineering, which is its application. Specialized engineering education, in particular, must be based on the firm grasp and sufficient knowledge of basic science.' The tenth section stated: 'The practice of autonomous learning and self-education should be promoted by allowing students to elect a wide range of subjects under the supervision of their professors and to study voluntarily and independently. Thereby, the individuality of each student must be cultivated sufficiently and conformism must be avoided.'

The same orientation manifested itself in the 'Outline for the reform of the Tokyo Institute of Technology' made public on 1 February 1946.[22] This document mapped out the aims of engineering education, suggesting that it should produce youth capable of independent thinking, expand the scope of freedom of study for students and enable them to master general application skills derived from basic science which forms the foundation of specialist techniques. The formation of civic-mindedness – which constituted a focal idea in postwar Japan and was emphasized in the report of the United States Education Mission to Japan – was reflected in this conception of desirable engineering education.

The Science Subcommittee approved the 'Standard for science-based faculties' on 18 July. On 25 November the Council revised the 'Standard for the establishment of universities' to include the condition that 'as to specialist courses, appropriate numbers of subjects should be taught in each area.' [23] The Council also set up two generic categories, namely, 'natural sciences' and 'applied sciences,' for each of which it provided broad lists of specific disciplines and fields. It also noted in an appendix that specific rules would be spelt out for medicine, dentistry, pharmacy and veterinary science.'

These statements differ from the positions that the university establishment criteria and the subcommittee deliberations had indicated until July. Precisely what brought about the changes remains unclear, but the *Daigaku kijun kyōkai jūnen-shi* (Ten year history of the University Accreditation Association)' suggests a belief that the specification of departmental education criteria should be limited to such new disciplines as social work and journalism and such disciplines as medicine, dentistry, pharmacy and veterinary science which were linked with national qualification examinations. The thinking at the time was that 'university education would be dangerously standardized if rigorous criteria were imposed on already established disciplines.'[24] Thus, in order to improve the level of specialist education, CIE apparently favored a wide range of freedom and autonomy for Japanese universities and academics rather than directing science education in a strictly planned fashion. Specifically, CIE expected to have accreditation systems voluntarily implemented.

ESS Scientific and Technical Division and the report of the Scientific Advisory Group to the United States National Academy of Sciences

There was mounting dissatisfaction amongst the members of the ESS Scientific and Technical Division (ESS/ST) with the existing Occupation education policy which attached great importance to the humanities and paid little attention to the development of science and technology. ESS/ST had initially given priority to the demilitarization measures but began to relax surveillance policies from about June 1946 with a view to creating a science and technology structure suitable to postwar Japan, a program oriented to the nation's economic recovery. However, the Government Section expressed concern over the ESS/ST's centralized policy style. Because of the internal conflict in GHQ, ESS/ST invited the Scientific Advisory Group from the United States in July 1947 and attempted to arrive at a policy consensus. The Group completed the 'Report of the

Scientific Advisory Group to the National Academy of Sciences on the Reorganization of Science and Technology in Japan' on 28 August.[25]

The Report criticized the Japanese university system for placing too much emphasis upon basic research. It stated: 'emphasis seems to fall upon work in pure sciences to an extent detrimental to applied science.' This point was not made in the 'United States Education Mission' report and implied a criticism of the higher education reform promoted by the CIE. The ESS submitted a paper entitled 'Scientific Advisory Group Report' dated 1 October 1947 with a proposal that it be made the basic policy document for science and technology in occupied Japan.[26] The proposed policy emphasized the importance of science for the purpose of Japan's economic recovery. It was an invitation to develop an awareness that the nation's rehabilitation should be based upon the achievements of science and technology.

In a document addressed to the GHQ Chief of Staff dated 8 January 1948, CIE opposed the proposal, stating: 'The responsibility of higher education is to train and develop leadership of all kinds necessary for a free society, only a small part of which is required to be scientific leadership.' The CIE refused to accept that universities should exist only for scientific research. Specifically, it took a negative position against the Scientific Advisory Group's recommendation proposing a new centralized control system for institutions of higher education. The CIE contended that 'It was precisely this type of administrative control that provided Japan with the ability to fully utilize its scientific resources for war purposes.' This statement reflects the fact that the CIE considered science and technology policy to be a part of the program to demilitarize and democratize Japan. The CIE argued that it should assume the primary responsibility for the reform of the educational system and that the Report of the Scientific Advisory Group should be used as the central manual to the extent that it did not contradict the established policy and implementation program. This argument was accepted on 5 March 1948.[27] The internal conflict in GHQ explains the fact that the Report was delivered to Kankurō Kaneshige, Chairman of the Renewal Committee for Scientific Organization, in 1948, half a year later than it was submitted to the US National Academy of Sciences.

This conflict was a precursor to alterations in higher education policies that began to attach increasing importance to specialized education at the expense of general education. With this confrontation happening within GHQ the trend that had begun immediately after the war accelerated, a process that cannot be reductively explained as a reflection of the so-called 'reverse course' (return to conservatism). ESS/ST invited the Second Scientific Advisory Group on 28 November 1948 in their efforts to strengthen applied science in Japan.[28]

Starting the New University System

This schism over the status of natural science education manifested itself within GHQ from 1947 to 1948, the same time that the new university system was making a start, the Cold War structure was consolidated and the repatriations policy faced a turning point. The designers of the new system had planned to include general education within science and technology education. However, the above-mentioned changes were already taking place at a concrete level when the system was put into motion. The debate focused on where general education should fit into the overall system and how it should be balanced against specialist education in science and technology. The balance issue was related variously to the questions of the shortened years of university education and also that of when students should complete general education and advance to specialist education. In most universities, the first two years were allocated to general education and the second two years to specialist education. However, there were exceptions. For example, the Faculty of Engineering at the University of Tokyo enabled its students to take specialist subjects in the second half of their second year,[29] while the Faculty of Engineering at Keio University decided to allocate one year to general education and three years to specialist education.[30]

Meanwhile, after the introduction of the new university system, the role of engineering research for economic recovery was taken more seriously. In August 1950, the Second United States Education Mission visited Japan, investigated and assessed the progress of educational reform, and made recommendations about future directions. The report filed by the second group made little reference to the importance of civic education in institutions of higher education which had been recommended by the first group. Instead, the new recommendation attached importance to the development of human resources which corresponded to the demands of local communities and industries and suggested that the sizes and disciplines of universities must be determined on the basis of those considerations. In reference to the financial implications of university education, the Group maintained that expenditures required for institutions of higher education would never be a waste and pointed out that research outcomes of agricultural studies and engineering produced fresh incomes, though of lesser importance. In September, the Ministry of Education established the Technology Education Conference to examine engineering education at the university level.[31]

In the meantime, on 23 October 1950, ESS/ST requested an advisory group from the American Society for Engineering Education to be dispatched. The advisory group arrived in Japan on 5 July 1951 and held a series of study meetings on engineering education around the nation from

12 July to 23 August.[32] The advisory group recommended that industry should be rebuilt on the basis of cooperation between universities and the industrial community; engineers and technicians should be educated and trained; applied research and experimental research should be encouraged; teaching methods should be ameliorated; and industrial educators should be organized.[33] However, this report did not make light of general education; it merely stressed the importance of universities satisfying the demands of the industrial world. The report assumed the significance of civil education, arguing that the desirable engineer is an 'individual who is a good citizen and a sensible member of society as well as an educated person.'

The End of the Occupation and Demands for Science and Technology Education

The twin process of specifying the nature of new-system universities and reviewing the reinforcement of industrial education was in progress. Yet, despite the many ideas and proposals presented, little was put into practice. The Economic Rehabilitation Plan which was prepared from summer 1947 to 1949 pointed out that the number of graduates from universities and specialized vocational colleges in the academic year of 1947 was excessive in the light of proper allocation of technical experts. The supply of graduates from the science and technical education system had expanded enormously during the war years and now exceeded the demand by industry.[34] During the war, the Navy recruited a large number of graduates as technical officers and engaged them in research on radar and weaponry. After the dissolution of the military, they were hired by private corporations such as NEC, Toshiba, Fujitsu, Mitsubishi Electric Company, and Sony, where they were involved in research on light electrical appliances and laid the foundations of this industry.[35]

Too many engineers and technicians meant no expansion of science and technology education. The report of the Deliberative Council for the Self-Sufficient Economy published in January 1951 suggested that the international balance of payment be restored and the level of self-sufficiency should be improved before 1953 by securing imports and accumulating capital. The report made no reference to technology policy. The report of the Deliberative Council of Industrial Rationalization (February 1952) and the Act for the Facilitation of Industrial Rationalization (March 1952) both placed an emphasis upon the encouragement of and assistance to technological research in corporations, provisions for securing research funds and tax concessions, and focused on the ways of fostering industrial technology within the private sector. In the educational sphere, industrial education was promoted mainly in high schools as a consequence of the enactment of the Law for the Promotion of Industrial Education (June 1951), legislation which eventuated under pressure from managers of private high schools.

However, the next phase of reform gradually attached importance to specialized education, with postwar changes being reexamined towards the end of the Occupation. Following the statement by M. B. Ridgway for the re-screening of the existing laws and ordinances, the Yoshida Cabinet appointed the Advisory Council for the Amendment of Government Ordinances. On 14 August the Council submitted a top secret 'Report on the reform of the administrative system'[36] and proposed the simplification of administrative work and the reduction of employees, commensurate with the size of government finance, with a specific suggestion that academic staff at national universities be reduced by ten percent. On 2 November, the Committee submitted a 'Report on the reform of the educational system' and proposed that vocational education be strengthened and the university structure be divided into the four-year general type and the two- or three-year vocational type. The report suggested that the reform should aim to 'establish a rational educational system which is genuinely effective in fostering all the many useful people required for various spheres in accordance with Japan's national strength and conditions.' Absent from this approach was the perspective of training engineers as citizens, a framework articulated in the 'Report of the Advisory Group for Japanese Industrial Education.'

In opposition to these moves that would require cutting the number of employees in the public sphere, the Ministry of Education argued for the necessity of science and technology education and the significant role that national universities played in this area. The Ministry maintained: 'It is a matter of urgency to strengthen the faculties of science and technology in the light of the present situation in which modern science and technology is swiftly developing to increasingly advanced levels. The management of the faculties of science and technology has been so difficult in private institutions that they should be developed in national universities.'[37]

At about this time, the business community began to actively make suggestions. The Japan Federation of Employer Associations published the 'Demand on the reexamination of the new educational system' (16 October 1952) to inform the Ministry of Education of the importance of science and technology education. The Federation made a series of proposals for the reinforcement of science and technology education, presenting in succession the 'Demand on the amelioration of the present educational system' (24 December 1954), the 'Opinion concerning technological education which meets the demands of the new age' (8 November 1956), and the 'Opinion regarding the promotion of science and technology education' (26 December 1957).

These moves influenced the University Accreditation Association which began reexamining the university standards and departmental education standards from the end of 1953. After the Association commenced its work, the interim report prepared by the Engineering Subcommittee stated: 'The substance of general education has come under severe criticism. However, it is not good to reduce the number of general education subjects. We should endeavor to let general education achieve its original objectives.'[38] On the whole, the Association took a stance to maintain the existing range of standards while addressing the necessity of raising special education standards.

However, upon receiving the 1954 document, the Association held an informal meeting with the Federation of Employer Associations in May 1955. After the Federation's 1956 document was released, the University Accreditation Association's Board attached more importance to technical education and began to study ways of restoring a balance in student numbers between law/arts departments and science/technology departments. This led to the formation of the 'Study Group of Technical Education' within the Association. The Group met six times from October 1957 to May 1958 and presented the Board a proposal to the effect that the basis of engineering education should be strengthened by making good use of the regulations about basic education units within the 'Standard for the establishment of universities.'

The 1956 Economic White Paper declared that 'the postwar period was already over' and stressed that Japan should strive to 'revise its economic structure in a bid to keep abreast of the waves of the technology revolution around the world.' The New Long-Term Economic Plan presented in December 1957 set forth a plan to increase student numbers in science and technology areas. Thus, from about 1955, university-level science and technology education was promoted in close connection with national economic planning and takes shape along these lines in the second half of the 1950s.

Notes

1 Shōichi Ōyodo, *Miyamoto Takenosuke to kagaku gijutsu gyōsei* (Takenosuke Miyamoto and science and technology administration), Tōkai Daigaku Shuppankai, 1989, p. 384.

2 'Kagaku gijutsu taisei kakuritsu yōkō ni taisuru shitsumon narabini iken (Questions and comments about the outline for the establishment of science and technology system [dated 25 September, 1940, proposal from The Cabinet Planning Agency]),' in *Yōji Minobe Monjo* (Yōji Minobe Records), 6602.

3 Tokyo Daigaku Hyakunenshi Henshū Iinkai (Editorial Committee for a

One-Hundred Year History of the University of Tokyo) (ed.), *Tokyo Daigaku hyakunen-shi, Tsūshi 2* (The one hundred-year history of the University of Tokyo, Complete history 2), University of Tokyo Press, Tokyo Daigaku Shuppankai, 1986, pp. 972, 974–5, 1103–5.

4 Tokyo Kogyō Daigaku (Tokyo Institute of Technology), *Tokyo Kōgyō Daigaku hyakunenshi, Tsūshi* (One hundred-year history of the Tokyo Institute of Technology, Complete history), 1985, p. 666.

5 Imoto, Minoru, 'Haisen zengo no nikki (A diary before and after the end of the war)', in *Osaka Daigaku-shi kiyō* (Journal of the history of Osaka University), vol. 1, May 1981, pp.111–16.

6 *Kanno Monjo* [7]

7 *Kanno Monjo* [7]

8 *Arimitsu Nikki* (Arimitsu Diary), entry on 3 September 1945, Daiichi Hōki Shuppan, p. 788.

9 'Bunka Nippon no kensetsu e (Towards the construction of Japan as a cultural nation),' in Nihon Kagakushi Gakkai (History of Science Society of Japan), *Nihon kagaku gijutsu-shi taikei* (A compendium of the Japanese history of science and technology), vol. 5, Historical Overview 5, Daiichi Hōki Shuppan, 1966 [18] p. 45.

10 Shuichi Katsuta, 'Sengo ni okeru shakaika no shuppatsu (Beginning of social studies in the postwar period),' in *Katsuta Shuichi chosakushū* (Collection of literary works by Shuichi Katsuta), vol. 1, Kokudosha, 1972.

11 'Kokumin gakkō, chūtogakkō ni okeru kagaku kyōiku narabini shidō ni kansuru ken (sōan) (Matters on science education and its teaching at national and secondary schools [draft]),' in *Sengo kenkyū shiryō* (Materials on postwar education), 1965, in the library of Kokuritsu Kyōiku Seisaku Kenkyūjo (The National Institute for Educational Policy Research) [1] II-5.

12 *Sengo kenkyū shiryō* (Materials on postwar education), 1965, in the library of Kokuritsu Kyōiku Seisaku Kenkyūjo (The National Institute for Educational Policy Research), 25 August 1945 [1] II-1.

13 Yoshio Fujioka, 'Kagaku tokubetsu kyōiku (iwayuru eisai kyōiku-ron) no kiroku (A record of science education [so-called elite education]),' in *Kagaku kyōiku-ron* (Essay on science education), Kawade Shobō, 1947.

14 Sub-Committee for Special Science Education, Special Research Committee for Science Education, 'Tokubetsu kagaku kyōiku no shori hōshin (Management policies for ending special science education [15 October, 1946]),' in *Sengo kenkyū shiryō* (Materials on postwar education), 1965, in the library of Kokuritsu Kyōiku Seisaku Kenkyūjo (The National Institute for Educational Policy Research), 1965 [1] II-8.

15 'Kagaku kyōiku jittai chōsa hōkokusho 1 (Report on the investigation into the actual situation of science education 1),' October 1946, in *Sengo kenkyū shiryō* (Materials on postwar education), 1965, in the library of Kokuritsu

Kyōiku Seisaku Kenkyūjo (The National Institute for Educational Policy Research), 1965 [1] II-101.

16 Yutaka Nishinarida, 'Senryō seisaku no tenkan to Nihon keizai (Change of the Occupation policies and Japanese economy),' in *Nihon dōjidai-shi* (Contemporary history of Japan), vol. 2, Aoki Shoten, 1990.

17 Shirō Sugai, *Shiryō, kokudo keikaku* (Materials, national land planning), Daimyōdō, 1975.

18 *Sengo kenkyū shiryō* (Materials on postwar education), 1965, in the library of Kokuritsu Kyōiku Seisaku Kenkyūjo (The National Institute for Educational Policy Research) [1] VII-8.

19 Terasaki, 1970 [19] p. 75.

20 Tanaka, 1981 [20] vol. 2, p. 168.

21 Tanaka, 1981 [20] vol. 1, pp. 96 and 98.

22 *Tokyo Kōgyō Daigaku hyakunen-shi, Tsūshi* (One hundred year history of the Tokyo Institute of Technology, Historical Overview), Tokyo Kōgyō Daigaku, 1985, pp. 672–4.

23 *Sengo kenkyū shiryō* (Materials on postwar education), 1965, in the library of Kokuritsu Kyōiku Seisaku Kenkyūjo (The National Institute for Educational Policy Research) [1] VI-121.

24 Daigaku Kijun Kyōkai (The Japan University Accreditation Association), 1957 [21] p. 124.

25 'Report of the Scientific Advisory Group to the National Academy of Sciences on the reorganization of science and technology in Japan,' in GHQ/SCAP Records [8]; Beikoku Gakujutsu Komondan, 1947 [13] is a Japanese translation. Refer to Tsuchimochi, 1991 [23] for a detailed discussion of the process of its creation.

26 'Report of the Scientific Advisory Group of the United States National Academy of Sciences,' Trainor Collection [9].

27 'Staff memorandum no. 10. Staff responsibility for scientific education research and technology,' Trainor Collection [9].

28 GHQ/SCAP, 'Reorganization of science and technology in Japan (1945 to September 1950),' in *History of the Nonmilitary Activities of the Occupation of Japan*, p. 75.

29 Tokyo Daigaku Hyakunenshi Henshū Iinkai, *Tokyo Daigaku hyakunen-shi, Bukyoku-shi 2* (One hundred year history of Tokyo University, Departmental history 2), Tokyo Daigaku Shuppankai, 1987.

30 Keio University, *Keiō Gijuku Daigaku Kōgaku-bu sanjūgonen-shi* (Thirty-five year history of the Faculty of Engineering, Keio University), Keiō Gijuku Daigaku Kōgaku-bu, 1974.

31 'Gijutsu kyōiku kyōgikai kankei shiryō (Related materials on the Conference for Technical Education),' in *Sengo kenkyū shiryō* (Materials on postwar education), 1965, in the library of Kokuritsu Kyōiku Seisaku Kenkyūjo (The National Institute for Educational Policy Research) [1] VI-527.

32 'Kōgyō kyōiku kenkyū shūkai jisshi yōkō (Guideline for a study conference for industrial education),' in *Sengo kenkyū shiryō* (Materials on postwar education), 1965, in the library of Kokuritsu Kyōiku Seisaku Kenkyūjo (The National Institute for Educational Policy Research) [1] VI-528.
33 'Report of the Engineering Education Mission to Japan' [12] 1952.
34 Hata, 1992 [24] p. 39.
35 Yasuzō Nakagawa, 'Kaigun erekutoronikusu jinmyaku (Electronics: personal connections with former naval colleagues),' in *Will*, June 1985.
36 *Tsujita monjo* (Tsujita Records), owned by Kokuritsu Kyōiku Seisaku Kenkyūjo (The National Institute for Educational Policy Research).
37 'Shinsei kokuritsu daigaku ni tsuite (On new-system national universities),' August 1951, in *Sengo kenkyū shiryō* (Materials on postwar education), 1965, in the library of Kokuritsu Kyōiku Seisaku Kenkyūjo (The National Institute for Educational Policy Research) [1] VI-34.
38 'Daigaku Kijun Kyōkai dai 72-kai rijikai ken dai 72-kai hyōgikai gijishōroku (Extract of the minutes of the 72nd meeting of the trustee-cum-the 72nd meeting of the directors of the Japan University Accreditation Association),' 29 January, 1954, in *Rijikai ken hyōgiinkai gijishōroku* (Extract of the minutes of the meetings by the Directors and Trustees of the Japan University Accreditation Association), owned by Daigaku Kijun Kyōkai (The Japan University Accreditation Association) [2].
39 'Daigaku Kijun Kyōkai dai 80-kai rijikai ken dai 80-kai hyōgikai gijishōroku (Extract of the minutes of the 80th meeting of the trustee-cum-the 80th meeting of the directors of the Japan University Accreditation Association),' 25 January 1955, in *Rijikai ken hyōgiinkai gijishōroku* (Extract of the minutes of the meetings by the Directors and Trustees of the Japan University Accreditation Association), owned by Daigaku Kijun Kyōkai (The Japan University Accreditation Association) [2]

Bibliographic Guide

Basic data

[1] *Sengo kenkyū shiryō* (Materials on postwar education), 1965, in the library of Kokuritsu Kyōiku Seisaku Kenkyūjo (The National Institute for Educational Policy Research). Refer to Chapter 15, 'Senryōka no kenkyūhi (Research funds during the Occupation)'.
[2] *Rijikai ken hyōgiinkai gijishōroku* (Extract of the minutes of the meetings by the Directors and Trustees of the Japan University Accreditation Association), owned by Daigaku Kijun Kyōkai (The Japan University Accreditation Association).
[3] *Kijun iinkai* (Standards committee), owned by Daigaku Kijun Kyōkai (The Japan University Accreditation Association).

[4] *Kagaku gijutsu kyōiku kenkyū iinkai* (Committee for science and technology education), 1957, owned by Daigaku Kijun Kyōkai (The Japan University Accreditation Association).
[2]–[4] are the records of the activities by Daigaku Kijun Kyōkai (The Japan University Accreditation Association) which had a great influence on the contents of university education during the reforms under the Occupation.

[5] *Kyōiku sasshin iinkai sōkai gijisokkiroku* (Stenographic records of the minutes of the general meeting by the Japan Education Reform Committee), 1946–51, owned by Noma Kyōiku Kenkyūjo (Noma Institute of Educational Research).
These are the minutes of the deliberative committee established by the Cabinet. The committee played a major role in the reform of the university system.

[6] *Gakujutsu taisei sasshin iinkai kaikei shiryō* (Related materials on the Renewal Committee for Scientific Organization), in Nihon Gakujutsu Kaigi Toshokan (Library of the Science Council of Japan).
The committee played a major role in forming The Science Council of Japan Law and the Scientific and Technical Administration Committee.

[7] *Kanno Monjo* (Kanno Records), The Tokyo Institute of Technology.

[8] *GHQ/SCAP Monjo* (GHQ/SCAP records), owned by Kokuritsu Kokkai Toshokan Kenseishiryō-shitsu (National Diet Library, Modern Japanese Political History Materials Room).
There are an enormous number of historical documents in existence, e.g. Higher Education Branch in the CIE Education Division and the ESS Scientific and Technical Division. In particular, there are policy documents by the Scientific and Technical Division in the following:
- Report of the Scientific Advisory Group of United States National Academy.
- Sciences (With summary of Recommendations and Brief), GHQ/SCAP.
- GHQ/SCAP Records, Box No. 7429, ESS (F) 1869–1870.

[9] *Trainor Collection,* owned by places such as Kokuritsu Kyōiku Seisaku Kenkyūjo (The National Institute for Educational Policy Research) and Kokuritsu Kokkai Toshokan Kensei Shiryō-shitsu (National Diet Library, Modern Japanese Political History Materials Room).
Historical documents owned by Joseph C. Trainor who was the Deputy Chief of Education Division. The collection is held at Hoover Institution, but a large number of copies were purchased and are stored in Japan. They are arranged according to the subjects and easier to use than the GHQ/SCAP Records. Regarding the scientific advisory group, there is a file named 'Scientific Advisory Group Report' in Box 49.

[10] Terasaki, Masao, 'Shiryō Tokyo Daigaku kyōiku seido kenkyū iinkai (Data on committee for Tokyo University education reform)', in *Tokyō*

Daigakushi kiyō (Journal of the history of the University of Tokyo), no. 7, March 1989.
This had a great influence on the planning of universities by Japanese intellectuals when Japan lost the war.

[11] *Beikoku tainichi kyōiku shisetsudan hōkokusho* (Report by the United States Education Mission to Japan), 1946, 2nd edn. 1950.
There are a number of translations, including: Masao Terasaki (tr.), *Sengo kyōiku kaikaku shiryō 10: Beikoku tainichi kyōiku shisetsudan ni kansuru sōgōteki kenkyū* (Postwar education reform data 10: Comprehensive research on the United States Education Mission to Japan), March 1991.

[12] Ministry of Education, Higher Education and Science Bureau, Technical Education Division (tr.), *Tainichi kōgyō kyōiku komondan hōkokusho* (Report of the Engineering Education Mission to Japan), August 1951.

[13] Beikoku Gakujutsu Komondan (The United States Scientific Advisory Group), *Nihon ni okeru kagaku to gijutsu no saihensei* (Reorganization of science and technology in Japan), 28 August 1947.

[14] Kindai Nihon Kyōiku Seido Shiryō Henshū Iinkai (Editorial Committee for the Historical data on Modern Japan's Education System), *Kindai Nihon kyōiku seido shiryō* (Historical data on modern Japan's education system), 35 vols., Dai Nippon Yūbenkai Kōdansha, 1956–64.

[15] Nihon Gakujutsu Shinkō-kai (Japan Society for the Promotion of Science), *Gakujutsu geppō* (Japanese Scientific monthly), monthly publication.

[16] History of individual universities:
These are important as they show the actual situation of science and engineering education at universities, but only a few can be useful. The following are important: *Tokyo Daigaku hyakunen-shi* (The one hundred year history of the University of Tokyo), 10 volumes, 1985–1987; *Tokyo Kōgyō Daigaku hyakunen-shi* (The one hundred year history of the Tokyo Institute of Technology), 2 volumes, 1985. The history of individual universities can be studied in *Gakkō enkakushi mokuroku* (Catalogue of school history), owned by Noma Kyōiku Kenkyūjo (Noma Institute of Educational Research), 1986, and in *Kakunendo zōka tosho* (Updated volumes published every year).

Related data

[17] Tetsu Hiroshige, *Kagaku no shakaishi – Kindai Nihon no kagaku taisei* (Social history of science – The structure of modern Japanese science), Chūō Kōronsha, 1973.

[18] Nihon Kagakushi Gakkai (History of Science Society of Japan), *Nihon kagaku gijutsu-shi taikei* (A compendium of the Japanese history of science and technology), 25 volumes, Daiichi Hōki Shuppan, 1996.

[19] Masao Terasaki, *Sengo Nihon no kyōiku kaikaku 9, 'Daigaku kyōiku'*

(Education reform in postwar Japan 9, 'University Education'), University of Tokyo Press, 1970.

[20] Masao Tanaka, 'Daigaku Kijun Kyōkai no keisei to "daigaku kijun" no seiritsu (jō)(ge) (Formation of the Japan University Accreditation Association and "university standards"),' in *Bulletin* (The Japan University Accreditation Association), nos. 44 and 45, 1981.

[21] Daigaku Kijun Kyōkai (University Accreditation Association), *Daigaku Kijun Kyōkai jūnenshi* (Ten year history of the University Accreditation Association), 1957.

[22] Hōichi Tsuchimochi, 'Senryōka no kyōiku kaikaku – Daiichiji beikoku tainichi kyōiku shisetsudan hōkokusho to kōtōkyōiku kaikaku (Education reform during the Occupation: The first report by the United States Education Mission to Japan and Reform of Higher Education),' in *Daigaku ronshū* (Research in higher education), no. 18, 1989.

[23] Hōichi Tsuchimochi, 'Beikoku gakujutsu komondan hōkokusho to sengo Nihon no kōtō kyōiku kaikakuan – Adams danchō bunsho o chūshin ni (The report of the United States Scientific Advisory Group and Japan's postwar reform of higher education with special reference to the Adams papers),' in *Daigaku ronshū* (Research in Higher Education), no. 20, 1991.

[24] Hata, Takashi, 'Sengo kyōiku kaikaku to kagaku gijutsu no shomondai (Postwar educational reform and various issues of science and technology)', in *Kōwa dokuritsu go no wagakuni kyōiku kaikaku ni kansuru chōsa kenkyū* (Research on Japan's educational reform after the conclusion of the Peace Treaty), Kokuritsu Kyōiku Kenkyūsho (National Institute for Educational Research), March 1992, pp.1–50.

11 The Science Council of Japan and the Scientific and Technical Administration Committee

Shigeru Nakayama

Great Anticipation in the Media for the Science Council of Japan

The Science Council of Japan (JSC) is an unprecedented 'parliament of scholars' (cf. ch. 9). Hence, at its outset there was great anticipation in the Japanese media. Major newspapers such as *Asahi Shimbun* and general science magazines like *Shizen* keenly observed and enthusiastically reported on its development.

It was not, however, as widely reported overseas as it was in Japan. At its birth, the Scientific and Technical Division of the Occupation Forces' Economic and Scientific Section (ESS/ST) wanted to publicize their role as the midwife but chose to wait until it was clear what the JSC would achieve. Occasionally, the JSC issued a politicized appeal in English, but otherwise there was no overseas media attention.

The Japanese media was not always friendly to the JSC. In fact it was often very critical. But one thing was certain; the JSC created the impression of an organization that was open to the outside world. The media focused its attention on the Council's political and ideological messages aimed at the outside rather than its purely academic internal problems. The members of Minka (Minshushugi Kagakusha Kyōkai [The Association of Democratic Scientists] cf. ch. 25) and the other leftwing members of the JSC, such as Shōji Ijiri, often pronounced demands for academic and ideological freedom and opposition to military research. In many such cases, the members of the basic science section supported them. But the majority of the members were without political inclinations and accepted the moderate liberal opinions of the law school professors of the Tokyo University as their guiding principle.[1]

Substantial Functions of the Science Council of Japan

Perhaps the most substantial function of the JSC was to deliberate the allocation of research funds – which all researchers had a keen interest in –

and to advise the government about same. For that purpose the Research Fund Committee was formed within the Council to discuss general principles of research fund allocation. It had three specialized subcommittees to deal with research funds for the Education Ministry, the other Ministries and the University Chairs.

Initially the representatives of the 'other' Ministries were suspicious that the JSC acted in the interest of the Education Ministry, but as time passed they became more trusting and cooperative. Two of the subcommittees were directly related to the Education Ministry. In actual practice, the JSC reviewed proposals sent from the Education Ministry in accordance with the guidelines for screening and allocating funds. The first two General Assemblies were totally absorbed in discussing these proposals. At the end of the second General Assembly, though, a special Committee for Research Fund Allocation was appointed to actually allocate funds. At the third General Assembly, research for the economic recovery of Japan, cooperative research, and basic research were given priority. The highest priority was given to technological research.[2]

Allocation of Scientific Research Funds

It was customary for the JSC to recommend the allocation of the Education Ministry Scientific Research Fund, and the order of priority, to the Science Promotion Council (Gakujutsu Shōrei Shingikai) of the Education Ministry. It was generally agreed that the JSC would provide the general guidelines while details were left to the Education Ministry which had more experience.

The relationship between the JSC and the Education Ministry was delicate. Pragmatists well versed in bureaucratic dynamics, like Kankurō Kaneshige, believed that the JSC should be part of the Education Ministry, while the General Headquarters, Supreme Commander for the Allied Powers (GHQ) did not want this and let it be placed under the direct control of the Prime Minister's Office. When the JSC was established, the 110 personnel of the old National Science Council moved to the JSC Secretariat. John W. O'Brien, the head of ESS/ST, was unhappy about this move, being opposed to any preservation of the old institutions.[3]

The Prime Minister's Office is primarily a coordinating body for the various government ministries, not an operating ministry. It was not in a position to request a large budget from the Finance Ministry and was therefore considered by scholars to be of no use to the JSC. Large numbers of JSC members were university people, so their propositions were primarily addressed to the Education Ministry. Thus it became obvious that without the intervention of the Education Ministry no budgetary allocations would be made and nothing could be achieved.

The Proposal to the Government for the Establishment of a New Research Institute

The major task for the JSC in its early stage was to recommend to the government the establishment of a new research institute. Significantly, the recommendations were all realized and implemented. When the JSC was founded in 1949 Japan was still suffering hard economic times. But during the JSC's second period (1951–53) recommendations were made for setting up various research institutes, such as the Nuclear Physics Laboratory (Genshikaku Kenkyūsho) and the Reflector at Okayama Astrophysical Observatory which were both affiliated with Tokyo University. Usually a proposition for a new research institute would be repeatedly discussed at the Long-time Research Investigation Committee (Chōki Kenkyū Chōsa Iinkai) of the JSC, and put to an open public symposium. Only after such deliberations would the General Assembly of the JSC approve it. The establishment of the Tokyo University Nuclear Physics Laboratory was proposed in 1953 and was adopted in 1955. Tokyo University Solid-State Physics Laboratory and Osaka University Protein Laboratory were both proposed in 1956 and approved in 1957 and 1958 respectively.

Liaison with International Scientific Organizations

In 1919, the International Research Council was set up within the Committee for Intellectual Cooperation of the League of Nations, in order to reorganize the international activity of scientists which had broken down due to the First World War. Specialized International Councils in astronomy and chemistry and so on were also set up within the International Research Council. For nation-states, the Liaison Committee of each discipline was founded as a sub-organization of the International Research Council. Japan founded its National Research Council (Gakujutsu Kenkyū Kaigi) in 1920. In 1932 the International Research Council was renamed the International Council of Scientific Unions (ICSU). It still survives today. The ICSU was classified as a Non-Government Organization (NGO) by the United Nations after World War II.

The JSC assumed the function of the National Research Council. In September 1949, Yoshio Nishina, a vice president of the JSC, represented Japan at the General Assembly of the ICSU in Copenhagen. While in other areas international hostility towards Japan's wartime activities still prevailed, Japan was accepted to join the international scientific organization without any problem. As a sub-organization of the JSC, a Research Liaison Committee (Kenkyū Renraku Iinkai, specialized liaison committee) for each discipline was set up corresponding to the International Union of

Specialized Disciplines within the ICSU. The ICSU was limited to natural science. But the JSC also had relationships with the international organizations for technology, medicine, the humanities and social sciences. The Research Liaison Committees were in charge of selecting representatives to the international conferences for each specialty. As a rule the JSC funded the cost of representatives participating in international meetings. At the same time, the JSC sponsored or partly supported the international conferences that were held in Japan. The JSC is still performing this liaison function now.

Until the early 1960s overseas travel was economically difficult for Japanese scientists and financial subsidies from the JSC were very important. But later, as the number of both international conferences and Japanese participants increased, the limited funds of the JSC could not meet the demands of scientists. Other sources of funding increased though, and, at the same time, it became increasingly possible for individual scientists to afford their own travel expenses. Thus the importance of the JSC for Japanese scientists dwindled.

ICSU conferences held in Japan in fields such as physics, mathematics and genetics were sponsored by the JSC. However, Government funding for international conferences was so limited that it became customary for the organizers of scientific societies to solicit funding from various corporations to cover the expenses of hosting international conferences.

On the Improvement of Scientists' Economic Life

Since the JSC is a professional organization whose members are elected by the entire Japanese scientific community, it became responsible for representing the financial interests of all scientists, much like labor unions such as the British Association of Scientific Workers. To this end, the JSC conducted an investigation of the living standards of scientists and assiduously petitioned for an improvement in scientists' living standards. According to a survey of scientists in 1950 entitled 'Questionnaire concerning the defense of the standard of living of scientists' a great many respondents expected the JSC to work for concrete improvements in their living standard. Thus, it can be said that most scientists expected the JSC to act as a trade union of scientists.[4]

Appeals and Declarations

The JSC became known to the general public through its appeals and declarations in the media on political affairs. Primarily, these were the initiative of Minka, often represented by Mitsuo Taketani. The best known of these was the declaration of 'Three Principles for the Research of

Atomic Power' (Genshiryoku Sangensoku) which said that nuclear research should be autonomous, democratic and open. These are simply the principles of academic science. These principles were accepted by liberal academics, such as Shin'ichirō Tomonaga and Fujio Egami, which comprised the mainstream of the JSC.

For the social scientists in Minka – who had anti-American and anti-capitalist views – science meant scientific socialism, while technocratic science and technology was regarded as the capitalist appropriation of science. Academic 'pure' research scientists were also offended by the applied science promoted by capitalism that dominated over pure academic research. Thus these two parties often agreed on political issues and public political declarations.

However, a silent majority of JSC members considered themselves to be representatives of the specialized discipline they belonged to. While they had no objection to the principles contained in many of these political appeals, such as 'The Declaration to Defend Science and Mankind from Warfare,' some members opposed such appeals on the grounds that the JSC was not the place to discuss politics. Thus political appeals were often voted down.[5]

The conservative government was reluctant to accept the principles of academic science, and objected to the pro-Soviet ideology of Minka. The Prime Minister, Shigeru Yoshida, wanted to strip away the official status of the JSC because it opposed his idea of a partial Peace-Treaty with the USA which would exclude the Soviet Union, but the Yoshida Government collapsed before he could do so.

Scientific and Technical Administration Committee and Liaison with Administrative Offices

The Science and Technical Administration Committee (STAC) was founded at the same time as the JSC, primarily at the request of technocratic members of the Engineering Section who wanted to have closer ties with, and influence on, the government administration. They were not content with a mere consulting body but wanted to be able to implement the policies developed by the JSC.

According to the STAC charter, the number of members was limited to 26, consisting of 12 selected from the JSC, the 12 highest Ministerial Administrative Officers (Jimu Jikan), the Prime Minister (chairman) and the Minister responsible for the JSC (vice-chairman). They were supposed to meet each month. In practice, though, the Prime Minister's chairmanship was only nominal; the Minister responsible for the JSC chaired the meetings. Similarly, the attendance of the Ministerial Administrative Officers was rare, with lower ranking officers usually attending in their place. The media criticized this arrangement, claiming that because the government ministries did not consider STAC as important enough, it had not been effective.[6]

STAC's prestige was based on the authority of the JSC but the Ministerial Administrative Officers were not well informed of the JSC's function or significance and the JSC had not yet produced any visible contribution. Since the Ministerial Administrative Officers did not thoroughly understand what STAC's mission was, they usually ignored it. Hence promoting the importance of its mission within the bureaucracy became STAC's highest priority. According to the Deputy Secretary of STAC, Kōhei Suzue, it did not attract bureaucrats' attention, since the JSC did not produce proper and concrete recommendations for the general administration. Hence it was only technocrats who found STAC useful as a medium for contacting ESS/ST.

For these reasons STAC was long considered to be a very weak and insignificant office[7] but recently technocrats have reevaluated it[8] as meaningful not only for its role as a liaison between JSC and government administrative offices but particularly as the liaison between the JSC and GHQ sections, particularly ESS/ST. Its importance in this regard can be seen from the many warnings that GHQ issued to Ministerial Administrative Officers who were not cooperative with STAC.

It has been said that when STAC was established there was no directive from GHQ,[9] but in relation to science policy ESS/ST probably had more interest in STAC than the JSC. ESS/ST was not satisfied with the report on science policies from the Education Ministry and called on Yoshio Nishina to ask for advice on the most effective way to vitalize STAC. Nishina answered that 170 of the 210 JSC members were university professors, who were ignorant of administration and industry and were only concerned about the research funds of the Education Ministry. Many members of the engineering section of the JSC participated in STAC but even they were mostly university men who thought of science policy in terms of the Education Ministry.

It was the intention of ESS/ST to promote applied science for economic recovery through allocating research funds to Ministries other than Education.[10] They suggested that STAC should allocate funds to other Ministries without consulting the JSC. They also suggested that the JSC should advise the Education Ministry to allocate funds to other operating Ministries. ESS/ST maintained that scientific research should not be the exclusive concern of the Education Ministry.[11]

During the war scientific research funding had been allocated to other Ministries, but this was abolished immediately after the war as part of the demobilization effort (cf. ch. 4). Thus, the Education Ministry acquired control of all scientific research funding, taking over the Technology Bureau's wartime research mobilization function (cf. ch. 15). The total quantity of funding was relatively insignificant; it could not keep up with the rapid inflation of the postwar economy. Hence, in spite of the intentions of ESS/ST, research funding was monopolized by the Education Ministry except for a nominal allocation to the Ministry of International Trade and Industry (MITI). However, following the JSC's recommendations – supported by ESS/ST – MITI, the Ministry of Agriculture and Forestry and other Ministries established their own technology subsidy systems in 1951, including the Subsidy for Engineering Research of Mineralogy and Technology, the Subsidy for Engineering Research at Middle and Small Industry and the Subsidy of Nuclear Power Engineering.[12]

The JSC had a special relationship with the Education Ministry; the Ministry consulted the JSC about the allocation of scientific research funds. Because of this close relationship, other Ministries had suspicions that the JSC was representing the interests of the Education Ministry. Thus, liaison between the other Ministries and the JSC was not very effective. ESS/ST believed that separating STAC from the JSC would be a good way to check the Education Ministry's monopoly of scientific research funding, but in reality it was extremely difficult to re-allocate funds from the Education Ministry to the other Ministries.[13]

As we have seen, the JSC was part of the Prime Minister's Office rather than the Education Ministry and STAC was the liaison between the JSC and the other Ministries. At a STAC meeting on 13 August 1949, Yoshio Nishina proposed that a motion be made in the Diet to give STAC responsibility for coordinating the science and technology budgets for all Ministries under the auspices of the JSC, but all of the Ministry members opposed it. Shunzei Inoue, the Director of the Agency of Industrial Science and Technology, justified his opposition on the grounds that it was impossible to determine the amount of budgeted funds for any Ministry before their negotiations with the Finance Ministry were complete. No conclusion was reached at this meeting.[14] Nishina's primary position was to represent the JSC's priority on purely academic science, but the distance between Nishina's academic position and the government's determination to coordinate various ministerial interests was too great to overcome.

At the time of budgeting negotiations for 1950, STAC (and its supporter ESS/ST) proposed to allocate 50 million yen for the Testing Laboratories of Ministries other than Education, but in the end it all went to the Education Ministry. The Education Ministry had already been allocating part of their budget to other research institutes such as the Central Meteorological Observatory, Railway Technical Research Institute, and Mechanical Engineering Laboratory of MITI. What the Education Ministry wanted, even against the wishes of GHQ, was to maintain the nominal right to allocate the entire research budget.[15]

In its first four years STAC held 50 meetings and deliberated on 71 recommendations from the JSC to the government as well as 28 requests from the government for advice from the JSC. The routine matters of STAC included annual deliberations on the allocation of the science and technology budget and applied research funds for the Ministries. STAC was not an administrative organization, but merely a consulting body. Thus it did not have the power to coordinate the Ministries' activities.

Demands on the Diet

In addition to the efforts made through the JSC, Japanese technocrats directly pressured the government to promote science and technology from immediately after the war. Hidetsugu Yagi, the President of Osaka Imperial University, and his associates formed the Science and Technology Policy Club (Kagaku Gijutsu Seisaku Dōshikai) as a political association in June 1946 in the Diet,[16] and conducted a weekly seminar for drafting science and technology policy.[17] The Club's membership – numbering more than 70 – came from all political parties. On 27 September the Club passed 'the declaration on the promotion of science and technology' at the House of Representatives. Yagi, the President of the Technology Bureau, claimed that Japan had lost the war because of its weakness in science and technology and therefore must strengthen its technological power. However, Yagi was purged from public office near the end of 1946 because his Bureau had been at the center of the wartime science mobilization. He also resigned from his position as University President for the same reason. In the middle of December GHQ blocked Yagi's movement on the basis that his plan was too premature to implement at the cabinet level.[18] It sounded too nationalistic for GHQ to approve.

These movements were terminated with the formation of the Renewal Committee whose mission was to prepare for the establishment of the JSC.[19] Yagi, however, was disillusioned with the JSC and STAC soon after their establishment, pointing out that 'Great expectations were not fulfilled due to political maneuvering based on left-wing ideology and the disillusionment of members who were manipulated by bureaucrats.'[20]

In 1952 the governing Liberal Party wanted to establish a stronger administrative body and budgeted to establish the Science and Technology Agency (STA) in the 1953 fiscal year. Both liberal and left-wing factions of the JSC opposed this move, seeing it as an attempt to revive the wartime Technology Bureau and promote the militarization of S&T.[21] Finally, the JSC committed to 'urgently strengthen STAC'; thus the establishment of the STA was set aside for the time being.[22]

When the STA was finally established in 1956, STAC was abolished and its function was transferred to the Science and Technology Council of the STA. The influence of the JSC on research administration was thus weakened. Technocratic members stopped depending on the JSC which had proven incapable of significant achievements because of the constant opposition from Minka and the indifference of the social science and humanities members. They began to implement direct contact between the administrative sections of government and the Science and Technology Council. Technocrats like Masao Maeda of the Ministry of Comm-unications (Teishinshō) began to plan for and promote the implementation of a national science policy, similar to the Americans' NASA, arguing for a ten percent increase in the R&D budget for S&T. Technocrats also tried to draft and pass through parliament a 'Basic Law of Science and Technology' that would oblige the government to support S&T.[23] Behind their actions was the desire to support S&T bureaucrats in government offices, who had until then been treated as subordinate to Law School graduate bureaucrats. In other words, they wanted a government ministry that would appoint top officials from amongst the Science and Technology School graduates. They were successful in fulfilling this desire.

Conclusion

The establishment of the Science Council of Japan as a democratic 'parliament of scientists' was unprecedented in world history. But most of the elected members were academics who had more concern with pure science and ideological position and less interest in technocratic policy implementation. Thus, its position as a Council with no real power proved relatively ineffective, bogged down as it was by internal ideological disputes. Thus, it was not long before it was totally marginalized and its functions assumed by a new technocratic Agency.

Notes

1 Jōji Ashida, 'Nihon Gakujutsu Kaigi no hansei: Dai 7-kai sōkai no hōkoku (Reflections on the Science Council of Japan: A report on its seventh general meeting)', in *Shizen* (Nature), no. 6, January 1951, p.23.
2 Masao Gorai, 'Nihon Gakujutsu Kaigi no dōkō (Trends of the Science Council of Japan),' in *Shizen*, vol. 4, no. 8, August 1949, p. 13; and Toshio Kitagawa, 'Nihon Gakujutsu Kaigi Dai 12-kai Sōkai okaerimite (Reflecting on the twelfth general meeting of the Science Council of Japan),' in *Shizen* (Nature), vol. 7, no. 7, July 1952, p. 34.
3 B. Dees' testimony.
4 Nihon Gakujutsu Kaigi Nijūgonen-shi Henshū Iinkai (Science Council of Japan, Editorial Committee for the Twenty-five Year History), 1974 [5] p. 28.

5 Jirō Date, 'Ichinen no Kaiko (Retrospective on Last Year),' in *Shizen* (Nature), vol. 6, no. 12, December 1951, pp. 30-1.
6 Editorial, *Mainichi*, 4 April 1950.
7 *Nihon Kagaku Gijutsushi Taikei, Tsūshi*, vol. 5, p. 129.
8 Kagaku Gijutsu Seisakushi Kenkyūkai (Workshop on the History of Science and Technology Policy), *Nihon no kagaku gijutsu seisaku-shi* (History of Japanese policy on science and technology), Mitō, 1990, p. 56.
9 Suzue, 1952 [15] p. 785.
10 Putnam, David L., 'Reorganization of Science and Technology in Japan (1945 to September 1950)', in *History of the Non-military activities of the Occupation of Japan*, Kokkai Toshokan (National Diet Library), microfilm number CHS-04304.
11 Kelly, quoted on 20 October 1949, in *Kagaku Bunka Shimbun*.
12 'Sengo kōgyzō gijutsu seisaku-shi no shōgen (Testimony for the postwar history of industrial technology policy),' in *Kōgyō gijutsu* (Industrial Technology), August 1986, p. 50.
13 ESS(B)-11611 11612, Box No. 7399 folder (g), Scientific And Technical Division, liaison group, Memo for File, 10 December 1949, Subject: Disposition of Research Budget [14]
14 *Kagaku Bunka Shimbun*, 20 August 1949.
15 B. Dees, Memorandum of a conference, 3 November 1949 between STAC officials and ESS/ST.
16 *Kagaku Bunka Shimbun*, 20 June and 13 July 1946.
17 *Kagaku Bunka Shimbun*, 31 August 1946.
18 Putnam, David L., 'Reorganization of Science and Technology in Japan (1945 to September 1950)', in *History of the Non-military activities of the Occupation of Japan*, Kokkai Toshokan (National Diet Library), microfilm number CHS-04304.
19 *Kagaku Bunka Shimbun* 5 September 1946.
20 Hidetsugu Yagi, 'Gakujutsu Kaigi ni kansuru Hansei (Reflection on the JSC),' in *Kagaku Bunka Shimbun*, 1 January 1951.
21 Hajime Matsuura, 'Gakumon no kiki (Science in crisis),' in *Shizen*, vol. 7, no. 10, October 1952, p. 12.
22 Kakuzō Maeshiba, 'Gakusha no ryōshin ha ikitteita (The conscience of scholars was still alive),' in *Shizen* (Nature), vol. 8, no. 7, July 1953, p. 14.
23 Teiri Nakamura, *Kiki ni tatsu Kagaku* (Science in crisis), Kawade, 1976, p. 47.

Bibliographic Guide

Science Council of Japan

[1] Sources issued by JSC:
- *JSC nyūsu* (JSC News), 1957–1960, thereafter *Nihon Gakujutsu Kaigi Geppō* (JSC Monthly).
- Science Council of Japan, *Annual Report*.

[2] Sources preserved at the JSC Library:
- 'Nihon Gakujutsu kaigi sōkai shiryō (Sources of JSC general assembly).' Prepared by the Secretariat and distributed to the members before the General Assembly. Mimeograph.
- 'Nihon Gakujutsu Kaigi sōkai sokkiroku (Shorthand records of the JSC General Assembly),' handwritten copy. Includes the names of discussants.
- 'Nihon Gakujutsu Kaigi Un'ei Kaigi Shiryō (Papers of the JSC Steering Committee).' The Steering Committee handles routine matters of JSC.
- The JSC Library also contains material on section meetings, committee meetings, branch meetings and the report of symposia, reports of members sent to international conferences, lists of members and committee members, etc.

[3] *Shizen* magazine featured articles on the activities of JSC. In the early days the report on the General Assembly was prepared by left-wing or liberal scientists.

[4] 'Nihon Gakujutsu Kaigi no Seiritsu (Establishment of the JSC),' in *Kenkyū Shōhō*, vol. 2, no. 2, 1950, p. 17.
- 'Nihon Gakujutsu Kaigi no Dai1 Iinkai (The First Committee of the JSC),' in *Kenkyū Shōhō*, vol. 2, no. 3, 1950, p. 22.
- 'Nihon Gakujutsu Kaigi nyūsu, Monbushō kagaku gijutsu shinkō ni hitsuyō na keihi no haibun ni tsuite (News about JSC, on the allocation of the Education Ministry fund needed for the promotion of S&T),' in *Kenkyū Shōhō*, vol. 3, no. 2, 1951, p. 43.

[5] Nihon Gakujutsu Kaigi Nijūgonen-shi Henshū Iinkai (Science Council of Japan, Editorial Committee for the Twenty-five Year History) (ed.), *Nihon Gakujutsu Kaigi nijūgonen-shi* (The twenty-five year history of the JSC), Nihon Gakujutsu Kaigi (JSC),1974.
An orthodox official history of the JSC, including all reports from the first period to the ninth period.

[6] Report on the Activities of the Science Council of Japan since its Inauguration (Science Council of Japan, 16 May 1949).
The activity of the first four months of the JSC written in English in order to be internationally known. It is also a report addressed to GHQ and the US National Academy of Sciences for their approval.

[7] Masao Gorai, 'Nihon Gakujutsu Kaigi sōritsu no koro (On the founding of the JSC),' in *Nihon no Kagakusha* (Japanese scientists), vol. 5, no. 12, December 1970, pp. 34–5.
A left-wing Account.

[8] Masae Ōmori, 'Kagaku kenkyūhi seido no henka to Nihon Gakujutsu Kaigi (Changes to the science research fund system and the JSC),' in *Nihon no Kagakusha* (Japanese scientists), vol. 21, no. 8, August 1986, pp. 443–9.

[9] *Kagaku*, vol. 40, no. 9, September 1970, pp. 492–3.
A discussion of the Common-use laboratories that were established following the JSC's recommendations.

[10] Masao Kotani, 'Kokusai Gakujutsu Rengō Kaigi to wa nanika (What is ICSU?),' in *Kagaku* (Science), vol. 22, no. 6, pp. 47–50.
On the Function of ICSU.

STAC Sources

[11] 'Kagaku Gijutsu Gyōsei Hyōgikai ni tsuite (On STAC),' prefaced in December 1949 by Kikuichirō Yamaguchi, Vice-President of the STAC, Minister.
Pamphlet published by STAC (44 pgs.). Record of the ten months following the first meeting on 30 March 1949.
Table of Contents: 1. A short history of the creation of STAC; 2. Purpose and organization of STAC; 3. Procedures of the meetings; 4. Outline of the deliberations.
Appendices: 1. The STAC Law; 2. List of STAC members; 3. List of STAC cadres; 4. The JSC Law; 5. Address by the Prime Minister to the first meeting of STAC; 6. Address by Dr. Kelly of ESS/ST to the first meeting.

[12] Gakujutsu Taisei Sasshin Iinkai (Renewal Committee for Scientific Organizations), 'Shin gakujutsu taisei no ritsuan ni tsuite (On the planning of new scientific organizations),' Report to the Prime Minister, 8 April 1948.
In August 1947 the Renewal Committee was established. Subsequently the eight members held discussions for eight months before recommending the establishment of JSC and STAC.

[13] ESS (B)-11612 STAC, 1949 (1948.9–1950.10). Mostly on the JSC election result. Also includes material on STAC.

[14] ESS (B)-11611 11612, Box No. 7399 folder (g), Scientific And Technical Division, liaison group, Memo for File, 10 December 1949, Subject: Disposition of Research Budget (3 pgs.).
On 7 December 1949, Kennoki of the Education Ministry submitted a report on STAC to ESS/ST. ESS/ST, in turn, invited Yoshio Nishina (a member of STAC) and Kunio Chiaki, the Secretary of STAC (and formerly

a civil engineer for the National Railway) to a meeting on 10 December to discuss what STAC's mission and structure ought to be. ESS/ST members include B. Dees, E.C. Allan, and D.L. Putnam. They admitted that it was not progressing as quickly as initially planned but clearly expressed their intention to allocate research funds to other Ministries (other than the Education Ministry).

[15] Kōhei Suzue, 'Kagaku Gijutsu Gyōsei Kyōgikai o kaerimiten (Reflecting on STAC)', in *Gakujutsu geppō* (Scientific monthly), vol. 5, no. 12, 1952, pp.784–7.
The author was the vice-executive secretary of STAC. He assesses STAC from the viewpoint of administration rather than university scientists.

[16] Masao Maeda, 'Kagaku gijutsu gyōsei kikō no kakuritsu (Toward the establishment of S&T administration)', in *Nissankyō geppō* (Japan Industrial Association monthly), July 1952.
The author was a Liberal Party Member of the Diet. He set up the Special Committee for the Promotion of S&T within his Party. He was the head of Keizai Antei Iinkai (Committee for Economic Stability) of the Diet. Critical that the S&T Promotion Fund had not resumed its prewar levels and that STAC was merely a coordinating organization, he argued for the establishment of a more powerful Science and Technology Agency.

[17] Kōsaku Shinkai, 'Kenkyū kikan no kakujū kyōka (Expansion and empowerment of research organizations),' in *Nissankyō geppō* (Japan Industrial Association monthly), April 1949.
The author was the President of Tokyo Shibaura Denki (Toshiba). He argued for research for the recovery of production.

[18] Eizō Ishige, 'Kenkyūkikan no seiri (Clearance of research laboratories),' in *Nissankyō geppō* (Japan Industrial Association monthly), May 1949.
The author is Senmu Riji (Executive Secretary) of Nihon Taaru Kyōkai (Japan Tar Association). He claims that during the time of economic difficulty private laboratories should be curtailed; they could be revived in the future.

[19] Yoshitaka Wakabayashi, 'Minkan gakujutsu kikan no josei (Subsidy for private research organizations),' in *Nissankyō geppō* (Japan Industrial Association monthly), July 1951; 'Kagaku gijutsu kenkyū kikan no kakuritsu (The establishment of S&T research organization),' in *Nissankyō geppō* (Japan Industrial Association monthly), May 1952.
A Liberal Party Member of the Diet.

12 The International Exchange of Scientific Information

Shigeru Nakayama

Prior to World War II, overseas scientific information was conveyed to Japan by ship. It took at least two months before an article published in a Western journal was available to Japanese researchers. Sending a manuscript from Japan to a European journal required the same amount of time. As long as international journals were centered in the West and the criteria for evaluation was the Western standards, the Japanese were distinctly disadvantaged in the competition for scientific discovery.

During the war acquiring scientific information via student travel and overseas study was nearly impossible, so the Education Ministry established the Kagaku-ka (Science Section) – later expanded to Kagaku-kyoku Chōsa-ka (Research Section, Science Bureau) – chiefly for the purpose of quickly introducing German scientific literature.[1] Astronomical ephemeredes and other necessary literature were transmitted via submarine from Nazi Germany but generally Japan was isolated from the Western world with about a ten-year lag in information.

In the postwar period, availability of science and technology (S&T) information rapidly improved. Airmail – and later email – minimized the geographical handicap. Scientific activity in the postwar era was thus able to narrow the time lag and fill the information gap.

Resuming the Reception of Scientific Information After the War

The situation of the Japanese S&T community during the war was comparable to the sakoku (closing the country) during the Edo period. There was no exchange of scientists or engineers between the West and Japan. The most significant effect of this was the absence of any S&T information exchange. The other side of this was that the Allied Forces were also cut-off from any information about what was going on in Japan during the war. The scientific survey teams who came to Japan immediately after the war had to develop their own overview of the wartime activities of Japanese science (cf. ch. 1 and 2). The Japanese scientists, though, did not have any systematic way to fill the lacuna of information that had developed during their wartime seclusion. Subscriptions to foreign scientific journals were disrupted during the war, and in the postwar period it was difficult to resume these because of the economic situation. There was no access to the US dollars necessary to defray the costs and the yen was not convertible to foreign currency at that time.

Until May 1948 the Japanese were forbidden from sending any mail abroad. Hence, with no officially sanctioned routes for accessing overseas scientific information, such information was available only through personal acquaintances like Louis Alvarez at the University of California, Berkeley, who donated back issues of *Physical Review* and *Review of Modern Physics* to Ryōkichi Sagane at Tokyo Imperial University. The Japanese scientific community fervently desired to know what was going on outside Japan in each field of specialty.

Visits by Foreign Scientists

The Advisory Mission members and scientific advisors that General Headquarters, Supreme Commander for the Allied Powers (GHQ) invited to Japan (cf. ch. 8) were indispensable sources of information for hungry Japanese scientists. Whenever the members of a new mission arrived, welcome committees were organized and seminars, symposia and get-togethers were planned. A number of Japanese scientists tried to develop personal connections with mission members in the hope that some day they might be invited to the US.

The extent to which these personal relationships developed and their impact on the S&T community is impossible to ascertain from official documents. However, there can be little doubt that extraordinary outcomes often emerged from seemingly innocuous personal contacts between particularly innovative Japanese researchers and foreign scientists. For example, a simple gift such as an electrical appliance from an Occupation officer to a Japanese scientist might instigate a significant breakthrough in the Japanese adaptation of some foreign technology.[2]

CIE Library

On 26 November 1945, GHQ established a Civil Information and Education Section (CIE) library in Hibiya, Tokyo. A number of American scientific journals and their back issues were displayed, together with some fashion magazines.

This was before the invention of the photocopier, so Japanese researchers would visit the library day by day and hand-copy journal articles. Many journals were also donated to the Tokyo University Library. But researchers in the provincial towns still had no easy access to this literature until August 1947 when CIE opened twenty-one branch libraries 'to recover and strengthen democracy and supply foreign literature.'

These libraries adopted an open display system that was suitable for accessing scientific journals, which thereafter became the standard design for libraries. Around the same time, microfilm services were established.[3]

The US had developed library and information services far more advanced than the rest of the world. CIE intended to form a network of information in provincial areas to promote and facilitate US-style democracy. After the Occupation ended these libraries survived under the name of 'American Cultural Centers.'

In addition to the CIE libraries, the Rockefeller Foundation and the Tokyo Branch of the Asia Foundation contributed literature and economic assistance to Japanese libraries.[4]

Microfilm Services

In July 1946 Gakujutsu Kenkyū Kaigi (the Japanese National Research Council) established Gakujutsu Bunken Chōsa Kenkyū Tokubetsu Iinkai (the Special Committee for Surveying Scientific Literatures) to promote the exchange of international literature and microfilm documentation. The Committee was abolished in March 1948 and Gaikoku Bunken Riyō Iinkai (Committee for Utilizing Foreign Literatures) was created to replace it. This Committee, in turn, was abolished when the Science Council of Japan (JSC) absorbed the functions of the Japanese National Research Council. Under the leadership of Ryōkichi Sagane, who was appointed head of the micro-document subcommittee, the subcommittee contributed to the utilization and promotion of microfilm for those Japanese scientists hungry for foreign scientific literatures.[5]

Importing Scientific Instruments

Outdated prewar machinery was not adequate for the project of economic recovery. But during the Occupation, the Japanese yen was in principle not convertible to US dollars and the allocation of exchange currency was very limited. From early 1949, GHQ's Economic and Scientific Section, Scientific and Technical Division (ESS/ST) began to defray the cost of importing the machinery necessary for the revitalization of the Japanese economy through the Economic Rehabilitation of Occupied Areas (EROA) Fund. This subsidization was later continued through the GARIOA (Government Appropriations for Relief in Occupied Areas) Fund (cf. ch. 2–6).

In fiscal year 1948–49, the EROA fund allocated 85,000 US dollars for purchasing S&T facilities for Japanese research laboratories. The Ministry of International Trade and Industry's (MITI) Agency of Industrial Science and Technology prepared a list of the items deemed most important for economic recovery. ESS/ST reviewed the purchasing program, approved (or not) the items on the list for EROA fund expenditure, and then sent it to Washington, D.C.

After the JSC was created in 1949, its associate organization, the Science and Technology Administration Council (STAC; cf. ch. 11) was responsible for deciding what sort of machinery should be imported and how to allocate the EROA fund. STAC established a special sub-committee for that purpose. Public sector scientific organizations collected information from universities and research laboratories, prepared a prioritized list of items required and submitted it to the sub-committee. As a general principle imported equipment was for common -use and the sub-committee determined which laboratory was the most appropriate for any particular piece of equipment before giving its approval.[6]

Characteristics of ESS/ST Science Policy: Importing Software

The Economic Stabilization Board (Keizai Antei Honbu, or Anpon) estimated the cost of the purchasing plan for technology importation for 1950 and submitted it to ESS/ST. ESS/ST then revised it, adding the cost of: 1. dues for participating in international organizations; 2. technological and medical literature, and; 3. international exchange of scientists and engineers. The different emphases in expenditure categories stem, perhaps, from the fact that Japanese planning officers were only interested in spending their limited funds on the importation of hardware (machinery) that could be listed on an official catalogue of acquisition, while ESS/ST emphasized the importance of software – i.e., information and the exchange of personnel. And ESS/ST was in a position to help obtain such software, including defraying the cost of foreign patents.

The Publication Board Report

During the final phase of World War II, US President Harry S. Truman ordered the creation of a US technology investigation team to collect documents from German wartime research. It was voluminous; for instance the documents of I.G. Farbenindustrie amounted to several hundred thousand pages. Moreover, under the auspices of the US Department of Commerce, the Publication Board was established to maximize the availability of S&T information that originated in the government-sponsored research of Germany, Japan and the Allied Forces by reviewing military secrecy classifications and declassifying documents for publication where deemed appropriate. Amongst the documents considered were the reports of the United States Strategic Bombing Survey of Japan prepared immediately after the war and the SCAP survey reports.[7] Reclassified documents were called 'PB Reports' and were numbered PB #'X' in the catalogue. Later, the Publication Board was renamed the 'Office of Technical Services' and the catalogue was renamed the *Bibliography of Scientific and Industrial Reports*, but they were still generally called PB Reports.

In July 1947 PB Reports were categorized and annotated. The resulting bibliography was then sent from the USA to the National Library at Ueno in the form of positive film.[8] This annotated bibliography amounted to a comprehensive survey of S&T carried out abroad during and after World War II. It was renamed the *Bibliography of Technical Reports* in July 1949.

Through this catalogue a Japanese private citizen became acquainted with the content of the PB Report, immediately flew to the US to obtain it and then made it available for Japanese production. Later, the scientific and business communities, as well as the government, became interested in it and it was often discussed in the media. The Diet accepted the need to purchase it for the Diet Library and, accordingly, the Finance Ministry allocated funds to purchase the entire PB Report.

After 1949 the PB Report in the Diet Library was made available for general use. Many researchers and laboratories were provided with copies of the microfilm from the Diet Library. They duplicated the relevant parts, distributed them to appropriate factories and laboratories and used it as a reference resource for replicating discoveries made overseas and for group research.

There were some criticisms of it for containing outdated wartime research data in a postwar nuclear age and because it was an unsorted mixture of good and poor work. Nevertheless, it provided tremendous benefits; by utilizing the PB Report the Japanese did not have to pay scarce and precious foreign exchange currency to visit the US in order to learn about the latest technology. Thus, it had an enormous impact on the Japanese S&T community.

A breakdown of the contents of the PB Report reveals that it was thirty percent chemistry, fourteen percent machinery technology, ten percent metallurgy, nine percent electric engineering, and nine percent weapons, etc.[9] Japanese interest primarily focused on organic chemistry with electronic engineering, dye chemistry and pharmaceuticals also prominent.[10]

The National Diet Library intended the PB Report to be widely known and utilized by provincial people. Thus, in 1953 they began to develop a network of PB centers in ten industrial towns nationwide. The microfilm preserved in the Diet Library in Tokyo was considered to be the primary edition. Reprints by contact printing on 35 mm microfilm were provided to the provincial centers as secondary editions and made available to the public. As the use of microfilm services became widespread, the development of domestically produced microfilm reading machines made extraordinary progress.[11]

Importing Books

The importation of Western S&T books was permitted from 1949. Initially this was limited to books in English; simply because after the war the European publishing industry was slow to recover – especially the Germans, who had been a primary source before the war – while the US industry remained intact. The regular importation of German books did not resume until around 1951.

Importing books was, like everything else, dependent upon the allocation of exchange currency by MITI. Book dealers had to submit an application for the allocation of exchange currency to MITI. MITI officials then assessed the application to determine whether the book was necessary for Japan's economic recovery. Therefore, science books had a higher priority than general literature. It took more than six months from the time a buyer placed an order with a book-dealer until the result of MITI's review of applications was published. If the application was approved, it took at least another six months to complete the various procedures necessary before the book could be delivered.[12]

In 1950 the demand for foreign books increased dramatically. Out of a total of fifty thousand books imported, eighty percent were in English (sixty percent American and twenty percent British). In the various S&T fields nearly ninety percent of books approved for import were American-made. These statistics explain the trend towards 'English (Language) Imperialism' in Japanese S&T. The dominance of German books before the war declined in the postwar era to a few percent of the total number of imported scientific books. That same year there were a total of 17,000 Japanese books exported, only one third the number that were imported. Of these, 6,000 books were sent to the US and 1,000 to Korea, the second largest market for Japanese books.[13]

UNESCO Coupons

Countries lacking in exchange currency generally prioritize the purchase of foodstuffs, crude materials and machinery, rather than cultural assets. To address this imbalance the UNESCO Coupon system was created so that domestic currency could be used to purchase foreign cultural assets.

In 1948 book coupons were experimentally introduced under the UNESCO Coupon system. By 1951 twenty countries participated in the book coupon program, thirteen participated in film coupons and ten in science coupons. Thus, an individual who did not have exchange currency was able to buy UNESCO Coupons with domestic currency and purchase whatever they chose. The international currency clearance was then conducted by UNESCO.[14]

Before the period of high economic growth in Japan, the allocation of exchange currency was strictly controlled and it was generally impossible to obtain exchange currency to freely purchase foreign books and machinery. To overcome this difficulty, participation in the UNESCO Coupon system was virtually mandatory for the revitalization of the scientific community. In 1952 the Minister of Foreign Affairs (with the approval of the Cabinet) submitted a request to UNESCO Headquarters. In response, UNESCO's Seventh General Assembly allocated 100,000 US dollars to Japan. In January 1953 Japan's Domestic UNESCO Committee was established as the chancery of UNESCO Coupons, while the Japan Society for the Promotion of Science (JSPS) lost its prewar official status, moved to the private sector and supported itself on the commissions derived from administering Coupons. Of the 100,000 US dollars allocated in the first year $50,000 was earmarked for books, $20,000 for film and $30,000 for scientific resources.[15] Five years later (in 1958) the fund was expanded dramatically: $75,000 was dedicated to books and $150,000 for scientific resources. Ninety percent of the scientific resources were ordered from the USA.[16]

Copyright and Pirating

Immediately after the war, when there was no way to obtain new or original Western books, scientists reviewed basic prewar literature. But these, too, were in short supply so, within their circles, they copied each other's collections and translated them into Japanese. The Science Bureau of the Education Ministry planned a systematic publication of translated scientific books in conjunction with private sector publishers. In January 1947, however, a directive from CIE prohibited the free translation and publication of British books within thirty-six years and other books within fifty-six years after the death of the authors.[17] Scientists, though, did not actually use such old books. Hence, it was necessary to obtain the permission of copyright holders. Some books were published with the agreement of the original authors – obtained with the assistance of GHQ – but these were often too expensive for postwar Japanese scientists to afford. Thus, in practice there were a number of pirated editions that had been either handwritten or typewritten and distributed among students and researchers. By the time the Occupation was over in 1952, such illegal publications were nearly eliminated. Still, even the legally translated books remained cheaper than the original editions until the 1960s.

The Boom in Russian Language Studies

During the Occupation scientific information from socialist countries was scarcely imported. Only after the Occupation ended – that is, after 1953 – did Russian scientific books become readily available.[18] The translations of Russian popular scientific books by Irin or Oparin, though, were available even during the Occupation and enjoyed readership among left-wing circles.

The evaluation of Russian science rose dramatically, though, with the Sputnik rocket launch in 1957. The study of the Russian language subsequently boomed amongst young researchers and Russian language S&T dictionaries were in great demand. According to a survey of science students at the College of General Education, Tokyo University, an overwhelming majority supported changing their second foreign language requirement (English was first) from German to Russian.

However, the US was severely shocked by the Sputnik launch and began to invest large sums for the systematic translation of Russian scientific literature into English.[19] Hence, Japanese scientists could soon read Soviet scientific literature in English – and sometimes in German (via East German translations). The boom in the study of Russian languages among scientists thus faded out. However a few universities that had created Russian language chairs at the time have maintained them ever since.

Publication of Japanese Scientific Research

The R&D conducted by Japanese scientists during WWII was thoroughly investigated by the US Forces in the first few months of the Occupation. This was a top priority for GHQ, contributing both to the demilitarization agenda and to preparations for reparations. Shortly thereafter, Harry Kelly from ESS/ST was asked to translate abstracts of Japanese wartime S&T journals into English by the editors of *Chemical Abstracts*, *Mathematical Reviews* and *Biological Abstracts*. Sixteen Japanese specialists were engaged in 1947 to abstract 4,700 articles.[20] During 1947–48, 7,438 articles were abstracted. In 1949 the journal *Japan Science Review*[21] was launched to make these abstracts available to overseas readers.[22] In 1950, the ESS/ST budget for abstracting articles was cut; the project was then continued by the Diet Library[23] with UNESCO support.

The Preparatory Committee for Editing the History of Science and Technology was formed as another initiative of ESS/ST, to edit a history of wartime Japanese scientific research. The texts were written in English by specialists in each field who were selected by the Committee. This plan originated as part of GHQ's efforts to gather Japanese military research, hence, the publication of a Japanese version was a secondary concern that was delayed or cancelled due to budget limitations.[24]

In 1947, *Journal of the Physical Society of Japan* and *Progress of Theoretical Physics* (edited by Hideki Yukawa) were launched. In 1948, *Mathematica Japonicae* and *Journal of the Mathematical Society of Japan* also appeared.

According to a 1952 survey of scientific journals, the journals published in Japan in European languages consisted of: seven basic science journals (constituting seventeen percent of the total of basic science journals published in Japan, including those in Japanese language), four technology journals (six percent of total technology journals), one medical and one economics journal. It means that Japanese scientific activity was still not well known internationally. These were all published by national scientific societies. The research institutes affiliated with national universities also published considerable numbers of their own institute bulletins in European languages, including thirteen in science (fifty-four percent of total institution publications, the rest being in Japanese), and fourteen in technology (forty-two percent).[25]

Another feature of Japanese scientific publishing is the widely accepted – although unverifiable – opinion that Japanese scientists started to distribute 'preprints' of their findings (and therefore pre-peer review) to a limited circle of researchers worldwide for the claim of priority of discovery. This practice became popular in certain rapidly progressing scientific fields. This trend reputedly began in prewar Japan in particle physics and continued during the war in order to overcome the Japanese geographical handicap of isolation from the Western centers of learning. The practice developed worldwide in the postwar world with the advent of airmail and photocopiers.

Conclusion

A ten-year lag in scientific information that was created by the war between the West and Japan was nearly filled within another ten years through the efforts and various channels discussed above. More importantly, through the development of technological means since then – means such as airmail and airline communications, microfilms and duplicating devices – the information gap was much less pronounced than was experienced before WWII, thus providing the possibility of the age of internationalized scientific communication that followed from the late 1950s.

Notes

1 Usaku Kakinuma, 'Gakujutsu Jōhōshitu no ninmu to shimei (The work and mission of the Scientific Information Branch),' in *Gakujutsu geppō* (Science monthly), vol. 5, no. 6, September 1952, p. 2.

2 Personal communication with Bowen Dees.
3 Nakamura, 1951 [10] p. 36.
4 Nihon Gakujutsu Kaigi Dokyumentēshon Kenkyū Renraku Iinkai (Documentation Section, JSC) (ed.), 1959 [12], pp. 2–3.
5 Nihon Gakujutsu Kaigi Dokyumentēshon Kenkyū Renraku Iinkai (Documentation Section, JSC) (ed.), 1959 [12] p. 4.
6 ESS (B) 11611-1, Box No. 73299, Folder (q), Economic and Scientific Section, Memo for File, 15 August 1949.
7 'PB repōto ni tsuite (On PB Report),' 1953 [5] pp. 510–11.
8 *Numerical Index to the Bibliography of Scientific and Industrial Reports*, vol. 1–10, 1946–48, Special Library Association, N.Y., 1949, preface.
9 Ōshima, 1952 [6] pp. 8–9.
10 Nishimura, 1956 [8] pp. 8–10.
11 Nishimura, 1952 [9] p. 17.
12 *Gakutō*, vol. 48, no. 3, 1951, p. 35.
13 F. H. Sei, 'Yōsho yunyū no kaiko (Recollections on the ways in which we used to import Western books),' in *Gakutō*, vol. 48, no. 3, March 1951, p. 45.
14 *Kagaku Bunka Shimbun* (Science culture newspaper), 16 February 1951.
15 *Gakujutsu geppō* (Science monthly), vol. 5, no. 8, November 1952, p. 58.
16 *Gakujutsu geppō* (Science monthly), vol. 11, no. 11, March 1959, pp. 36–7.
17 'Gaikoku Tosho no Shuppan ni tsuite (On the publication of foreign books),' in *Nihon butsurigaku kaishi* (Journal of the Japan Physics Association), vol. 1, 1947.
18 Jun'ichi Nagamura, 'Kagakugijutsusha notameno Roshiago (Russian for scientists and engineers)', in *Gakkensha*, 1968.
19 *Nauka no Mado* (Journal of Nauka Publishers) [4] vol. 7, 1957.
20 *Kagaku Bunka Shimbun* (Science culture newspaper), 17 December 1947.
21 *Gakujutsu geppō* (Science monthly), vol. 8, no. 12, March 1956, p. 790. The first printing was scheduled for March 1949 but it did not appear until December 1949.
22 Genkichi Hara, 'Gakujutsu bunken no katsuyō (Utilization of scientific literature)' in *Gakujutsu geppō* (Science monthly), vol. 7, no. 4, July 1954, p. 24; cf. [11]
23 Testimony of Bowen Dees.
24 *Kagaku Bunka Shimbun* (Science culture newspaper), 25 December 1947. This work is a collection of records of wartime activity in different fields written by practicing scientists rather than historians.
25 *Gakujutsu geppō* (Science monthly), vol. 6, 1953, p. 335.

Bibliographic Guide

[1] Takahisa Sawamoto, 'Kagakugijutsusha no kenkyū jōhō chōsa (Survey of

research information for scientists and engineers),' in *Library Science,* vol. 5, July 1967, pp. 137–52.

[2] Tomoo Odaka, 'Yunesuko kūpon seido ni tsuite (On the UNESCO Coupon system),' in *Gakujutsu geppō* (Science monthly), vol. 3, no. 6, 1950, p. 1.
The Education Ministry was in charge of the UNESCO Coupon system, so the miscellaneous section of *Gakujutsu geppō* (a magazine affiliated with the Education Ministry) often reported on it. Imported machinery was also listed in this journal.

[3] ESS (A) 09889.
ESS/ST was responsible for assessing the requests of international organizations (through UNESCO) for strategic scientific information, e.g., Japanese geological data.

[4] *Nauka no mado.*
A magazine published by Nauka Shobō, the publisher that imported and distributed books from the Soviet Union. Reports on S&T information from socialist countries.

On PB Reports

[5] 'PB repōto ni tsuite (On PB Reports),' in *Gakujutsu geppō* (Science monthly), vol. 6, 1953, pp. 510–11, 572–4.
Includes a synopsis of literature on the PB Reports on p. 574.

[6] Jinpei Ōshima, 'PB repōto,' in *Dokusho shunjū* (Reading chronicle), vol. 3, no. 11, 1952, pp. 8–9.
The author is a staff member at the Diet Library. A tabulation of the frequency of utilization of PB Reports for each specialty is attached.

[7] Tokujirō Kanamori, 'PB Repōto no katsuyō o nozomu (Hoped-for utilization of PB Reports),' in *Dokusho shunjū* (Reading chronicle), vol. 4, no. 10, 1953, p. 1.
The author is the Director of the Diet Library. Publicity for the PB Reports acquired by the Diet Library.

[8] Kō Nishimura, 'PB Repōto to sono riyō (PB Report and its utilization),' in *Dokusho shunjū* (Reading chronicle), vol. 7, no. 5, 1956, pp. 8–10.
A table of provincial PB Report centers is appended.

[9] Kō Nishimura, 'Kansai chiku PB sentaa no zōkyō (Strengthening PB Centers in the Kansai area),' in *Dokusho shunjū* (Reading chronicle), vol. 3, no. 2, 1952, p. 17.

On CIE Libraries

[10] Hatsuo Nakamura, 'Shinjuku CIE Toshokan (CIE Library at Shinjuku),' in *Dokusho shunjū* (Reading chronicle), vol. 8, no. 2, 1951, pp.36–8.
Overseas Dispatch Of Japanese Scientific Research Information.

On Japanese Scientific Research

[11] *Japan Science Review*, *Biological Science*, *Medical Science*, etc.
Edited and published by the Education Ministry and the Science Council of Japan. Listed below are the volumes I could find.
- *Biological Sciences*, no. 1, 1949–50, Tokyo. Nihon Nōgakkai, Nihon Dōbutsu Gakkai, Nihon Iden Gakkai, Nihon Shokubutsu Gakkai in charge.
- *Medical Science*, vol. 2, 1953, edited by the Science Council of Japan and the Ministry of Education.
- *Engineering*, vol. 4, Nihon Kōgakkai.

[12] Nihon Gakujutsu Kaigi Dokyumentēshon Kenkyū Renraku Iinkai (JSC Documentation Section) (ed.), *Nihon ni okeru kagaku bunken no riyō oyobi saabisu no genjō* (The present situation of the use and service of science publications in Japan), April 1959, 25 pages.
A Japanese translation of a report submitted in English to the International Conference on Science and Information held in Washington, D.C., November 1958.

13 Sending Scientists Overseas

Shigeru Nakayama

From Germany to the USA

Before World War II (WWII) most Japanese students traveling abroad to study on government funds went to Germany. There was a political change in Japan in 1881, when the Government turned towards Germany as its model of modern civilization. Students began to travel to Germany rather than England, France or the US to study. For Japanese science students, at least, the choice was academic rather than political; German science was the most advanced and flourishing in the late nineteenth century.

In the 1920s the world center of science and technology moved from Germany – suffering from the reparations of WWI – to the booming US economy. Japan was rather late to recognize that the center of the world of science had moved; the tradition of visiting Germany continued from generation to generation until the end of WWII. This preference for Germany was particularly strong in medical schools, for example, which remained German-oriented for a decade after the end of WWII, and continued to use German technical terms until the late 1950s.

But it wasn't just the world of science that recentered; the center of higher education itself moved from Europe to the US during this period. Before the war colonial students studied in their suzerain state, mainly because of language familiarity. But during the reconstruction of war-torn Europe foreign students from around the world increasingly studied in the US. According to the Institute of International Education (IIE; an American institute providing assistance to foreign students) the number of foreign students jumped from less than 10,000 before WWII, to more than 26,700 in 1948–49. There were 75 Japanese students in the US in 1948–49 and 265 in 1949–50; very few in comparison to other Asian countries.[1]

During the Occupation, it was impossible for a Japanese to go abroad except under very special circumstances. The Japanese were too poor to afford travel expenses. Even after the Occupation, it was virtually impossible to change Japanese yen to exchange currency (US dollars). The penalty for purchasing exchange currency on the black market was hard labor on Okinawa where the US Occupation still continued.

For scientists and engineers, the end of the war meant the second Kaikoku (opening the country from wartime seclusion; the first Kaikoku was the Perry incident in the mid-nineteenth century). It was their professional instinct to eagerly seek current S&T information from overseas and if possible to confirm the information by visiting abroad.

In the early period of the Occupation, the Japanese could not obtain a passport and visa unless they had a relative abroad who would provide an affidavit of support. It was 1949 before scientists were officially allowed to travel overseas again. Beginning in 1950, a number of programs were created to send them overseas, but there were strict selection procedures and many applicants. Most of them went to the US. Thus across two World Wars and the devastation of Europe – and especially Germany – both through the physical damage of the war but also the economic damage of the reparations and the political influences of the Occupations, the center of world science moved from Germany to the US, and the US model of scientific research became the dominant paradigm.

Overseas Travel through Personal Invitations

The first postwar invitation from an international learned society arrived in Japan on 22 June 1947. It was an invitation from the United Nations' World Health Organization (WHO) to the Japanese Government to attend a conference on the 'international classification of the cause of death.' Under the economic and political circumstances it was virtually impossible to send a delegate. Hence Japan's only participation was sending papers.

In 1948 Hitoshi Kihara, a world-renowned geneticist, received an invitation from the Eighth International Conference of Genetics. He had no funds for travel expenses until the Rockefeller Foundation sent him 1,000 US dollars. This enabled him to depart for overseas on 2 July[2] and made him the first Japanese scientist to travel overseas after WWII. He was the Professor of genetics at Kyoto Imperial University. After attending the Genetics Conference, he traveled around with further sponsorship from UNESCO, returning home on 19 September.

Hideki Yukawa was engaged in elementary particle physics research and was the Japanese scientist whom the international scientific community was most eager to meet. Western scientists had been either war-damaged or mobilized into wartime research. Wartime seclusion had affected all of the parties in the war and the Western scientists were ill informed about what Yukawa and his group had been doing both before and during the war. In the physics community, particularly the elementary particle physics community, there was great demand to meet Yukawa and discuss subjects of common interest.

Similarly, Western mathematicians were ill informed about the activities of Japanese mathematicians. They were afraid that isolation could result in the duplication of work and keenly felt that communicating with Japanese mathematicians was a matter of priority.[3] Ernest O. Lawrence, the director of Lawrence Berkeley Laboratory of UC Berkeley initiated the invitation of Yukawa.[4] Lawrence's intention was realized after Robert Oppenheimer, the director of the Princeton Institute of Advanced Studies, sent a letter to Douglas MacArthur, the Supreme Commander of the Allied Powers (SCAP), inviting Yukawa to go to Princeton as a visiting professor.[5] Although it was generally impossible for Japanese to travel to the US, Yukawa was able to travel on 2 September 1948 with special consideration from SCAP's General Headquarters (GHQ).

At the same time Shizuo Sumitani was invited back to Princeton. Sumitani was an Assistant Professor of Osaka University and an ex-member of the Princeton Institute who had returned from the US to Japan in a wartime exchange boat. Yukawa and Sumitani both spent a year at Princeton on a contract basis. Before his departure for the US, Yukawa cleared the security investigation of the US Atomic Energy Commission on the grounds that he had never been engaged in military or nuclear research.[6] He extended his stay for another year and then, during the summer of 1950, returned home to consult with Japanese university authorities about whether he should stay permanently in the US or eventually return home. In those days, many Japanese scientists who stayed in the US were concerned about whether or not they were contributing to a 'brain-drain' at home.

In more practical engineering fields, Shūichi Tashiro, the head of the geology section of Mitsui Mines Co., participated in the Eighteenth International Geological Conference held from 25 August to 1 September 1949, accompanied by Dr. H. G. Schenk of GHQ's Natural Resource Section (NRS). After the conference he conducted studies of the safety technology in US coal mines before returning home. Special permission for this research was granted on the grounds that coal-mining technology was particularly important to the priority production policy being pursued by the Japanese with GHQ's assistance.[7]

Following the precedents set by Kihara and Yukawa, a number of Japanese scientists wished to visit abroad. There were a few scientists who were invited by American universities and research institutes on individual contracts. However, apart from exceptionally well-known scholars such as Yukawa, most Japanese scientists tried hard to find visiting or more permanent positions by writing directly to American scientists in their field. The invitation of Japanese scholars to lecture in Japanese and Oriental studies had begun before WWII and soon resumed in the postwar era. According to an Occupation Forces' evaluation, Japanese scholars' ability to speak English had declined, perhaps because of the wartime seclusion.[8] Because their subject matter is international, scientists and engineers were less handicapped than scholars in the humanities and social sciences, who had practically no chance to visit abroad.

ESS/ST helped scientists to get permission and visas to travel abroad on an individual basis but could not allocate exchange currency to them. An individual American scientist at the university or research institute they intended to visit had to guarantee their travel and living expenses. Conditions varied greatly but no records of these conditions were compiled.

In June 1948, Harry C. Kelly of ESS/ST requested the Ministry of Foreign Affairs to provide a list of Japanese scientists visiting abroad. The list that was compiled was not exhaustive, but it records 897 scientists studying abroad from 1 June 1948 to the end of 1951, of whom 807 went to the US.[9] However, even though an individual Japanese scientist had been invited by a US University or an International Conference, the institution usually could not cover all of his expenses; so from 1950 GHQ managed to help scientists purchase sufficient exchange currency to defray the cheapest travel expenses – a Japanese freighter – for studying abroad or attending international conferences.[10]

Official Programs for Studying Abroad

From the beginning of the Occupation, Dr. G. W. Fox of ESS/ST had proposed to invite Japanese scientists to the US for a three-month period to foster contact with the US scientific community. He also proposed to start an exchange program for students with a fund of ten million US dollars that had been appropriated in reparations.[11] This proposal was not implemented until the later phase of the Occupation.

The US Exchange Education Mission arrived in Japan at the invitation of GHQ in August 1948, conducted investigations for a month and then advised that an exchange program be established for students and professorial researchers between the US and Japan. This was intended to create a route to encourage fellowships for students and short-term appointments for researchers to work at US universities. They also advised that civil servants, including most university faculty and researchers, should be able to study abroad with their salary fully paid and suggested that Japanese leaders in their fields should be sent to the US to survey their field.[12] This final suggestion was implemented steadily as discussed in the following section.[13]

National Leaders Project

GHQ started the National Leaders Project in 1949 with the Government Appropriations for Relief in Occupied Areas Fund (GARIOA). The program was created to select one or more leaders in each specialty field, invite them to the United States and let them study the installations and equipment in their field for several weeks. There was opposition to the program within the Far Eastern Commission, with representatives from the Soviet Union arguing that the invitations should not be limited to the US, but GHQ ignored them.

In the first year, from the end of 1949 to the end of 1950, one hundred and sixty projects were organized with Japanese representatives sent to the US. The projects included various S&T installations. The fields chosen for participation were primarily socially significant areas of research rather than those of 'purely' academic interest. The participants traveled to the US on board US Army vessels.

The Science Council of Japan (JSC) was one of the participants in this project. Naoto Kameyama, Sakae Wagatsuma and Yoshio Nishina were selected from its members. They visited the National Academy of Sciences, major research institutions and universities for forty-five days. Another representative visited research institutes, universities and utilization offices for ninety days as part of a project to train specialists in the development and utilization of forestry research.[14] In the same program representatives of the Ministry of International Trade and Industry (MITI) also visited the US.[15]

Student representatives were sponsored at the various sites by the Institute of International Education while the Conference Board of Associated Research Councils sponsored scholars and researchers. It was intended that both the strong points and the weak points of the US science system would be demonstrated. Nevertheless, the US was the only large and advanced nation that had survived WWII virtually undamaged; thus all of the Japanese who visited were deeply impressed by the mighty power that was so far removed from the reality of postwar Japan.

The program continued in 1951. At a time when there was no other opportunity to see and study overseas industries, the program provided an opportunity for Japanese leaders to observe the leading-edge installations of the US and provided them with concrete targets to strive for.[16] They continued to travel on US army vessels and aircraft. Their budgets remained extremely limited; they were advised to eat at cafeterias and stay at the least expensive hotels like the YMCA. Nevertheless, the curiosity and goal-oriented focus of the Japanese who had the opportunity to look at the US for the first time enabled them to overcome their financial difficulties.

GARIOA-Fulbright Study Abroad Program

The GARIOA Fund lasted three years, from 1949 to the end of the Occupation. One thousand Japanese were sent to the US on this program with full stipends for a year. Thus it was the largest study abroad program of its time in Japan. The Japanese students ranged in age up to thirty-six years, in consideration of providing the opportunity to the generation that had been brought up in wartime seclusion and had – since 1937 – missed the chance to study abroad. The selection process was very severe with 6,500 applicants for 300 positions in the first year.[17]

The Fulbright Act came into effect in the US in 1946 with a fund of 140 million US dollars obtained from selling obsolete military equipment. Its aim was to provide funding to enable the exchange of professors, scholars and students between the US and other nations for the next twenty years. More than ten countries were participating in this program before it was available to the Japanese. Thus, in April 1951 a motion was unanimously passed at the Tenth General Assembly of the Science Council of Japan (JSC) for the President of the JSC to push the Japanese Government to actively petition the US Government to include Japan in the Fulbright Program.[18] They were successful in their petition and in its first ten years 2,072

Figure 13-1: Changes in the numbers of scholarship students under the GARIOA and Fulbright Programs (1949–1987)

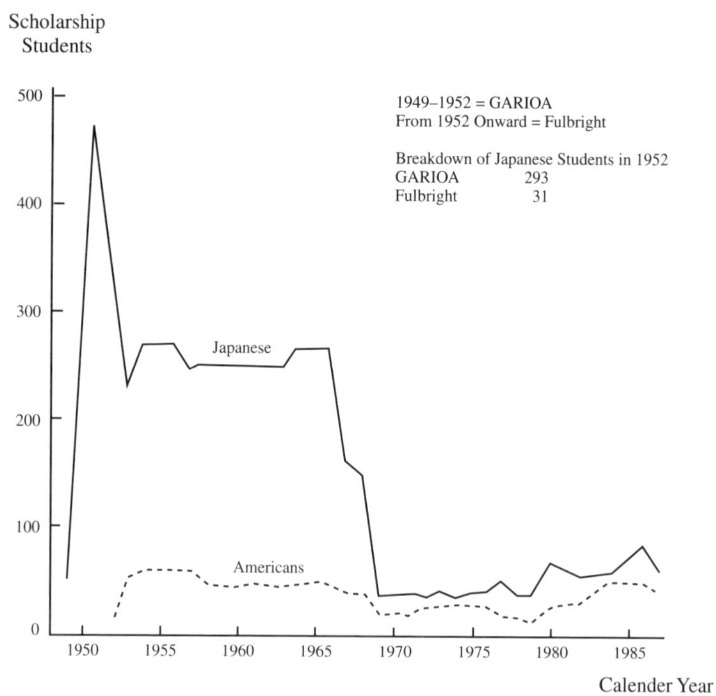

Source: Shigeru Nakayama, *Amerika diagaku e no tabi* (trips to universities in the United States), Recruit, 1988, p.223.

Japanese went to the US.[19] Subsequently, the program continued to send more than 200 Japanese each year until 1969 when, following the budgetary cuts brought about by the Vietnam War, the budget of the Fulbright Program was halved and the program for natural science ceased altogether.[20]

Through this program Japanese students were freed from strict isolation. Japan was ranked twenty-second in the international order of the number of students traveling to the US with this program in the first year of the GARIOA Fund (1949–50), jumping to tenth place in the following year (1950–51). The presence of Japanese students then began to be recognized amongst the academic community abroad.[21]

The selection criteria of the Fulbright Committee were not strictly academic. Their preference was for those who occupied a solid social position in their society and could benefit their workplaces through what he or she learned from studying in the USA.

In the beginning, the Committee selected Japanese who worked for the public benefit and sent them to the US. In the first year, 142 were selected from 6,491 applicants. Amongst the prerequisites were English proficiency and an age limit of thirty-six years. Most of the Japanese selected were humanities majors, with few natural scientists;[22] this is partly explainable by the fact that the period of studying abroad was limited to one year and it was difficult for natural scientists to complete their research in that period of time.

Fulbrighters were divided into two categories, those who were fully supported by the Committee and those who were provided only with travel expenses and whose living expenses were covered by the universities and institutions where they visited and worked. Most scientists belonged to the latter category, perhaps because their subject matter was in more demand internationally than humanities scholars and thus they had more opportunities for finding positions abroad.

STAC's Project for Sending Scientists Overseas

On 1 January 1950, the first Japanese initiated project for overseas travel was created. The Scientific and Technical Administration Committee (STAC; cf. ch. 11) initiated a project to send Japanese scientists and engineers overseas with the aim of bringing the relatively inferior Japanese S&T sector up to a world-class standard, thereby contributing to an economic recovery. Out of the Economic Stabilization Office's non-trade exchange currency budget 250,000 US dollars was allocated to this project. With this budget, STAC's Special Committee selected thirty-five S&T personnel in the first quarter (January to March) of 1950 and followed with further quarterly selections.[23]

The participants in this project were primarily scientists and engineers from the private sector. They were allocated exchange currency and paid their own way. In the first three quarters of 1950, 220 scientists traveled overseas through this project, of whom 39 were from the public sector and 181 were from the private sector. The largest field was applied chemistry with fifty-eight participants, followed by forty mechanical engineers; most of whom were in areas of applied technology.[24] Most of them visited the US for three months.[25]

ESS/ST also received a budget for the importation of technology and allocated it primarily to sending Japanese scientists to the US (cf. ch. 12).[26]

Allocation of Exchange Currency for Participation in International Conferences

The currency control policy of the Occupation period involved a severe examination process for the allocation of exchange currency. Hence, this policy was crucial to the Japanese S&T sector; it could determine the direction of the development of Japanese S&T.

The exchange currency for sending researchers abroad steadily increased. In 1951 it amounted to 40 million yen (with 23 million distributed by STAC and the Ministry of Education and 17 million by the JSC) and in 1952 it was 46 million yen (11 million for STAC, 18 million for the Ministry of Education and 17 million for the JSC).[27] As the Table below indicates, more than half of all invitations to international conferences were declined because of the limitation of exchange currency but the situation gradually improved.

Table 13-1: Participation in international conferences

Year	No. of conferences invited to	No. participated in	No. of participants
1950	70	24	28
1951	62	26	33

Source: *Kagaku Bunka Shimbun*, 25 November 1952.

Political Purge

Those applying to visit the US had to visit the US embassy prior to departure to swear a statement that they were not communists and to have their fingerprints taken. A medical examination was also required; those with symptoms of tuberculosis, for example, were excluded.

As the cold war intensified and the McCarthy purge of communists reached deeper into the US intellectual community, those affiliated to the Japanese Communist Party or Minka (Association of Democratic

Scientists; cf. ch. 25) were denied visas. For instance, Hajime Matsuura, Professor of Hokkaidō University was recommended by the biology liaison committee of JSC to be one of three representatives for an international conference. He wanted to visit the US on his way back to Japan but his visa was denied on the claim that two representatives were sufficient.

Japanese feelings about communism were somewhat different, though. According to GHQ documents, for example, when Walter C. Eells of GHQ's Civil Information and Education Section (CIE) gave an anti-communist speech at Hokkaidō University, the students protested against him. This became known as the Eells incident. Eells objected to issuing a visa to Matsuura since, at that time, Matsuura did not maintain a neutral position. GHQ was, however, cautious to ensure that this incident did not create tension with the Japanese Government and JSC or disturb the general situation of Japanese scientists traveling overseas.[28]

Approaches Toward International Scientists

Prewar Japanese students were, especially in the field of science and technology, mostly affiliated with national universities and publicly supported. They were thus public servants representing the Japanese nation and promised posts in the public sector after their return home. The period of their overseas life was therefore limited and they did not develop close contacts with the Western scientific community, always preparing instead for their life after returning to Japan.

In contrast, Japanese scientists visiting overseas in the postwar era, especially university affiliated researchers, were much more like intellectual immigrants, without the financial support of the Japanese Empire, but rather invited by foreign universities and research institutes as competent specialists. Many of them developed the behaviors and norms of international scientists, and were evaluated 'purely' on their competence and research achievements. They became part of the international community of scientists who speak English as a *lingua franca*. In particular, pure mathematicians and physicists were heavily involved.

Mainstream Japanese scientists were, however, more concerned about their position and social status within the Japanese scientific community than were fellow Asian scientists. The Japanese scientific community thus remained a relatively self-sufficient society, closed from the international community of scientists that was forming around a US center in the postwar era.

Notes

1 Institute of International Education, 1950-51 [6]

2 *Kagaku Bunka Shimbun*, 5 June 1948.
3 The Occupation documents include records about Tsurusaburō Takasu, mathematics professor of Tōhoku University.
4 The Lawrence Papers preserved at Bancroft Library, University of California, Berkeley, include documents expressing a desire to invite Yukawa.
5 GHQ/SCAP Records, ESS(A)09891.
6 GHQ/SCAP Records, ESS(A)09846.
7 *Kagaku Bunka Shimbun*, 20 October 1948.
8 *Report of the Education Exchange Survey to the Supreme Commander for the Allied Powers* [1]
9 CIE, *Postwar Developments in Japanese Education*, vol. 1, 1952, p. 377.
10 Dees, 1950 [10]
11 G. W. Fox, 'Japanese Reparations to the United States' to Chief, ESS, 9 April 1946. GHQ/SCAP Records, ESS (B) 11723-34.
12 *Asahi Shimbun*, 25 October 1949.
13 Dees, 1950 [10]
14 GHQ/SCAP Records, ESS (F) 06346.
15 *Nippon Times*, 5 April 1950.
16 Taizō Yakushiji, 'Sengo kōgyō gijutsu-shi no shōgen (Testimonies on the history of postwar industrial technology),' in *Kōgyō gijutsu* (Industrial Technology), July 1986, p. 35.
17 CIE, *Postwar Developments in Japanese Education*, vol. 1, 1952, p. 380.
18 Hajime Matsuura, 'Nihon Gakujutsu Kaigi Dai10kai Sōkai kara (From the Tenth General Assembly of the JSC),' in *Shizen* (Nature), vol. 6, no. 7, July 1951, p. 14.
19 'Tainichi Furuburaito keikaku (Fulbright Program for Japan),' in *Gakujutsu geppō* (Science monthly), vol. 14, no. 5, August 1961, pp. 268-9.
20 Nichibei Kyōiku Iinkai (Japan-US Education Committee), Chronological Table of the GARIOA-Fulbright Program.
21 Institute of International Education, 1950-1951 [6] p. 11.
22 Tomio Ogata, *Kagaku* (Science), vol. 20, no. 3, March 1950.
23 *Kagaku Bunka Shimbun* (Science culture newspaper), 10 March 1950.
24 *Gakujutsu geppō* (Science monthly), vol. 3, no. 3, June 1950, p. 31.
25 *Kagaku Bunka Shimbun* (Science culture newspaper), 30 May 1950.
26 Dees, 1950 [10]
27 *Kagaku Bunka Shimbun*, 1 January 1953.
28 'Travel abroad special,' ESS(B)11645.

Bibliographic Guide

During the Occupation, GHQ controlled all Japanese overseas travel. Hence GHQ documents are the primary source of information on the subject. The general policy of sending students abroad was the responsibility of CIE, while ESS/ST took care of individual scientists. The project of sending engineers to the US began in 1949 with the NRS responsible for the mining industry and related research and everything else under the jurisdiction of ESS/IND (Industry Division).

[1] *Report of the Education Exchange Survey to the Supreme Commander for the Allied Powers* in J. C. Trainor Collection at Hoover Institution, Stanford University.
J. C. Trainor was a deputy director of CIE. A copy is preserved in the Japanese National Institute of Educational Research, Roll 18.
Since the arrival of the First Education Mission, the exchange of personnel was recognized as being important. Their advice was implemented to some extent. At one stage, SCAP wanted to assess the educational exchanges completed to date and requested a survey from the Department of the Army. The Dept. recommended the survey proceed, and a team of US educationalists arrived in Japan in August 1949. Their report suggested that visiting Japanese scientists be shown how short the distance was between US university research and industrial research.

[2] ESS/ST, travel abroad by Japanese.
All records of Japanese scientists traveling abroad were maintained by ESS/ST beginning from 1948. The earliest one is ESS (A) 116433-5 'Travel abroad, special, April 1948 – May 1951' which documents the cases of Hideki Yukawa, Hitoshi Kihara, Shin'ichirō Tomonaga, Kunihiko Kodaira, Yūsuke Hagihara and others. Later and more general cases are recorded in ESS (A) 09843-7 'Travel abroad general, April 1947–May 1951.'

[3] ESS/ST (A) 09846.
Kelly's request that records be kept on all Japanese scientists traveling abroad, dated 25 June 1948. The Department of the Army allocated a budget of 300,000 US dollars to invite US scientists to Japan and send Japanese scientists and students to the US. Funds for bringing US scientists to Japan were distributed as follows: 170,000 to CIE, 60,000 to NRS, 30,000 to ESS, etc. Sixty-five thousand dollars were allocated to sending Japanese research specialists to the US and 73,300 dollars were allocated for sending students. The money arrived in Japan late, so they only started to send students in the second semester of the following year ('Procedures and Requirements for Japanese Traveling to the U.S. as Temporary Visitors').

[4] ESS (A) 09847.
ESS/ST started an 'Import of Technology Program' in 1949 and listed those Japanese sent to the US in 1950 and 51.

[5] ESS (E) 06345.
National Leaders Program, September 1949. Lists the participants and the sites visited in this program.

[6] Institute of International Education, *Education for One World, Annual Census of Foreign Students in the United States*.
IIE (Institute of International Education) is a non-profit organization based in New York to provide aid to foreign students visiting the USA. With the cooperation of the Committee on Friendly Relations Among Foreign Students, they sent questionnaires to more than 2,500 higher education institutions in the US and received responses from 1,210 organizations (surveyed in 1949–1950). Since then, statistics have been published every year based on surveys of foreign student advisors and secretariats, and including only the students regularly registered. Since 1954 the title has been *Open Doors*.

[7] Kaigai Haken Seido Kenkyūkai (Study Group of the Overseas Dispatch System) (ed.), *Kaigai haken, shisatsu sōran* (Survey of overseas studies and inspections), 1972.
Edited for civil servants. Revised every year. Covers all of the Japanese Government programs and funds for overseas study.

[8] Hajime Saitō, *Furuburaito ikkisei* (First year Fulbright students), 1984.
Selecting several students, it recorded the influence of their US experience on later life.

[9] Ministry of Education, Science and Library Bureau, UNESCO International Department (Section of Students Abroad), *Wagakuni no ryūgakuseiseido no gaikan: Ukeire oyobi haken* (An Outline of Japanese students exchanges: Receiving and sending), 1977.

[10] B. C. Dees, 'Checklist: Travel of Japanese Nationals Abroad; Japanese Traveling between Japan and the U S', 1950.

14 The Scientific Community Post-Defeat

Shigeru Nakayama

Before GHQ's Arrival

Immediately after they heard the Imperial Edict of surrender many scientists set about burning secret military documents. On the previous day, 14 August 1945, the Deputy Minister of Army issued a directive to eliminate official documents. Military related documents in particular were to be burned before the incoming Occupation Forces could confiscate them and use them as evidence of Japanese war crimes. On 15 August, the smoke of burning documents issued forth from every research institute and many universities.

Ordinance resources were released even before the surrender. The Osaka Military Arsenal started to transfer their essential resources to private sector factories on 9 August when they realized that the day of surrender was approaching. The Kantarō Suzuki Cabinet believed that when the Occupation Forces came, they would confiscate everything; hence before then all resources had to be covered up and dispersed. Covered up resources amounted to one third of the Army's total resources and 100 billion yen worth (in current prices) of Navy resources. These resources have been estimated to be the equivalent of four years supply of the peacetime economy or one year of wartime military production. GHQ (General Headquarters, Supreme Commander for the Allied Powers (SCAP)) tried to prevent this hoarding and ordered the Japanese government to stop it. The Naruhiko Higashikuni Cabinet deliberately delayed taking action for eight days, but on 28 August, the Cabinet called back dispersed military resources.[1]

Until the end of the war, the science sections of universities were engaged in mobilization research, and laboratories had been evacuated to the countryside in spite of their staff's continuing teaching commitments. The period between the surrender and the Occupation Forces' arrival was very confusing; rumors were rife that the Occupation troops would rob, rape and kill Japanese people. So a lot of people, including teachers and students, stayed home to wait-and-see what would happen. The secretariat of the university requested home-staying teachers and students to come back to school.[2] Professors could not foresee what sort of life awaited them and many passed the time in a state of apathy. Many professional military engineers who taught war-related disciplines in universities resigned before the Occupation Forces arrived in anticipation of a purge of positions.

When the Occupation Forces arrived in September the Japanese realized that this peacetime invasion was not so disorderly. Thus, in the autumn, students who had been evacuated to the countryside returned to school and course work resumed.

Problems with Demobilization

The arrival of the Occupation Forces, though, did not alleviate the confusion. The first priority of GHQ – as discussed in Part 1 of this book – was the demilitarization of Japan. Whilst disbanding military units and closing down the weapons and ordinance industries was confusing enough, demilitarizing military-related S&T research facilities created special problems. Most of these facilities were to be reassigned to non-military institutions rather than disbanded. But the command channels were confused and where a particular laboratory was to be transferred to was not clear.

The Japanese Government tried to preserve ex-military facilities and personnel by transferring entire institutions to peacetime bureaus; for example, the facilities and personnel of the Navy were to be transferred to the Ministry of Transportation.[3]

To further complicate matters, inventories and records of facilities and equipment had been burned at the time of surrender and it was thus impossible to check them. All ex-military facilities and equipment were supposed to be confiscated by the Occupation Forces and put on reparations lists. If all of their mobile resources were to be confiscated by GHQ, though, they figured that they might as well take them home for their private use. With no adequate records, those who lived nearby and could carry munitions or equipment items could get almost anything released from their workplaces. These items were often sold and circulated on the black market. During the period of confusion a lot of police activity focused on the black market operations and criminal cases were often reported in the newspapers. Police frequently uncovered stolen goods.

The literature of science and technology suffered a similar fate. The documents and books related to wartime military action were all to be inspected by GHQ. Fearing censorship by the Occupation Forces, secret documents were either concealed or transferred to the non-military sector for preservation – e.g., the Imperial University of Tokyo or Hibiya Public Library – in order to avoid GHQ searches. Navy documents had been evacuated during the war to protect them from US air raids. After the San Francisco Peace Treaty these resources were partly returned to the Navy College (Etajima) or the University of the Coast Guard (Kaijō Hoan Daigaku) located at Kure.[4]

Reparations and Restitutions of Military-Affiliated S&T Facilities

Edwin Pauley, the head of the Reparations Mission to Japan, arrived on 13 November 1945 as a Special Ambassador. He listed machine tools, etc., as reparations items. On 6 December he decided to dismantle half of Japan's production capability. In November 1946, the final inventory of the Pauley Reparations Mission was released; it was designed to reduce the Japanese production capacity to the 1935 level.[5]

Parallel to this program, in January 1946 SCAP issued a directive to confiscate military arsenals, aircraft factories and affiliated research laboratories. This directive was in keeping with a decision to meet the reparations demands of the Far East Command.

The Technical Intelligence Company of the US Army conducted site evaluations and confiscated facilities. Their task was so vast that the assessment work progressed very slowly. By January 1947, 218 site surveys were completed and another 205 sites remained untouched.[6] In the meantime, some research equipment was illegally sold on the black market, some of which had to be purchased back by the institutions from which it was stolen.

GHQ's initial orders to the Eighth Army were to dislodge the parts worthy of confiscating for reparations from the ex-military equipment and then destroy the rest. Immobile facilities were listed for reparations at the site. Mobile scientific instruments were assembled at the Tokyo First Arsenal and exhibited to those nations demanding reparations. Non-functioning or obsolete equipment was all scrapped. Educational institutions that had not been involved in the production of military goods were generally not considered to be subject to reparations. The Aeronautical Research Institute (Kōkū Kenkyūsho) affiliated with Tokyo Imperial University was exceptionally listed for reparations in August 1946 because of its military affiliation and some instruments therein were destroyed.[7] Very few items, however, were considered to be for military use only; most of the equipment there was confiscated for reparations. Later, when it was decided the Institute would survive – with a name change from the Aeronautical Research Institute to the Science and Technology Research Institute – the Japanese scientists submitted petitions for some facilities to be lifted from the reparations list. G. W. Fox of ESS/ST supported them.[8] Harry C. Kelly also considered lifting them from the reparations list.[9]

Disrupted by the US-Soviet rivalry, the reparations plan was quite slow to be realized. Not until early 1947 could the Far Eastern Commission define the type of scientific facilities to be listed for interim reparations.[10] For the US scientists at ESS/ST, like Kelly and Fox, Japanese facilities did not appear to be good enough to warrant sending them to the US; hence many items were considered to be not worth listing for reparations.[11] They also thought that research facilities without military application should be removed from reparations so that Japanese researchers could use them. But the Far Eastern Command and military organizations – especially Australia represented by John W. O'Brien, head of ESS/ST,[12] and China – were dissatisfied with this assessment and demanded an 'independent' inspection of Japanese S&T facilities for themselves.[13] They claimed that they themselves badly needed such equipment and argued that, in general, in order to maximize reparations, no equipment should be returned to the Japanese except for what had already been scrapped. By the end of 1947, thirty percent of the interim reparations plan was in effect. The detailed allocation ratio of reparation value was: three for China and one each for the Philippines, Dutch-Indonesia and Australia. Small machine tools were removed from the First Navy Arsenal and sent to the Philippines by the end of the year as reparations.[14]

It is apparent that the reparations value of scientific instruments declines as the facilities get older. Moreover GHQ sources[15] reveal that around 1947 GHQ changed their position on reparations, deciding that some scientific instruments should be left for the economic recovery of Japan. Around May 1947, a revision of the reparations lists began. After reinvestigation, equipment that had been mistakenly listed for reparations were removed from the list. The standard procedure for the Japanese to object to the listing of an item for reparations was to petition the local military commander who would then forward the petition to ESS/ST for assessment and advice; but with these changes in reparations policy, many such issues were handled directly by ESS/ST.

The Japanese scientists were incredibly frustrated by the fact that some of their research facilities that had not been physically removed remained 'confiscated' and were thus sitting idle, inaccessible for anyone to use. ESS/ST documents reveal that while these scientists maintained close contact with ESS/ST, they realized that they needed Kelly's understanding if they were to be successful in their petitions for salvaging equipment from the reparations program in the interests of conducting the research necessary for a Japanese economic recovery. When one university succeeded in 'rescuing' its facilities from the reparations list, the news quickly spread to other universities and one after another they petitioned for the removal of their own facilities from the lists. In most cases, these petitions were successful and instruments were lent to the requesting scientists. Even Occupation Forces organizations, such as the Atomic Bomb Casualty Commission (ABCC) at Hiroshima, petitioned ESS/ST for the release of equipment from the reparations lists, such as some measuring instruments to be used at the site.[16]

There was also a proposal that certain machine tools should be struck from the reparations list and given to Japanese educational institutions purely for educational purposes. On 19 January 1948, the Education Ministry surveyed these proposals and submitted a list of this equipment to ESS/ST who, considering educational use to be a top priority, approved it.

The wind tunnels at the Aeronautical Institute of Tokyo University (renamed the Science and Technology Institute) and machine tools from the Second Aeronautical Arsenal were particularly desired. In September 1946, the Railway Technical Research Institute requested that the wind tunnels be removed from the reparations list for use in basic research for a new bullet train,[17] but O'Brien of ESS/ST refused this request[18] being eager to get hold of this equipment for Australia. No record of what subsequently transpired with the wind tunnel has been found.

Also at the Aeronautical Institute were duplications of a number of scientific instruments. In 1947 a proposal was submitted to ESS/ST to provide equal access to this equipment for universities other than Tokyo University. Initially Kelly was not very enthusiastic about the use of these instruments, considering them to be somewhat obsolete. But they were more than sufficient for the student laboratories at universities and polytechnics.

On 28 June 1948, Tatsuo Morito, the Minister of Education, petitioned GHQ to transfer non-reparations instruments from the Aeronautical Institute to the (First and Second) Engineering Schools of Tokyo University, Tokyo Kōtō Kōgei Gakkō (Tokyo Higher Polytechnic School) at Matsudo, Chiba Prefecture, and the Engineering Schools of Keiō and Waseda Universities.[19] This petition was approved, but since these were public resources, they were provided to the institutions in the form of a long-term loan, rather than a purchase acquisition.

A number of machine tools had been transferred from university engineering schools to military arsenals as part of the wartime mobilization. University researchers, therefore, rightfully made claims for their return. Then, there was a widespread rumor among university researchers that the reparations would soon be removed. Upon hearing the rumor, Yasumasa Tani, of the Second Engineering School of Tokyo University at Chiba began negotiating to get back the items confiscated at Chiba Aeronautical Arsenal, but GHQ explained to him the difference between reparations and non-reparations and refused it. Non-reparations resources were a matter to be discussed between Japanese offices. All of this suggests that there was some confusion at the time about the Occupation policy for the removal of reparations in addition to the strife between GHQ and the other members of the Far Eastern Command.

The reparations program effectively started on 1 July 1948 and ended in March 1950.[20] The program was not carried out smoothly because of the US government's growing determination to mitigate reparations.

In 1949–50 when the reparations program was relaxed, GHQ planned to lend reparations items for a short time – six months or a year – to Japanese research institutions for the economic recovery of Japan. But even some Japanese considered the wartime facilities to be outdated. Director Sakurai of the Production Engineering Institute of Tokyo University, for example, was not interested in the salvaged resources of the Tokyo First Army Arsenal,[21] although Yoshio Nishina, the President of Riken Laboratories, petitioned for reparations items for use in his laboratory.[22] In the negotiation between GHQ and Japanese scientists, the Japanese often expressed a desire for surplus US-made equipment rather than the salvaged Japanese-made equipment.[23]

The State-of-Affairs of Postwar Research

During the war most scientists had been engaged in war mobilization related work. After the surrender, they no longer had an objective for their research. They were forbidden from conducting research in strategic fields like aeronautics, electromagnetism and nuclear energy. The heavy-chemical industry was severely war-damaged, too, and reparations had ruined their research facilities. Thus, many Japanese scientists were in a state of apathy. Even at the celebrated Riken, still remembered as the most productive research institution of the prewar era, the famous charismatic scientists had either died or retired, and the middle level scientists had lost their affinity for the institution because research was virtually impossible. The traditions of excellence at Riken were discontinued, as evidenced by a directive from the Riken administration to the effect that any project that could earn money should be taken up as a priority to overcome the severe economic crisis.[24]

Within a year after the surrender, Japanese scientists had completely switched their research from wartime to peacetime. A number of distinguished (physical) scientists turned to problems of agricultural production and biology. Scientists had no privilege to escape from hunger and there were many projects to test their capability for solving urgent foodstuff problems. Yoshio Nishina applied radiation to plants and researched high yielding varieties. Bunsaku Arakatsu, a Kyoto University physicist, collaborated with chemists and biologists to investigate the effects of radiation for high yielding farm products, improving species, food processing and sterilization.[25]

With no resources available for experimentation, theoretical work with paper and pencil prevailed in academia. But even then, according to Shōichi Sakata of Nagoya University, in the absence of experimental data from the US, they could do only passive work such as resolving theoretical contradictions.[26]

Professor S of experimental physics at Tokyo University had studied in the US before the war and had become closely acquainted with American scientists. During the Occupation, with his fluent English, he maintained close contacts with GHQ, which apparently gained him special 'privileges' concerning reparations items. He was provided with abundant research resources that made him the envy of his fellow scientists. He was thanked for providing them with necessary resources but did not reveal how he got them. Minka researchers and students tried to disclose the truth about S and picked up some rumors about him, but the case was too big for them to deal with. However, reacting to these moves against him, S departed for the US without completing a lecture course he was giving at Tokyo University. Subsequently he worked for the Nuclear Energy Laboratory and never returned to Japan.

Reality of Scientists' Life and Employment

GHQ ordered a purge of all military personnel from public positions as part of the demilitarization program. This applied to all who had military positions even if they were scientists. With no decent work available for scientists, the university laboratories were, for a time, full of disappointed people, who had no regular salaries, but nowhere else to go. Tokyo University in particular had many of these unpaid scientists.

Overall unemployment in Japan at that time is estimated to have been about ten million people. But the unemployment rate was higher in the scientific community,[27] although, since there were no reliable statistics compiled at the time, there is no concrete evidence of this. However, it has been estimated that there were around one hundred thousand unemployed engineers in the spring of 1946, with an overwhelming representation from the field of aeronautical engineering.[28]

In the beginning of the Occupation it was estimated that the official purge would apply to one third of natural scientists in Imperial Universities, which might have dealt a fatal blow to the Japanese scientific community. At that time GHQ was afraid that secret military research would be continued in Japan, but they soon realized that the Japanese were not so inclined. The terms specified for the official purge turned out to have very little impact on scientists in teaching positions. Initially the purge was supposed to apply categorically, after close examination, to all who had participated in scientific expeditions or excavation in the territories occupied by the Japanese military from 1 January 1928. The Education Mission, however, issued a directive on 12 August 1946 to limit the purge to those who had engaged in work with purely military purposes and those who had confiscated precious resources and items.

Some scientists, such as Toshima Araki, Professor of Astrophysics at Kyoto Imperial University and a spokesman of the extreme right wing, resigned proactively before the purge began. People were not purged due to their work as scientists but for other reasons, such as ideology or politics. Purged scientists were prohibited from teaching positions but were allowed to engage in research at universities or research institutes.[29]

Japan had sent numerous scientists and engineers to Keijō (Seoul) and Taipei Imperial Universities as well as other Institutions in their prewar colonies. After the surrender, these professionals were repatriated, but without any anticipation of appropriate jobs. In fact, many of them must have found jobs outside of their area of specialty. Many enterprises were severely damaged without any sign of recovery. Initially the National Railway was almost the only large institution to absorb returned scientists and engineers. A little later, new universities created under the Occupation educational reform also absorbed returnees.

Governmental organizations like the National Railway Service and National Universities survived relatively unscathed, while private corporations had to convert from military to peaceful industry. Peacetime businesses struggled, however, under difficult economic conditions. A small number of private research laboratories that had existed since before the war barely managed to continue their existence.[30] In private sector industries researchers were often transferred to production lines, with few remaining in research laboratories. In general the managers of private corporations believed that they did not have the luxury of running research laboratories.[31]

According to the inaugural issue of *Kagaku Bunka Shimbun* on 29 April 1946, the monthly salary for a new graduate from primary school was 200 yen and graduates from high school received 300–400 yen, while professors of Imperial Universities only earned 200–400 yen, associate professors 120–130 yen and research/teaching assistants received only 80 yen.

According to a questionnaire the Education Ministry sent to researchers in July 1946, while natural scientists always demand more research facilities and better salaries, the demand for more food rations among research workers who lived in the cities rather than provincial areas was particularly noteworthy.[32]

University scientists had the privilege of continuing research funds allocated to the professorial chairs and the Science Research Fund but in fact these funds did not keep up with the skyrocketing inflation. Hence most research funds were spent on utilities (electricity, gas, water, etc.), not for research. While national universities continued to receive research funding from the Education Ministry, the new universities established after the war were not allocated any such funds. Since the Science Research Fund of the Education Ministry could not keep up with inflation, the actual funding for research was inferior to prewar standards.[33]

During the war, scientists and engineers had often enjoyed slightly more generous rations than average citizens since they were considered to be key personnel for continuing the war effort. After the war these privileges were lost; they were poor and suffered in all aspects of life just like ordinary citizens. They wore military surplus clothing and foot-soldier's shoes. For nutrition they cultivated sweet potatoes in the gardens of universities and research laboratories. Chemistry professors made alcohol and saccharine – which were in short supply in the market – in their laboratories. These were marketed with the university's name on the label. Under the pseudonym of 'food social science,' one professor was a broker for the barter exchange of foodstuff.[34] Scientists with nothing else to offer the black-marketeer must have survived by selling their books in exchange for food. Other scientists engaged in small-scale production for the black market as a side business.[35] They all keenly desired to sustain their livelihood by engaging in decent work as scientists.

There were housing problems for many whose houses were bombed and burned. Many scientists returned to their university posts and lived in the laboratory until they were able to find proper housing. Railway travel was also difficult, as the railroad had been bombed and damaged. Special permission was required to acquire a long distance railway ticket. Thus, for scientists to attend the meetings of JSC, Liaison Committees and Renewal Committees, they needed special permission from GHQ. Without GHQ's support it would have been impossible to have meetings with members from the provincial universities.[36]

In postwar Japanese society, where this crisis in living conditions cried out, numerous labor unions were organized. Scientists in the laboratories of both the public and private sectors were organized into labor unions. Some of them were modeled after the Association for Scientific Workers[37] but professional labor unions failed to materialize. Thus, even proud professors from Tokyo Imperial University joined the workers' unions. In 1948, parallel to the general labor union movement which organized the Economic Recovery Conference, the unions representing research laboratories were united to form the Research Recovery Conference. They lost power after the Red Purge was carried out (cf. ch. 25).

Research and Publications

During the war, Zenkoku Kagaku Gijutsusha Dōmei (Nation-wide Union of Scientists and Engineers) allocated paper stock for scientific journal publishing. Dōmei, however, was abolished after the war. Gakkyōkai Rengō (Federation of Learned and Scientific Societies) was established to replace it, but it was only a nominal organization without any real power. It was difficult at this time to publish research articles through regular means, because of the shortage of paper stock.[38] But old military stocks of paper became available on the black market.

During the first year after surrender, there was a publishing boom that was reliant on this wartime stock but it had little to do with the publication of scientific journals (cf. ch. 27). The stock was exhausted, though, after a year. The shortage of paper produced an unimaginable obstacle to the proper functioning of scientific societies. Bulletins of national universities and research institutes were officially allocated paper stocks as well as funds for publication costs since they were part of the public sector. Scientific societies dependent upon membership fees, however, were voluntary associations, which was a disadvantage in obtaining paper for their publishing activities.

For two years after the surrender, scientific publications could barely survive, and those that did published thinner journals with much less frequency than before.[39] For example, before the war *Teikoku Gakushiin Kiji* (Journals of the Imperial Academy) functioned as a bulletin of quick notification like the Western magazines *Nature* and *Science*, but in the postwar environment it ceased to exist; hence, Japanese researchers started to make 'preprints' of their own articles and distribute them to interested parties (cf. ch. 12). It has been said that preprinting, the new mode of scientific communication, originated in Japan at this time.

While academic scientific journals did not recover for a couple of years after the surrender, new popular scientific magazines flourished. Freed from the wartime censorship regime, dozens of new science magazines sprang up, driven by the appeal of (re)constructing a new society through science. The left-wing journalist affiliates of Minka were quite active in this regard. We can now judge the 'hot' climate of the postwar scientific community through the essays and commentaries that appeared in such magazines.[40]

Transfers in Education from Science to Humanities

During the war university students with science majors were privileged with exemptions from or postponement of compulsory military service. Hence, during the war the number of science students increased. The prewar ratio between the humanities and sciences had been 6:4. That was more than reversed during the war. Especially in the last phase of war, as all humanities students should be drafted, students entering Kōtō Gakkō (elite higher schools to prepare for Imperial Universities) as humanities majors were sent to the battlefield without exception. Hence, humanities studies were nearly annihilated, with only thirteen percent of the total student enrolment opting for such majors. Almost all students transferred to science majors.

After the war, those who had chosen science majors for the privileges changed their major to humanities. For 1946, the entrance quota for Kōtō

Gakkō was determined by the Education Ministry to be humanities: science = 1:2 but in practice the ratio returned to approximately the prewar ratio of 55:45.

Polytechnic students also changed to humanities subjects. Comparing student numbers before and after the war, agriculture majors increased slightly and medical students decreased.[41] The School of Engineering had been very popular during the war but because of the slump in industry in the postwar period, it did not attract any new students for a while.

The period immediately following the surrender on 15 August 1945 was the most extraordinary period of poverty, misery and confusion in Japanese history. The Japanese people had no experience before and do not wish to repeat it ever again. The Japanese scientific community experienced extraordinary circumstances during this period that were not officially recorded to any great extent. To fill this vacuum, I have tried to document the situation from my personal recollections, for it was out of this confusion that Japanese science and technology was rebuilt from the 1950s onwards.

Notes

1 Seizaburō Shinobu, *Sengo Nihon Seiji-shi* (A history of politics in postwar Japan), vol. 1, 1965, pp. 131-3.

2 *Tokyo Daigaku Dai2 Kōgakubu-shi* (History of the Second Engineering School of the University of Tokyo), p. 76. A document dated 21 August 1945 calling for teachers and students to return to the Second Engineering School, University of Tokyo; written by the Secretariat of the School.

3 Kaigun Shisetsukei Gijutsukan no Kiroku Kankō Iinkai [6] May 1972.

4 Interviews with librarians from both libraries.

5 Hitachi Seiki Kabushiki Gaisha Shashi Henshū Iinkai (Editorial Board of Company History for Hitachi Steel), *Hitachi Seiki nijūgonen no ayumi* (Twenty-five year history of Hitachi Steel), 1963, pp. 117-21.

6 *Summary of Reparations Activities of the Scientific and Technical Divisions – December '46 to March '47, inclusive.*

7 'Aeronautical equipment for reparations,' ESS/ST, 7 August 1946. A memo of a meeting held on the previous day.

8 'Institutions of a Scientific Nature for Reparations,' 14 June 1946; 'Aeronautical Laboratories and Reparations,' Col. G. W. Fox, 12 June 1946.

9 'Laboratory Reparations,' Memorandum from H. C. Kelly, 16 January 1947.

10 'Reparations Removal' from Lt. E. C. Allan, 13 January 1947.

11 'Japanese reparations to the United States,' by G. W. Fox, 9 April 1946.

12 'Removal and storage of laboratory equipment from present locations to storage,' Notes by J. W. O'B. 27 September 1946. O'Brien proposed to the Pauley Reparations Committee that in order to curtail Japanese military capabilities Japanese scientists, engineers and their families

should be permanently removed to the Allied Forces countries (B. Dees). On the Japanese science and technology that O'Brien wanted to import for Australia, especially in the fields of magnet and metallurgy research, see Morris S. Low 'Mosquito Coils, Ball Bearings and Batteries; The Australian Scientific Mission in Occupied Japan, 1945-1947,' Paper presented at the Japanese Studies Association of Australia, Sixth Biennial Conference, 6-9 July 1989, p. 9.

13 GHQ, 'Transfer of Japanese aviation research facilities to the Chinese government,' 8 May 1946. 'Application of Maj. Gen. Lee Dai-chin and Mr. Z. W. Chang for inspection of some Army & Navy Arsenals,' to G-2 section, GHQ, SCAP, 1 July 1946. A personal letter from the Chinese Mission in Japan (The-chang Koo) to Dr. Kelly, 6 Nov. 1947.

14 Seizaburō Shinobu, *Sengo Nihon seiji-shi* (History of politics in postwar Japan), 1974, p. 622.

15 Conference memo, 20 November 1947 by Donald J. Pletch, Fundamental Research Branch.

16 'Use of Japanese Sanction Apparatus,' from the Atomic Bomb Casualty Commission to ESS, 12 September 1947.

17 Tetsudō Gijutsu Kenkyū Shochō (Director of the Railway Technical Research Institute of the Ministry of Transportation), 'Baishō wariate yori kikai o jogaisuru ken (On the removal of machines from reparations assignment),' ESS(B)11723-34, 6 September 1946.

18 'Memo for record,' JWO'B, 5 February 1947.

19 ESS(B)11723.

20 'Reparations of Science and Technology in Japan,' in GHQ/SCAP, *History of the Non-military Activities of the Occupation of Japan*, 1945-51, vol. 54, pp. 22-3.

21 Memo for file, Y. E. K. 15 January 1950, ESS(B)11723, 'Laboratory equipment listed as reparations materials.'

22 Petition addressed to GHQ in March 1950.

23 K. Kazato (The Electron Optical Laboratory, Inc.) to Kelly, 12 December 1949, ESS(B)11723.

24 Yoshimi Sakayanagi, 'Sengo no ichimen: Rikagaku Kenkyūsho rokujūnen no ayumi (One aspect of the postwar period. The footsteps of Riken for sixty years),' in *Shizen* (Nature), vol. 33, no. 13, special issue, p. 117.

25 *Kagaku Bunka Shimbun*, 7 September 1946.

26 'Kagakukai no dōkō kataru (Speaking of the trend in the scientific community),' in *Kagaku Bunka Shimbun*, 15 January 1948.

27 Yasoji Kazahaya, 'Minka sōritsu no koro (When Minka was established),' in *Kagakusha undō no shōgen* (Testimony of the Scientists' Movement), 1978, p. 26.

28 Estimated from Rinji Kokumin Tōroku (Interim Poll of Japanese People). Humitaka Nakamura, 'Sengo Nihon no gijutsu dōnyū seisaku no keisei to

sono seisakuteki na kiketsu (The formation of technology importation policy in postwar Japan and its consequence),' in *Meiji Daigaku Shakai Kagaku Kenkyūsho Kiyō* (Bulletin of the Research Institute of Social Sciences, Meiji University), October 1989, p. 118.

29 *Kagaku Bunka Shimbun*, 6 September 1946.

30 Nakamura, 1947 [11] p. 38.

31 'Kenkyū Kikan no Seiri (The closing down of research laboratories),' in *Nissankyō geppō*, May 1949.

32 *Kagaku Bunka Shimbun*, 16 November 1946, p. 2.

33 'Sōsetsu iraino Monbushō Kakenhi (Science research funds of the Education Ministry since its founding),' in *Gakujutsu geppō* (Science monthly), vol. 5, no. 8, November 1952, pp. 59-62. Includes annual variation with price index. Also see GHQ-CIE, *Postwar Developments in Japanese Education*, vol. 1, p. 361.

34 Kaneseki, 1947 [10] p. 32.

35 Nakamura, 1947 [11] p. 37.

36 There are a number of such permits recorded in GHQ documents.

37 J. D. Bernal's *Social Function of Science* was translated into Japanese in 1946 and provided a model for the Association for Scientific Workers.

38 'Shiryō to Dōkō (Resources and Trend),' in *Shizen kagaku* (Natural science), no. 9, September 1947.

39 Masuo Usui, 'Shizen kagaku, jiron, gakkaishi ni tsuite (On natural science, commentaries and scientific journals)' in *Shizen kagaku* (Natural science), no. 9, September 1947, p. 1.

40 Kaneseki, 1947 [10] pp. 30-6.

41 *Kagaku Bunka Shimbun*, 3 August 1946, p. 2.

Bibliographic Guide

[1] ESS(B)11720-40.
The military facilities confiscated by GHQ included a lot of science and engineering facilities and resources. This document primarily addresses the issues of how to deal with them. Japanese scientists petitioned for the return of confiscated goods and ESS/ST responded. Lists of equipment, facilities and resources were attached to these petitions. Petitions for the equipment at the Aeronautical Institute of Tokyo University were especially numerous. From the lists of facilities confiscated from military laboratories, we can estimate military S&T activity during the war. There are also documents recording a controversy between a scientist of ESS/ST who advocated returning confiscated items to the Japanese scientific community and the Far Eastern Command, which wanted them as reparations.

[2] Nihon Gakujutsu Kaigi (Science Council of Japan), *Kagakusha seikatsu hakusho* (White paper on the living standard of scientists), 1959.

[3] 'Shizen to tomoni nijūgonen (Twenty-five years with *Shizen* Magazine),' in *Shizen* (Nature), vol. 25, no. 5, May 1970, pp. 84–9.

[4] 'Zadankai, Kenkyū minshuka ni kansuru zenkoku butsurigakusha no kondankai: Minshu sensen ni taisuru kagakusha no taido ni tsuite (Panel discussion, a workshop of nationwide physics community on laboratory democracy: On the attitude of scientists towards a People's Front),' in *Shizen kagaku* (Natural science), inaugural issue, July 1946, pp. 22–33.
The workshop was held on 30 April 1946 at Sanjō Goten of Tokyo Imperial University. One hundred members of the General Assembly of the Japan Physics Society attended it. Shoichi Sakata was the chairman. They freely discussed topics such as the democratization of laboratories and workplaces, union activity and science for 'the people.' Participants include not only left-wing but also liberal scientists in the establishment. A good source for understanding the atmosphere of that time.

[5] *Kagaku Bunka Shimbun* (Science and Culture Newspaper).
Issued weekly or every five days from 29 April 1946. It was later renamed *Kagaku Shimbun* (Science Newspaper) and continues to be published now. It was valuable for its unique ability to explain the reality of the scientific community particularly in the early Occupation period.

[6] Kaigun Shisetsu-kei Gijutsukan no Kiroku Kankō Iinkai (Committee for the Publication of 'A Record of Naval Facility Engineers and Technicians') (ed.), *Kaigun shisetsu-kei gijutsukan no kiroku* (A record of naval facility engineers and technicians), limited press, May 1972.
Valuable and interesting records on what happened to the Navy technological facilities after the surrender. The list of the Kasumi Kai (Club of ex-Navy engineers) reveals the course that Navy engineers followed in the postwar world.

[7] 'Gunkankei gakuto (Military-affiliated students),' in *Kokumin no kagaku* (Science for national people), vol. 1, no. 5, 1946, p. 18.

[8] The following are examples of short-lived magazines and articles of the postwar left-wing scientists' movement.

'Kagaku senpan tsuikyū ni tsuite (On the search for science war criminals),' in Minshushugi kagaku (Democratic science), vol. 1, no. 3, June 1946, p. 2.

Tatsuichi Hishiyama, 'Warerano gikai o tsukurō (Let's make our own Parliament),' in *Warerano kagaku* (Our science), vol. 1, no. 1, March 1946, p. 3.

Masao Nakamura, 'Kagaku Akademii no shuyō na mondai (Main problems of the Science Academy),' in *Kokumin no kagaku* (Science for national people), vol. 2, no. 1, 1947.

[9] Genkichi Hara, 'Nihon ni kagakugijutsusha wa donokurai iruka (How many scientists and engineers in Japan?),' in *Kokumin no kagaku* (Science for the people), no. 3, May 1958, p. 4.
[10] Yoshinori Kaneseki, 'Sokoku saiken to kagakugijutsusha no mondai (Recovery of Japan and the problems of scientists and engineers),' in *Shizen kagaku* (Natural science), no. 9, August 1947.
[11] Sachio Nakamura, 'Shiryō to dōkō, Kenkyu hukkō kaigi no kaidō (Sources and trends: Conference on the way to research recovery),' in *Shizen kagaku* (Natural science), no. 9, August 1947.

15 Research Funding in Occupied Japan

Takashi Hata

Prewar Science Mobilization and Research Funding

According to the 'Science and Technology Report' prepared by the Japanese Government's Statistics Bureau, Management and Coordination Agency, 'research funding' is defined as 'funds that are used internally by companies, research agencies and universities to cover the costs of labor and raw materials as well as the costs involved in acquiring tangible fixed assets.' While this is perhaps an adequate definition of research funding as a general concept, it is not adequate for describing the situation of research funding during the period when Japan was occupied by the Allied Powers following World War II, a period when statistics were not reliably compiled.

Several studies have already examined university and science funding during this period,[1] so this chapter will focus on general trends in scientific and technical (S&T) research funding and funds allocated for support and assistance during the Occupation. Many different terms were used to designate various schemes – terms such as 'scholarship,' 'financial aid,' and 'financial supplement' – but they were all simply different ways to indirectly increase the funding available for organizations and individuals to conduct research. Through an examination of these different schemes we can identify a general trend in the administration of scientific and technical research. This trend began before World War II, expanded during the war and developed further after the war.[2] One of the earliest examples of this type of indirect funding was the Imperial Academy Research Assistance Subsidy established in 1913. This was primarily a means for assisting illustrious families rather than the beginnings of what we can call 'science mobilization' and, as such, was used for only three projects, including copying Mannen Ueda's *Ruiju Koshū*.

From the end of World War I the opportunities for financing scientific research gradually expanded as more and more funding schemes were established, including: The Invention Encouragement Fund of the Ministry of Agriculture (established in 1917); the Science Research Encouragement Fund (est. in 1918 and administered by the Ministry of Education); and the Industrial Research Encouragement Fund (est. in 1921 under the administration of the National Census Agency; when this Department was later abolished the fund was administered by the Ministry of Agriculture and Commerce). These funds mark the beginning of government financial assistance to private organizations.

However, the economic downturn following World War I retarded the development of research funding sources. In 1930 government research funding totaled about 400,000 yen, which was not enough to even cover the research program of the Aviation Research Institute at Tokyo Imperial University. However, as the nation moved into a general war mobilization mode – in an effort to increase its military capabilities – a variety of scientific research funds were established.

The Japan Society for the Promotion of Scientific Research Fund (est. in 1933) was born from a movement for the general promotion of science education. It commanded a budget of approximately 510,000 yen in its first year, rising to 1.32 million yen five years later. The Science Research Fund (est. 1939 in the Ministry of Education) was initially established with a budget of 3 million yen, and also expanded very rapidly. These funds were intended to finance specific research projects for the mobilization effort, rather than to provide general funding for private research. The Japan Society for the Promotion of Scientific Research Funds were allocated to research projects determined by the Society to be in the national interest. These projects were broken down into basic research tasks that were then allocated with appropriate funding to various universities who were required to report back the names of the primary researchers engaged in the project.[3]

After 1942 there was a significant increase in funding for airship research. Existing research facilities were expanded and new facilities were established through the Cabinet Technology Agency. They also provided financial assistance for research institutions and additional research funds were available from the Ministry of Army and the Ministry of Navy. The Cabinet Technology Agency distributed enormous funding to universities and research institutes in the form of 'sponsored research funds.'[4] The exact amount of research funding administered by the Cabinet Technology Agency in 1944 is not clear. Similarly, ascertaining the precise sum designated for research funding in the national budget in 1944 is difficult, but it has been estimated to be in excess of 230 million yen. If that sum is correct, the research funds available to the Ministry of Education were less than ten percent of the total national research budget for that year (see Table 15-1).

Reorganization of Research Funding after WWII

From 1943 the 'employees of research bodies at high schools and technical colleges were prioritized for screening' for natural science subsidies.[5] Higher normal schools (teacher training secondary schools) in Tokyo, Hiroshima and Kanazawa, Tokyo higher normal schools for women

Table 15-1: Distribution of research funds, 1944 & 1947

	yen (x 1000)	percentage
1944		
Ministry of Education (cf. *Table 15-2*)	21,332	9.2
Ministry of Commerce and Industry	3,362	1.4
Ministry of Welfare	33	0.0
Technology Institute	12,758	5.5
Ministry of Munitions	1,530	0.7
Special Military Fund	193,294	83.2
Total	232,309	100.0
1947		
Ministry of Education	71,501	96.0
Ministry of Commerce and Industry	2,950	4.0

Source: 1944 details compiled from the Japanese Government's 'Research White Paper' and '1944 Settling of Accounts.'

(1944), and Kyoto Prefectural No. 1 Junior High School (1945) each established special science classes to try to develop the talents of students who demonstrated strong science ability. To further expand their science programs after the war, special science education research societies were established and the budget allocated to the promotion of scientific research was increased so that teachers at secondary schools could be assisted in the scientific research being conducted by teachers at secondary schools.[6]

Following Japan's defeat in WWII the Government promptly set about dismantling and then reorganizing the science mobilization framework. Research funding was maintained as much as possible.[7]

Scientific research funding subsidies were taken over by the Ministry of Education as a result of the Occupation policy of demilitarization. In 1947 scientific research funding within the Ministry of Education was divided into 'Natural Science Research Funds' and Humanities and Social Sciences Research Funds' with a clearly defined ratio allocated for each.[8] With the abolition of the Cabinet Technology Agency, its Research Mobilization Unit was integrated into the Ministry of Education's Science Education Bureau and its funds for Utilization of the Results of Research and Practical Pilot Study became the Subsidy for the Promotion of Utilization of the Results of Research in the Ministry of Education.

In 1947 it became the Grant for the Promotion of Utilization of the Results of Research [9] and focused primarily on applied research. To encourage greater cooperation and interaction between research bodies, organizations that could independently operate in accordance with the research plan of the governing authorities in Japan were excluded from these funds. Financial subsidies that had been paid to corporate research institutes by the Cabinet Technology Agency were transferred to the Ministry of Education and continued to be paid.

In 1936 the Committee for the Promotion of Various Sciences in Japan was established within the Bureau of Education Reform (Kyōgaku-kyoku) with research funding designated for this purpose. When this Bureau was abolished on 15 October 1945 its funds were transferred to the newly established Science Education Bureau. The Japanese Studies Promotional Committee was also abolished on 19 March 1946 by Directive no. 3 of the Ministry of Education with a proposal that it could be remodeled into the Humanities and Social Sciences Promotional Committee.[10] In September 1946 it was re-established as the Humanities and Social Sciences Committee. The financial subsidies it administered became the Humanities and Social Sciences Research Financial Subsidy.[11] This was later renamed the Humanities and Social Sciences Research Grant.[12]

During this period the Ministry of Education's research funding increased dramatically, accompanied by the creation of new funding bodies such as the Science Council's Enterprise Assistance Fund. Thus, the dismantling of military-oriented research bodies resulted in elevating the importance of the Ministry of Education in the field of scientific research funding. In fact, the vast majority of such funds were concentrated in the Ministry of Education (cf. Table 15-2) which as we will see, became a matter of concern to GHQ and other Japanese Government ministries.

Table 15-2: Breakdown of Ministry of Education Budget, 1944 & 1947

	yen (x 1000)
1944	
Science Research Fund Subsidy	18,700
Science Research Promotion Subsidy	20
Japan Science Promotion Council Research Fund	2,583
Science Research Encouragement Fund	29
1947	
Natural Sciences Research Fund	54,264
Science Research Fund	31,785
Science Research Promotion Subsidy	249
Science Experimental Research Fund	15,000
Resource Science Council's Association Subsidy	1,520
Civil Science Research Agencies Subsidy	3,710
Council Magazine Publication Assistance	2,000
Humanities and Social Science Research Fund Subsidy	11,238
Science Research Fund	7,868
Science Research Promotion Subsidy	150
Humanities and Social Science Research Fund Subsidy	350
Ethnology Association Assistance	180
Civil Research Agencies Subsidy	2,190
Council Magazine Publication Assistance	500
Japan Science Promotion Council Assistance	6,000

GHQ's Research Funding Policy

As the S&T promotional strategies became centralized in the Ministry of Education, it was natural enough that the question of encouraging all kinds of applied and pilot research began to be pointed out to the education administration which had heretofore emphasized basic research in universities. A typical perspective on this can be found in the Scientific and Technical Division of GHQ's Economic and Scientific Section (ESS/ST). ESS/ST had primary responsibility within GHQ for S&T policy although the Civil Information and Education Section (CIE), the Civil Communication Section (CCS), the Natural Resources Section (NRS) and the Public Health and Welfare Section (PHW) each had some responsibility for particular areas of S&T policy. ESS/ST's regulatory function, though, generally took precedence over these other bodies.[13]

In the early stage of the Occupation S&T research funding policy was dictated by the Potsdam Declaration, the 'United States Initial Post-Surrender Policy for Japan'[14] and the 'Basic Initial Post-Surrender Directive to Supreme Commander for the Allied Power for the Occupation and Control of Japan.'[15] All military research was prohibited in accordance with the policy of demilitarization.[16] GHQ directed that any allocation of research funds to the Ministry of Education and the Ministry of Commerce and Industry in the 1946 budget required prior approval from GHQ. The screening of all applications for such research funds was reviewed by NRS. Under this process, in 1946 applications for funding were rejected for: 'manmade fertilizers' by the Government Chemical Industrial Research Laboratory, Tokyo (under the jurisdiction of the Ministry of Commerce and Industry); 'natural and artificial rubbers' by the Government Chemical Industrial Research Laboratory, Osaka; the 'carbonization of coal, its gasification and liquefaction,' 'the liquefaction of coal tar using extremely high pressure' and 'manmade oil' by the Fuel Research Institute. Furthermore, expenditure on 'oil refining and related facilities' was cut-back.[17] Subsequently research funding was screened every year.

In 1947 GHQ invited a US Scientific Advisory Mission to come to Japan and advise on the drafting of long-term S&T policies. Its advice was presented in a report entitled 'The Reorganization of Science and Technology in Japan' on 28 August 1947. The report pointed out the importance of S&T for economic reconstruction and advised:

- that a Science Research Committee should be established to examine the formulation of S&T policies;
- the committee should be empowered to allocate research funds;

- university funding should be allocated for specific research proposals rather than being uniformly distributed to all chairs of national universities;
- priorities should be established between basic and applied research, between the natural sciences and the social sciences and between research focused on primary production (e.g., agriculture and mining) and research directed towards manufacturing production.

The report also criticized the higher education system reforms in progress by CIE and the Ministry of Education. Although the report was submitted to GHQ in August 1947, it was not published – i.e., released for public perusal – until March 1948 (cf. ch. 10).

ESS/ST also criticized the ongoing education reforms in a report on S&T research funding dated April 1948, pointing out that Japan's S&T research activities focused on basic research and neglected applied or developmental research.[18] It analyzed the Ministry of Education's function and concluded:

> It is possible to show, however that the Monbushō has reached a position, within the past decade, where it can exercise direct control over a large fraction of Japanese natural science research.
>
> Many of the powers which have come to reside in the Ministry of Education during its history have apparently been placed there without much consideration being given to the appropriateness or inappropriateness of the Monbushō having such powers. It would, therefore, appear likely that if a complete re-apportionment of duties were to be made in the Japanese central government, a number of the duties now detailed to the Monbushō might be limited or given to other agencies. (Thus, for example, the duty of supervising the Physical Institute of Radio Waves might be transferred to the Ministry of Communications; or the duty of exercising surveillance over the zaidan hōjin might be passed to the Finance Ministry.)

As can be seen in the quote above, the report considered it not only acceptable, but also advisable to allocate many of the Ministry of Education's duties to other government agencies. For example, the authority to supervise the Electric Wave Research Institute might move to the Ministry of Communications (whose portfolio includes postal services and telecommunications, cf. ch. 18) and the authority for overseeing research foundations (zaidan hōjin) might be relocated to the Ministry of Finance. The Ministry of Education's authority in the field of S&T was thus weakened. Furthermore, ESS/ST expressed concerns that the management of research funds was concentrated in the Ministry of Education who maintained a preference for basic research. However, it seems that this report did not lead to the drafting or implementation of a new S&T research funding policy.

CIE, the GHQ section responsible for education, was particularly concerned about the tardiness of the formulation of research funding policies for the Humanities and Social Sciences and their promotion. CIE critically responded to the Science Advisory Mission's policy outline, pointing out that 'A persistent problem in the financial support of research has been the disparity between cultural and the natural-technological sciences' and stated that funding 'in the cultural sciences too often fails to provide for essential field trips and empirical observations.'[19] CIE cooperated in the rectification of these shortcomings, primarily through the provision of financial support.

Changes in Research Funding Policy

GHQ's internal disagreements concerning changing priorities for the management and allocation of research funds arose concurrently with overall changes in Occupation policy; changes from demilitarization to economic reconstruction which began about 1948. As discussed in other chapters, these changes arose in response to the escalating 'Cold War' (cf. ch. 4, 8).

Accordingly, S&T policy shifted from simple observation and restriction to providing guidance for and encouraging economic dev-elopment. The relative importance of the natural sciences increased, particularly in regards to applied research. In particular, after the 'Nine-Point Economic Stabilization Program (FEC-329)' was set out, ESS/ST began a concerted effort to promote applied research through reorganizing the S&T research funds that had hitherto been controlled by the Ministry of Education.

One concrete example of this can be seen in the preparation of the 1950 budget which was drawn up in accordance with the 'Dodge Line,' which pursued policies for a truly balanced consolidated budget.[20] ESS/ST consulted with the Science Council of Japan's (JSC) Science and Technology Administration Commission (STAC) (cf. ch. 11) about S&T policy changes before advising ESS Finance Division on the budget preparations.[21] STAC referred the preparation of a draft funding proposal to a special committee constituted of selected members of JSC. After conferring with ESS/ST, the committee's draft proposal included a recommendation that approximately 50 million yen of the research funding slated for allocation to the Ministry of Education for distribution to other Government ministries should instead be allocated directly to the concerned ministries.

Consideration of the special committee's proposal sparked vigorous and heated debate within STAC. Mr. Katsuhiro Kennoki, the Deputy Director of the School Education Bureau argued that the Ministry of Education budget promoted the type of research conducted in universities and research institutions for the benefit of industry, studies which are not necessarily categorized as applied research. He did not accept that these funds should be curtailed at a time when university research was increasingly development-oriented. Furthermore, he countered critiques of the concentration of funding in the Ministry of Education with the claim that splitting these funds up amongst the different ministries would undermine the only regulatory process that existed for research and would therefore be inefficient, resulting in duplicated and nonsensical research. Nevertheless, with a number of committee members reflecting the intentions of ESS/ST and making strong counter-arguments, he remained in a minority within STAC, who strongly supported the redistribution of funds to the various ministries concerned. In the end, there was a majority agreement that overall research funds should be increased by 100 million yen.[22]

ESS/ST considered STAC's position – situated in between the Government ministries, and thus with no effective authority – as a problem. It is reasonable to assume that the reason ESS/ST considered this to be a problem was because STAC was powerless to overcome the Ministry of Education's resistance to changing the research emphasis from basic to applied science and – perhaps more importantly – to overcome the Ministry's efforts to maintain total control of research funding.[23]

Nevertheless, ESS/ST incorporated the special committee's recommendation into their advice to the Ministry of Finance. This advice included ESS/ST's objections to the Ministry of Education's control of S&T research funds designated for other ministries' research programs. More specifically it recommended that the Ministry of Industry and Trade should be directly allocated the 40 million yen that it would be seeking from the Ministry of Education.[24]

The actual budget that was drawn up did not include the additional 100 million yen that had been requested, but it did allocate the 50 million yen in research funding to ministries other than Education, to be allocated by STAC. This redistribution with no accompanying increase in the overall budget necessitated a corresponding reduction in the Ministry of Education's budget.[25] However, there was no decision included as to how the distribution or expenditure of these funds was to be supervised. Thus on 3 March a Prime Ministerial Inquiry was launched to investigate the appropriate role of the JSC for the distribution of the 1950 Funds for Utilization of the Results of Science Research.[26] This inquiry finally resolved the debate about the concentration of research funds in the Ministry of Education. Subsequently scientific research funding ceased to be concentrated in the Ministry of Education and increased very rapidly.

Breaking the Ministry of Education's control of scientific research funding appears to have initiated a greater emphasis being placed on applied scientific research. Thus the 1949 'Economic Science Bureau's Annual Report' provided the following evaluation:

> ESS/ST tried in all ways possible to shift the research emphasis from basic science research to applied science oriented towards technology. It now appears that this sort of change in emphasis has occurred. Many scientists who were previously interested in basic scientific research are now actively assisting in finding solutions to important problems confronting industry.[27]

Science Technology Research Promotion and Applied Research Funding

The application of science technology to industry was advocated by the Japanese, too. Industrial rationalization was promoted with a view to economic revival and independence. A financial assistance policy was implemented for the practical application of technology to this end. The Economic Stability Head Office Economic Revival Committee adopted a long-term economic plan in May 1949 that advocated greater emphasis on the practical application of research and proposed a range of policies to promote practical research in private laboratories. Similarly, JSC began lobbying for an increase in research funding soon after its inauguration.[28]

JSC established its eleventh committee in February 1949 to investigate the practical application of the Nine-Point Economic Stabilization Program for science technicians. Following the deliberations of the eleventh committee, JSC vigorously lobbied the government for increased funding. On 5 March 1949 they requested an increase in industrialization oriented research funding to be administered by the Industrial Technology Agency. The committee's third plenary session in April decided there should be an increase in the sponsored research funds for private research institutes and that a portion of the trade accounts should be allocated to a fund for the acquisition of overseas technology. On 1 July they agreed to request 146 million yen for steel technology research from collateral funds. On 11 July they voted in favor of organizing a loan from the collateral funds for vital fields of industrialization.

The request for increased research funding received bipartisan support in the Upper House of the Diet.[29] This was, at least in part, a response to the massive cuts in science research funding in the 1950 draft budget – a reduction from 1,680 million yen to 500 million yen – which resulted from the balanced budget policy of the Dodge Line. In accordance with the Dodge Line, the increased funding did not come to fruition, but the movement for it continued to grow throughout 1950. On 11 March 1950 the Lower House also voted, again with bipartisan support, in favor of increasing the S&T budget and funding loans to improve facilities.[30] On 13 June the JSC requested an increase in research funding and recommended that the various ministries and government offices establish applied research funds and coordinate between themselves more closely.[31]

At this juncture the government embarked on a course of industrial rationalization and gradually increased funds for applied research. The request of the Science Council, mentioned earlier was also granted.[32] In September the Cabinet decided in favor of 'the promotion of practical oriented research and the diffusion of superior technology'[33] and in December the Industrial Rationalization Council of Review was created to examine various policies aimed at bringing about an economic revival. In December 1950 Cabinet decided to establish the Industrialization Experimentation Subsidy and the Mining Industry Technology Research Subsidy for the application of technology to industry, under the Agency of Industrial Science and Technology. In 1951 the Labor Health Experimental Research Fund Subsidy (Ministry of Labor), the Welfare Science Research Subsidy (Ministry of Health and Welfare), the Science Technology Applied Research Fund Subsidy (Ministry of Transport), the Construction Technology Research Subsidy (Ministry of Construction), and the Farming, Forestries and Fisheries Industries Technical Experimentation Fund Subsidy (Ministry of Agriculture and Forestries) were all established.

In February 1951 the Industrial Rationalization Council of Review produced its first report, advocating 'fostering private research institutes, expanding and strengthening national and publicly managed experimental research institutes and promoting co-operation amongst these institutes.' It also referred to the 'coordination of pilot research funding for industrialization' and recommended 'an increase in funding from the public purse for this.' The reorganization and increase of applied research funds proceeded accordingly. Article 3 of the Enterprise Rationalization Promotion Law, proclaimed in March 1952, provided for granting subsidies to researchers for practical oriented research. In accordance with the Promotion Law, the Industrialization Experimentation Subsidy and the Mining Industry Technology Research Subsidy were created under the jurisdiction of the Ministry of International Trade and Industry. The Subsidy for private research institutes was put into place according to the law relating to the fostering of private research institutes (1 June 1951, Law no. 272).[34]

The availability of applied research funds from new sources – that is, from sources other than the Ministry of Education – raised questions about the relationship between these new funds and the Ministry's Funds for Science Experimental Research. After much deliberation it was decided that the Ministry of Education's Promotion of Utilization of the Results of Research Funds were for basic and applied research at basic level, including interdisciplinary studies, in the natural and social sciences at prefectural, municipal and private universities, non-profit foundations and corporations (excluding national universities).[35] STAC would establish an 'Applied Research Funds Allocation Regulation Section' to oversee the administration of these funds. Applications for funds from universities, non-profit organizations and companies were first lodged with the various ministries. STAC then reviewed the allocations of research funds amongst the various ministries based on advice from specialists recommended by JSC.[36]

From about 1950 a high proportion of scientific research funding in Japan has been allocated for applied and developmental research. However, it is important to note that the concerted efforts of GHQ, JSC and others to break the Ministry of Education's stranglehold on research funding and to shift the scientific community's research priorities towards more applied research were not intended to undermine the importance of basic research. During the late 1940s the shifting emphasis inevitably cut into basic research funds, but the Ministry of Education remained the primary government office for funding and conducting S&T research (see Table 15-3, Distribution of Research Funds, 1952). As more funding was directed towards applied research, efforts to strengthen basic research funding increased as well. In September 1949, for example, the Ministry of Education conducted an economic evaluation of the outcomes of research carried out with the Promotion of Utilization of the Results of Research Funds and began efforts to make the significance of basic research more widely understood.[37] Similarly, on 6 May 1950 a plenary session of JSC reached a decision that research funding for chairs at national universities and affiliated organizations should be increased and that budgetary measures needed to be taken to ensure that the new graduate schools were fully funded (cf. ch. 10).

Before concluding this chapter it is important to point out that the problems facing S&T research during the period we have been discussing were not limited to the questions of how funding should be administered or allocated. Detailed investigations of universities conducted by the Research Aid Division (of the Ministry of Education) in 1949 revealed that forty-two percent of the war damage to university facilities had not yet been rectified due to lack of funds. There were no funds available for upgrading out-dated equipment and facilities either, and only six percent of applications for imported equipment subsidies had been approved. And while there were little or no budget cuts to university operating and research funds, their budget allocations remained (relatively) constant in a time of spiraling inflation. Hence, in real terms, by 1949 their purchasing power had fallen below twenty-five percent of the prewar (1928) levels.

Table 15-3: Distribution of research funds, 1952

Fund Name	yen (x 1000)
Under the Jurisdiction of the Ministry of Education	
Science Research Funding Subsidy	340,000
Science Experimental Research Fund	110,000
Research Outcome Fostering Subsidy	30,000
Machinery and Book Importation Subsidy	45,000
Civil Research Agency Grant	46,500
Japan Science Promotion Council Assistance	3,500
Ministry of Educational Sub-total	610,000
Under the Jurisdiction of the Ministry of Health and Welfare	
Science Experimental Research Subsidy	20,000
Under the Jurisdiction of the Ministry of Agriculture and Forestry	
Farming, Fisheries and Research Subsidy	282,112
Under the Jurisdiction of the Ministry of International Trade and Industry	
Science and Technology Research Aid	23,950
Resource Development Subsidy	153,426
Under the Jurisdiction of the Industrial Technology Institute	
Science and Technology Research Aid	426,000
Industrialization Experimental Fund Subsidy	244,000
Under the Jurisdiction of the Ministry of Transport	
Science and Technology Applied Research Funding Subsidy	50,000
Under the Jurisdiction of the Ministry of Labor	
Labor Hygiene Experimental Research Subsidy	3,000
Under the Jurisdiction of the Ministry of Works	
Construction Technology Research Subsidy	15,000
Total	**1,583,488**

Source: compiled from the 'Final Account Details, 1952'

Under these circumstances universities appropriated faculty and student research funds simply to cover administration expenses. Thus, for example, fully sixty percent of research funds for the Science and Engineering Faculty at Tokyo University were from external sources. It was under these circumstances that JSC and ESS/ST had strenuously argued that limited research funds should be redirected towards applied research. And it was under these circumstances that the Ministry of Education had vigorously defended their budget and their control of S&T research funds. The results of these power struggles had long-lasting repercussions for the Japanese University system and for the S&T research community.

From the mid 1950s, research funding gradually increased. The Ministry of Finance accepted that scientific research funding – with scientific research defined as 'integrated joint research' – was of paramount importance. Its emphasis was not on private research, which it considered to be supplementary to research funding for chairs, but rather on institutional research and integrated research.[38] However, the Ministry of Education continued to oppose the system of prioritizing applied research over fundamental research that had developed during a period when research funding was very limited. Conflict between the different parties to this dispute continued, as did the search to determine precisely how research funding should be distributed.[39]

Notes

1 Gakujutsu Taisei Kenkyū-kai (Science System Research Group), 1953 [15]; Nihon Gakujutsu Shinkō-kai Kagaku Gijutsu Seido-shi Henshū Iinkai (Japan Society for the Promotion of Science, Editorial Committee on the History of Science and Technology System), 1960 [16]

2 Tetsuo Hiroshige's pioneering research provides a basic outline of this trend and provides a basis for much of this chapter. Hiroshige, 1973 [17]

3 *Hatsuka* (Memorandum issued by the Bureau of Science Education), no. 25, 1 April 1943, 'Shōwa 18-nendo kagaku kenkyū-hi ni kansuru ken (Matters relating to the 1943 science research funding),' in Tokyo Daigaku Hyakunenshi Henshū Iinkai (Editorial Committee for a One-Hundred Year History of the University of Tokyo) [18] *Tsūshi 2* (Historical overview, vol. 2), 1985, pp. 675–6.

4 MITI, 1979 [20] vol.13, 'Kōgyō gijutsu (Industrial technology).'

5 *Hassen* (Memorandum issued by the Director-General of the Bureau of Special Educational Affairs), no. 15, 'Shizen kagaku kenkyūhi shōreikin kōfu ni kansuru ken (Matters relating to natural science research promotion subsidies),' in Tokyo Daigaku Hyakunenshi Henshū Iinkai (Editorial Committee for a One-Hundred Year History of the University of Tokyo) [18] *Tsūshi 2* (Historical overview, vol. 2), 1985, p. 674.

6 Science Education Bureau, Research Division, 'Science Education Bureau Jurisdiction Articles (1946)', in *Sengo kyōiku shiryō* [1] VI – 567, 1965.

7 Nihon Gakujutsu Shinkō-kai Kagaku Gijutsu Seido-shi Henshū Iinkai, 1960 [16]

8 *Hatsuka*, no.115, 1 April 1947 [4] vol. 28, 1964.

9 *Hatsuka*, no.15, 22 January 1947 [4] vol. 28, 1964.

10 Ministry of Education, Bureau of Science Education, Research Division, 'Shokan jūyō jikō (Important jurisdictional matters) (1946),' in *Sengo kyōiku shiryō* [1] vol. 1, 1965, p. 566.

11 *Hatsuka*, no.169, 29 November 1946 [4] vol. 28, 1964.

12 *Hatsuka*, no.16, 17 January 1948 [4] vol. 28, 1964.

13 ESS/ST, 'Phase-out Report,' 29 May 1951, in *History of Division to Personnel Action (Japanese Nationals)*, GHQ/SCAP Records, Box no.7412 ESS (A) 9794-(A) 9795.

14 SWINCC-150/4/A, 22 September 1945.

15 JCS 1380/15, 3 November 1945.

16 Discussions between ESS/ST, the ESS Finances Division and CIE produced a directive banning any scientific research whose objective was waging war. SCAPIN 9841, 24 May 1946. 'SCAPINS and SCAPIN A'S Produced by Scientific and Technical Division 1945.9–1950.4' [7b] ESS (A) 9776-9778.
17 *Check Sheet* [7c], ESS (A) 9780.
18 'S&T in Japan Report No.13,' April 1948 [7e]
19 Education Division, 'Development in Japanese Education in terms of the Report of the United States Education Mission to Japan,' in *Wanamaker Papers*.
20 Nihon Gakujutsu Shinkō-kai Kagaku Gijutsu Seido-shi Henshū Iinkai, 1960 [16], pp. 196–7.
21 'Record of meeting with the Vice-President and Government Members of STAC and JSC Officials (October 1949)' [7a] ESS (B) 11613.
22 Nihon Gakujutsu Shinkō-kai Kagaku Gijutsu Seido-shi Henshū Iinkai, 1960 [16], p. 196.
23 Nihon Gakujutsu Shinkō-kai Kagaku Gijutsu Seido-shi Henshū Iinkai, 1960 [16], p. 196.
24 'Memorandum of Interim Oral Report by Mr Senshu Re: Reexamination of Research Budget (29 October 1949)' [7a] ESS (B) 11613.
25 'Proceedings of Ninth Meeting, STAC, 2 November 1949' (2 November 1949) [7a] ESS (B) 11613.
26 Nihon Gakujutsu Shinkō-kai Kagaku Gijutsu Seido-shi Henshū Iinkai, 1960 [16] no.243, p. 196.
27 '1949 Annual ESS Historical Report, Scientific and Technical Division,' January 1950, GHQ/SCAP Records, Box no.7377, ESS (C) 9362 – 9364.
28 Nihon Gakujutsu Kaigi [19] 1974.
29 'Kagaku gijutsu shinkō ni kansuru ketsugi (Resolution on the promotion of science and technology),' on 28 November 1949 at the plenary session of the House of Councilors.
30 A resolution of the same title as above on 11 March 1950 at the plenary session of the House of Representatives of the Diet.
31 'Kagaku gijutsu kenkyū shinkō ni yōsuru keihi no yosan hensei hōshin ni tsuite (On the budget preparation policy of the costs relating to science technology research promotion).'
32 Hiroshige, 1973 [17] and MITI 1972, 1985 [20] vols. 10 and 19.
33 'Sangyō gōrika ni kansuru ken (Matters relating to industrial rationalization),' Cabinet Decision made on 13 September 1949.
34 Takeharu Miyashita, 'Kaisetsu: Minkan gakujutsu kenkyū kikan no josei ni kansuru hōritsu-an ni tsuite (Interpretation on draft laws relating to the fostering of private research institutes)' [5] 1951, vol. 4, no.1.
35 *STAC News*, 13 March 1951, *Gakujutsu geppō* (Science monthly) [5] vol. 4, no. 2, 1951.

36 *STAC News*, 13 March 1951, *Gakujutsu geppō* (Science monthly) [5] vol. 4, no. 2, 1951.
37 Ministry of Education, Bureau of Higher Education and Science, Aid Division, 'Daigaku kenkyūhi no jittai chōsa kara daigaku zaisei no jittai chōsa (Survey on university finances from the survey on university research funding). Also see: Heihachirō Miyayama, 'On Surveys of University and other Research Funds,' in *Gakujutsu geppō* [5] vol. 4, no. 3, 1951; Research Fostering Unit, 'Survey on University Finances from the Survey on University Research Funding,' in *Gakujutsu geppō* [5] vol. 5, no. 11, 1953.
38 Hideyuki Aizawa, '29-nendo yosan o chūshin to shita kenkyūhi no shomondai (Problems with research funding centering on the 1954 Budget)' in *Gakujutsu geppō* [5] vol. 7, no. 1, 1954.
39 Heihachirō Miyayama, 'Kagaku kenkyūhi no jūten haibun ni tsuite (On the prioritized distribution of scientific research funding),' in *Gakujutsu geppō* [5] vol. 3, no. 4, 1950; Research Fostering Division, 'Kenkyūhi o chūshin to shite (Centering on research funding),' in *Gakujutsu geppō* [5] vol. 7, no. 12, 1955; Yūsuke Sumiki, 'Kōza kenkyūhi to kagaku kenkyūhi (Research funds for chairs at national universities and science research funds),' in *Gakujutsu geppō* [5] vol. 8, no. 2, 1955.

Bibliographic Guide

Basic Data

[1] *Sengo kyōiku shiryō* (Postwar education materials), 1965, held by the National Institute for Educational Policy Research.
A collection of materials relating to postwar educational reforms of those involved with the Ministry of Education (Kyōsuke Yamazaki, Teiyū Amano, and Junnosuke Haruyama). Includes indispensable materials relating to legislative processes. For an inventory, see *Sengo kyōiku shiryō sōgō mokuroku* (Postwar Education Data Comprehensive Table of Contents).
[2] *Honda Hiroto kyūzō shiryō* (Materials of Hiroto Honda), 1990, held by the National Institute for Educational Policy Research.
Materials held by the Mr. Honda, the former Secretary-General of the Science Council of Japan. Includes materials relating to the Economic Recovery Planning Committee. A useful inventory is *Sengo kyōiku kaikaku shiryō 9: Honda Hiroto kyūzō shiryō* (Materials on postwar educational reform series 9: Hiroto Honda).
[3] Ministry of Education, Records Unit, *Tsūtatsu-tsuzuri* (Files of official notices), held by the National Educational Research Institute.
Includes communications missing from [4] and is an important reference for instructions concerning scientific research funds.

[4] Kindai Nihon Kyōiku Seido Shiryō Henshū Iinkai, *Kindai Nihon kyōiku seido shiryō* (Historical data on modern Japan's education system), 35 vols., Kōdansha, 1956–64.

[5] *Gakujutsu geppō* (Science monthly), Nihon Gakujutsu Shinkōkai (Japan Society for the Promotion of Science).
Includes numerous items relating to scientific research funding and comments on each of the research funds. Valuable information about trends in the Science Education Bureau of the Ministry of Education, the Science, Technology Administration Council and the Science Council of Japan.

[6] *Yosan* (Budget), *Kessan* (Settlement of Accounts).
Held in the Hōrei Gikai-shitsu (Data rooms for laws, ordinances and the Diet deliberations) of the National Diet Library. These items were brought before the Diet: the draft proposals which were submitted and the accompanying reference materials are held in the Ministry of Finance.

[7] GHQ/SCAP Records. These are such a huge volume of records that the present author was able to scan only parts of the higher education records of CIE and the ESS/ST. As for research funds, it is highly probable they are included as concrete instructions in the materials of the Economic and Scientific Section's Finance Division. A more systematic search should be conducted in the future.
In relation to the ESS/ST, I was able to identify the following:

a) *Science Reorganization-Research Restoration Council, 1947 to JSC no. 5 (Inquiries, Reports), 1951, GHQ/SCAP Records*, Box No. 7399, ESS (B) 11589 through 11632.

b) *History of Technical Intelligence-Southwest and Western Pacific Areas, 1942–1945*, vol. II to SCAPIN 400 A; Appropriation of Funds for Scientific and Technical Activities 1946.5–1946.6, GHQ/SCAP Records, Box No. 7410, ESS (A) 9776-(A) 9778.

c) *SCAPIN 1425-A: Application for Removal of High Voltage Testing Sets to Consolidated Accession List #2, GHQ/SCAP Records*, Box No. 7411, ESS (A) 9779-(A) 9788.

d) *Chronological Listing of Arrival of Radioisotopes to (International) Pacific Science Congress, GHQ/SCAP Records*, Box No. 7417, ESS (B) 11780-(B) 11782.

c) *Associations, Japan to Minutes on the Meeting of the Special Committee for the Investigation and Study of the Atomic Bomb Disaster (National Research Council), GHQ/SCAP Records*, Box No. 7423, ESS (F) 1773-(F) 1775.

[8] Beikoku Gakujutsu Komondan (The United States Scientific Advisory Group), *Nihon ni okeru kagaku to gijutsu no saihensei* (Reorganization of science and technology in Japan), 28 August 1947.

[9] 'Daigaku tō no kenkyūhi no jittai chōsa (Survey reports on research funding of universities and other institutions), six 'occasional issues' in *Gakujutsu*

geppō (Science monthly), 1949–59.
The investigation of greatest scope carried out by the Ministry of Education in relation to the public finances of postwar national universities.
[10] Kōgyō Gijutsu-in (Agency of Industrial Science and Technology), *Kagaku gijutsu-in nenpō* (The Science and Technology Agency annual), from 1951.
[11] Kōgyō Gijutsu-chō (Agency of Industrial Science and Technology), *Kenkyū hakusho* (Research white paper), Nikkan Kōgyō Shimbunsha, 1951.
[12] Kōgyō Gijutsu-chō (Agency of Industrial Science and Technology), *Gijutsu hakusho: Wagakuni kō-kōgyō gijutsu no genjō* (Technology white paper: Mining and manufacturing technology in Japan), Nikkan Kōgyō Shimbunsha, 1949, 1951, 1956.
[13] Kōgyō Gijutsu-in Sōmu-bu Gijutsu Kenkyū Shinkō-ka (Agency of Industrial Science and Technology, General Affairs Department, Technical Research Promotion Division), *Shiken kenkyū josei seido no kaisetsu* (Explanation of the experimental research assistance system), Kagaku Gijutsusha, 1964.
[10]–[13] are valuable in terms of checking trends in applied research funding.
[14] It is not possible to cite the names of concrete materials but the clerical data record books for the science research funds of each university are useful in terms of learning about the nature of research funding. However, unlike university running costs these are dealt with by each faculty. Accordingly, it is virtually impossible to grasp an overall summary of the amount of, for example, the total distributed to universities. In the Archives Section of the University of Tokyo where one can access internal university data, there are a small number of record books relating to science research funding and science research promotional funds. In recent years, the statistical information in this area has been vastly improved, with data on research funds for each university being made available.

Related Data

[15] Gakujutsu Taisei Kenkyū-kai (Science System Research Group), *Gakujutsu kenkyū no haikei* (A background to science research), Nihon Gakujutsu Shinkō-kai (Japan Society for the Promotion of Science), 1953.
[16] Nihon Gakujutsu Shinkō-kai Kagaku Gijutsu Seido-shi Henshū Iinkai (Japan Society for the Promotion of Science, Editorial Committee on the History of the Science and Technology System) (ed.), *Kagaku gijutsu seido-shi: Kagaku kenkyū-hi (Shōwa 21 – 34-nen), Shiryō-hen* (A history of the science and technology system: Science research funding (1946–1959), volume on data), Nihon Gakujutsu Shinkōkai (Japan Society for the Promotion of Science), 1960.

[17] Tetsu Hiroshige, *Kagaku no shakaishi – Kindai Nihon no kagaku taisei* (A social history of science – The structure of modern Japanese science), Chūō Kōronsha, 1973.

[18] Tokyo Daigaku Hyakunenshi Henshū Iinkai (Editorial Committee for a One-Hundred Year History of the University of Tokyo) (ed.), *Tokyo Daigaku hyakunen-shi* (The one hundred-year history of the University of Tokyo), 10 vols., University of Tokyo Press, 1985–1987.

[19] Nihon Gakujutsu Kaigi Nijūgonen-shi Henshū Iinkai, *Nihon gakujutsu kaigi nijūgonen-shi* (Twenty-five year history of the Science Council of Japan), Nihon Gakujutsu Kaigi (Science Council of Japan), 1974.

[20] Ministry of International Trade and Industry (MITI) (ed.), *Shōkō seisaku-shi* (A history of the policies of commerce and industry), 24 vols, Shōkō Seisaku-shi Kankō-kai (Publication Group for the History of Commercial and Industrial Policies), 1965–1985.

[21] Masamichi Rōyama (ed.), *Daigaku seido no saikentō* (A re-examination of the university system), Minshu Kyōiku Kyōkai (Democratic Education Association), 1962.

[22] Genkichi Hara, *Kagaku kenkyūhi* (Science research funds), Kagaku Shimbunsha, 1982.

[23] Ministry of Education, Science International Bureau, *Wagakuni no gakujutsu* (Science in Japan), Nihon Gakujutsu Shinkōkai (Japan Society for the Promotion of Science), 1975.

[24] Yasuo Kuroda, 'Kenkyū-hi taikei to daigaku (The research funding system and universities),' in *Kōza Nihon no daigaku kaikaku, 4* (Series on Japanese university reform, vol. 4), Aoki Shoten, 1982.

16 Reform of Medical Education

Hiroyuki Fujii

The State of Medical Education at the end of WWII

Medical Treatment and the Health System at the end of WWII

The deteriorating public health situation, brought about by poor working conditions, shortages of material things and medical supplies, and the devastation of health facilities during the war, was exacerbated by food shortages and spiraling inflation after the war. There was also a tremendous increase in the number of unemployed people resulting from the flood of disbanded military personnel, repatriated servicemen and the demilitarization of industry.

Infectious diseases went on a rampage and the medical treatment and health maintenance strategies necessary to combat them were insufficient. Many doctors had been pressed into military service and medical treatment facilities had been destroyed or otherwise rendered non-functional.[1]

A GHQ evaluation said: 'At the time of surrender Japanese public health conditions showed every indication of becoming desperate.'[2] Whilst certainly there were problems caused by the devastation wrought both directly and indirectly by the war – i.e., both war damage and general neglect rendered by the prioritization of military needs – there were also problems with the medical system itself, both before and during the war. GHQ was well aware of this. It reported that 'The larger hospitals, operating with a staff of full-time doctors, were closed to the private practitioner. With this closed-staff policy, private practitioners desiring to retain the care of their patients were forced to provide their own hospital facilities. This resulted in the establishment of thousands of inferior hospitals in which it was frequently impossible to provide facilities necessary for the proper care of patients with serious illness.'[3]

The Number of Established Medical Departments and Medical Universities and the Number of Students

At the end of WWII the number of medical schools and students enrolled therein was as follows: thirteen national universities (1626 students), one public university (117), four private universities (565) and fifty medical specialist schools (8225).

In terms of the Medical Faculties of national universities, there were the so-called 'Seven Imperial Universities': Tokyo Imperial University (established in 1877) Kyoto (1899), Tōhoku (1880), Kyūshū (1903), Hokkaidō (1919), Osaka (1915), and Nagoya (1920). There were also six medical universities: Niigata (1922), Okayama (1922), Chiba (1923), Kanazawa (1923), Nagasaki (1923) and Kumamoto (1922). The public university was Kyoto Prefectural Medical University (1921) and the private universities were Keiō University's Medical Faculty (1916), Jikei-kai Medical University (1921), the Medical University of Japan (1926) and Nippon University's Medical Faculty (1925).

The medical universities had each developed from specialist medical schools and had been incorporated as universities by the University Order of 1918.

Other specialist medical schools at the end of the war included: Tokyo Ladies' Medical Specialist School (est. 1900), Imperial Ladies' Specialist School (1915), Tokyo Medical Specialist School (1918), Osaka Senior High School Specialist Medical School (1927), Kyūshū Senior High School Specialist Medical School, Iwate Specialist Medical School, Osaka Ladies' Specialist Medical School and Shōwa Specialist Medical School (1928). These were recognized for training practicing doctors and became universities during the postwar reforms. Other than these the great majority were specialist medical schools established after the proclamation of the 1938 National General Mobilization Law.

In response to the increased demand for military doctors as the war developed, in 1939 specialist medical faculties were established as provisional affiliations of the seven Imperial Universities and the six government medical universities. In 1943 the 'Wartime Emergency

Table 16-1: Increase in Medical Schools and Students, 1939–1945

	Number of schools		Number of students	
	1939	1945	1939	1945
Universities and medical colleges (a)				
Government	13	13	5,100	5,640
Prefectural	1	1	320	480
Private	3	4	1,920	2,320
Senmon Gakkō (b)				
Government	0	20	0	8,485
Prefectural	0	19	0	4,576
Private	9	12	5,970	6,720

Notes (a) Medical departments of universities and medical colleges.
(b) Technical and vocational schools.
Source: Ministry of Health and Welfare.

Table 16-2: Number of medical students as of 1945

Type of institution	Names or numbers of institutions	No. of students
Imperial universities	Tokyo, Kyoto, Tōhoku, Kyūshū, Hokkaidō, Nagoya	
Medical universities (national):	Niigata, Okayama, Chiba, Kanazawa, Nagasaki, Kumamoto	
imperial universities + national medical universities		5640
Medical universities (prefectural):	Kyoto Prefectural	480
Medical universities (private):	Keiō University, Jikei-kai Medical University, Nihon Medical University, Nihon University	2320
Medical Specialist Schools (national)	20	8485
Medical Specialist Schools (metropolitan and prefectural)	19	4576
Medical Specialist Schools (private)	12	6720

According to Chart 14 of the hygiene statistics of 'The eighty year history of the medical system' the number of medical schools as of 1945 stood at 63 – there were 17 universities (13 government, 1 public and 3 private) and 46 specialist schools (19 national, 18 public and 9 private). The number of graduate students stood at 3859. The difference between these figures and GHQ data (see Bibliographic Guide [3]) is thought to be due to the fact that there were some schools which, although established, had not produced any graduates.

Measures Policy Relating to Education'[4] was adopted to establish specialist medical schools and medical specialist faculties of a provisional nature in all geographical regions of Japan and Manchuria, and to reduce the length of courses in order to ensure the training of a large number of doctors.[5]

Thus, in 1945 the total number of university places offered for studying medicine was 10,533, which was greater than at any other time after the war. A majority of these places were provided by special medical schools.[6]

The Educational Content of Medical Schools

It is important to recognize that reforms to the medical education system in the US, following the Flexner Report of 1910, provided the background to GHQ's evaluation of the educational content of Japan's medical schools. The US reforms included abolishing medical schools that could not maintain acceptable standards, emphasizing the importance of pre-medical education and biological medical science and fully implementing an internship system. The specialization of American medicine accelerated after this and, with the war increasing the opportunities, the specialist medical system settled into place.[7]

The Gap between different Medical Schools

There were different course requirements between the various types of educational institutions – especially the Imperial Universities' Medical Faculties, medical universities and the specialist medical schools on many different fronts. The students enrolled at university had come through the old high school system medical stream (three years of high school) and then completed a four-year specialist course at University. By contrast, students at the specialist medical schools were admitted straight after graduation from five years of junior high school and were only required to complete one year of basic education and three years of specialist study at tertiary level. The specialist schools were reportedly inferior to the universities in terms of both teachers and facilities. According to GHQ's analysis: 'Laboratory space was rarely adequate to accommodate the whole class at one time and frequently there were only enough microscopes for a few of the students to utilize at one sitting.'[8]

Lack of Importance Placed on Experiments

In both the universities and specialist schools, educational content centered on lectures. Pre-clinical education (anatomy and physiology – in other words, basic medicine) nowadays centers on practical work involving the human body and involves the use of specimens and animals, but in those days the importance for students to experience things for themselves was not recognized and the great majority of instruction occurred through students observing the teachers' demonstrations. Furthermore, in the hierarchical faculty system of the university there was more prestige derived from instructing post-graduate students and, accordingly, professors were averse to teaching the 'general student' populace.[9] 'Students were rarely properly guided in performing satisfactory research problems under controlled conditions.'[10]

Lack of Clinical Guidance

The focus on providing education through lectures delivered to large numbers of students applied to clinical subjects as well. GHQ estimated that 'Throughout the four years of medicine, about 70 percent of the time was devoted to lectures and about 30 percent to experimental and clinical work combined.'[11] By way of example if we look at the weekly lecture timetable for the third and fourth year students at a certain medical university (about 1930–1939) we see that whereas there were twenty-five to thirty hours of lectures there were only between two and nine hours of outpatient clinic lectures (refer to Table 16-3).[12] In other words, 'Japanese medical students received little clinical instruction.'[13]

There were also great discrepancies between medical universities, the leading medical specialist schools and the provisional medical specialist schools with regard to the use of university hospital educational diagnostic facilities.[14]

Table 16-3: Chiba University Medical Department curriculum circa 1930–1940

Subject	Term I	II	III	IV	V	VI	VII	VIII	IX	X	XI	XII	no of terms
Anatomy													
Systems Anatomy	8(10)	12											2
Organizations	5(2)												1
Anatomy Practice		4 x	3 x										2
Organizational Studies Prac	3 x		3 x										2
Use of Microscopes	3 x		3 x										1
Viviparity Studies #			2										1
Topography #				2									
Physiology													
Physiology	5	5(6)	3										3
Physiology Practice				2x(3)	1x(0)								2(1)
Medical Science													
Medical Science	5(4)	2(3)											2
Medical Science Practice	2x2x(1)												1(2)
Microbes And Serology													
Microbiology	2	2											2
Serology			2										2
Serology Practice				2x	2x								1
Pharmacology													
Pharmacology			2(0)	2(4)	3								3(2)
Pharmacology Practice						2x							1
Prescriptions #					1								1
Pathology													
Introduction to Pathology			4(6)	3(0)									2(1)
Pathological Anatomy			3(6)	4(8)									2
Pathological Structures Prac					2 x	2x0							2(1)
Abnormal Material ID									2	2	2	0(2)	3
Pathological Anatomy Techniques #					1(2)	1(0)	1(0)						3(1)
Hygenics													
Hygenics and Practice				(4)	(2)				2(0)	2(0)	2(0)		3(2)
Social Health Studies #										(1)			(1)
Legal Medicine													
Legal Medicine				(4)	(2)					3(0)	3(0)		2
Internal Medicine													
Diagnostics and Treatment			3(8)	3(0)	3(0)								3(1)
Internal Medicine													3(1)
Discussions				4(0)	4(0)	4	4(3)	4(3)	4				6(4)
Internal Medicine Clinical Lectures							4	4	4	4	4	0(4)	5
Internal Medicine Day Patient Clinical Lectures						*(0)	*	*	*	*	*	0(*)	5
Diagnostic Practice #						*							
Physical Therapy													
Physical Therapy #						0(1)				2			1(2)
Introduction to Surgery			3	3(2)	3(2)								3
Surgery Discussions				4(0)	4(0)	4	4(3)	4(3)	4				6(4)
Surgery Clinical Lectures							4	4	4	4	4	0(4)	5
Surgery Day Patient Clinical Lectures						*(0)	*	*	*	*	*	0(*)	5
Surgical Operation Practice #								0(*)	0(*)				1
Bandage Practice #						*							
Dentistry													
Dental Clinical Lecture											2		1
Obstetrics and Gynecology													
Gynecology						2	2	2					3
Obstetrics							2	2	2				3
Obstetrics Clinical Lectures										2	2		2
Obstetrics Day Patients Clinical Lectures										*	*		2
Maternity Model Practice #											*		1

Subject	I	II	III	IV	V	Term VI	VII	VIII	IX	X	XI	XII	no of terms
Opthalmology													
Opthalmology						3(2)	3	2					3
Opthalmology Clinical Lecture									1	1	1	0(1)	3
Ophthalmology Day Patient Clinical Lecture									*	*	*	0(*)	3
Eye Examination Practice #								*					1
Otorhinolaryngology													
Otorhinolaryngology						2	2	2					3
Otorhinolaryngology Clinical Lecture									1	1	1	0(1)	3
Otorhinolaryngology Day Patient Clinical Lecture									*	*	*	0(*)	3
Otorhinolaryngology Investigative Methods Practice #								*					1
Skin and Urology													
Skin Studies and Minor Poisons						2	2	2					3
Skin Studies Clinical Lectures									1	1	1	0(1)	3
Urology						1	1	1					3
Skin and Urology Day Patient Clinical Lecture									*	*	*	0(*)	3
Pediatrics													
Pediatrics							2	1					2
Pediatrics Clinical Lecture									1	1	1	0(1)	3
Pediatrics Day Patient Clinical Lecture									*	*	*	0(*)	3
Psychiatry													
Psychiatry								2	2				4
Psychiatry Clinical Lecture										2	2		2
Psychiatry Day Patient Clinical Lecture										*	*		2
Special Lectures #													
Total no of hours for lectures	25(23)	21(23)	20(14)	25(22)	23(16)	19(29)	30	30	30(26)	25(21)	25(20)	0(14)	
Total no of pracs	3(4)x	6(5)x	6x	4(5)x	5(6)x	4(0)x							
Total no of day patient clinical lectures						4(2)*	2*	4*	7*	8*	9*	0(6)*	

# indicates subjects without examinations.
Figures are from around 1930–1940.
The numbers in ( )s represent the number of hours that were required during the war.
'x' represents the frequency of practical sessions.
Source: 'The eighty-five year history of the Medical Faculty at Chiba University' 1964.

This lack of emphasis on practical or clinical experience reflects a tradition of neglecting the technical education aspects of medical education before the war. As GHQ pointed out, 'There was considerable emphasis in Japan upon medicine as an abstract science. The researcher felt that completion of a piece of research work, however impractical it might be, was essential to a position of respect in the community.'[15] Strong traces of this survived both the war and the postwar reforms and are still with us today (we will return to this later).

The Qualification System and Practical Training

The national examination system for medical qualifications was first established in 1875, one year after the Government announced the regulation of the medical system. Some three decades later, the Medical Doctors Law was enacted in 1906. These two sets of regulations paved the way for the full specifications of qualifications for medical practices and the national licensing scheme controlled by the Minister for Home Affairs. However, graduates from the medical departments of the Imperial Universities and graduates from government and public universities as well as specialist private medical schools designated by the Minister for Education were considered to have qualified without sitting the national medical practitioners' examination.

The People's Medical Law of 1942 established a minimum of one year's practical training as a prerequisite to a license to practice medicine. However, the war generated a need for large numbers of doctors and this requirement lost its impetus; the amount of medical treatment practice received in the course of the formal school education was then deemed to be sufficient,[16] despite the fact that the provision of practical training in these courses was negligible; it didn't really go beyond lip service.

The Road Forward after Graduation

As previously mentioned, the most respected Japanese medical doctors were those who had achieved success in research. Thus, graduates from the more prestigious universities generally sought employment in the medical office of their university and would 'work in a sub-ordinate position for a number of years then be given a research theme by their professor and spend a number of years further conducting research in a basic medical educational setting. The aim was to make use of one's research as soon as possible and write one's doctoral thesis. The whole point was to work towards a medical doctorate.'[17]

Opportunities to study in Germany and Austria ceased when diplomatic ties were cut with Germany as a consequence of World War I. To maintain the supply of doctors, the Government amended the requirements for academic degrees in such a way that 'the privilege of conferring degrees is awarded to government, public and private universities other than imperial universities.'[18] This led to the establishment of the Japanese *ikyoku* system. *Ikyoku* literally means the department which deals with the medical affairs or the room where doctors are stationed in hospitals. In practice, under the *ikyoku* system, chair professors exercise enormous power, not only over the academic, administrative and financial matters of the universities and hospitals to which they belong, but also over the appointment and distribution of doctors in and between the medical institutions under their control. The system is built upon several centralized medical hierarchies at the apex of which full professors – at the top universities – have a wide discretionary and covert power over low-ranking medical institutions, including their personnel matters. The practice has had implications for the independence of Japanese medical studies and the formation of the closed labor market for Japanese doctors.

Most of the graduates from the medical specialist schools went straight into practice without sufficient training, either before or after graduation.[19] In contrast, the graduates of universities were considered to be first class. 'Most of these first class doctors stayed in the big hospitals of the major cities and the environs of Tokyo. Consequently the poorer education provided by the specialist medical schools...was necessary to answer the demand for doctors in the country...The result was that the people in the major cities enjoyed the services of first class doctors but people in the country areas...were treated by second rate medical practitioners.'[20]

The Commencement of the GHQ Medical Education Reforms: The Reforms of the Sams Era

Colonel Crawford F. Sams, the head of the American Pacific Forces General Headquarters Military Administration Bureau's Health, Education and Social Welfare Section from May 1945, arrived in Yokohama on 30 August, the same day that MacArthur landed at Atsugi. Sams was a graduate of the University of Washington and did his internship at a military hospital in California. His career as a military doctor had taken him from military school to service as a military doctor in the Middle East and Panama. Immediately prior to his assignment to the US Pacific Forces GHQ he had been the Head of the Department of the Army's Planning Division, No. 4 Planning Section. In this capacity he had participated in the strategic planning for the invasion of the Japanese home islands. On 2 October he was appointed Head of the Public Health and Welfare Section (PHW) of the General Headquarters, Supreme Commander for the Allied Powers (GHQ/SCAP). The staff that he personally appointed to PHW numbered in the thousands and included not only those he knew from his days as a military doctor and those he had trained at the military medical school but also medical school graduates who had been conscripted after the war plus one hundred and fifty civilian specialists.[21] GHQ's reform of medical education was conducted with him at the helm until his retirement in May 1951.[22]

Periodization of the Reforms

The 'Eighty Year History of the [Japanese] Medical System' divides the postwar reforms of the medical and health system into three time periods: from late 1945 to the middle of 1947 when several strategies were adopted; mid-1947 until mid-1948, during which various measures were taken in accordance with the establishment of the Japanese Constitution; and finally the period from 1948 forward, when steps were taken to implement extensive reforms.[23] However, serious reforms had been implemented in the medical education system from relatively early in the Occupation. Hence, in the discussion that follows, while maintaining a division into three time periods, the periodization is marginally different than that outlined above.

Emergency Measures Immediately Following WWII

On 22 September 1945 GHQ's PHW issued a 'memo on public health' [24] which ordered that a survey be conducted to determine appropriate emergency reform measures for the public health system. In January 1946 a temporary system[25] was established to recognize the medical qualifications of those members of the medical profession who were returning from overseas. To the best of my knowledge, this was the only emergency measure taken.

The 'Medical Education Commission of Inquiry': Direct Reforms by GHQ – August 1945 to October 1948

The 'Medical Education Commission of Inquiry'

On 11 February 1946 Sams issued a policy memo that ordered an official inquiry into medical education in Japan.[26] On 23 February the Medical Education Commission of Inquiry Formation Preparation Group was convened in PHW which, in turn, decided to hold the Medical Education Commission of Inquiry Inaugural General Meeting at GHQ on 27 February.[27] The commission was established under the auspices of PHW, and consisted of five members – Yoshio Kusama (Head of the Medical Faculty at Keiō University/Public Health Studies), Chūjirō Nishino (Keiō), Yoshinori Takagi (Jikei), Takeo Tamiya (University of Tokyo) and Ryōjun Kinoshita (Osaka University) – plus observers from the Ministry of Health and Welfare, the Ministry of Education and the Japan Medical Association.[28] Yoshio Kusama was appointed the Chairman of the Commission.

In his address to the inaugural meeting, Sams criticized what he called the 'Teutonic institutions' – i.e., the alma mater cliques, the existence of 'a medical scientific elite class' and the autocratic rule of professors – of university medical faculties.[29] These criticisms were later repeated by Major Moolton of GHQ's Medical Education Section in a press interview given on 6 May 1946.[30] Sams and Moolton both claimed that these institutions were responsible for compromising the academic standards and the social standing of practicing doctors.

At the meeting, Sams proposed a plan for general reforms which included a re-organization and improvement of the medical schools, education courses, facilities and textbooks, plus the introduction of a system of internships and national examinations.

These proposals were modeled on the 1910–20 medical education reforms implemented in the US, but they also attempted to incorporate the findings of the investigations conducted into the Japanese situation, including the Japanese medical educators' hearings that had been underway since the previous year.[31] Further, it should also be recognized that PHW/GHQ did not merely pass judgment from the position of an assumed superiority based on their military victory, but in fact, were reflecting on problems similar to those faced in the US medical system. To that extent, PHW was both cautious and fair (as will be discussed below).

As early as 30 August 1946 an Imperial edict was issued entitled the 'Partial Amendment to the National Medical Law Execution Order' which declared that a national doctor's examination and a system of practical training were made prerequisites to *all* medical graduates who wanted to be qualified as doctors. These requirements were now enforced across the board, uniformly and without exceptions. In October 1946 the Ministry of Education created an 'Essential Points Proposal for Reforming Medical Education' based on advice received from the Ministry of Education's Commission of Inquiry. On 1 November 1946 the first national doctors' examination was administered (but only for students who had graduated in March 1946 from Keiō University and the Specialist Medical School attached to the Jikei Medical University).[32] In 1947, the March '46 and March '47 graduates commenced the first internships together. In 1948 the Medical Practitioner's Law, which incorporated these stipulations, was proclaimed together with the Dental Practitioner's Law, the Public Health Nurse, Mid-Wife and Nurse Practitioner's Law, the Dental Technician Practitioner's Law and the Medical Treatment Law.

On 12 April 1947 the Commission of Inquiry submitted its report recommending a seven-year medical course which would make the completion of a three-year degree a condition for enrolment in a medical university in the new system. The report also recommended abolishing the lecturing system and adopting a classroom based approach to learning as well as appointing deputy professors.[33] These were quite radical reform proposals which followed on from the implementation of the Education Basic Law and the School Education Law of March that year. They paved the way for the new system of education that was gradually being established.

The Preparation and Re-organization of Medical Schools

Reforms of medical schools commenced in 1947 and continued into 1948. Depending on the existence or otherwise of the requisite facilities, experimental equipment, affiliated hospitals and professorial profiles the schools were divided into 'A class' and 'B class.' 'B class' schools were granted provisional licenses. Those schools that were deemed to have no potential to meet the new standards in the future were closed down.[34] Thus of the fifty-one specialist medical schools extant at the beginning of the reform process, twenty-four were closed (a number of these were revived as new medical universities in the 1960s and '70s). On 10 February 1948 all of the government specialist medical schools were promoted to university status, followed by others in rapid succession so that by the end of 1948 there were thirty-nine such schools. This number increased to forty-six by the end of 1950.[35] No new medical faculty was established for fifteen years from 1947.

The Establishment of a National Examination System

A Doctors Examination Commission of Inquiry was established and chose the members for an Examination Group comprised of seventeen educators from a variety of medical fields. The commission proposed policies for a national examination and supervised the committee responsible for preparing the examination and administering it twice a year.[36] The first national examination of all medical faculty graduates was administered in April 1947.

Initially the examinations took four days and consisted of approximately twenty hours of written tests. The pass rate for the first six national examinations was in the order of sixty percent. In 1949, after the sixth exam, the process was reviewed, and subsequently, it was not unusual, in the case of the spring examinations, for pass rates to approach the ninety percent mark.[37]

The Establishment of the Internship System

The amendments to the National Medical Law Execution Order of 30 August 1946, as mentioned above, made practical training a prerequisite for sitting the national examinations.[38] In October 'A Synopsis of Practical Training as it pertains to Diagnosis, Treatment and Public Health' was published by the Ministry of Health and Welfare, who also determined that a committee would be established to determine the length and contents of the practicums as well as what facilities would be required.[39] Initially, two hundred and sixty national, public and private hospitals and sixty-nine health centers were designated nationwide as training facilities.[40]

The first internships commenced in 1947, undertaken by those who had graduated in March of that year. However, the 'students who had graduated from the specialist medical schools attached to Keiō and Jikei in March 1946 were also required to undertake practical training – commencing in April 1947 – as an administrative measure prior to the Imperial edict amendment.'[41]

Figure 16-1: Numbers of successful and unsuccessful candidates in the national examination for a licence to practice medicine

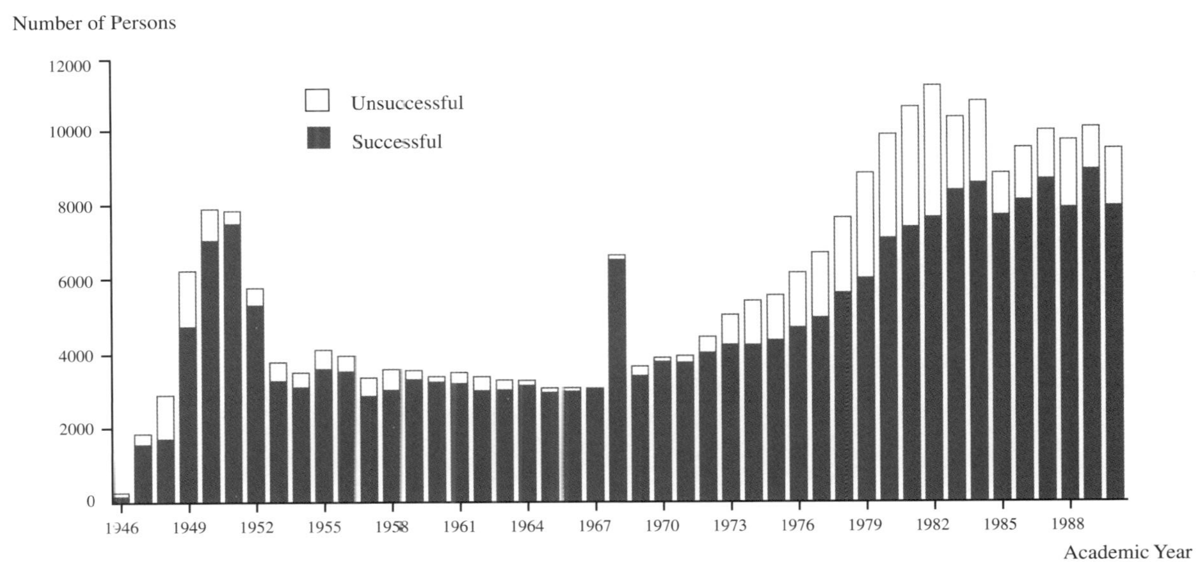

Sources: *Igaku kyōiku hakusho* (White paper on medical education), 1986 and 1990 editions.

The custom of graduates joining a university hospital and receiving coaching existed in some quarters before the war. However, this was an elitist custom open to only a select few; only those who hailed from the leading universities were able to receive such practical training and the number of facilities offering such training was limited.[42] The internship system was therefore epoch making in terms of the practical training it offered to graduates from all medical faculties and by virtue of the fact that it promoted a new course structure that included: looking after a large number of patients; a system of rotation; emphasis on pathological anatomy; and regularly conducted group discussion sessions. The absence of public health training from the medical education system had been a failing of the prewar medical education system, and hence, its introduction was epoch-making. However, this aspect of training was removed from the curriculum in 1968 and today, at the beginning of the twenty-first century, there is again this serious shortcoming in the education of medical practitioners.

The Japanese system of internships was, however, different from the US system inasmuch as the interns had no form of license and were unpaid.[43] Furthermore, most of the treatment facilities and health centers designated as training institutes lacked the requisite infrastructure and experience in training. As a result, particular designated training facilities were swamped with applications and those applicants with no personal connections were effectively excluded from the hospitals.[44] In 1949 the forecast was that more than 1800 hopefuls would not be placed in hospitals nationwide. Due to this the Training Committee gradually affected a great increase in the number of additional designated facilities. There were seven additional hospitals in March 1947 and nine more in July that year.[45] In the autumn of 1948 there were an additional one-hundred and two,[46] and fifty-three more in 1949.[47]

The ensuing chaos was an acute problem for medical students. In 1948 the National Universities Medical Students' Association sought legal guarantees and submitted a draft law on internships to the Training Committee of the Ministry of Health and Welfare.[48] The question of internships became a social issue and the demands of the interns (known as the 'Tri-Ask' demands; three demands on the guarantee of satisfactory salaries, stable positions and good educational content) led to the subsequent internship campaign.

The 'Medical Education Committee' Period – Post-October 1948

In October 1948 the Medical Education Commission of Inquiry became the Medical Education Committee, an organ of the Japan Medical Association. The chairman of the Commission of Inquiry, Yoshio Kusama, was appointed chairman of the Committee. The seventeen committee members were joined by two observers from the Ministry of Health and Welfare and two from the Ministry of Education.

Kusama received his medical education at Stanford and had developed a very close relationship with the Rockefeller Foundation. Before the war he had gone to Johns Hopkins University with financial assistance from the Foundation to conduct research on a course he was preparing in preventative medicine at Keiō University. His US education, his contacts and his fluent command of American English probably all contributed to his appointment as the Chair of the Commission and later the Committee. While serving as Chairman of the Medical Education Committee, he became the 'face' of medical education in postwar Japan.[49] In this role he traveled to the US and Canada in October 1949 and stayed there until February 1950 to observe the educational methodology [50] and the specialist doctor system. Upon his return to Japan he produced a tentative plan for introducing a similar system for Japan.[51]

In 1950 and '51, a 'medical education delegation comprised of twelve US professors' came to Japan to demonstrate educational techniques and pass on new teaching materials to the Japanese professors. The American Unitarian Service Committee sponsored this delegation. American professors gave lectures to Japanese professors and demonstrated their educational methodologies.[52] There were two meetings conducted, one in Tokyo and one in Osaka. Both were proclaimed a success, with professors who had not been programmed to attend participating at their own expense.[53]

It is widely believed that this delegation was the result of approaches made by Yoshio Kusama while he was in the US. It was fairly common during the Occupation for delegations (or 'Missions'; see ch. 8) of US specialists to travel to Japan to offer advice and provide training to their Japanese counterparts. The US Education Mission, which had visited Japan in 1946, greatly influenced the investigations and reform proposals of the Medical Education Commission of Inquiry.[54]

According to GHQ, the Medical Education Committee's proposed '...new medical curriculum emphasized practical methods of instruction, including demonstrations, laboratory experiments, anatomical dissection and attendance at autopsies. (Practical instruction of students at the bedside and dispensaries was particularly stressed.) Fifty-three percent of the required hours were devoted to clinical teaching and lectures, primarily to supplement the clinical subjects. The other 47 percent of the required hours were devoted to preclinical subjects with a preponderance of these hours spent in practical laboratory work.' [55]

However, the Committee wielded little influence over the Ministry of Education and thus there were no concrete outcomes from these proposals beyond the servicing and re-organization of medical schools (primarily the elevation of medical schools to medical university status).[56]

The Changes and Continuities Resulting from the Reforms

From the German Model to the US Model

The Japanese medical system's dependence on Germany began when the Meiji government adopted German medical practices. This is clearly evident in the tradition of graduates from the Imperial Universities taking up professorial posts at medical universities after traveling to Germany to study and by the German language textbooks, lectures and medical reports used in Japan. As previously mentioned, opportunities to study in Germany ended after WWI.[57] The defeat in WWII, the US Occupation and the medical education reforms that were implemented, fixed the orientation of Japanese medical practice towards the US.

The transition away from German models and towards US models is very clearly evident in language studies. After WWII, textbooks were from the US and patients' charts were modified to accommodate the English language as the postwar generation of medical students moved through the system. The US was very clearly the main destination for medical students studying overseas,[58] the students who came to be the elite core of medical researchers.

However, the explanation for these changes in focus cannot be reduced to the US Occupation of Japan and GHQ's role in reforming the medical education system. We must also recognize the economic strength of the US after WWI, which was further enhanced after WWII, and the general relocation of the world center of higher education from Europe to the US during the first half of the twentieth century (as discussed in ch. 10). Furthermore, the reforms to the US medical system during the 1910s provided Sams and his medical education advisors with both the competence and the confidence to make observations and offer advice. We must also recognize that the US came to the forefront in the fields of medical technology, with developments such as antibiotics, transfusions, anesthetics, etc. For all of these reasons, Japan abandoned the German models of medicine and turned towards the US model.

The Medical Office Lecturing (*Ikyoku*) System

As previously mentioned, the dominant career path for medical graduates was to apply for an appointment to the University medical office where they would work under the supervision of a professor for several years. In addition to practicing medicine in an affiliated hospital – also under the professors' supervision – the young doctor would conduct research towards completing a medical doctorate, and this research topic would be of the professor's choosing.[59] Thus their professors were the absolute authority as far as education and research were concerned, as indeed they were in terms of finding employment. This is what is known as the Japanese style medical office lecturing system (the *ikyoku* system).

Although GHQ criticized this system from the earliest days of the Occupation,[60] and despite the fact that the Medical Education Commission of Inquiry lodged a radical reform plan,[61] the response from the University Standards Association Medical Education Deliberation Committee,[62] the body of inquiry of the Ministry of Education, was silent contempt. And since the Ministry of Education was the governing authority for the university system, the medical office lecturing system lived on,[63] blocking reforms to the doctor training and employment system.

The Specialist Doctor Authorization System

The US system of medical treatment, which provided the model for the reforms, already had an established system of specialist doctors and amongst the GHQ advisers there were those who recommended introducing a similarly specialized system. Before the war, in Japan too, the Medical System Improvement Policy expressed in 1940 by the Medical System Commission of Inquiry, proposed 'a national system of certification in relation to the professing of specialist treatment' in order to effect improvement in the provision of medical treatment.[64]

However, such a system was never introduced in Japan. It was only through later medical technology reforms that advances were made in the area of medical doctor specialization. However, even today medical specialization has not been incorporated into law, which suggests a great historical significance to the fact that GHQ did not introduce a system of medical specialists.

One reason why this system was not introduced lies in the fact that Sams himself felt that there were problems with the specialist doctor system. At the root of this was his view of medical techniques which held that a medical specialist must not only be a respected doctor in all regards but, beyond that, he must have specialist skills. These must be the qualifications required of a specialist.'[65] He appears to have been fully aware of abuses of the US medical specialist system at the time, saying: 'I myself am not sure whether or not the sort of system operating in the US is really appropriate' and 'I felt that an extreme system that trained all Japanese doctors to become specialists should not be introduced to Japan, at least until, through economic incentives, a nice balance is obtained in the ratio of specialists and general practitioners' (which accounted for approximately eighty-five percent of private medical treatment).[66] More than anything we must not forget the fact that he himself was a military administrator and not a medical specialist and could thus view the situation from a broad perspective.

Co-medicine

Our discussion so far has been about the Japanese home islands where the Occupation Forces exercised indirect control after WWII. The situation was different in Okinawa, however, which was under direct control during the Occupation. The shortage of doctors was even more acute there than in the home islands. In response to this shortage they developed their own system of doctor assistants known as 'medical officer helpers.' The idea of having medical officer helpers was conceived at the close of the war when there were a mere sixty-four doctors on hand. 'The term referred to people who had local doctor's licenses in Taiwan and Korea and who had experience working in hospitals, as well as the so-called pharmacists' assistants, non-commissioned health officers and medics. They were all recruited, registered and engaged as doctor's assistants.' Several of them not only did the work of doctor's assistants but also were appointed as heads of medical centers in country areas where doctors didn't go. 'In Okinawa, where the number of doctors was exceedingly small, we certainly should not underestimate the value of the work the medical office helpers did.'[67]

The medical situation under US military control in Okinawa needs to be studied as a separate issue. There are problems there which might be informative for the study of Occupation policy in the home islands of Japan.[68]

I have restricted my comments to the training of doctors, which represents medical education in a narrow sense of the term. However, the significance of the Occupation reforms in relation to co-medical education, namely that of the medical specialist professionals other than doctors (nurses, pharmacists, physiotherapists and occupational therapists) must also be recognized. In particular the reforms to nursing education were, in a sense, more significant than the proposals for reforming medical doctor training.[69]

Conclusion

The GHQ-initiated reforms to the Japanese medical education system were successfully adopted, introducing substantial elements of the United States medical education system – a system which maintained the highest standards in the world – into Japan. However, the reformers failed to either correct or abolish the *ikyoku* chair system, which had significant influence over the entire medical system in Japan by controlling the labor market of doctors. It appears that Colonel Sams' reluctance to impose the specialist medical system so deeply entrenched in the US medical system to Japan reflected his personal medical philosophy, but was perhaps also an attempt by him to avoid imposing a new hierarchical structure of specialization into an already tightly controlled community.

Notes

1 Ministry of Health and Welfare, Medical Affairs Bureau, 1955 [1] p. 60; Kawakami, 1965 [2] pp. 493–4.
2 GHQ, 1950 [3] (Chapter I – 'Problems and Politics: Public Health Conditions in 1945'), p. 1.
3 GHQ, 1950 [3] (Chapter XII – 'Hospitals and Clinics: Conditions at the beginning of the Occupation'), p. 144.
4 Ministry of Health and Welfare, Medical Affairs Bureau, 1955 [1] p. 165.
5 Kawakami, 1965 [2] p. 464.
6 GHQ, 1950 [3] p. 133; Ministry of Health and Welfare, Medical Affairs Bureau, 1955 [1] p. 825.
7 Ogishima and Shimomura, 1970 [5] pp. 1–23.
8 GHQ, 1950 [3] (Chapter XI – 'Medical Education and Standards: Pre-surrender Medical Education'), p. 130.
9 Sams, 1986 [6] p. 231.
10 GHQ, 1950 [3] (Chapter XI – 'Medical Education and Standards: Pre-surrender Medical Education'), p. 130.
11 GHQ, 1950 [3] (Chapter XI – 'Medical Education and Standards: Pre-surrender Medical Education'), p. 130.
12 Chiba Daigaku Igaku-bu Sōritsu Hachijūgo-shūnen Kinen-kai, 1964 [7] p. 111.
13 GHQ, 1950 [3] (Chapter XI – 'Medical Education and Standards: Pre-surrender Medical Education'), p. 130.
14 GHQ, 1950 [3] (Chapter XI – 'Medical Education and Standards: Pre-surrender Medical Education'), p. 130.
15 GHQ, 1950 [3] (Chapter XI – 'Medical Education and Standards: Pre-surrender Medical Education'), p. 130.
16 Ministry of Health and Welfare, Medical Affairs Bureau, 1955 [1] p. 166.
17 Kawakami, 1965 [2] p. 162.
18 Kawakami, 1965 [2] p. 322.
19 Sams, 1986 [6] p. 231.
20 Sams, 1986 [6] p. 233.
21 Sams, 1986 [6] pp. 419–25.

22 Nihon Kagakushi Gakkai, 1967 [10] p. 320, (main staff of the Occupation Forces' Public Health and Welfare Section).
23 Ministry of Health and Welfare, Medical Affairs Bureau, 1955 [1] p. 61.
24 GHQ, 1950 [3] (appendix P30 AG710 SCAPIN 48, Public Health Measures).
25 Ministry of Health and Welfare, Medical Affairs Bureau, 1955 [1] p. 64.
26 Sugiyama, 1995 [11]
27 Sams, 1986 [6] (The Occupation Medical Welfare Chronological Table), p. 13.
28 Sams, 1986 [6] p. 244.
29 Sams, 1946 [12]
30 Moolton, 1946 [13]
31 GHQ, 1950 [3] p. 134.
32 Ministry of Health and Welfare, Medical Affairs Bureau, 1955 [1] p. 166.
33 'Igaku Kyōiku Shingi-kai Tōshin (Report of the Deliberative Council on Medical Education),' 1947 [14] p. 279.
34 GHQ, 1950 [3] p. 134.
35 Nihon Igaku Kyōiku Gakkai (Japan Medical Education Society), *Igaku kyōiku hakusho* (A white paper on medical education), 1990 [15] (University Medical Deptartment/Medical Universities, Table 2), pp. 7–9.
36 GHQ, 1950 [3] p. 139.
37 Yonezō Nakagawa, 'The Doctors' National Examination' (chart 1, 'The Doctors' National Examination Pass Ratio') [16] 1972, p. 61.
38 Ministry of Health and Welfare, Medical Affairs Bureau, 1955 [1] p. 166.
39 Sams, 1986 [6] pp. 244–5.
40 Higashida, 1946 [17] p. 505.
41 Ministry of Health and Welfare, Medical Affairs Bureau, 1955 [1] p. 166.
42 Sams, 1986 [6] p. 242.
43 'Taylor-hakushi kōen...(Dr. Taylor's Lecture...),' 1948 [18]
44 'News: jitchi kunrensha no haichi...(Partial Amendment to the Rules Governing...Trainees...),' 1948 [19]
45 'Shūren shisetsu tsuika shitei (Additional Designations as Training Facilities),' 1947 [20]
46 'Intaan shitei byōin 102-kasho tsuika (Intern hospitals: An additional 102 locations),' 1948 [21]
47 'Intaan shitei byōin 53-kasho tsuika (Intern hospitals: An additional 53 locations),' 1949 [22]
48 'Intaan hōki-an sakusei...(Framing draft intern legislation...),' 1948 [23]
49 Kusama, 1983 [24]
50 Kusama, 1950 [25]
51 Kusama, 1950 [26]
52 GHQ, 1950 [3] p. 134.
53 Chiba Daigaku Igaku-bu Sōritsu Hachijūgo-shūnen Kinen-kai, 1964 [7] p. 154.
54 GHQ, 1950 [3] p. 134.

55 GHQ, 1950 [3] (Chapter XI – 'Medical Education and Standards: Pre-surrender Medical Education'), p. 130.
56 Chiba Daigaku Igaku-bu Sōritsu Hachijūgo-shūnen Kinen-kai, 1964 [7] p. 154.
57 Kawakami, 1965 [2] pp. 322–4.
58 Kusama, 1945 [27]
59 Kawakami, 1965 [2] pp. 322–5.
60 Sams, 1946 [12]
61 'Igaku Kyōiku Shingi-kai Tōshin (Report of the Deliberative Council on Medical Education),' 1947 [14] p. 279.
62 'Iiku kikan shisetsu hyōjun…(Facility standards for medical education…),' 1948 [28]
63 Ministry of Education Ordinance No. 28, 1956 [29]
64 Ministry of Health and Welfare, Medical Affairs Bureau, 1955 [1] p. 166.
65 Sams, 1948 [30]
66 Sams, 1986 [6] p. 248.
67 Teruya, 1987 [31] p. 24.
68 Teruya, 1987 [31]
69 Shimazaki, 1990 [32]; Kawashima, 1992 [33]

Bibliographic Guide

Basic Literature, particularly regarding public hygiene and treatment at the end of WWII

[1] Ministry of Health and Welfare, Medical Affairs Bureau, *Isei hachijūnen-shi* (The eighty year history of the medical system), Ministry of Health and Welfare, 1955.
Subsequently, *Isei hyakunen-shi* (The onc-hundred year history of the medical system, 1976) appeared encompassing a greater scope but most of the chronological commonality was copied from 'The eighty year history.'
[2] Takeshi Kawakami, *Gendai Nihon iryōshi* (A history of the medical treatment of modern Japan), Keisō Shobō, 1965.
[3] GHQ/SCAP, *History of the Nonmilitary Activities of the Occupation of Japan,* 1945 through 1951.

Medical Education Reforms in the United States of America

[4] A. Flexner, 'Medical Education in the United States and Canada,' in *Bulletin*, no. 4, Carnegie Foundation for the Advancement of Teaching, New York, 1910.
[5] Hideo Ogishima and Katsurō Shimomura, *Amerika no igaku kyōiku to iryō jijō* (America's medical education and medical situation), Igaku Shoin, 1970.

Medical Education Reforms Enacted by GHQ and their Influence in Medical Departments and Medical Universities

[6] C. F. Sams, *DDT kakumei: Senryō-ki no iryō fukushi seisaku o kaisō suru* (The DDT revolution: Reflecting on medical welfare policy during the Occupation), Eiji Takemae (tr.), Iwanami Shoten, 1986.
[7] Chiba Daigaku Igaku-bu Sōritsu Hachijūgo-shūnen Kinen-kai (Chiba University Medical Faculty Eighty-fifth Anniversary Commemorative Society), *Chiba Daigaku igaku-bu hachijūgonen-shi* (The eighty-five year history of the Medical Faculty of Chiba University), 1964.
[8] *Ishihō* (The Medical Act), proclaimed 2 May 1906.
[9] *Kokumin iryōhō* (National Medical Treatment Act), proclaimed 25 February 1942.
[10] Nihon Kagakushi Gakkai (Japanese Society of Science History) (ed.), *Nihon kagaku gijutsu-shi taikei 25: Igaku* (A compendium of the Japanese history of science and technology, 25 volumes: Medical 2), Daiichi Hōki Shuppan, 1967.
[11] Akiko Sugiyama, *Senryō-ki no iryō kaikaku* (Medical reforms during the Occupation period), Keisō Shobō, 1995.
[12] C.F. Sams, 'Igaku kyōiku kaikaku ni tsuite: Igaku kyōiku shingikai dai 1-kai kaigi ni okeru kunji (On medical education reforms: Address at the first meeting of the Deliberative Council on Medical Education, 17 February 1946)', in *Nihon iji shinpō* (Japan medical affairs bulletin), no. 1190, July 1946.
[13] Major S.E. Moulton, 'Nihon ni okeru igaku kyōiku kaizen-an (Proposals for medical education reform in Japan),' in *Nihon iji shinpō* (Japan medical affairs bulletin), no. 1191, July 1946.

[12] and [13] are difficult to obtain and peruse today. (On pp. 344–347 of [10], [13] is supplied as data).

[14] 'Igaku Kyōiku Shingi-kai Tōshin (Report of the Deliberative Council on Medical Education),' in *Nihon iji shinpō* (Japan medical affairs bulletin),' no. 1221, June 1947, p. 279.
[15] Nihon Igaku Kyōiku Gakkai (Japan Medical Education Society) (ed.), *Nihon igaku hakusho* (A white paper on medical education), Shinohara Shuppan, 1990.
[16] 'Tokushū: Nihon igaku kyōiku no genjō (Feature edition: The present state of medical education in Japan),' in *Igaku kyōiku* (Medical education), vol. 3, no. 1, February 1972.
[17] Iwao Higashida, 'Intaan hando bukku (The intern handbook (1)),' in *Nihon iji shinpō* (Japan medical affairs bulletin), no. 1203, November 1946, p. 505.

[18] 'Taylor-hakushi kōen: Intaan-sei wa ishi kokka shiken mo shūren mae ni (Dr Taylor's lecture: Intern students should take the National Doctors' Examination before placement),' in *Nihon iji shinpō* (Japan medical affairs bulletin), no. 1269, 21 August 1948, p. 12.

[19] 'News: Jitchi kunrensha no haichi kettei yōryō ichibu kaisei (Partial Amendment to the Rules Governing Decisions Concerning the Allocation of Trainees on the Ground),' in *Nihon iji shinpō* (Japan medical affairs bulletin), no. 1265, July 1948, p. 834.

[20] 'Shūren shisetsu tsuika shitei (Additional designations as training facilities),' in *Nihon iji shinpō* (Japan medical affairs bulletin), no. 1227, August 1947, p. 15.

[21] 'Intaan shitei byōin 102-kasho tsuika (Intern hospitals: An additional 102 locations),' in *Nihon iji shinpō* (Japan medical affairs bulletin), no. 1280, November 1948, p. 30.

[22] 'Intaan shitei byōin 53-kasho tsuika (Intern hospitals: An additional 53 locations),' in *Nihon iji shinpō* (Japan medical affairs bulletin), no. 1289, January 1949, p. 14.

[23] 'Intaan hōki-an sakusei – Kokuritsu Daigaku Igakusei Rengōkai (Framing draft intern legislation – The National Universities' Federation of Medical Students),' in *Nihon iji shinpō* (Japan medical affairs bulletin), no. 1249, March 1948, p. 32.

[24] Yoshio Kusama, 'Kusama Yoshio sensei kikigaki (Yoshio Kusama: His verbatim account),' in *Igaku-shi kenkyū* (Medical history research), no. 57, 1983, pp. 43–8.

[25] Yoshio Kusama, 'Amerika no igaku kyōiku (American medical education),' in *Nihon ishikai zasshi* (Journal of the Japan Medical Association), vol. 24, no. 5, May 1950, pp. 418–26.

[26] Yoshio Kusama, 'Amerika no senmon'i scido to wagakuni no seido shian (The American medical specialist system and Japan's draft plan)', in *Nihon ishikai zasshi* (Journal of the Japanese Medical Association), vol. 24, nos. 10–12, Oct–Dec 1950 (serialized).

[27] Yoshio Kusama, 'Amerika no igaku kyōiku (American medical education), in *Nihon iji shinpō* (Japan medical affairs bulletin), no. 1170, 15 October 1945, p. 184.

[28] 'Iiku kikan shisetsu hyōjun: Igaku Kyōiku Shingikai Tōshin (Facility standards for medical education: Report of the Deliberative Council on Medical Education), in *Nihon iji shinpō* (Japan medical affairs bulletin), no. 1268, August 1948, p. 24.

[29] Ministry of Education Ordinance No. 28, *Daigaku setchi kijun* (University accreditation standards), 22 October 1956.

[30] C. F. Sams, 'Igaku Kyōiku Iinkai no jigyō ni taisuru shishin (A guide to the undertakings of the Medical Education Committee), in *Nihon ishikai*

zasshi (Journal of the Japanese Medical Association), vol. 22, no. 11, December 1948, pp. 439–41.

Post-war Medical Reforms in Okinawa

[31] Hiroyoshi Teruya, *Sengo Okinawa no iryō: Watashi no ayunda michi kara* (Postwar medical treatment in Okinawa: The path I took), Medikaru Furendosha (Medical Friend Co.), 1987.

Nursing Education Reforms

[32] Reiko Ryder Shimazaki, 'Hi-senryōka (1945–1951nen) ni okeru Nihon no kango seisaku 1–7 (Japanese nursing policies during the Occupation [1945–1951], 1–7),' in *Kangofu kyōiku* (Nursing education), vol. 31, nos. 2–8, 1990.
[33] Midori Kawashima, 'Hitotsu no kango kyōiku-shi 1946–53: Tokyo kango kyōiku mohan gakkō de mananda hitobito, 1–3 (A nursing education history, 1946–53: People who learned at the Tokyo Model School for Nursing Education, 1–3), in *Kangofu kyōiku* (Nursing education), vol. 33, nos. 1–3, 1992.

PART III

Reorganization of Industry and Social Systems

17 The Reorganization of the Electric Power Industry

Takuji Okamoto

Introduction

As of 31 March 1991 there were sixty-six electricity businesses in Japan licensed by the Electricity Business Law. Ten of them were private 'general electric power companies' permitted to sell electricity to general consumers: including nine electric power companies [1] established in 1951 with a sum generating capacity as of 31 March 1991 of approximately 151.7 GW; and the Okinawa Electric Power Company, which was established in 1972 when Okinawa returned to Japan and whose generating capacity as of 31 March 1991 was about 1.0 GW.[2] The other fifty-six companies are 'wholesale electricity industries' which sell electricity to general electricity businesses but not to general consumers and are far smaller and less influential than the general electric power companies. The sum of their generating capacities as of 31 March 1991 was some 29.9 GW. The ten general electric power companies had a monopoly on selling electricity in their own 'supply areas,' which combined, covered the entire country.

Compared to European countries which had been dominated by state run electricity companies until deregulation in the 1990s,[3] and the US where a mixture of national, state, municipal and private electricity companies operate, Japan's electricity industry is characterized by large scale and well organized privately owned and operated companies monopolizing supply areas. The electricity industry is undoubtedly unique when compared to the national management regimes of other public utilities such as the railways and the telecommunication industries in Japan.

It was not until the so-called 'reorganization of the electricity industry' in the early 1950s that the existing framework of the electricity industry was established as a unified system in Japan. This system is unique for three characteristics: different areas being individually administered; being privately run; and each area having its own means of electricity generation, transmission and distribution.

An understanding of the postwar reorganization of the electricity industry is essential to an understanding of the history of the electricity industry and the development of electrical technology in the postwar years.

This chapter outlines the developments of the industry reorganization with critical comments regarding various reorganization plans proposed by the different groups with vested interests in the industry.

The Electricity Industry immediately after World War II

The business system at the end of the war

Following WWII, Japan was confronted with the challenge of reorganizing and sometimes reconstructing its industrial sectors. The particular challenge for the electricity industry was what to do about the system of national control that had developed before and during the war.[4]

The question of centrally controlling the nation's electric power industry first arose in the late 1910s. However, it was only after the tremendous competition between the five biggest electricity companies during the so-called 'electricity war' in the 1920s that moves – in the form of the amendments to the Electric Enterprise Law of 1931 – were made to implement changes to this end. In an effort to prevent any unnecessary duplication of facilities, this amendment created regional supply monopolies for the electricity businesses, a measure that was exceptionally favorable for the electric power companies.[5] At the same time it aimed to institute government intervention in the industry in the form of: a system of fee approvals; establishing policies regarding the generation and transmission of electricity; and establishing a government supervisory organization (the Electricity Committee).

The Ministry of Communications further entrenched the state's control of the industry as the war effort intensified. In March 1938 four laws relating to the industry came into effect, including the Japan Electric Generation and Transmission Company Limited Law. This law established the Japan Electric Generating and Transmission Company (*Nippon Hassōden*) which came into being in 1939. It required all of the private and local governmental electric power industries to invest the hydroelectric power generation facilities with a capacity larger than 5 MW that they had already planned or were developing, all of their existing thermal electric power generation facilities whose capacities exceeded 10 MW, and all of their main substations in *Nippon Hassōden*. Meanwhile, the Agency of Electricity was established within the Ministry of Communications, with the aim of promoting state control of the electric power industry. Furthermore in April 1941 the Electricity Adjustment Law Enforcement Ordinance was revised to invest all of the existing hydroelectric power facilities with a capacity in excess of 5 MW in *Nippon Hassōden*. The Electric Distribution Control Ordinance was effective from August 1941, bringing about the creation of nine regional Electric Distribution

Companies (*Haidens*).[6] Both *Nippon Hassōden* and the nine *Haidens* were privately financed and run by the government.

The national control system was characterized by a low rate of power supply, the enforcement of a uniform national power rate, and government subsidies to ensure these. The government subsidies were pooled in *Hassōden* and then the electricity fee – which was paid by the nine *Haidens* to *Hassōden* – was adjusted with payments from these pooled subsidies such that the uniform national rate could be guaranteed. Although the adjustment of the revenue of the transmission companies was made possible in this way, the concerns of the ten electricity companies, that is to say the nine *Haidens* and *Hassōden*, concentrated only on the portion given them from the pooled subsidies. Later, the Electricity Agency of the Ministry of Communications, the authority in charge of the electricity industry, was renamed the Electricity Bureau. Later again, it was renamed the Electric Power Bureau and moved to the Ministry of Munitions. However the accounting methods that constituted the foundations of the system of national control never changed.

During the Occupation of Japan by the Allied Powers after WWII the Ministry of Munitions was abolished and responsibility for the administration of the electricity industry shifted to the Electric Power Bureau of the Ministry of Commerce and Industry. Government subsidies to *Hassōden* stopped in March 1946. In September and October the Electric Distribution Control Ordinance and the Electricity Adjustment Ordinance were repealed and a revised Electric Enterprise Law came into effect. State control of the electricity industry seemed to have come to an end. However, the revised law retained many elements of the state control that had been legislated by the two repealed laws. For example, the government retained the right to regulate electricity usage and supply, and the Minister in charge maintained the authority to determine power supply rates. Over and above these continuities there were newly introduced regulations regarding the power to control the disposal of profits, the power to appoint and dismiss executives and so on. Furthermore the Electric Power Control Law and the Japan Electric Generation and Transmission Company Law were regarded as a continuation – even *reinforcement* – of state control.

Hassōden's monopoly on main power stations, transmission lines and substations remained virtually unchanged, as did the arrangement giving regional monopolies to the nine distribution companies. Moving the Electric Power Bureau to the Ministry of Commerce and Industry also had minimal effect on the Bureau's administrative function, with the exception of some authorizing rights, such as price controls that were moved to the Price Board, and planning electricity allocations, which went to the Economic Stabilization Board. The Bureau retained responsibility for work approvals, work transfers and merger approvals. Its powers ranged from financial management – including profit disposal, company bonds and changes to capital amounts – to business planning – such as planning the construction of electricity facilities, power supply and the method of power transmission – and personnel management – such as the appointment and dismissal of directors.[7] This was the structure of government control of the electric power industry in Japan immediately before the reorganization began.

Electricity supply and demand just after the war

The electricity industry did not suffer very much direct damage during the war when compared to other Japanese industries. Although the thermal power stations were mostly located in city suburbs and suffered damage resulting in about a fifty percent reduction in power output, the industry had long favored hydroelectric to thermal power and the damage suffered by hydroelectric stations was minimal (Table 17-1).[8]

The end of the war heralded a sharp decline in all industrial activities. In August 1945, the demand of electricity dropped to one third of what it had been in the same month of the previous year.[9] For a few months after the war there was actually surplus power supply.[10]

However, this period was followed by a speedy recovery in demand (Table 17-2). The postwar shortage of fuel accentuated the reliance upon electricity. Electric power, which was rated at the same low price that it had been during the war, was used for heating and cooking. Furthermore, illicit use of electricity was widespread. Due to the insufficient numbers of meters and inadequate supervision of distribution lines, it was not difficult to tap into the distribution systems and obtain unauthorized, unmetered electricity.[11]

The electricity companies, with *Hassōden* as a key player, were troubled by the deteriorating condition of the aging power stations. They were run down from overuse and a lack of repairs during the war, they were using an inferior quality of coal, and recruiting new and qualified labor was difficult, since the food ration of the electric power workers was less

Table 17-1: War damage status (unit: 10, 000 kW)

		All facilities	War damaged facilities
Generation	Hydro-power	598	11
	Thermal power	288	146
Transformer stations		1,586	475

The power generating capacity of the thermal power stations immediately before and after the war was 800,000 kW and 120,000 kW respectively. The decreased capacity of the transformer stations was 1,160,000 kW.

Source: Electricity Bureau, Ministry of Commerce and Industry, 'Electricity Business Data' 1948, pp. 37–40.

Table 17-2: Outline of electricity demand (unit: 1 million kWh)

(Numbers in brackets – e.g., (16.6) indicate the percentage of total demand.)

Industry		Major electric power	Minor electric power (≤500 kW)	Lights and heating for homes (≥500 kW)	Illicit use	Total
1930–34 average	Consumed electricity	8,884 (62.1)	2,416 (16.9)	3,000 (21.0)		14,300 (100)
1939	Consumed electricity	20,788 (73.2)	4,242 (15.0)	3,330 (11.8)		28,360 (100)
1943	Consumed electricity	25,729 (81.3)	3,454 (10.9)	2,404 (7.6)		31,587 (100)
1946	Consumed electricity	13,273 (56.1)	4,092 (17.3)	4,266 (18.0)	2,016 (8.6)	23,647 (100)
1947	Consumed electricity	14,637 (58.0)	5,031 (19.9)	4,106 (16.3)	1,449 (5.8)	25,223 (100)

Source: Planning Section, Research Division, Japan Electric Power Generation and Transmission Company Ltd 'Trend in Electricity Demand' 1949.
Note: the figures have been rounded-off so their sum is not always equal to the number of the corresponding total.

than that granted workers in more favored industries. They were also experiencing a shortage of funds and materials plus recurring labor disputes. Fearing the possibility of the thermal power stations being removed for reparations,[12] they tried to answer the soaring demand with repairs and other measures.[13] The results of these efforts started to become apparent in the three years after the end of the war. Thus, by 1948 electricity output had almost returned to 1943 levels when output had previously peaked (Table 17-3).

However, 1946, '47 and '48 were three very difficult years. The winter of 1946 has been described as a period in which 'an unprecedented starvation of electricity' was experienced. Blackouts, power cuts and drastic restrictions on power supplies to factories occurred at least every

Table 17-3: Change in the output of electricity generation for industries

(unit: 1 million kWh)

	Hydro-power	Thermal power	Total
1943	28,641	6,193	34,834
1944	28,345	3,734	32,079
1945	19,521	539	20,060
1946	27,247	903	28,150
1947	28,409	1,956	30,365
1948	31,870	2,743	34,613

Source: Electricity Section, Power Agency, Economic Stabilization Board, 'Recent Situation on Electricity' 1949.

second day and often daily.[14] Aiming to use the law to limit the use of electricity, the 'Electricity Supply and Demand Adjustment Rule' was enacted in 1947. Its regulations continued to be enforced until 1952.[15]

To solve the 'electricity crisis,' *Hassōden* and the Economic Stabilization Board independently developed their own five-year plans in 1948.[16] However, as will be discussed later, the Counterpart Fund [17] that the companies expected for development projects were used as a bargaining tool in the restructuring negotiations and were not satisfactorily dispensed. Thus, even these modest restoration plans – plans too modest to adequately address electricity shortages – could not be realized.

Although the primary objective of the so-called 'reorganization of the electricity industry' was to change the way the industry was structured, a solution to the electricity crisis was integral to every proposal.

Details of the Reorganization of the Electricity Industry

Initial moves towards and various plans for reorganization

Nippon Hassōden and Labor

Before the surrender there was already some movement towards restructuring the management system of the electricity industry. Towards the end of the war, an Investigation Team from *Hassōden* conducted 'an investigation into the unification of *Hassōden* and the *Haidens*.' The investigation's report stated that the unification of *Hassōden* and the nine *Haidens* was possible; if achieved, business rationalization through a simplified organizational structure, a reduction in the number of workers, lower operating costs and reducing the distance between producers and consumers could be expected.[18]

On 7 November 1945 *Hassōden* sent a report to the Electric Power Bureau of the Ministry of Commerce and Industry entitled 'A Report on the Improvement of the Management System of the Postwar Electricity Industry.' It stated that the main aim of reorganization was the creation of a single national company through the unification of the *Hassōden* and nine supply companies.[19] Some of the advantages the investigators anticipated would ensue from unification were: the increased possibility of developing hydroelectric power, the expansion of supply networks, the elimination of uneven pricing structures in different areas and the eradication of the irrational pooling system of subsidies. But the primary objective of the creation of a unified national company, they claimed, was to facilitate an 'advanced socialization of the industry.'

At the end of the report was a section 'in response to the argument that management by a nationally unified company would be against the democratic movements of the economic structure.' The investigators rebutted this argument with the following three claims: as during the war the electricity industry had already been 'released from capitalist administration under state management' there was no 'fear of causing a disturbance from the enforcement of a capitalist monopoly'; the unification into one company was in accordance with 'socialization' intended to contribute to 'the development of the national economy and social welfare, which is also in harmony with the direction that public enterprise is taking in general'; and that management policy should be set 'under the direction of the state and society' in accordance with 'the nation's will' and operate 'independently and democratically.' Furthermore, it argued that such a scheme 'is not contrary to democratizing the economic structure.' It is unclear whether the investigators intended for the unified company to be owned by the state or privately. However, it claimed that unification into one company under state control was indeed a 'democratizing' of the enterprise, and, moreover, this would be the complete opposite of a 'capitalist monopoly.'

This restructuring proposal had many points in common with proposals made by the labor union who maintained that unification was the primary objective because it would create the possibility of a national flat rate for the provision of electricity, of simplifying the management structure of the business, and clearly establishing where responsibility lay. The union movement had been active from around 1945, but discussions about the industry's management structure surfaced only after 4 July 1946 when the Japan Electric Power Industry Labor Union Council (*Densan Rōkyō*) was established as a joint federation of the enterprise-based unions at the ten different electricity companies. The three common objectives proclaimed by the union in July 1946 included 'the democratization of electricity enterprises.'[20] The labor movement further developed this aim the following autumn.

For the labor movement, 'democratization' meant 'socialization,' and such socialization required democratic management by the state. In other words, the socialization of enterprises could only be realized in a unified enterprise under 'social public' management, and 'democratic management' by the state would be required to make this possible. But under the then non-democratic government, 'state management' meant 'bureaucratic management.' Therefore, the union adopted a compromise 'structure in transition' – that is, they aimed for 'a nationally unified, privately owned electricity enterprise.' This would achieve an interim unification of the industry into one company while freeing the industry from bureaucratic government controls. To maximize democratization under private ownership the union suggested supervision by the 'Electricity Industry Supervisory Committee' (composed of capitalists, workers and consumers) and that an 'Electricity Committee' – chosen democratically by the industry managers and workers – should be created to run the company.

The union articulated four concrete policies that would be required to make such a social-democratic privately owned monopoly a reality: 'the abolition of bureaucratic state management laws'; supervision and guidance by 'the social public'; the integrated management of generation, transmission and distribution by a single company; and establishing an industry socialization law concurrently with establishing a democratic planning body. Accordingly the bloc plans, state-ownership plans and plans for publicly owned electricity generators (which will be discussed later) were rejected. The union also rejected any attempt at achieving such democratization by establishing several companies on the grounds that such a system would be 'a return to free competition' and would therefore run contrary to the principles of socialization.[21]

The proposals of *Hassōden* and *Densan Rōkyō* were consolidated for economic and technical rationalization reasons. They shared a belief that the abolition of capitalist and liberal tendencies meant democratization and socialization. The only point of difference was that some practical management proposals presented by *Densan Rōkyō* were not suggested by *Hassōden*.

In the Electric Power Industry Labor Union's so-called first industrial dispute and mediations, in the autumn of 1946, these shared beliefs enabled capital and labor to reach agreement relatively easily. On 16 September, the *Densan Rōkyō* dispute began when the union demanded three things including the democratization of the activities of the ten electricity companies. The negotiations developed to the point where they involved the Ministry of Commerce and Industry, and the situation deteriorated when the union, as a result of its requests being refused, conducted a nationwide stoppage of electricity supplies. On 22 October the Central Labor Relations Commission, an independent mediating body, commenced mediations.[22] We can develop an understanding of the position taken by the ten electric companies and the Ministry of Commerce and Industry (the main ministry involved) by examining the issues addressed in the dispute and the negotiation processes.

The ten electric companies, *Densan Rōkyō* and the Ministry of Commerce and Industry had signed a paper of intent on the four items relating to democratization put forward by the union as part of the negotiations conducted on 7 October during union action.[23] However, the Minister for Commerce and Industry, Jirō Hoshijima, indicated in a statement on the following day that while the Ministry did not have any objection to the process of democratization, no detail should be discussed before the establishment of an Electricity Committee. The ten electric companies later agreed to this. The union, however, objected to this apparent equivocation on matters previously agreed to, and the discussion was postponed until mediation by the Central Labor Relations Commission could be sought.

After three rounds of negotiations, the mediations concluded with the formal signing of an agreement by management and the union on 21 December 1946. As for the democratization of business, the four items of 7 October were acknowledged in the agreement. To promote this matter further, a plan to establish the 'Electric Industry Democratization Council' was added.[24]

Upon acceptance of the Agreement, the first meeting of the Electric Industry Democratization Council was held on 22 and 23 January 1947 at which thirteen company members and fifteen union members were appointed to the Council. Discussions included practical concerns relating to the 'Plans for an Electric Industry Socializing Law' and a study of the provisional measures suggested for a merger into one company. As for the management structure, it was confirmed that the government would bear responsibility for this in accordance with the public company arrangement under the guidance and supervision of the 'Electricity Committee' and the 'Electricity Industry Audit Committee.'[25] However, the two sides did not agree on the timing of a proposal for the socialization laws. At the eighth meeting, on 15 September 1947, as soon as a system for combining the accounting systems of the ten electric companies' management was proposed by the union members who were pushing to bring about the unification of the ten electric power companies, the management's attitude quickly hardened. As a result, it was decided that matters relating to the question of democratization as well as a sliding wage system and management/union agreements should be handled in conjunction with mediation by the Central Labor Relations Commission.[26] In the meantime (in May 1947), the Japan Electric Industry Labor Union (*Densan*) was formed as the sole industry based labor union, with the Electric Power Industry Labor Union (*Densan Rōkyō*) as its mother body.[27]

In spite of there being agreement on the principles involved, the ten electric companies and *Densan* could not agree to any revised joint plans. This was partly because, as stated in the mediation plans [28] drafted by the Central Labor Relations Commission, the Ministry of Commerce and Industry was not party to the discussions. No matter what legislation or reorganizing plans relating to electricity administration the industry and the union may have agreed upon, the talks could not move forward without the Ministry's participation. At the end of the second Electricity Industry dispute on 25 March 1948, it was confirmed that the union and management would proceed with a discussion about democratization before the establishment of the 'Democratization Committee' prepared by the Government.[29] However, the discussion did not start until the 'Electric Industry Democratization Committee' was established in April 1948 (we will return to a discussion of this Committee in the next section).

Ministry of Commerce and Industry

The Ministry of Commerce and Industry, the electricity administrator, was engaged in negotiations with GHQ in the hope of continuing a national system of management. At the same time it studied several options in the process of planning a company structure based on the integration of the electric generation, transmission and distribution system under state management. In November 1946 the Ministry of Commerce and Industry started negotiations with GHQ to establish a Central Electric Power Committee within the Ministry in order to lay out practical plans for the reorganization of the electricity industry.[30] However, not only did GHQ not grant permission for this, it demanded that the Electric Power Bureau be abolished, thereby providing for a change in company structure. It also demanded the establishment of a 'five-member committee' as a replacement for the electric administration organization.[31]

Although the Ministry of Commerce and Industry had strong doubts about the administrative position and level of authority the 'five member committee' might command, it attempted to establish the committee within the Ministry, concluding that GHQ's demands could not be avoided. However, the Ministry failed to establish the committee within the Ministry.[32] As for the business structure, it reached the conclusion that the unification of the electricity generation, transmission and supply management could be supported as this would allow for the eradication of duplicated facilities, flexible interchange of electricity between local areas and nationwide unified rates. Moreover, with state management it was thought that an abuse of the monopoly would be eliminated.

Local Governments

Local governments, who had lost the power supply business to the nine power suppliers (under the Electric Distribution Control Ordinance which was based on the National Mobilization Law), started to hope for a retrieval of this business. After the 'Report Paper on Views on the Power Supply Business Administration Transfer' was adopted by the Tokyo Metropolitan Assembly on 20 February 1946 [33] the push for this spread to Sendai City, Kyoto City and Fukushima Prefecture. However it was not until 1948 that strong ties among the local governments were achieved and practical and statewide business reorganization plans were produced.

From the Application of the Elimination of Concentration of Excessive Economic Power Law to Replies by the Electric Industry Democratization Committee

In accordance with GHQ's early Occupation policy of 'demilitarization and democratization' the Elimination of Concentration of Excessive Economic Power Law was proclaimed and enforced on 18 December 1947. Its intent was to facilitate economic democratization through the dismantling and/or restructuring of large companies. This law created an opportunity for the reorganization of the electricity industry.[34] On 8 and 22 February 1948, three hundred and twenty five companies, including *Hassōden* and the nine *Haidens*, were designated as excessive concentrations of economic power that needed to be deconcentrated.

However, it was about this time that what later became known as the Cold War was intensifying. In response to this developing situation the focus of GHQ's Occupation policy shifted from demilitarization to fostering 'independence and revival' in the Japanese economy. This resulted in overturning the orders to reorganize businesses and industrial sectors in 1948 and 1949. Thus, in the final analysis, there were only eighteen companies outside of the electric power industry that were terminally effected by the Elimination of Excessive Economic Concentration Law.[35] The dismantling of *Hassōden*, however, would eventually be executed in 1951 on the ground of deconcentration of excessive economic power.[36] In light of these decisions we can surmise that GHQ considered the reorganization of the electricity industry to be a separate issue to the democratization of other industries in general.

Entities that embraced their own reorganization in accordance with the Elimination of Concentration of Excessive Economic Power Law developed plans and made them public. *Densan*'s reorganization plan released on 9 March 1948 corresponded to the structures outlined by the Democratization Council. It proposed a business structure whereby the government would manage a corporation handling electric generation, transmission and distribution nationwide under the supervision of an appropriate guiding and supervising organization.[37]

Local governments were aiming for the transfer of authority over power supplies and formed the National Alliance for Local Government Management of Electric Distribution on 23 April 1948. They publicized their basic campaign policies the following month, stating that a nationally unified and managed state or public company was appropriate for electricity generation and transmission, and that local governments should monopolize electric distribution within their proper supply areas (namely, prefectures).[38]

On 22 April 1948 *Hassōden* and the *Haidens* – who were all subject to the elimination law – independently produced and submitted their own plans for reorganization. The plans produced by *Hassōden* re-articulated the proposal for business management by an integrated electric generation, transmission and distribution entity in the form of a nationally unified joint-stock company under the administrative supervision of the Electricity Committee.[39] However, the nine power supply companies – who had sided with *Hassōden* since the negotiations with *Densan Rōkyō* – changed their position dramatically and proposed that power generation, transmission and supply should be managed by nine privately owned and managed regional companies.[40] The plan to establish regionally based companies combining electricity supply, generation and transmission is, on the one hand, hard to understand when we consider the general thrust of the law aimed at eliminating economic concentration. On the other hand, though, the companies' rationale was that by abolishing the pool calculation and establishing privately run companies with independent profit schemes, the electricity industry could get rid of state protection and at the same time render possible the dismantling of the monopolistic *Hassōden*.

The Ministry of Commerce and Industry also modified their reorganization proposal. In March 1948 the Ministry held a study meeting with the Economic Stabilization Board that considered a business plan fundamentally oriented towards maintaining the status quo as well as a plan for a nationally unified management of transmission and generation. They also proposed a study of a plan for a system of nine regional companies with integrated management of electric generation, transmission and distribution in their supply areas.[41] They attempted to establish a committee within the Ministry to develop a plan for reorganization. It was to be known as the 'Electric Industry Democratization Committee.'

With the consent of the Ashida Cabinet on 16 April 1948, it was decided that a Democratization Committee would be established with Matsujirō Ōyama as Chairman. This committee, the first commission of consultation for reorganization created by this government, consisted of nineteen members from *Hassōden*, the Kantō Electric Distribution Co., *Densan*, local governments and electricity consumers as well as one member from each House of the Diet and three government officials from the Ministry's Electricity Bureau (Table 17-4).

As can be expected from such a diversely constituted committee, discussions were very heated. However, the scope of the discussions was restricted by the GHQ announcement outlined below. E. C. Welsh, Manager of the Anti Trust and Cartels Division of GHQ's Economic and Scientific Section (ESS), announced his view that the nine existing distribution companies would not be divided. Responding to this, the Holding Company Liquidation Commission, a Japanese government body responsible for elimination of excessive economic power, publicly announced that it would not restructure the nine companies.[42] These two

Table 17-4: Members of the Electric Industry Democratization Council

Chairman	Matsujirō Ōyama (President, Institute of Electrical Engineers of Japan; and Professor at the University of Tokyo)
Members	
Hassōden	Eiichi Ōnishi (President, Nippon Hassōden)
Kantō Haiden	Ryōtarō Takai (President, Kantō Haiden)
Densan	Nobuhide Uematsu (Central Committee member, Densan)
	Atsuyoshi Suzuki (Member, Densan)
Municipal Institutions	Kanshirō Ishikawa (Governor, Fukushima Prefecture)
	Nagaaki Ishihara (Chairman, Tokyo Metropolitan Assembly)
	Torao Hayashi (Governor, Nagano Prefecture)
	Seiichi Fujiwara (President, Electric Power Council, Tokyo)
Users	Ichirō Ishikawa (President, Japan Industrial Council)
	Harutaka Toyota (President, Japan Federation of Medium and Small Businesses)
	Takashi Miki (President, Japan Steel Industry Association)
Academics	Seiichi Tōhata (Chairman, Agriculture Revival Association; Professor, University of Tokyo)
	Ichirō Nakayama (Member, Central Labor Committee; Professor, Tokyo University of Commerce)
Labor	Seita Itō (All Japan Industrial Unions Council)
	Yoshitarō Takado (Head, Research Department, Japanese Confederation of Labor [Sōdōmei])
	Toshihide Mizukami (Secretary-General, All Japan Coal Industry Unions)
Other	Kijio Kishi (President, Japan Industrial Bank)
	Ryōichi Yamakawa (President, Japan Coal Association)
Parliamentarians	Einosuke Maeda (Member of the House of Representatives)
	Ryōsaku Sasaki (Member of the House of Councilors)
Secretariat	Keizō Tamaoki (Director-General, Bureau of Electric Power, Ministry of Commerce and Industry)
	Shinjirō Mitsui (Technical Officer of Commerce and Industry, Bureau of Electric Power, Ministry of Commerce and Industry)
	Ichirō Nakajima (Head, Division of Electric Affairs, Bureau of Electric Power, Ministry of Commerce and Industry)

Source: Ministry of Commerce and Industry, Bureau of Electric Power, *Denki jigyō minshuka iinkai gijiroku* (Minutes of the Electric Industry Democratization Committee), 1948, p. 1.

announcements must have proven a shock for *Hassōden*, *Densan* and the local governments. *Hassōden* Vice-President Shindō sent a letter to H. Scheier of the Anti Trust and Cartels Division, stating that *Hassōden*'s reorganization plan was to maintain the status quo and hinted that they would support the rather conservative proposal of the Chairman of the Democratization Committee, Ōyama,[43] considering it to be a policy 'one step ahead' of the alternatives.[44]

Against this background the Democratization Committee also proposed to support Ōyama's plan.[45] The central tenets of this proposal were that the business structure would maintain *Hassōden* as a joint-stock company under state control, establish an integrated operation of electricity generation, transmission and supply in the Hokkaidō and Shikoku areas and retain the existing structures in Honshū and Kyūshū. It was clearly their intention to minimize damage to *Hassōden* and the *Haidens*. Furthermore, it proposed establishing a Central Council, local Electricity Councils and an Electricity Agency in the government for the administration of the electricity industry. Essentially, state control was to be preserved.

The above proposal was never acted upon, though, because the Ashida Cabinet resigned on 7 October, a week after the proposal was made public, and because, given GHQ's reorganization policy outlined below, such a system was not acceptable to the Occupation Forces.

GHQ's reorganization policy

GHQ had already considered the reorganization of the electricity industry before the Elimination of Concentration of Excessive Economic Power was proclaimed. At a meeting on 15 April 1947 I. T. Bush of the Anti Trust and Cartels Division announced a plan for reorganization that called for the 'establishment of a new self-organized electricity administration' and a 'change towards a unified management of the electricity generation, transmission and supply network divided into nine areas.'[46] GHQ staff were basically of the view that the bureaucratic administration of the Ministry of Commerce and Industry was preventing the business from being rationally managed and were in agreement on a reorganization policy, consisting mainly of two points: revamping the administrative structures and a division of enterprise. It was from this perspective that GHQ rejected the Ministry of Commerce and Industry's attempt to establish an Electricity Committee. Moreover, as has been mentioned, GHQ suggested the abolition of the Electric Power Bureau and the establishment of an integrated management of the electric power industry by private regional companies.

One of the two main points of GHQ's reorganization policy – a revamping of the administrative structures – would remain unchanged. However, opinions were divided concerning definite plans for dividing the industry into regional management areas.[47] A report released by the Industry Division (ESS/ID) in July 1948 [48] recommended five divisions while about the same time, as previously mentioned, Welsh, of the Anti Trust and Cartels Division announced a structure involving nine divisions.

In February 1949 ESS/ID prepared a report based on comprehensive research into the Japanese electricity industry that proposed a reorganization dividing the country into seven areas and the establishment of a self-managed coordinating body.[49] The Anti Trust and Cartels Division agreed to this seven-division plan. In June the Deconcentration Review Board, formed by US experts at the request of SCAP, submitted a report to General MacArthur outlining this proposal.

On 10 May E. J. Burger of the Deconcentration Review Board unofficially announced the seven division plan to *Hassōden* trustee Mori.[50] This plan was thus conveyed to the Japanese.[51] However, the Japanese showed a strong aversion to it. In response, International Trade and Industry Minister Heitarō Inagaki [52] presented W. F. Marquat, head of ESS, with a list of three points for consideration: he opposed the seven division plan, sought a postponement of the reorganization and wanted to establish a Japanese committee for planning the reorganization.[53]

But, in fact, as mentioned above, the seven-division plan was not universally supported within GHQ. Furthermore, the Holding Company Liquidation Commission, whose activities generally corresponded to GHQ's policies, opposed the plan, fearing that it might create an even larger entity than the existing supplier.[54]

A few months later, the Deconcentration Review Board was wound-up due to changes in Occupation policy and Burger returned to the US. It seems that this situation led GHQ to decide to leave the reorganization of the electricity industry until the arrival of T. O. Kennedy who had experience as the head of a small electricity company. Thus the seven-division plan was not imposed upon the Japanese, and in the autumn of 1949 GHQ permitted the establishment of a Trade Ministry Advisory Committee empowered to negotiate definite plans for reorganization with Kennedy. At a Cabinet Meeting on 4 November the Yoshida Cabinet established the Electric Industry Reorganization Council to that end.[55]

GHQ considered de-bureaucratization of the industry to be a primary objective of reorganization as can be seen in its demand that a regulatory body entirely divorced from the Government be established for the administration of the industry. The plan for a company structure based on divisional areas also seems to have been aimed at abolishing bureaucratic management.

As we have seen, *Hassōden* and *Densan* considered that, from the perspective of democratization and breaking up an excessive monopoly, the electricity business could not be managed and monopolized by a private nationwide company. They argued, rather, that these objectives could only be achieved through a unified public enterprise, either directly or indirectly under state management. However, GHQ continually opposed state management and argued that a unified company was more bureaucratic and thus less democratic. Private companies operating in their own given areas, they argued, would lead to securing competitive electricity rates in each area and, hopefully, the rationalization of management, long term budget plans and the development of management plans.

The Electric Industry Reorganization Council's report

Members were appointed to the Electric Industry Reorganization Council on 21 November 1949 (Table 17-5). People associated with the electricity business were removed and Yasuzaemon Matsunaga – who had been President of Tōhō Electricity Company before the war – was appointed as the Chairman. He was known for having cloistered himself in Odawara during the war in protest at government control of the electric power industry. He had also published *The Question of Electricity Control: My Personal View* in 1928 which proposed an integrated system of management for the generation, transmission and supply of electricity by nine regional private companies, to be regulated and advised by a committee.[56] The appointment of Matsunaga – who ardently supported division into nine areas and privatization – perhaps indicates that the Yoshida Cabinet and MITI were, by this stage, prepared to allow privatization to some extent.

Kennedy and G. R. Roames of ESS contacted the Council to encourage them to push for the implementation of GHQ's reorganization plans. But there was hardly any difference of opinion between Chairman Matsunaga and GHQ regarding private companies operating an integrated system of electricity generation, transmission and supply in different areas and the establishment of a regulatory body, so the negotiations in this early stage went very smoothly.[57]

Matsunaga's management of the Advisory Council, however, was so overbearing that it excluded the government officials of MITI who had been appointed to the Council Administration Office. Matsunaga's reorganization plans were prepared, instead, mainly by the private staff he had known since his days in the Tōhō Electric Power Company.[58] His plan for the nine regional divisions, though, was not accepted by the other Council members and thus, the report was tabled in an irregular manner on 1 February 1950; it includes Chairman Matsunaga's proposed reorganization attached as a minority view to the proposal – known as the

Table 17-5: Members of the Electric Industry Reorganization Council

Chairman	Yasuzaemon Matsunaga (Former President, Tōhō Electric Power Co. Ltd.)
Deputy Chairman	Ryūichi Koike (Dean, Faculty of Law, Keiō University)
Members	Shōshirō Kudō (Vice-President, Reconstruction Finance Corporation)
	Takashi Miki (President, Japan Steel Co. Ltd.)
	Shigeru Mizuno (Vice-President, Kokusaku Pulp Co. Ltd.)

Source: Denki Jigyō Saihensei-shi Kankō Iinkai (Editorial Committee for the History of the Reorganization of the Electric Power Industry)(ed.), *Denki jigyō saihensei-shi* (A history of the reorganization of the electric power industry), 1952, p. 342.

Miki Plan – that was supported by the other four members (including Takashi Miki).[59]

The Miki Plan proposed the creation of nine bloc companies to cater for the already existing electricity supply areas plus a company for generating and transmitting electricity to buffer those nine companies after dismantling *Hassōden* and the nine *Haidens*. The nine bloc companies were not only to supply electricity but also to generate and transmit it, each in its own area. The report stressed three reasons that made it necessary to establish a generation and transmission company: to alleviate shortages of electricity caused by regional divisions, to avoid a duplication of facilities and to remedy discrepancies in power supply rates that differed from region to region.

Under this plan the Electric Power Bureau and the Gas Section of the Bureau of Coal Supervision and Management (in MITI) would be abolished and a Public Utilities Commission would be established; not to operate and manage, as such, but to administer the industry in an adjusting and supervisory capacity. Although this Commission was to have an independent status, it would be subject to Cabinet decisions and would have obligations for communication and cooperation with relevant government offices regarding supply and demand adjustments and price settlements. Also, its budget and personnel management came under the Prime Minister's portfolio and were thus subject to the constraints of the Prime Minister's Office. In other words this supposedly independent committee could only act in accordance with government instructions.

Although the Miki Plan represented a compromise with GHQ's area divisions, one of its main objectives was, in fact, to ensure the continued existence of *Hassōden*, even if greatly reduced in scale (the proposed electric generation and transmission company would keep forty-two percent of *Hassōden*'s electric facilities and about thirty-six percent of all electricity generating facilities in Japan). In fact, the report is primarily a justification for establishing an electric generation and transmission company, while in terms of the regional division it describes only the negative aspects of the policy – for example, the shortage of electricity in certain areas, price differences and duplicated facilities. It argued that even the Public Utilities Commission would not be able to make companies in different areas interchange electricity satisfactorily without depending upon a generation and transmission company.

The Matsunaga plan also proposed the establishment of nine bloc companies. However it did not include the generation and transmission company. Instead, it proposed that the nine bloc companies would be allowed to own generation and transmission facilities outside of their assigned supply areas to supplement any shortage in electricity supply (a system known as the kite method). The aim of the Matsunaga plan was to depart from government control, achieve management rationalization and make supply and demand adjustments easier. This could be accomplished, it said, by abolishing the system of pooling calculations, establishing the nine companies on a self-funding basis and assigning supply areas of a suitable size to each of them. These nine companies were supposed to be responsible for electric generation, transmission and distribution in each area and to develop power resources on their own. The plan did not include a lending company in the expectation that the Public Utilities Commission would coordinate the exchange of electricity among the bloc companies.

The report also contains an argument against a ten-division plan – a plan that would make Shin'etsu independent from Chūbu – which suggests that GHQ's inclination at the time was in favor of such a scheme. Kennedy conveyed this preference to the Reorganization Council at a meeting on 19 January 1950 with a recommendation to establish a Public Utilities Commission.[60] This scheme was based on an idea that no electric power company should operate equipment outside its own supply area (the territorial principle).

We can only surmise that Kennedy was responsible for GHQ embracing this ten-division plan in place of its previously preferred seven-division plan. Whatever the reason was, though, upon receiving the Matsunaga Committee's report on 6 February – after it was translated by MITI – Roames indicated that he disagreed with the Matsunaga plan on the basis of the territorial principle, although he considered it to be better than the Miki plan. He strongly disagreed with the Miki plan's attempt to secure the continued existence of *Hassōden*.[61] Furthermore, he expressed discontent at the failure to guarantee the independence of the Public Utilities Commission from the Government.[62] Marquat, the head of ESS, sent a memo dated 11 February to Inagaki, the Minister for Trade and Industry, asking the Japanese Government to draw up an appropriate reorganization plan – in the form required for submission to the Diet – and to indicate a program for implementing this plan to GHQ by 15 February 1950.[63]

The Application of the Potsdam Government Order

MITI received Marquat's memo on 13 February and called an urgent meeting at which it was decided to negotiate with GHQ on the Matsunaga plan – the plan for which GHQ had expressed less resistance. However, regarding the Public Utilities Commission, the meeting confirmed that the restrictions stipulated in the report should be retained: the Committee's actions were to be subject to cabinet decisions on important matters of policy and it would have some duties to communicate and cooperate with the related ministries. Inagaki reported these decisions to Marquat on 15 February.[64] In addition to adopting the plan for dividing the country into nine regions and special measures for adjusting prices in different areas, on 21 February the Cabinet decided that the existing organization for administration would be maintained and that a public utilities committee would be established as an advisory body to the Minister for International Trade and Industry. These decisions were reported to GHQ on 23 February.[65]

GHQ's note to MITI dated 10 March indicated that it was no longer attached to the ten-division plan and the territorial principle. Matsunaga's nine division plan and its accompanying special price adjustment measures were approved.[66] However, GHQ suggested that the Public Utilities Commission should be responsible to the Diet and repeated its views on the abolition of the Electric Power Bureau and the Gas Division of the Agency of Natural Resources' Bureau of Coal Supervision and Management. Kennedy provided the Government with a Summary Plan for a Public Utilities Law on 18 March.[67]

The Government and GHQ reached a compromise on the following points: the Public Utilities Commission would be an extra-ministerial board attached to the Prime Minister's Office, the Prime Minister would appoint committee members from the private sector and the Commission should be placed under no restriction of the Government but would virtually be subject to Cabinet decisions in its performance.[68] Following the Ministry's resolution of 6 April and the Cabinet's resolution of 8 April the Yoshida Cabinet submitted the Electric Industry Reorganization Bill and the Public Utilities Bill to the Diet on 20 April.[69]

Getting these Bills passed by the Diet, though, was by no means plain sailing. *Hassōden* played a leading role in staging a nationwide protest movement. Both the Electricity Committee and the Trade and Industry Committee in the Lower House had heated discussions about the question of regional differences in electricity charges and policies aimed at solving electricity shortages. Members of both the Opposition and the ruling party voiced strong objections. In the end, both Bills were put on the shelf on 2 May, the last day of the seventh session of the Diet.[70]

The Yoshida Cabinet was hesitant about submitting the Bills to the eighth session of the Diet. Kennedy's response was to withhold Counterpart funds. This was the largest source of funds for the development of electricity resources. The Yoshida Cabinet quickly discovered that switching around the President and Vice President of *Hassōden* and making amendments to the Bills were not sufficient to satisfy GHQ.

Immediately before the end of the Occupation GHQ took strong measures to hasten a reorganization of the electricity industry. On 22 November 1950, the day after the opening of the ninth session of the Diet, General MacArthur, the Supreme Commander of the Allied Powers (SCAP), sent a strongly worded letter to Prime Minister Yoshida demanding the implementation of the reorganization measures outlined in the Bill presented to the seventh session of the Diet. In response, the Government decided to enforce both Bills in accordance with the Potsdam Government Order which invoked the authority of SCAP and allowed the Cabinet to override the Diet. It proclaimed the Electricity Industry Reorganization Law and Public Utilities Law on 24 November and decreed that they should be implemented on 15 December 1950.[71]

The Yoshida Cabinet appointed members to the Public Utilities Commission on 15 December (Table 17-6). This Committee was inaugurated as an extra-ministerial board attached to the Prime Minister's Office. The Electric Power Bureau of MITI's Natural Resources Agency was abolished. Only a few of its management staff moved to the administration office of the Public Utilities Commission. The Electricity Facility Division of the Resource Agency was established for the maintenance and management of electricity facilities and the supervision of independent electric power plants and took on many of the Electricity Bureau's former staff.[72]

The immediate issues facing the Public Utilities Commission were the dismantling of *Hassōden* and the nine electric distribution companies whilst at the same time preparing for the establishment of nine regional electric power companies. In the face of strong resistance by *Hassōden* it was expected that this work would be hard-going.[73] However, Committee Chairman Matsumoto and Acting Committee Chairman Matsunaga pushed ahead and finally succeeded on 1 May 1951.[74] During this process, however, because of the aggressive attitudes of the Committee, it became very unpopular, not only with the (former) staff of *Hassōden* but also with the nine *Haidens*, the Government and GHQ, who had all hoped for a smoother transition of operations. The above can be seen as a clear indication of the fact that the Public Utilities Commission acted quite independently from the Government.

Evaluation of the Reorganization of the Electricity Industry

During the unfolding of this process two clearly identifiable preferences were articulated and fought for. One aimed at a nationally unified system under government control. This was the preference of *Hassōden*, *Densan Rōkyō*, the Ministry of Commerce and Industry and the Ministry of International Trade and Industry. The other aimed for regionally based privatization of the industry and was advocated by the nine *Haidens*, Matsunaga and GHQ. Both sides embraced the principle of supply area monopolies in order to avoid undesirable facility duplication. Although as things progressed *Hassōden* and the two ministries were increasingly inclined to maintain the status quo, they all supported the unification of

Table 17-6: Composition of the Public Utilities Commission, Candidates and appointees

Candidates	Appointees	Period of appointment (years)
Names Affiliations	*Names* Affiliations	
Shōji Arai Former President, Hassōden; Former President of Kantō Electricity Distribution Co.	*Jōji Matsumoto* (Chairman) Former Minister for Commerce and Industry; Doctor of Law	5
Seitarō Iida Former Vice-Minister for Railways; Former member of the House of Councilors	*Yasuzaemon Matsunaga* (Deputy Chairman) Former Chairman, Electric Industry Reorganization Council	4
Chūbei Itō Chairman, Itōchū Store	*Kiyoshi Miyakawa* President, Kamishima Chemicals Co.	3
Naritarō Okamatsu Former Vice-Minister for Commerce and Industry	*Kōichi Kawakami* President, Export and Import Bank	2
Yōzō Kaitō President, Tōa Synthetics Co.	*Chūbei Itō* Chairman, Itōchū Store	1
Shōshirō Kudō President, Reconstruction Finance Corporation		
Junzō Kosaka President, Hassōden		
Michisuke Sugi President, Osaka Chamber of Commerce and Industry		
Aiichirō Fujiyama President, Nippon Sugar Manufacturing Co.		
Jirō Masuda Ex-President, Hassōden		
Tarō Matsuda Former Vice-Minister for Commerce and Industry		
Tasuku Watanabe Ex-President, Yokohama Gas Co.		

Source: Takeo Kikkawa, 'Denryoku-gyō (The power industry)', in the Ministry of International Trade and Industry Policy History (ed.), *Tsūshō sangyō seisaku-shi* (A history of international trade and industry policies), vol. 3, 1992, p.375.

electricity generation, transmission and distribution. All of the parties involved had begun on the common ground of the ideology of democratization and opposition to a monopoly by a large statewide private company. The development of proposals, debates, negotiations and compromises, however, produced divergent paths.

If a nationally unified company were to manage an electricity industry under the principle of supply area monopoly, a huge monopolistic body would be required. According to the first view, if the objective of democratization were to be achieved – the central tenet of the Occupation

Forces' initiated postwar reforms – such a company could not be privately managed. *Hassōden* and the Government Ministries believed that the rationalization and democratization of the industry could only be achieved by a public concern managed by the state on behalf of the public.[75] Thus, for these parties, realizing nationwide unification meant the completion of a system of state management that had been established in the prewar years in an unsatisfactory form.

From the perspective of GHQ, Matsunaga and the *Haidens*, however, an essential aspect of democratization and rationalization was to move away from state-management. As they also embraced the principle of supply area monopoly, they maintained that the industry could not be nationally unified. Dividing the country into different regions, they believed, would weaken the rationale for state management and guarantee competition between private regional companies. Legalizing supply area monopolies was, they maintained, a necessary compromise to avoid excessive competition that might damage the public nature of the industry.

As we have seen, the forces aligned in favor of the private management of regional monopolies finally won the day. The Public Utilities Law enabled a reorganization based on the private management of regional divisions, with Clause 3, Article 28 legalizing monopolies of supply areas.[76] According to the prevailing position expressed by GHQ and Matsunaga, establishing such monopolies did not justify intervention by a public authority. Thus the role of the newly established Public Utilities Commission was not to control, but rather to regulate the industry.

The new industry system had all the features necessary for the type of management advocated by GHQ and Matsunaga. It is, however, very difficult to satisfactorily answer whether the new system achieved the expected results because the electricity industry experienced new developments within two years after this reorganization was implemented. However, in the struggle between the new companies over the ownership of undeveloped electricity sources, a struggle that has been described as 'reminiscent of the era of the prewar fierce electricity conflicts among the five major private electric power companies,'[77] we can ascertain the competitive stance taken by the new electricity companies in the early stage of their fight towards securing an independent management base.

Let me finish this chapter by describing the subsequent moves made by some of the parties defeated in the power struggle to define the nature of the reorganization. *Densan* was perhaps the hardest hit by the re-organization. *Densan* had been proud of the strength of the industry-based unions but proved incapable of recovering after the regional divisions were established. Some people believe that this was one of GHQ's primary objectives for the reorganization.[78] Actually, though, the collapse of *Densan* began in a struggle between its own Communist and Democratic Alliance factions.[79] It should therefore be regarded as part of the general decline in the union movement seen in the 'reverse course' (return to conservatism) that accompanied GHQ's change in Occupation policy. However, in a sense it can also be seen as symbolizing the struggles for reorganization that ended in victory for the prewar capitalist electricity industry managers.

The local governments that had aimed for establishing the public management of electricity distribution had to reduce their ambitions for retrieving the municipal and prefectural electric power businesses that they had run in the prewar years due to the establishment of the new regional companies. The movement was later subsumed by new organizations.[80]

As previously mentioned, although the Ministry of Trade and Industry was stripped of the authority to regulate and control the electricity industry, it maintained a foothold through the Electric Facility Section. A small number of former government officials who had been engaged in the administration of the industry moved to the administration office of the Public Utilities Commission, and each new electricity company took on many of the former directors and employees of *Hassōden* after it was dismantled.[81] The future direction of the electricity industry would be determined not only by those whose views on reorganization had prevailed, but also by these survivors from the other side of the conflict.

Notes

1 Electric Power Companies of Hokkaidō, Tōhoku, Tokyo, Chūbu, Hokuriku, Kansai, Chūgoku, Shikoku and Kyūshū.

2 Ministry of International Trade and Industry, Resources and Energy Agency, Public Utilities Division, [1] (Heisei 3rd edition), pt. 2, pp. 372–6.

3 Recently, changed to be divided and privatized.

4 Nori Sakurai, 1964, 'The Electricity Industry and State Control,' in Kurihara (ed.) [2] pp. 235–356.

5 However, not mentioned in the legal document.

6 Hokkaidō, Tōhoku, Kantō, Chūbu, Hokuriku, Kansai, Chūgoku, Shikoku and Kyūshū.

7 Koike [3]; Kanayama, 1947 [4]

8 Ministry of Commerce and Industry, Electric Power Bureau, [5]; Ministry of Commerce and Industry, Electric Power Bureau, 1945 [6]; Ministry of Commerce and Industry, Electric Power Bureau, 1948 [7] pp. 37–40; Yamamoto, 1950 [8]

9 Yamamoto, 1950 [8] p. 7.

10 The Ministry of Commerce and Industry instructed the electricity industry to manufacture salt using surplus electricity (Electric Power Bureau, Ministry of Commerce and Industry, 1945, 'How to make salt with surplus electricity'). The Ministry of Commerce and Industry and

Hassōden planned a long expected nationwide unification of electric power frequencies at this time when works at war-affected sites were easy to incorporate into the program. However, the plan came to naught, as the Ministry and *Hassōden* prioritized the measures to meet the rapid growth in demand for power (The Preparation and Investigation Committee for Frequency Unification, 1946, 'Reports on Electricity Frequency Unification'; Ministry of Commerce and Industry, 1946, 'Data on Electricity Frequency Unification').

11 The Gas and Electricity Section of GHQ regarded the many cases of stealing electricity seriously and questioned the management of the electricity industry (Electricity Committee, Lower House, 1947, 'Electricity policy research data no 5').

12 GHQ planned the removal of a thermal power station as part of reparations, but it was only partially completed due to the change in Occupation policy. However, it caused the delay in recovering the supply capacity because restoration works were forbidden on some thermal power stations that were designated to be removed for reparations. Arai [9]

13 Ministry of Commerce and Industry, Electric Power Bureau, 1947 [10]

14 Nippon Hassōden, Research Division, 1947 [11] p. 1.

15 Ministry of Trade and Industry (MITI), Public Utilities Bureau, [1] 1961 edition, pp. 104–7.

16 Nippon Hassōden, 1948 [15]. Economic Stability Board, et. al., 1948 [16]

17 The Counterpart Fund consisted of monies the government had accumulated to pay for US aid goods. The government canceled this accumulated savings to spend on economic restoration. Sokuichi Odake, 'One-hundred year history of electricity, vol. 2,' 1980, Seikei-sha, p. 141.

18 Research Room, Nippon Hassōden [22]

19 Nippon Hassōden, 1945 [23]

20 Ministry of Labor [17] 1946 edition, pp. 490–2.

21 Nippon Denki Sangyō Rōdō Kumiai Kyōgikai, Chūō Kyōdō Tōsō Iinkai (Japan Labor Union Federation of Electrical Industries, Central Joint Struggle Committee), 1946 [24]

22 Chūō Rōdō Iinkai (Central Labor Relations Commission Office), 1946 [18] p. 3.

23 Chūō Rōdō Iinkai (Central Labor Relations Commission Office), 1946 [18] pp. 7–8.

24 Chūō Rōdō Iinkai (Central Labor Relations Commission Office), 1947 [18] p. 9.

25 Denki Jigyō Minshuka Kyōgikai (Electric Power Industry Democratization Council), 1947 [25]. Denki Jigyō Minshuka Kyōgikai (Electric Power Industry Democratization Council), 1947 [26]

26 Chūō Rōdō Iinkai (Central Labor Relations Commission Office), 1948 [18] pp. 13–14.

27 Ministry of Labor [17] 1947 edition, p. 652.
28 Ministry of Labor [17] 1948 edition, p. 488.
29 Ministry of Labor [17] 1948 edition, p. 108.
30 Kikkawa, 1992 [20] p. 328.
31 Kikkawa, 1992 [20] p. 329.
32 Kikkawa, 1992 [20] pp. 330–3.
33 Kōei Denki Fukugen Undō-shi Henshū Iinkai 1969 [34] p. 69.
34 Nippon Hassōden, Documents Division, 1948 [41]
35 On GHQ's Occupation policy change and the Elimination of Excessive Economic Concentration Law, see Ministry of Finance, 1981, 1976 [40]
36 The Holding Company Liquidation Committee responsible for enforcing the Elimination of Excessive Economic Concentration Law determined in July that nine distribution companies out of the ten electricity companies did not need to be reorganized.
37 Nihon Denki Sangyō Rōdō Kumiai (Japan Electric Power Industry Labor Union), 1948 [30]; Densan Chūō Jōnin Tōsō Iinkai (Central Standing Struggle Committee of the Labor Unions of the Electric Industry), 1948 [31]
38 In other words, the movement for the restoration of local government supply businesses was working on the assumption that the supply area monopolies would be approved. Later, local governments strongly opposed the supply area monopoly principle. However, this should probably be seen as employing a convenient argument for the restoration movement of local government supply under the nine company system rather than an in principle objection to monopolies *per se*. Haiden Jigyō Zenkoku..., 1948 [32]; Haiden Jigyō Zenkoku..., 1948 [33]
39 Nippon Hassōden, 1948 [27]
40 At that time, it was only the nine distribution companies that insisted on a privately owned and unified management of electricity generation, transmission and supply divided into separate areas. Therefore, it is quite important to make it clear how those nine transmission companies' changed their stance, although not enough work has been done in this area – partly due to the limited availability of data. Electric Distribtion Compaines..., 1948 [28]; 1948 [29]
41 Kikkawa, 1992 [20] pp. 341–5.
42 Nippon Hassōden, 1950 [42] pp. 16–17.
43 Nippon Hassōden, 1950 [42] p. 19.
44 Ministry of Commerce and Industry, Electric Power Bureau, 1948 [45] p. 33.
45 Ministry of Commerce and Industry, Electric Power Bureau, 1948 [45] p. 33.
46 Kikkawa, 1990 [39] p. 301.
47 Behind the scenes there was conflict between the Anti Trust and Cartels Division/Holding Company Liquidation Committee, which were about to proceed with eliminating the concentration of economic power

according to GHQ's early Occupation policy, and the Deconcentration Review Board whose members were appointed according to the later (revised) Occupation policy.

48 Kikkawa, 1992 [20] p. 352.
49 Kikkawa, 1992 [20] p. 353.
50 Nippon Hassōden, 1950 [42] p. 26; MITI, 1950 [43] pp. 2–3.
51 The Japanese were worried about the possibility that the seven-region plan would immediately be implemented. The Kantō Haiden Labor Union separated from *Densan* and became independent on 23 December 1949. Among its reasons for doing so was an analysis of the situation that concluded that division into seven areas would be unavoidable. Ministry of Labor, 'Labor Movement History-Data,' 1949, p. 404.
52 On 25 May 1949, The Ministry of Commerce and Industry was restructured and renamed the Ministry of International Trade and Industry (MITI). The Electric Power Bureau was relocated (again) to the Natural Resources Agency, a body external to the Ministry.
53 MITI, 1950 [43] pp. 3–4.
54 Kikkawa, 1992 [20] p. 354.
55 MITI, 1950 [43] p. 4; 1975 [44] pp. 38, 48.
56 Tōhō Denryoku-shi Henshū Iinkai, 1962 [35] p. 542.
57 Denki Jigyō Saihensei-shi Kankō Iinkai, 1952 [19] pp. 340–530.
58 Sangyō Seisaku-shi Kenkūsho (Research Institute of Industrial Policy History), 1985 [47] p. 88.
59 Denki Jigyō Saihensei Shingikai (Electric Industry Reorganization Council), 1950 [46]
60 Nippon Hassōden [42] p. 47.
61 Kikkawa, 1992 [20] p. 365.
62 GHQ/SCAP, 1951 [38] pp. 41–2.
63 MITI, 1975 [44] p. 178; MITI, 1950 [43] pp. 5, 11–12.
64 MITI, 1950 [43] pp. 13–16.
65 MITI, 1950 [43] pp. 17–19.
66 MITI, 1950 [43] pp. 20–2
67 MITI, 1950 [43] pp. 23–38.
68 Denki Jigyō Saihensei-shi Kankō Iinkai, 1952 [19] p. 545.
69 MITI, 1950 [43] pp. 7–8.
70 Denki Jigyō Saihensei-shi Kankō Iinkai, 1952 [19] pp. 599–806.
71 Kikkawa, 1992 [20] pp. 371–4.
72 Research Institute for the History of Industrial Policy, 1981 [52] p. 134; Kōeki Jigyō Iinkai, 1951 [53] pp. 163–5.
73 The main points of dispute were the ratio of the new electricity companies' share allotment to the stockholders of *Hassōden* and nine electric distribution companies, and how the directors of the new companies would be appointed. *Hassōden*'s resistance persisted for a long time and even

caused the adjournment of the Tokyo Electricity Company's shareholders' General Meeting.

74 Nippon Hassōden, 1951 [48] [49] [50]. Kōeki Jigyō Iinkai (Public Utilities Commission), 1951 [51].

75 The movement for the establishment of local government supply business – whose policy was to combine supply area monopolies and public management – presented the same reasoning, although it did not pursue a unified nationwide management.

76 The Public Utilities Law was not the first to legalize a monopoly of supply areas; the Electric Distribution Control Ordinance had already enabled this.

77 Yasushi Kobayashi, 'Saihengo no denryoku sangyō (Electricity industry after reorganization),' in Kurihara, 1964 [2] p. 389.

78 Yasushi Kobayashi, 'Saihengo no denryoku sangyō (Electricity industry after reorganization),' in Kurihara, 1964 [2] p. 387.

79 We cannot, however, ignore the impact that GHQ's seven-division plan had on the departure and independence of the Kantō Haiden Labor Union.

80 On 29 October 1951, the Association for the Restoration of Municipal Management of Electric Power Business was set up as an alliance between municipal governments seeking the restoration of their power businesses. Kōei Denki Fukugen Undō-shi Henshū Iinkai, 1969 [34] p. 228.

81 Of the one hundred and fourteen directors of the new regional electricity companies, twenty of them were from *Hassōden* and forty-two from the Haidens. Kōeki Jigyō Iinkai, 1951 [53] p. 107.

Bibliographic Guide

On the Overall History of the Electric Industry

[1] Ministry of International Trade and Industry, Public Utilities Bureau (later the Ministry of International Trade and Industry, Agency of Natural Resources and Energy, Public Utilities Department), *Denki jigyō no genjō* (The present situation concerning electric industry), published annually.

[2] Tōyō Kurihara (ed.), *Gendai Nihon sangyō hattatsu-shi, III Denryōku* (A history of the development of modern Japanese industry: III Electric power), Kōjunsha, 1964.

On the Electric Industries Law Amended in 1946

[3] Shinzō Koike, *Kaisei denki jigyō-hō kaisetsu* (An Explanation of the amended Electric Industry Law), Dai Nippon Denki-kai (The All-Japan Electric Association) (year of publication unknown).
The author was then Head of the Electric Power Bureau of Ministry of Commerce and Industry.

[4] Muneo Kanayama, *Kaisei denki jigyō-hō chikujō kaisetsu* (An article-by-article explanation of the amended Electric Industry Law), Nippon Hassōden Co. Ltd., Chūgoku Branch, 1947.
The author was then an employee of Nippon Hassōden's Chūgoku Branch.

Damage Incurred by the Electric Power Facilities as a result of Air Raids

[5] Ministry of Commerce and Industry, Electric Power Bureau, *Kūshū higai songai kingaku sōkatsu-hyō* (Comprehensive table of air raid damage), (year of publication unknown).
[6] Ministry of Commerce and Industry, Electric Power Bureau, *Hatsuden setsubi higai jōkyō meisai-hyō* (Detailed chart giving damage sustained by power producing facilities), 1945.
[7] Ministry of Commerce and Industry, Electric Power Bureau, *Denki jigyō shiryō* (Electric industry data), 1948.
[8] Senji Yamamoto, 'Sengo no hatsuden-sho no fukkyū ni tsuite (Postwar restoration of power plants),' paper submitted to the Fourth World Power Conference, Nippon Hassōden Co. Ltd., 1950, pp. 2–21.
The author was an employee of Nippon Hassōden.

On Reparation Removal of the Thermal Power Electric Facilities

[9] Shōji Arai, President of Nippon Hassōden, *Denryoku setsubi no baishō ni taisuru yōbō* (Pleas relating to electric power facility reparations), (year of production unknown).
This paper contains correspondence addressed to Ambassador Edwyn Pauley from Arai, requesting a softening of reparation measures.
[10] Ministry of Commerce and Industry, Electric Power Bureau, *Hatsuden setsubi kinkyū fukkyū seibi yotei-hyō* (Schedule for restoration and maintenance of power producing facilities), 1947.

On the Electric Industry Immediately Following Defeat in the War

[11] Nippon Hassōden, Research Department, *Wagakuni denki jigyō no genjō* (The present situation of the electrical power industry in Japan), 1947.
[12] Ministry of Commerce and Industry, Electric Power Bureau, *Denki jigyō shiryō* (Electric industry data), 1948.
[13] Nippon Hassōden Co. Ltd., Research Department, Planning Division, *Denryoku juyō no sūsei* (Trends in electric power demand), 1949.
[14] Economic Stabilization Board, Electric Power Bureau, Energy Division, *Saikin no denryoku jijō* (The recent energy situation), 1949.

On the Immediate Postwar Electric Power Restoration Plan

[15] Nippon Hassōden Co. Ltd., *Denryoku setsubi kakujū oyobi hoshū kairyō kōji 5-kanen keikaku gaiyō* (An outline of the five year plan for the expansion, repair and improvement works for electric power facilities), 1948.

[16] Economic Stabilization Board, Production Bureau, Supply and Demand Division; Power Bureau, Electric Power Division; and Construction Bureau, Industrial Facilities Division, *Keizai fukkō keikaku dai 1-ji shian no urazuke to naritaru denryoku 5-kanen keikaku-an* (Five year initial draft plan for electric power which served as the basis for the first draft of the Economic Recovery Plan), 1948.

On the so-called Densan Dispute

[17] Ministry of Labor, *Shiryō rōdō undō-shi* (Reference data on the history of the labor movements), published annually.

[18] Chūō Rōdō Iinkai (Central Labor Relations Commission Office), editorial supervision, 'Densan sōgi no chōtei keika, (The arbitration process in the Densan dispute),' 3 installments; 'Densan sōgi chōtei keika (The arbitration process in the Densan dispute'; 'Densan sōgi assen keika gaiyō (An outline of the Densan dispute mediation process)', in *Chūō rōdō jihō* (Central Labor Review), vol. 1, nos. 9–10 (Dec 1946); no. 11 (Dec 1946), nos. 14–15 (Feb 1947); no. 56 (March 1948); no. 60 (May 1948) pp. 14–24.

Research into the History of the Re-organization of the Electric Power Industry.

[19] Denki Jigyō Saihensei-shi Kankō Iinkai (Editorial Committee for the History of the Reorganization of the Electrical Industry) (ed.), *Denki jigyō saihensei-shi* (A history of the reorganization of the electrical industry), 1952.
This is a definitive history of the reorganization history containing numerous primary source materials. References to GHQ trends and trends prior to the process of eliminating the concentration of economic power, however, are not dealt with in any depth.

[20] Takeo Kikkawa, 'Denryoku-gyō (The power industry),' in The Ministry of International Trade and Industry Editorial Committee on International Trade and Industry Policy History (ed.), *Tsūshō sangyō seisaku-shi* (A history of international trade and industry policies), vol. 3, Tsūshō Sangyō Chōsa-kai, 1992, pp. 317–95.
Kikkawa, who specializes in business history, has in recent years assiduously dug out materials relating to GHQ, the Ministry of Commerce and Industry and MITI and published the results.

[21] Takeo Kikkawa, 'Denki jigyō saihensei ni okeru min'ei chiiki-betsu 9-bunkatsu-an no keisei katei: Saihensei no shuyaku wa dareka (The formulation of the 9-bloc plan in the reorganization of Japan's electric power industry in 1951: Who was the opinion leader?)', in *Shakai keizai shigaku* (Socio-economic history), vol. 57, no. 6, March 1992, pp. 735–61.
According to Kikkawa, the leading figure in the re-organization was Yasuzaemon Matsunaga. However, at the present stage, we have not been able to obtain any data that clarify the actions of the nine electric distribution companies. Consequently, the present author has not been able to determine who the leading figure was.

Several Japanese Proposals for the Reorganization of the Electric Power Industry – Before the Electric Power Companies were Designated as Excessive Concentrations of Economic Power by GHQ.

Nippon Hassōden

[22] Nippon Hassōden, Inquiry Section, *Hassōden kaisha oyobi haiden kaisha no tōgō ni kansuru chōsa* (Survey into the amalgamation of the electric generation, transmission and distribution companies). This was not dated but appears to have been drawn up immediately after the war.
[23] Nippon Hassōden Co. Ltd., *Sengo denki jigyō no un'ei keitai no kaizen ni kansuru tōshin-an* (A report on improvements to the administrative structure of the postwar electrical industry), 1954.

The Labor Union Perspective

[24] Nippon Denki Sangyō Rōdō Kumiai Kyōgikai, Chūō Kyōdō Tōsō Iinkai (Japan Labor Union Federation of Electrical Industries, Central Joint Struggle Committee), *Densan yōkyū kyūyo minshuka-an zen* (The complete draft of the electrical industry's salary democratization demands), 1946. In the material obtained by the author, the word 'complete' has been erased.

The Electrical Industry Democratization Council

[25] Denki Jigyō Minshuka Kyōgikai (Electric Power Industry Democratization Council), *Denki jigyō minshuka jōnin iinkai giji yōroku (shūroku)* (Proceedings digest of the standing committee on the electrical industry's democratization – Record), 1947.
[26] Denki Jigyō Minshuka Kyōgikai (Electric Power Industry Democratization Council), *Denki jigyō no minshuka wa dō susume rarete iruka – zōho* (How is the democratization of the electrical industry proceeding?), revised and enlarged edition, 1947.

After the Electric Power Companies Were Designated As Excessive Concentrations of Economic Power

Nippon Hassōden

[27] Nippon Hassōden, *Saihensei keikakusho* (A reorganization plan), 1948.

Electric Distribution Companies

[28] The Electric Distribution Company of each of the Hokkaidō, Tōhoku, Kantō, Chūbu, Hokuriku, Kansai, Chūgoku, Shikoku and Kyūshū Regions, *Saihensei keikakusho* (A reorganization plan), 1948.

[29] The Electric Distribution Company of each of the Hokkaidō, Tōhoku, Kantō, Chūbu, Hokuriku, Kansai, Chūgoku, Shikoku and Kyūshū Regions, *Denki jigyō saihen keikaku-an* (A draft plan for the reorganization of the electricity industry), 1948.
A pamphlet explaining the reorganization plan of the distribution companies for general readers.

The Japan Electric Industries Labor Union

[30] Nihon Denki Sangyō Rōdō Kumiai (The Japan Electric Industries Labor Union), *Kado keizai-ryoku shūchū haijohō tekiyō ni yoru denki jigyō saiken keikaku* (A plan to rebuild the Japanese electric industry based on the application of the Elimination of Excessive Economic Concentration Law), 1948.

[31] Densan Chūō Jōnin Tōsō Iinkai (Central Standing Struggle Committee of the Labor Unions of the Electric Industry), 'Densan-an ni tsuite (On the Densan plan)', in *Denkijigyō saiken keikaku* (Reconstruction plan of the electric industry), 1948.

Local Governments

[32] Haiden Jigyō Zenkoku Todōfuken-ei Kisei Dōmeikai (All Japan Alliance for Prefectural Administration of Electricity Distribution), *Denki jigyō saihensei ni tomonau haiden jigyō todōfuken-ei mondai ni tsuite* (On the question of prefectural administration of electric distribution, accompanying the re-organization of the electric power industry), 1948.

[33] Haiden Jigyō Zenkoku Todōfuken-ei Kisei Dōmeikai (All Japan Alliance for Prefectural Administration of Electricity Distribution), *Haiden jigyō todōfuken-ei kihon hōshin setsumei-sho* (An explanation of the basic policy of the prefectural administration of electric distribution), 1948.

[34] Kōei Denki Fukugen Undō-shi Henshū Iinkai (Editorial Committee on the History of Movements to Restore Local Government Management of the Electric Power Industry), *Kōei denki fukugen undōshi* (A history of movements to restore the local government management of electric power industry), 1969.

Yasuzaemon Matsunaga

[35] Tōhō Denryoku-shi Henshū Iinkai (Editorial Committee on History of Tōhō Denryoku), *Tōhō denryoku-shi* (A history of Tōhō Denryoku), 1962. Includes an outline of Matsunaga's 'Denryoku tōsei shiken (My personal view on the regulation of electric power industry)'.

[36] Yasuzaemon Matsunaga, *Denryoku saihensei ni tsuite* (On the reorganization of electric power industry), 1950.

[37] Yasuzaemon Matsunaga, *Nippatsu o kokonotsu no genhaiden kuiki ni bunkatsu shita riyū o nobu* (Stating the reasons for dividing the Hassoden into the present nine local regions of power distribution).

On the Reorganization Plan of the GHQ

[38] GHQ/SCAP, 'Expansion and Reorganization of the Electric Power and Gas Industry,' in GHQ/SCAP, *History of the Nonmilitary Activities of the Occupation of Japan,* XV (1951), Fuel and Power, Part B, pp. 29–44. Part of the Official History of the Occupation produced by GHQ.

[39] Takeo Kikkawa, 'The Reorganization of Japan's Electric Power Industry and General Headquarters/Supreme Commander of the Allied Powers (1)–(4)', in *Aoyama keiei ronshū* (Aoyama Journal of Business), vol. 25, no. 3, Nov. 1990; vol. 25, no.4, Mar. 1991; vol.26, no. 1, June 1991; vol. 26, no. 3, Dec. 1991.
Internal GHQ materials are presented, including correspondence from MacArthur to Shigeru Yoshida urging an enforcement of the reorganization. Kikkawa collected these materials at the Washington National Records Center, a branch of the National Archives located in Suitland, Maryland. It is believed that almost all of these are presently (Spring 1994) available in the GHQ/SCAP data in the Constitutional Data Section of the National Diet Library.

[40] Ministry of Finance, Public Finance History Section (ed.), *Shōwa zaisei-shi*: *Shūsen kara kōwa made* (A history of public finances during the Shōwa period: From the end of the war to the peace treaty), Tōyō Keizai Shinpōsha, vol. 2 (1981), vol. 3 (1976).
A comprehensive history of the Occupation period.

On the Reorganization Process

[41] Nippon Hassōden Co. Ltd., Documents Division, *Kado keizairyoku shūchū haijo shitei-go no keika oyobi tetsuzuki tō no gaiyō* (An outline of the progress and processes following the designation as an excessive concentration of economic power), 1948.

[42] Nippon Hassōden Co. Ltd., *Sai-hensei kankei nisshi* (A diary on matters relating to reorganization). This is not dated but is thought to have been published in 1950.

[43] Ministry of International Trade and Industry, Agency of Natural Resources, Electric Power Bureau, *Denki jigyō saihen no kon'nichi made no keika* (The course of the reorganization of the electric power industry to date). This is thought to have been published in 1950.

[44] Ministry of International Trade and Industry, Minister's Secretariat, Public Relations Division, *Shōgai shūhō* (Weekly bulletin of negotiations with GHQ).

On the Electric Power Industry Democratization Committee

[45] The Ministry of Commerce and Industry, Electric Power Bureau, *Denki jigyō minshuka iinkai gijiroku* (Minutes of the Electric Industry Democratization Committee), 1948.

On the Electric Industry Reorganization Council

[46] Denki Jigyō Saihensei Shingikai (Electric Industry Reorganization Council), *Denki jigyō no saihensei tō ni kansuru tōshin* (Report on the electricity industry reorganization and other matters), 1950.

[47] Sangyō Seisaku-shi Kenkyūsho (Research Institute of Industrial Policy History), *Sangyō seisaku-shi kaisōroku* (Reflections on the history of industrial policy), 32 volumes, 1985.
Recollections of the situation concerning the council given by Tsuneo Komuro, the former head of the Electricity Policy Division, Electric Power Bureau, Resources Agency. He also held the position of secretary general of the council.

On the Dissolution of Nippon Hassōden Co. Ltd. and the Nine Electric Distribution Companies

[48] Nippon Hassōden Co. Ltd., *Igi mōshitate-sho* (Statement of objections), 1951.

[49] Nippon Hassōden Co. Ltd., *Kaku haiden chōmon-kai gaiyō* (An outline of the public hearings of each of the electric distributors), year of compilation unknown.

[50] Nippon Hassōden Co. Ltd., *Matsumoto Kōeki Jigyō Iinchō kaiken kiroku* (Record of the interviews with Mr. Matsumoto, Chairman of the Public Utilities Commission), year of compilation unknown.

[51] Kōeki Jigyō Iinkai (Public Utilities Commission), *Kigyō saihensei keikaku ni kansuru kettei shirei* (The Order relating to the enterprise reorganization plan), 1951.

On the Changes to the Administrative Mechanisms and Personnel

[52] Research Institute for the History of Industrial Policy, Toranomon Annex of the Research Institute, MITI, *Kaitei-ban: Shōkōshō – Tsūshō sangyōshō gyōsei kikō oyobi kanbu shokuin no hensen* (Revised edition: The Ministry

of Commerce and Industry/The Ministry of International Trade and Industry – Changes to the administrative structure and executive staff, 1925–1980), 1981.

[53] Kōeki Jigyō Iinkai (Public Utilities Commission), (editorial supervision), *Shin denryoku jigyō meikan* (New directory of electric power enterprises), Nikkan Kōgyō Shimbunsha, 1951.

18 GHQ and Changes in the Postwar Telecommunications Structure

Kiyoshi Yamauchi

The Postwar Telecommunications Revival

Japan marked the centenary of the introduction of the telephone on 16 December 1990. As of 1 January 1988 Japan was second only to the USA in terms of subscriber numbers, with 49,980,000 subscribers compared to 118,400,000 in the US.[1] The third highest number of subscribers was in the Soviet Union, followed by West Germany, France and Britain. There were 40.8 telephones per 100 people in Japan placing it fifteenth in the world on the scale of per capita distribution. The large population of Japan accounts for the different rankings.

By the end of World War II virtually all of the telephone facilities in place prior to the war had suffered damage in the hostilities. The telephone exchanges in the major cities were almost completely destroyed. As *Asahi Shimbun* reported on 2 August 1945 '...of the total of 1.08 million telephone subscribers nationwide, approximately 500,000 lost their telephones in air raids.' [2]

Of course, '...the destruction of Hiroshima by the atomic bomb had a devastating effect on telecommunications. At the end of the war there were no telephone or telegraph lines to the areas beyond Hiroshima. Neither the Shikoku route nor the Sanyo route had resumed operations yet. It goes without saying that the bomb wiped out the Hiroshima repeater station.' [3] The damage sustained before the cease-fire is outlined in Table 18-1. It was 1950 before the number of subscribers returned to the prewar figure of 1,070,000.

It seems appropriate to consider the period between the end of the war and the establishment of Nippon Telegraph and Telephone Public Corporation on 1 August 1952 as a time of reconstruction and the subsequent period as one of development.

This chapter traces the process of the reforms to the postwar telecommunications structure in Japan. In particular, it aims to evaluate the continuities and changes of the postwar reforms effected during the seven year period between 15 August 1945, when the war ended, and 1 August 1952 when the Nippon Telegraph and Telephone Public

Table 18-1: Damage to the telecommunications infrastructure.

	Equipment total	Damaged	Percentage
Local telephones			
Local telephone lines	1.18 million lines	522,575 lines	44%
Cables	25,300 km	5170 km	20%
Open wires	247,800 km	10,560 km	4%
Telegraphs and toll telephones			
Toll telephone lines	17,200 lines	6687 lines	39%
Telegraph lines	2110 km	2027 km	96%
Cables	7390 km	20 km	0.3%
Open wires	305,500 km	600 km	0.2%
Telephone offices	97	26	27%
Telegraph offices	10	3	30%
Post offices	590	72	12%
Radio stations	24	3	12%
Repeater stations	98	2	2%
No of local telephone subscribers	1,080,000	540,000	50%

Source: *The Postwar Telecommunications Industry*, Federation of Electrical Telecommunications Industries Associations of Japan, 1959.

Corporation was established. The principal events of this period are summarized in Table 18-2.

Telecommunications Administration during the Allied Occupation

The Allied Forces arrived in Japan on 28 August 1945, and on 2 October 1945 the General Headquarters of US Army Forces in the Pacific (GHQ/AFPAC) arranged the dissolution of the Military Government Section and in its place established the General Headquarters, Supreme Commander for the Allied Powers (GHQ/SCAP, hereinafter referred to as GHQ) which set about organizing the administration of occupied Japan. The Economic and Scientific Section (ESS) was established within GHQ to assume the responsibility for guiding the reconstruction of industry and the economy.[4] One of its primary responsibilities was to assist in converting 'the munitions based industry to a consumer-based industry.'[5]

The Civil Communications Section (CCS) was also created on 2 October with sole responsibility for the administration of telecommunications. The establishment of this single-purpose organization indicates the great importance that GHQ placed on the reconstruction of Japan's telecommunications system and its role in the successful execution of Occupation policy.

GHQ instructed the Japanese government that the security and maintenance of telecommunications should be given the highest priority for the reconstruction of Japanese industry.[6] There were two primary objectives of GHQ's telecommunications policy, securing the telecommunications capability of the Allied Forces' equipment and the restoration of telecommunications services to the general public.[7] Fulfilling these required: restoring the telecommunications systems destroyed in air raids, re-establishing municipal and suburban telecommunications services, revitalizing the telecommunications equipment manufacturing industry, training personnel and re-constructing international telecommunications facilities.

CCS's responsibilities included postal services, telegraphs, telephones and radio services. Its duty, first and foremost, was to demilitarize Japan's telecommunications (cf. Tables 18-1 and 18-2). Army and Navy radio facilities, telegraph stations and telephone services were either to be closed down or converted, and the radio broadcast stations that had been used for transmitting military propaganda to the Japanese public were to be reorganized for civilian use. The International Telecommunications Co. Ltd. and the Japan Telegraph and Telephone Construction Company Ltd. were dismantled, the postal system was democratized and modernized guidelines were issued for increasing the capacity for producing telecommunications equipment.

In June 1947 the Japanese Government responded to the establishment of CCS – to cooperate in fulfilling its objectives – by establishing the Electrical Telecommunications Equipment Bureau within the Ministry of Commerce and Industry as a telecommunications administration body and the Communications Bureau was established within the Economic Stabilization Board. In June 1949 it established the Ministry of Telecommunications, a telegraph and telephone administration to promote a revival of public telecommunications. At the same time, the Ministry of Postal Services was established to handle postal services. The creation of these two new ministries was accompanied by the abolition of the Ministry of Communications.

The State of the Telecommunications Industry

In January 1946 the Communications Equipment Manufacturing Association was established with membership consisting of companies such as Nippon Electric Company Ltd, Fuji Communications Equipment Manufacturing Company Ltd (now Fujitsu Limited), Hitachi Ltd and Oki Electric Industry Company Ltd. CCS utilized the association as a control organization and exerted an indirect influence over the Japanese manufacturers of telecommunications equipment through it. This arrangement continued until November 1947 when GHQ directed the closure of all controlling organizations. In December 1947, though, it demanded that a replacement organization be created. Thus, on 20 April

1948, three industrial associations were founded: the Wired Telecommunications Equipment Industry Association, the Radio Telecommunications Equipment Industry Association and the Telecommunications Cable Association. At the same time the Federation of Electrical Telecommunications Industries Associations of Japan was established as an umbrella affiliation.[8]

Thus, under the leadership of CCS, the framework for revitalizing the telecommunications industry was in place about two years after the ceasefire. The revival of the telecommunications industry and market proceeded relatively smoothly and quickly under the guidance of CCS.[9]

The telecommunications equipment manufacturers were in a very difficult situation at the end of the war. They had to cope with the disposal of their factories in the formerly occupied areas overseas and the concomitant loss of production. They also had to deal with returning employees and the reconstruction or disposal of domestic facilities that had been destroyed by air raids. Their surviving production facilities were aging and there was an urgency to convert military production to civilian production. The experience of the Nippon Electric Company Limited – just one of many in this situation – helps to illustrate the extent of this turmoil:

> The personnel, machinery, tools and materials that had been deployed immediately before the end of the war to the various rural districts in Japan to avoid air raids now had to be returned to their former plants...At the end of the war nearly five thousand personnel and several hundred freight carriages of goods and equipment were to be returned...The withdrawal was, by and large, completed in November, although the entire operation was not concluded until December. The hardships and dangers experienced by those people sent to Japan's overseas occupation countries was far greater than anything experienced by people who were returning from deployments within Japan itself...[10]

Under these conditions the guidance provided by CCS was far more effective than could ever have been imagined.

During this chaos, a severe shortage of raw materials was a major problem for the manufacturers of telecommunications equipment. The Electrical Telecommunications Equipment Bureau was established in the Ministry of Commerce and Industry to provide support for the telecommunications equipment industry which until then had little influence on the Ministry. CCS played a significant role in securing the priority rationing of materials for the telecommunications industry. On 23 October 1948 CCS gathered together representatives of the Ministry of Communications and the manufacturers of telecommunications equipment to deliver a special directive on the terms and conditions of Ministry of

Figure 18-1: Organization of the Civil Communications Section (CCS)

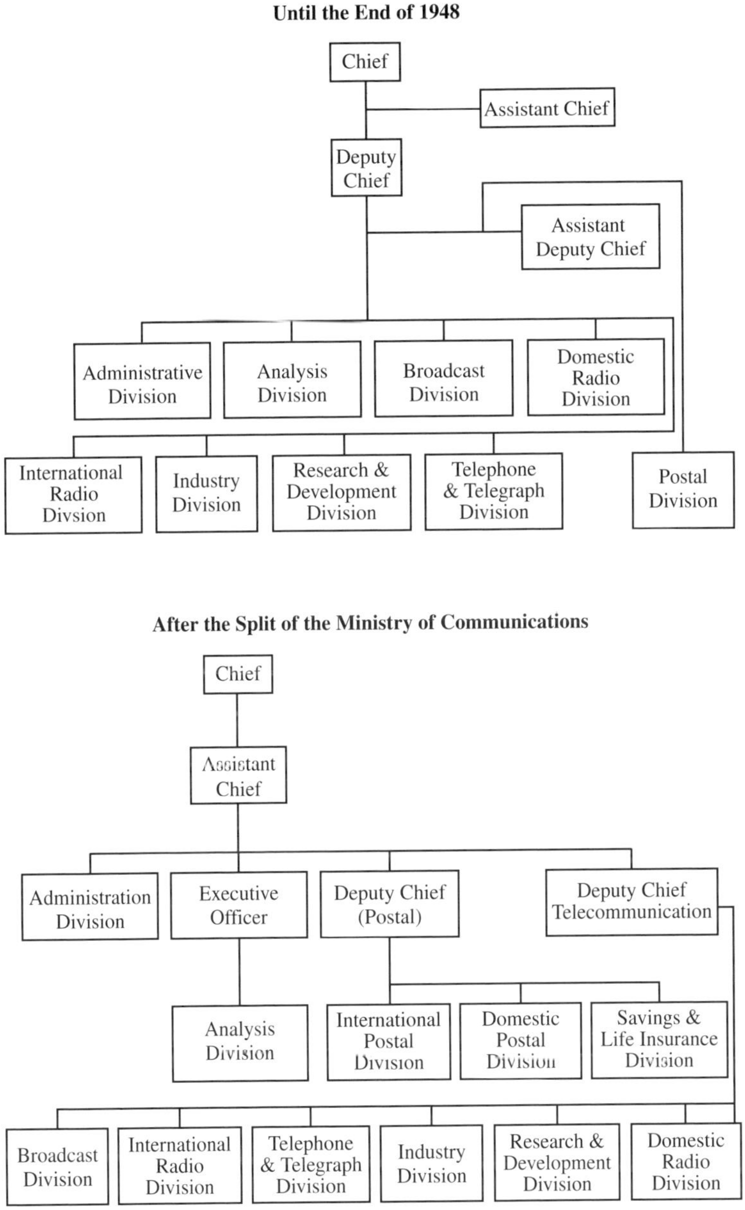

Source: *The Postwar Telecommunications Industry*, 1959, p. 10, edited by the Postwar Telecommunications Industry Editorial Committee.

Communications supply contracts.[11] These new contract conditions were designed to ensure cooperation between the government and the private sector for the manufacture of high quality telecommunications equipment and that manufacturers who fulfilled their contractual responsibilities would be guaranteed a certain profit.

The essence of the 12-point special directive was that:

1 The Ministry of Communications was to establish a comprehensive inspection system for telecommunications equipment.

2 Detailed specifications for the manufacture of all telecommunications equipment were to be clearly spelled out by the Ministry of Communications. All equipment manufactured must comply with these specifications.

3 All technical requirements and specifications must be set forth in covering supply contracts so that they may bind both the Ministry and the manufacturer.

4 The Ministry would reject any equipment that did not comply with the specifications set forth in the contract.

5 A manufacturer would have the right to refuse any contract if the Ministry disallowed payment of the costs necessary to ensure conformity to the specifications.

6 A manufacturer could reject contracts that called for materials not available via the proper supply routes and contracts that demand guarantees where unacceptable to the manufacturer.

7 A manufacturer could reject contracts that included penalty clauses that were unacceptable to the manufacturer.

8 Specifications in a contract could not be altered without the mutual consent of the Ministry and the manufacturer.

At the foundation of this new directive was CCS's intention to establish parity between the government and private enterprise. In other words, while there was no requirement to pay for goods that do not conform to the specifications, if the Ministry of Communications could reject goods that did conform to the specification then the two parties were not equal. There can be no doubt that this new system stimulated manufacturers to take part in the revival of the postwar Japanese telecommunications industry.

Changes in the Public Telecommunications Service

Unification of Telecommunications Services

CCS's model for reorganizing the Japanese telecommunications industry was undoubtedly the US industry. This is totally understandable given that many CCS personnel had come from the Bell Telephone Laboratories and other US telecommunications companies. CCS strongly suggested that in comparison to the vitality of private enterprises in the US, the Japanese telecommunications industry was non-entrepreneurial and inefficient, even when the differences in organizational structure – i.e., government-operated in Japan and privately operated in the US – were taken into consideration. In order to 'correct' this situation the Ministry of Communications and CCS jointly established the Telecommunications Coordinating Committee in February 1947. This committee put forward organizational reform proposals after considering: 1) what enterprise management 'should' be like; 2) the specific circumstances in Japan; and 3) US examples. One of these proposals was for the unification of the electrical telecommunications service.

At that time in Japan the police and national railways each had their own telephone networks independent from the public telecommunications system. When the Ministry of Telecommunications was established as the controlling authority in June 1949, all means of public telecommunications were placed under its jurisdiction – with the exception of the national railways, but including the police telephones (absorbed in June 1950).

Birth of the Nippon Telegraph and Telephone Public Corporation

The question of who should be responsible for telephone operations has been debated in Japan ever since the introduction of telephone services to Japan. The issue elicited considerable debate immediately after the Meiji restoration in 1868. Opinions were sharply divided between those in favor of private enterprise and those who advocated government enterprise. In 1885 both arguments were rejected because there was no decisive advantage from either side. Since, however, the introduction of the telephone service was fiercely needed, it was finally decided that the telephone service in Japan would be operated directly by the government, on the grounds that, with the exception of the US, all of the advanced countries had government-operated telephone services. Subsequently, the push towards privatization was a continuous undercurrent in Japan. For example it manifested again as the Telephone Privatization Bill which was announced by the Ministry of Communications in 1930, but remained a plan only, with no concrete action or implementation.

When CCS's request for a change in the organization responsible for telecommunications service in Japan is considered against this background, the question of what should be done to develop an 'enterprising and rational' business operation arises. It was considered vitally important that the management of the telephone system be transformed from a merely supervisory administration to an entrepreneurial organization.

To this end the Government established the Telegraph and Telephone Restoration Review Committee in July 1949 with Ichirō Ishikawa, Chairman of the Japan Federation of Economic Organizations, as its head. It deliberated over the best management structure for the telegraph and telephone systems as well as concrete matters such as pricing, long-term planning, the capacity for producing necessary equipment and payrolls.

The Committee revisited the relative merits of private versus government enterprise as well as a dual management system in which part of the business would be privatized and part remain under government ownership. The Committee reached a consensus that to achieve an 'enterprising and rational' telephone service 'it would be necessary to have a form of enterprise other than nationally operated.'

Taking full account of the public nature of the telecommunications service, though, the Committee's report to the Prime Minister on 31 March 1950 recommended that 'It is appropriate to have a government corporation so that the advantages of a private enterprise' may be realized to the greatest possible extent.

Based on this report, the 'Proposal for the Rationalization of the Telecommunications Service' was prepared and submitted to GHQ on 27 May 1950. It was never accepted, though, probably due to the outbreak of the Korean War.

However, on 14 August 1951 the Cabinet's Government Ordinance Inquiry Committee proposed that 'telecommunications service should be handled by a government corporation for the time being on the presumption that it will be transferred to the private sector in the future.'

The Ministry of Telecommunications subsequently drafted the 'Nippon Telegraph and Telephone Public Corporation Bill' and the 'Inter-national Telegraph and Telephone Company Bill,' which both received Cabinet approval and were submitted to the thirteenth Diet Assembly on 10 May 1952. Both Bills were approved on the evening of 31 July 1952. On 1 August 1952 the Nippon Telegraph and Telephone Public Corporation Law was proclaimed and took effect.[12] It was 1 April 1985 before privatization (private ownership and private management) came about.

The CCS Management Course

One of the most significant contributions CCS made to the restructuring of the Japanese telecommunications sector was the provision of management guidelines that were developed from the findings of a survey of six Japanese telecommunications equipment manufacturers. The survey revealed that:

1 There had been insufficient research concerning the relationship between chief executive officers (CEO) and management.
2 CEOs were too busy to spend any time on creative work.
3 Objectives, policies, criteria and job descriptions were not clearly defined; nor were they adequately communicated to the rank and file.
4 Insufficient research had been conducted into optimal corporate structures.
5 The responsibilities and authority of managers were not clearly defined.
6 There was a lack of cooperation and coordination between divisions.
7 There was a lack of supervisory and leadership skills.
8 There was an absence of training and education for managers.
9 There was insufficient research about, and practice of, quality control measures, cost management and technology management.

To overcome these weaknesses the Federation of Electrical Telecommunications Industries Associations of Japan cooperated with CCS and sent the top management of affiliated manufacturers to a management course that CCS developed and conducted (from September 1949 to January 1950).[13] The tremendous contribution that this course made to the development and rationalization of the Japanese telecommunication industry is undisputed.

CCS's introduction of statistical quality control methods was equally as significant. CCS's head of manufacturing, G. D. Combs, visited a Nippon Electric Company vacuum tube factory in May 1946 with a team of advisors and demonstrated how such methods should be practiced. As a result, this factory became the first in postwar Japan to implement statistical quality control. The effect was very considerable. The introduction of a total quality control system through which product quality checking was carried out at each stage of the production process resulted in a remarkable improvement in output and ultimately contributed to a reduction in costs.

Changes in the Postwar Public Telecommunications Service

The management system of the Japanese telecommunications services and industry was treated as a government administrative mechanism, rather than as a government enterprise before, during, and for approximately seven years after World War II. CCS pointed out the lack of adaptability and efficiency of this system. To remedy this situation and improve telecommunications services, the telecommunications sector was restructured in 1952 under the Japan-US Joint Committee initiative. One of the outcomes of this scheme was the revival and modernization of telecommunication facilities, including the introduction of new technologies.

After the outbreak of World War II, Japan was isolated from foreign technology. The US announced its withdrawal from the Japan-US Commerce and Navigation Treaty in 1939. Access to new technology had already become quite difficult before then, but it is reasonable to consider 1939 as 'the year in which the introduction of technology ceased.' In the postwar restoration it was therefore imperative for the telecommunications industry to overcome the deterioration in technology that had occurred during the war. Thus, there was a hunger for new technologies in postwar Japan.

One of the sources of new technology was Japan's own 'development' efforts. Today the expression 'development' is quite popular in the Japanese language, but in those days it was a totally new term. The 'development' referred to by the Bell Telephone Laboratories was different from the development referred to when using the Japanese term 'kaihatsu,' giving full consideration to the marketability of a technology and implying research activities that aimed at taking a new technology to the point where it fulfilled market demands. It seems obvious that this concept contributed to the remarkable development of industry in the US. Those concerned in Japan assigned the expression 'Jitsuyōka' to the English word 'development,' making this 'Jitsuyōka' a core concept for improving telecommunications services.

In the rapidly changing technology of the telecommunications industry, rapid adaptation to world-class technology was an important element in improving technological expertise in Japan, even if expensive license fees were required. The restoration of prewar technical agreements with overseas companies and the introduction of new technologies under new technical agreements brought about not only improved technology, but also improved manufacturing processes, factory management and access to overseas information. Japan's postwar telecommunications industry is thus characterized by a dramatic reorganization of enterprise structures and an emphasis on technological innovation.

Notes

1 AT&T, (ed.), *The World's Telephones: A Statistical Compilation as of 1 January 1987 and 1 January 1988*.
2 *Asahi Shimbun*, 2 August 1945.
3 Sengo no Tsūshin Kōgyō Henshū Iinkai (Editorial Committee on the Postwar Telecommunications Industry), 1959 [10] p.13.
4 Established 15 September 1945. Takemae, 1983 [5] p. 45.
5 SCAPIN 47 (order no 3), 22 September 1945.

Table 18-2: Chronological summary of events: 1945–1952

1945	
19 May	Board of Telecommunications abolished and the Board of Communications was established.
15 August	World War II officially ends: telecommunications facilities incurred enormous damage from air raids.
30 August	Establishment of GHQ/AFPAC (Army Forces in the Pacific).
8 September	The Telecommunications Restoration Bureau was established in the Board of Communications.
22 September	Announcement of GHQ's science policy for Japan.[14] General research activity was permitted. A report detailing research activity to be submitted via the Central Liaison and Coordination Office to GHQ/SCAP monthly (later biannually).[15] Research related to atomic energy is prohibited.[16]
2 October	Civil Communications Section (CCS) established within GHQ. Responsible for guidance and administration of telecommunications operations and telecommunications manufacturing industries. Abolished 3 October 1951.
24 December	GHQ directive prohibiting six categories of research at the Electrotechnical Laboratory[17] including television, radar and radar counter measures, pulse modulated multi-channel communication systems, electronic proximity devices, voice secrecy system (speech scramblers), electronic code and cypher systems.
1946	
19 January	Radio and radar aids to navigation added to the list of prohibited research at the Electrotechnical Laboratory. [18]
January	Abolition of Telecommunications Restoration Bureau.
January	Communications Equipment Manufacturing Association is established.
1 July	The Board of Communications abolished and the Ministry of Communications restored.
24 July	GHQ demands improvement in local telephone services.
25 July	GHQ relaxes limitations on research at the Electrotechnical Laboratory.[19]
October	GHQ directs the Ministry of Communications to thoroughly overhaul the telecommunication laws and regulations.[20]
1947	
5 May	Inauguration of the Japan-US Joint Committee[21] to study what a research organ directly related to telecommunications should be (a committee meeting was held about once a week, for a total of around fifty meetings by September 1948).
November	GHQ orders the closure of the Communications Equipment Manufacturing Association.
1948	
April	Establishment of the Federation of Electrical Telecommunications Industries Associations of Japan, comprised of three organizations: the Wired Telecommunication Equipment Industrial Association, the Radio Telecommunication Equipment Industrial Association and the Telecommunications Cable Association.
May	CCS proposes that the Ministry of Communications be reorganized (to include establishing the Electrical Communications Laboratory).
22 July	General MacArthur, in a memorandum to Prime Minister Hitoshi Ashida, suggested that the Ministry of Communications be split into two ministries and that the Japanese national railways and monopoly corporations be reorganized as government corporations.
1 August	Inauguration of the Electrical Communications Laboratory in the Ministry of Communications for encouraging 'development.'
9 September	GHQ directs the dismantling of the Ministry of Communications to establish the Ministry of Telecommunications.[22]

1949	
1 June	Abolition of the Ministry of Communications. Establishment of the Ministry of Telecommunications and the Ministry of Postal Services.
26 August	CCS releases the findings of the Corporate Management and Quality Control Survey to Japan's telecommunication companies.
26 September	CCS commences the Top Corporate Management Course (ended 18 November).
1950	
January	CCS commences Specialist Course.
31 May	The Telegraph and Telephone Revival Deliberation Committee (established July 1949) submits recommendation that telephone service be operated by a government corporation.
31 May	The number of telephone subscribers reaches the prewar level (1.07 million).
April	The House of Representatives decides that telecommunications service will be operated by a government corporation.
May	Cabinet approves the telecommunications industry management rationalization plan and submits it to GHQ for approval. GHQ rejects the plan.
June	The Ministry of Telecommunications assumes control of the police telephone system.
1951	
August	Cabinet Order Inquiry Committee reports on move of telecommunications service to government corporation with the understanding that they will be privatized in the future.
4 September	San Francisco Peace Treaty.
1952	
April	Cabinet decides telecommunications network to be owned and operated by a government corporation.
28 April	Declaration of the end of the state of war. San Francisco Peace Treaty becomes effective. GHQ is abolished.
31 July	The Ministry of Telecommunications is abolished.
1 August	Nippon Telegraph and Telephone Public Corporation comes into existence.

6 Sengo no Tsūshin Kōgyō Henshū Iinkai (Editorial Committee on the Postwar Telecommunications Industry), 1959 [10] p. 8.

7 *Nihon senryō GHQ seishi*, 1990 [1] vol. 55, p. 1.

8 Sengo no Tsūshin Kōgyō Henshū Iinkai (Editorial Committee on the Postwar Telecommunications Industry), 1959 [10] p. 9.

9 CCS was abolished on 3 October 1951. The US Army Pacific Telecommunications Division assumed its responsibilities.

10 *Nippon Denki Kabushiki kaisha nanajūnen-shi* (The seventy year history of the Nippon Electric Company, Limited), 1972, [8] p. 223.

11 On 23 October 1948, Shinohara and Koizumi, Heads of the Engineering Bureau and the Purchasing Bureau, respectively, and representatives from the manufacturers were summoned to the CCS offices by Combs for a detailed briefing on the new contract conditions.

12 Nippon Denshin Denwa Kōsha Nijūgonen-shi Henshū Iinkai (Editorial Committee on the Twenty Five Year History of the Nippon Telegraph and Telephone Public Corporation), 1997 [9] vol. 1, pp. 10–18.

13 Lectures given on the ideology and techniques of US-style management concepts and technologies by Frank A. Polkinghorn, Homer M. Sarasohn and Charles W. Protzman of CCS on 9 and 26 August 1949. *Nihon Denki Tsūshin Kōgyō Rengōkai kaihō* (The Federation of Electrical Telecommunications Industries Associations of Japan Bulletin), 15 November 1949. The minutes of these lectures were published under the title *CCS Management Course* in two volumes translated and edited by the Japan Electrical Telecommunications Industries Association, Management Research Group. 1952 [6] Diamondōsha.

14 SCAP Directive No 3, 22 September 1945.

15 SCAP Directive No 3, 8-c.

16 SCAP Directive No 3, 8-d.

17 SCAPIN 494, 24 December 1945.

18 SCAPIN 626, 19 January 1946.

19 SCAPIN 1084, 25 July 1946.

20 Nippon Hōsō Kyōkai Hōsōshi Henshū-shitsu (Japan Broadcasting Corporation, Editorial Office for Broadcasting History), 1966 [4] vol. 1, pp. 690–1.

21 GHQ/SCAP, *Nihon senryō GHQ seishi*, 1990 [1] vol. 55, p. 74.

22 SCAPIN-A5985, Report of the Joint Investigative Committee on Organization Matters within the Ministry of Communications, 9 September 1948.

Bibliographic Guide

On GHQ Telecommunications Policy

[1] General Headquarters, Supreme Commander for the Allied Powers, *History of the Non-military Activities of the Occupation of Japan, Introduction*, 1945 through 1951. Japanese translation: *Nihon senryō – GHQ seishi* (The Occupation of Japan: GHQ's official history), Nihon Tosho Sentaa, 1990.
Describes GHQ occupation policy, directives and instructions. The reconstruction of the telecommunications industry is addressed in Vol. 55, 1990.

[2] Ministry of Foreign Affairs, Political Affairs Bureau, Special Data Section, (ed.), *Nippon senryō oyobi kanri jūyō bunshoshū* (The Occupation of Japan and a collection of important documents, vol. 1: 'Fundamentals'), 1949.

[3] Nihon Kagakushi Gakkai (ed.), *Nihon kagaku gijutsu-shi taikei 5, tsūshi 5* (A compendium of the Japanese history science and technology: vol. 5, Historical overview 5), Daiichi Hōki, 1964.

[4] Nippon Hōsō Kyōkai Hōsōshi Henshū-shitsu (Japan Broadcasting Corporation, Editorial Office for Broadcasting History) (ed.), *Nippon*

hōsōshi (History of Japanese broadcasting), vol. 1, Nippon Hōsō Kyōkai (Japan Broadcasting Corporation), 1966.

On the Civil Communications Section of GHQ (CCS)

[5] Takemae, Eiji, *GHQ*, Iwanami Shinsho, 1983.
A discussion of the positioning of CCS within GHQ is provided in Chapter 3: 'GHQ no soshiki to hito (The organization and people of GHQ).'

[6] General Headquarters, Supreme Commander for the Allied Powers, CCS, *C.C.S. Keiei kōza* (CCS Management Course) *Vol. 1, Toppu maneejimento no hōshin to soshiki* (The policy and organization of the top management) and Vol. 2, *Toppu maneejimento no taisei to unei* (Control and operation of top management), Japan Electrical Telecommunications Industries Association, Management Research Group (tr.), Diamond-sha, 1952.
These two books are a record of the lectures (including Q & A sessions) of the management course conducted for executives of manufacturers of telecommunications equipment to assist in reconstructing the Japanese telecommunications equipment industry. This record of lectures reveals a sense of mission within GHQ to introduce and disseminate a scientific management system in Japan.

[7] *Nihon Denkitsūshin Kōgyō Rengō kaihō* (The Japan Federation of Telecommunications Industries Association of Japan Bulletin), inaugurated 15 August 1948, published bimonthly. News about the CCS Industrial Management Course was reported from issue to issue.

[8] The history books of Japanese telecommunications manufacturers – such as *Fujitsu-shashi* (History of Fujitsu), 1964, *Nippon Denki Kabushiki-kaisha nanajūnen-shashi* (The seventy year history of Nippon Electric), 1972 and *Oki Denki hyakunen no ayumi* (The one-hundred year history of Oki Electric), 1981 – include descriptions of the activities of the CCS.

On the Restoration of Postwar Telecommunications

[9] Nippon Denshin Denwa Kōsha Nijūgonen-shi Henshū Iinkai (Editorial Committee on the Twenty Five Year History of the Nippon Telegraph and Telephone Public Corporation) (ed.), *Nippon Denshin Denwa Kōsha nijugonen-shi, Jōkan* (1977), *Chūkan* (1978), *Gekan* (1977) *and Bekkan* (1978) (The Nippon Telegraph and Telephone Public Corporation: A twenty-five year history), vol. 1, (1977), vol. 2 (1978), vol. 3 (1977), Special vol. (1978)).
In Chapter 2 of volume 1, 'Denden Kōsha no hossoku (the establishment of the Nippon Telegraph and Telephone Public Corporation)', changes in the postwar communications system in Japan are explained in detail in

order to show the processes and circumstances leading to the establishment of the Nippon Telegraph and Telephone Public Corporation.

[10] Sengo no Tsūshin Kōgyō Henshū Iinkai (Editorial Committee on the Postwar Telecommunications Industry) (ed.), *Sengo no tsūshin kōgyō* (Postwar telecommunications industry), Nihon Denki Tsūshin Kōgyō Rengōkai (Federation of Telecommunications Industries Associations of Japan), 1959.

[11] Tsūshin Kōgyō Nijūnen no Ayumi Henshū Iinkai (Twenty Years of Progress in the Telecommunications Industry's Editorial Committee) (ed.), *Tsūshin kōgyō nijū-nen no ayumi* (Twenty-years of progress in the telecommunications industries), Tsūshin Kikai Kōgyōkai (Communications Equipment Manufacturing Association), 1968.

19 GHQ and the Patent System of Japan

Tetsuo Tomita

There are primarily four issues that arise when we consider the system of ownership of industrial property in Japan during and immediately after World War II (WWII). First, the restoration of the property rights of foreigners that had been forfeited and made invalid during the war. Secondly, the problem of illegal products manufactured – i.e., products that violated patents, copyrights and trademarks, etc. – by Japan before the war. Thirdly, the secret patents granted for technology during the war. Finally, the dismantling of the *Zaibatsu* (cross-industrial business conglomerates).

The first of these issues was a continuation of problems that arose in the aftermath of World War I (WWI) because the method of restoration established in Japan at the time created confusion, but the other three issues were new. The first and third issues – especially the core of the third issue – until now have not been very clearly addressed, so we will begin with a survey of ownership rights in Japanese industry during the two world wars.

The System of Industrial Property Rights Before World War I

Historical Summary

The origins of the Japanese system of industrial property rights traces to the patent regulations of 1885. These rights were extended to foreigners under the modern law system through negotiations to amend the unequal treaties after the Sino-Japanese War. In 1896 legislation to protect the rights to various types of assets was established in civil law. In 1899 Japan signed the Paris Convention – an alliance treaty to protect the rights to industrial property – which determined equality between national citizens and foreigners (guarantees to the citizens of other countries the same rights in patent and trade matters that the law provides to its own citizens). Also in 1899, as a prerequisite to joining the Paris Convention, the relationship between basic laws – such as civil and criminal law – was clarified and laws concerning patents, designs and trademarks were revised. Also, legislation was enacted to ensure that common law would protect trade names.

Building on the foundations of this legislation, contracts were entered into to transfer technology with respect to property and patent rights. Examples include the initial Japan-US joint venture to establish Nihon

Electricity Co. Ltd. and contracts to import new technology such as the Parsons (British) and Curtis (U.S.) turbine engines and the Bessemer converter. From this perspective, the system of industrial property rights in Japan before WWI carried out the role of technology transfer very well.

Commercially, though, the imitation of products, brand names and foreign business names that had not been trademarked or patented multiplied in Japan after 1900. The British Government was the first to react to these practices and in 1908 demanded that legislation be enacted by the Japanese Government to prohibit such behavior. However, Tokyo was reluctant to take steps to rectify the situation on the pretext that torts were very narrowly defined in the Japanese civil law. Consequently, Britain's discontent grew. Thus the end of the Anglo-Japanese Alliance after WWI was not simply due to the League of Nations as is commonly assumed, but rather the imitation of products by Japanese industry and the export of these products to foreign colonies resulted in turning British national sentiment against Japan.

Administration of Patents During World War I

With the outbreak of WWI, nations around the world were confronted with the question of how to treat the terms and provisions of the Paris Convention; more specifically, how to treat the property rights of people and companies from enemy nations. There were two main issues arising in this regard: first, was the morality of employing the technology available – irregardless of property rights – as tools for conducting war, and; secondly, how to deal with the property rights of individuals from enemy countries.

Initially, most countries opted to protect the assets of individuals from enemy countries. But Germany decreed the rights of people from enemy countries forfeit, and in response, Britain negated the copyrights of Germans. Subsequently the general practice was to confiscate all enemy property. In 1917 the Japanese Supreme Court ruled to suspend the Paris Convention amongst the warring countries, and therefore, equality amongst patent rights belonging to German people were forfeited.

In Japan these standards of conduct were enshrined in the Wartime Laws on Industrial Property Rights (1917; Law no. 21) as outlined below.

1 The people of enemy countries cannot sue (or bring an action to the courts), nor apply for judgment on any kind of claim. Those inventions etc. that become widely used during the war cannot be granted patents after the war.

2 If there is demand from the military as a public utility, the patents and trademarks of people from enemy countries can be extinguished (but a utility model and design cannot be extinguished). [A 'utility model' is

a small patent, typical of the French, German and Japanese industrial property rights systems.]

3 The people who hold a license from a property owner from an enemy nation have sole use of the product or technologies so licensed as determined by an Imperial Ordinance.

4 The trademarks of people from enemy countries that have been extinguished cannot be registered by Japanese people.

5 Applications on patents and trademarks by people from an enemy country, even when said trademark or patent belongs to someone from another (non-enemy) country, cannot be registered while the rights deliver benefits to the people from an enemy nation.

6 The measures to be taken after the war will be determined by an Imperial Ordinance.

These same laws and regulations, in place since the Sino-Japanese incident, also applied during WWII. Following WWII, claims by citizens of the victorious countries were restored, the term for claims was extended to include the period when they had been forfeited during the war, and claims by defeated countries were forfeit. All of these decisions were agreed to by all countries in accordance with the Treaty of Peace with Japan.

In accordance with the third provision, Japan employed a lot of German technology. It has often been observed, though, that this served to strengthen the links between Japan and Germany in later years.

The Establishment of the Law to Prevent Unfair Competition

Another significant issue is how to avoid international friction arising from unfair industrial competition. The general problems of interpreting Japan's civil law regarding 'unlawful acts' (§709) is beyond the scope of our discussion, but it is important to note that the imitation of famous foreign trademarks and designs that were not registered in Japan was not prohibited by Japanese law. During WWI Germany illegally exported imitated goods through neutral countries. This problem of illegal exporting through third countries became an international issue after WWII.

During WWI Japan was isolated from European trade and, therefore, from the imported goods that it relied on. Thus the need for domestic industry rapidly developed. Soon manufactured products from Japan were being exported to various colonies of the European countries. However, at the close of WWI, when a network of international cartels began to envelope the world, products manufactured for export became enmeshed in a network of international patent cartels. Disputes over the practices of these cartels – their capacity to impose restrictions on international trade, etc. – arose in every geographical region and trading relationship.

In 1926 the Japanese Supreme Court re-interpreted the rules on unlawful behavior, and a law to prevent unfair competition (domestically) was enacted by the Japanese Government in 1934 based on this new interpretation. However, after the Sino-Japanese Incident of 1937 – i.e., the beginning of WWII – there was a directional change in Japan from free trade to a controlled economy and society. Subsequently this anti-monopoly legislation was not enforced.

International Patent Cartel

Between WWI and WWII international cartels dominated most industries and divided the world markets between themselves. In most industrialized areas, international cartels were formed. The US was the only nation with effective anti-trust legislation – elsewhere, international cartels were incredibly powerful and virtually unbreakable. However, as the division of world markets became more and more extensive, permission to use patents and trademarks became central to contracts.

A cartel consisted of companies that held the patents on particular technologies or products and associated companies who entered into contractual alliance. The contracts included a license – i.e., an exclusive right – to manufacture and/or sell the goods in a particular market – usually a country or geographic region. This arrangement frequently included the associated company acquiring the relevant patents in the country(ies) in which they had a license to operate. These alliances acted to protect the mutual interests of the member companies, with local members lobbying local governments to prevent the passage of any legislation which might impede their monopolies (such as the US anti-trust legislation) and taking action in the local courts to prosecute violations of their patents, trademarks, etc. by imitators. Even when they could not stop a particular company/country from manufacturing competing products, through their cartels they could often prevent these products from being imported by other countries – and thereby block the exports of the 'transgressor.'

Cartels were never limited to single products, but produced and controlled a broad range of products. And, of course, their domination of any market was never absolute or total, nor were all governments and courts totally sympathetic to their cases. So they also employed other means of obstructing their competition. For example, they lobbied international labor conferences to exclude Japan on the grounds that the Japanese were engaging in unfair trade through employing 'cheap' labor and poor working conditions, such as working night shifts. Such pressure was unrelenting.

In 1932, the State of Manchukuo (formerly Manchuria) was founded as a sovereign state, and a set of laws relating to industrial property rights was established. The Sino-Japanese Incident of 1937 instigated the Military Mobilization of Industry Law which was the beginning of an era of a planned domestic economy and the regulation of commodities. Since this economy did not entail free competition, the system of patents and trademarks instituted to support such a system became unnecessary.

The Administration of Patents During World War II

With the outbreak of WWII the Wartime Laws on Industrial Property Rights again came into effect. Since then, Japanese policy has been similar to what European countries practiced in WWI.

Forfeiture of Enemy Property

When WWII began Japan already had many enemies. The assets of enemy countries were frozen during WWI, and depending on the case, they may have been confiscated. The war time law for rights to industrial property – concerning applications for patents, utility models, designs and trademarks – suspended registration, and since it was determined that it could establish exclusive licenses and or cancel rights, all patents, utility models, etc. were accommodated.

Concerning exclusive licenses for rights to aptitude patents, there were 1,829 applications from 141 companies. We do not know enough about what became of these applications now.

These international rights were not restored to the Japanese people after WWII as a result of Japan's losing the war. All of the property rights by Japanese people in foreign countries met the same fate.

Compulsory Enforcement Provisions

Following the precedent set in WWI foreign peoples forfeited their industrial property rights in every country involved in WWII. However, the precise manner in which each country treated domestic property rights in regards to military mobilization differed somewhat.

In Japan the National Mobilization Law was created to maximize the productivity of industry. It addressed obstacles to productivity such as the monopoly that certain enterprises had on various technologies. For example, one *Zaibatsu* (a cross-industrial business conglomerate) might have exclusive rights to the use of alternating current, which seriously hindered the productivity of the manufacturing plants of other *Zaibatsu*. In other words, one steel manufacturer (for example) had the use of this technology while others did not. For the war effort to succeed it was necessary to break these monopolistic practices.

In order to do that, the order to enforce patent inventions, etc. (1943, Imperial Ordinance no. 159) was officially proclaimed. Via this Imperial Ordinance, any enterprise involved in the mobilization could apply to the Prime Minister for access to any technology, regardless of its patents, etc. The Prime Minister was empowered by the ordinance to grant and enforce access to the requested technology. The extent to which this ordinance was employed is difficult to ascertain, as all records and data pertaining to technology institutions were destroyed by fire at the end of the war. However, after the war there were many technicians who reported having spent their holidays working in competitors' factories training the technicians there how to implement and operate various technologies.

While such evidence is only hearsay and therefore not completely reliable, the fact that at the end of the war enterprises across all the *Zaibatsu* were employing the same technology indicates that wholesale technology exchange did, in fact, occur. In precisely the same way that this assisted the war effort, it facilitated rapid industrial and economic reconstruction in the postwar era.

War Time Investigation – Secret Patents, etc.

In the early stage of the war, the processing of applications by the Patent Bureau proceeded as usual. However on 31 October 1943 the war office created special provisions for wartime rights to industrial property (Law no. 109) which stopped the applications of designs, suspended appeals of applications, prohibited lawsuits and demands for judgment, and suspended the publication of patent information. Subsequently, only applications for trademarks, utility models and patents were processed.

Patents for military technology were granted, but under the secrecy provisions enshrined in the 1899 Law the contents of those patents were not open to the public. The first secret patents were granted to individuals in 1903, while the secret patents of the Army and Navy were registered little by little in the 1910s. After WWI the secret patents of the Army increased. The Navy was facing disarmament at the time and the number of secret patents was few, but after 1928 they suddenly increased.

The secret patents of enterprises began from military industrial enterprises, and gradually expanded to national policy companies (i.e., companies engaged in the wartime mobilization). The Mitsubishi Ship Building Company received its first secret patent in 1923, Mitsubishi Aircraft and Japan Optics in 1928, and Yokokawa Electricity in 1929. Forty-nine enterprises had received secret patents by 1941.

The first secret patents were limited to weapons and other armaments, but before long they included the chemical industry, internal combustion engines, electricity, communications, measuring machines and, eventually, all technology fields. By the end of the war 1,571 secret patents had been issued including 355 to the Army Minister, 583 to the Navy Minister and 87 utility model patents.

The exercise of the Mobilization Law and the 1943 patent invention enforcement order based on it has inspired an assumption that the acquisition of rights to patents had no meaning (the Japanese, today, are still sometimes accused of disregarding patent rights), but the people responsible for protecting the patents of enterprises were actually quite busy during this period. Under the provisions of the Mobilization Law, the enterprises engaged in ammunition production needed military permission in order to apply for general patents. However, a patent application requires a detailed description of the item to be patented, so, in most cases, when an enterprise applied to the military authorities for permission to apply for a patent the permission was denied on the grounds of military secrecy. Unfortunately, this does nothing for the morale of research engineers for whom a successful patent application is considered to be recognition of the quality and originality of their work. Hence, the military authorities opted for secret patents, or other similar means of registration. Similarly, among the military patents were applications for copying technology from foreign countries, for which secrecy was also required. Consequently, secretly patented technology was a mixture of good and bad.

Administration of Patents after WWII

When WWII ended in defeat for Japan, all of the foreign property management that had been conducted during the war was invalidated. Foreign claims that had been declared invalid were recovered, and rivalry arose amongst the victorious countries for preferential access to technology information.

GHQ's Patent Administration

During the Occupation GHQ had two primary interests in the Japanese patent system. One was the need to recover the rights of foreigners whose property had been confiscated and to secure ordinary property rights in Japan. The other was the need to uncover and acquire information about technology that had been developed during the war. Information concerning the technology that Japan had imported from Germany was of greater interest to GHQ than the technology that Japan had developed itself, but in both cases the acquisition of secret patent information was a top priority.

The Private Property Administration Bureau (CPC) was established by GHQ to oversee the management of patent issues, but before long it allocated responsibility for investigating secret patents to the Economic and Scientific Section, Scientific and Technical Division (ESS/ST). ESS/ST's highest priority was to investigate the whereabouts of radioactive materials and Japanese atomic technology (cf. ch. 2), but they gradually acquired a wider portfolio. The instructions from CPC to ESS/ST were issued on 25 March 1948 and on 27 March a summary of the information compiled to date was sent to Dr. Harry C. Kelly, the head of ESS/ST. This memo states that: 'Of 1,280 secret patents, 80 are secret utility models, 213 secret expired patents but still held as secret, 286 secret patent applications, 39 secret utility model applications. All these patents are written in Japanese...They are now in a safe in the Japanese Patent Office.'

By this step, the exclusive possession by America of the information related to the technology developed by Japan became an issue at the Far Eastern Commission. The Outgoing Message ESS/ST WFM/WJA/mm, from SCAP on 12 April 1948, signed by Marquat (US Army) outlined the terms and conditions for representatives of other countries of the Far Eastern Commission to have access to information about Japanese technology. The memo reported that there were 1,898 secret patents, and that there were radioactive 'stockpiles' stored in 78 different locations. This diversity of materials and locations created difficulties with site-security as well as intelligence gathering.

At the Far Eastern Commission, the US representatives were in a difficult position because the official terms of the Occupation gave all members of the FEC equal access to intelligence information. As discussed in Part 1 of this text, though, the Occupation was effectively by the US Forces alone, and they strictly controlled all intelligence gathering and resource materials. Thus Marquat's recommendation for alleviating the difficulties of the US representatives to the FEC was to remove the 'unscreened secret Japanese patents to the US prior to' allowing technical representatives from other FEC nations access to industrial and military sites. The memo suggests that the relevant materials could be removed in time for the FEC representatives to commence their investigations on 1 July 1948 and that they should be given (relatively) free access for nine months.

One of the problems facing GHQ was that the rights to industrial property, whether patents or trademarks, are primarily individual rights rather than national rights, and with the end of the war claims on these rights would appear and should be honored. This is true whether they were foreign claims or Japanese claims. Handling secret patents separately, including military technology and general concerns about the Japanese system would create confusion. Therefore, CPC and ESS decided to divide the work between them. Staff memo APO 500 of FEC and SCAP dated April 1948 (industrial property rights and technology information) outlined the different responsibilities for these operations as follows:

a The drafting and implementation of policy for the restoration of the rights of United Nations nationals in industrial property, which rights were cancelled or which lapsed because of non-payment of fees or were terminated by expiration of their terms during the war, in accordance with directives from high authority, and all transactions pertaining thereto. (CPC has responsibility)

b The drafting and implementation of policy relating to filing of applications by United Nations nationals for industrial property rights in Japan where such filing was prevented by the cessation of communication during the war and the period of occupation. (CPC has responsibility)

c The protection of the rights and interests of United Nations nationals in their industrial property in Japan or in such property likely to be restored or revalidated. (CPC has responsibility)

d The custody and control of industrial property rights of governments and nationals of the United Nations; enemy nations other than Japan; nations whose status has changed as a result of the war and the industrial property rights owned by the Japanese Government. (CPC has responsibility)

(In this connection the CPC has no responsibility for the control or development of scientific and technical information or research in Japanese Government owned or controlled laboratories or research institutes; such responsibility lies with ESS…)

e Licensing the use in Japan of all rights in industrial property under the custody and control of Supreme Commander for the Allied Powers and all transactions pertaining thereto. (CPC has responsibility)

f Amendment of the Japanese Industrial Property Laws and procedures relating to any of the items above. (CPC carries this out under the acknowledgement of ESS)

g Amendment of the Japanese industrial property Laws and procedures with the exception of laws and procedures relating to items a to e above. (ESS has responsibility)

h The obtaining of the proper distribution of technical information derived from Japanese sources. (ESS has responsibility; with G-2 concurrence in all cases and, additionally, with CPC concurrence where the information forms the subject of an industrial property right comprehended by d above.)

i Technical investigation, with a view toward development and industrial application, of any inventions, utility models or designs, whether subject to industrial property rights or not, knowledge of which is received by the Supreme Commander for the Allied Powers. (ESS has responsibility)

j Drafting and implementation of policy for the publication by the Japanese Bureau of Patents and Standards of past and currently registered patents, utility models and designs. (ESS has responsibility)

Thus the responsibilities of CPC and ESS were divided such that CPC handled the patent rights and other claims of foreigners, while ESS oversaw domestic industrial property rights, as well as matters related to Japanese industry, commerce and technology.

When the first industrial property rights issues arose during the Occupation it became apparent that these GHQ sections did not have much knowledge about the patent system. Thus there was uncertainty about how to distinguish secret patents concerned with technology information from the patent applications that appeared every day. To address this shortcoming and explain their policies to others an article detailing procedures for managing the patents system was developed.

GHQ's administration of Secret Patents and Other Military Technology

ESS/ST's views on secret patents are expressed in a memo dated 6 May 1948 in response to a telegram instruction (W-97791) concerning the destruction of military equipment and associated items. The definition of 'military equipment' explicitly included patents, but as ESS/ST observed, the only way to 'destroy' a patent is to de-register it. Further, they note that it is possible to de-register patents without actually examining the invention that is covered by the patent and that any examinations could be conducted subsequently if deemed necessary. The telegram instruction also states that 'all measures necessary should be taken to prevent any revival of the Japanese Army, Navy, Gendarmerie, Secret Police and their administrative organs.' ESS/ST's memo responds that the only measure necessary in this regard concerning patents is to revoke those Japanese ordinances that enable the issuance of secret patents. It also observes that this measure is not absolutely necessary provided that the equipment covered is itself destroyed and no further manufacture or R&D of weaponry and other instruments of war is allowed, but considers that it would be inconsistent if all existing secret patents were deregistered and made public while provision still existed for creating new secret patents.

Further on the issue of secret patents, the following draft instruction was drawn up as a memo to the Japanese government from GHQ/SCAP.

1 The Japanese Government will cause amendments to be made to the domestic legislation relating to patents and utility models, revoking all provisions whereby inventions, patented or unpatented, may be dealt with by the government for military reasons or for military purposes.
2 The Japanese Government will cause all current patents and utility models relating to weapons and means of war to be cancelled forthwith.
3 The Japanese Government will cause such legislation to be enacted that will prevent the future granting of any patents or utility models in respect of weapons or means of war.

4 The Japanese Government will before 30 June 1948, advise General Headquarters, Supreme Commander for the Allied Powers of the steps it has taken to implement the provisions of this memorandum.

This proposal was recommended by the memorandum to make the rights of secret patents owned by the Japanese Government on non-military applications open to the public (SCAPIN 1923, 29 July 1948).

Meanwhile Marquat's suggestion regarding how to mollify the FEC was accepted. A memo dated 8 May 1948 reported that the 1800 secret patents uncovered had been bound in volumes of 100 patents each. Of these eighteen volumes of originals, fifteen volumes had been multiply printed (approximately 20 copies each, each volume reportedly 'about the size of a *National Geographic* magazine'). All original volumes and most of the copies would be air freighted by CPC to the Dept. of the Army in the US for screening, while CPC would confiscate all copies of the secret patents from the Japanese Patent Office and secure them in a vault under CPC supervision.

The US subsequently withheld information about technology covered by the secret patents from the other Allied Powers as much as possible, and thus monopolized these technologies.

On 28 May CPC reached the conclusion that under the present circumstances, while it was valid to impede the Japanese from re-militarizing and to prohibit them from profiting from their military technology, there was little risk involved in making their military oriented secret patents public and, therefore, in the interests of administering property rights it advised that the records of the Patent Bureau should be preserved and made public.

Thus it became the responsibility of the Japan Patent Bureau to have the full texts of secret patent applications published.

General Property Issues

A general matter for the Occupation Forces (conditions of peace, etc.) was the recovery of industrial property assets that had been forfeited during the war. This recovery was carried out in the same way as the measures taken after the end of WWI as previously discussed. In respect of this procedure, it was presented as an agenda item to the Far Eastern Commission, and those instructions were carried out as stated.

In respect of the methods for recovering 'enemy country' property assets, patents and patent applications, etc., that were cancelled by the wartime law, industrial property rights were restored and the term of patents was extended beyond the wartime period.

A very significant issue relating to the handling of property rights was GHQ's determination to dismantle the *Zaibatsu* in the interests of deconcentration of industrial/economic activity. To this end it recommended that the use of trademarks, trade names, company names and marks of designated stock holding companies be prohibited. The designated companies were: Mitsui, Mitsubishi, Iwasaki, Sumitomo, Yasuda, Nomura, Ōkura, Asano, Nissan, Nakajima and Koga. The trademarks of these *Zaibatsu* group enterprises were cancelled in accordance with GHQ's orders.

Unfair Competition

The previously mentioned concerns about Japan's 'unfair' competition became an issue again during the Occupation. Great Britain raised this as a matter of concern during the negotiations of the San Francisco Peace Treaty. The preamble of the Treaty stated: 'in public and private trade and commerce to conform to internationally accepted fair practices.' Great Britain's opposition to the Treaty was not against this statement in the preamble, but they argued that unless the condition was incorporated into the provisions of the Treaty, Japan would perhaps again engage in unfair trade.

To address this point, GHQ demanded that the Japanese Government establish a proper protection policy, but because it was a matter of civil law (§709) the Government resisted introducing this bill. GHQ therefore issued SCAPIN 2042/1 (9 September 1949; concerning the designation of trademarks, trade names and businesses in Japan), which ordered the enactment of an injunction provision in the unfair competition law to prevent such an action whether intended or sheer error.

Subsequent Development

An overall revision to the system of industrial property rights in Japan was completed in 1959, entailing changes to the judicial system and appeals to higher courts (i.e., appeals no longer went straight to the Supreme Court). Since then, there have been some significant amendments and revisions, but basically the system has remained unchanged.

Amongst the revisions several are noteworthy. First were efforts to acquire leading-edge technology to catch-up in the development of a nuclear power industry. Second were revisions in the 1970s that allowed for patenting new materials (previously only the methods of manufacture could be patented). In the 1980s changes were made to enable territorial protection of seeds and seedlings and the copyright of computer software programs was originated.

At an international level, in recent years international technology transfer provisions have been made uniform, although the Japanese system has not undergone substantial changes in this respect. One can see the greatest alterations in the area of industrial rights that used to be unregulated but now are subjected to the anti-monopoly law. Furthermore, Japan has also increasingly protected individual rights that do not require government licensing; rights to prohibit unfair competition, and rights to defend the industrial secrets of companies. With the development of software technology, the areas to be protected have widened.

We must also take into consideration that the legal systems of the socialist states, including the former Soviet Union and China, have been changing qualitatively. The North/South division of the global economy is the central issue in international law regarding technology transfers, given that the technological discrepancies between the developed and the developing nations are so great. The need for innovative legislation that surpasses the national treatment of property rights (provisions not to discriminate against foreigners in accordance with the industrial property rights [Paris Convention for the Protection of Industrial Property]) has never been greater.

Bibliographic Guide

This is a list of some commentaries on the Japanese patent system, and does not include the many legal documents available on the topic.

Materials Collection

[1] Hatsumei Kōhō Kyōkai (Society for the Official Reporting of Inventions), 1942, *Eibei sonota tekikoku shoyū no tokkyo-ken shūroku* (Recorded collection of the patent rights of the property of the other enemy countries of America and Britain), Hatsumei Kyōkai (Society for Inventions), 1942.

[2] Patent Agency, *Kogyō shoyūken kankei rengōkoku saikō shireibu oboegai-shū* (Collection of memoranda of SCAP relating to industrial property rights), 1951.

Publications of the Japanese Patent Office

For the history of the Patent System see *Tokkyo seido shichijūnen-shi* (Seventy year history of the patent system), and *Shōkō seisaku-shi: Tokkyo* (History of commerce and industry policy: Patents), but more generally:

[3] Patent Agency (compilation), *Kōgyō shoyūken seido hyakunen-shi: jō, ge, bessatsu* (One-hundred year history of the system of industrial property rights), First volume, Last volume, Separate volume, Hatsumei Kyōkai (Society for Inventions), 1984–85.

[4] Tokkyo-chō Kōgyō Shoyūken Seidoshi Kenkyū-kai (Society for the Study of the History of the Industrial Property Rights System of the Patent Agency) (ed.), *Tokkyo seido no hassei to hensen* (Creation of and changes

to the patent system), *Ōkura-shō Insatsu-kyoku* (Printing Bureau of the Ministry of Finance), 1982.
[4] is the product of the preparatory work to draw up [3]. It briefly compares the systems of Great Britain, the USA, Germany and France.
[5] Riemann, Günther, *Hittoraa no tokkyo senryaku* (Patents for Hitler), Hideo Tanaka, Masato Matsujō, and Toshimichi Komine (tr.), Daiyamondosha (Diamond Company), 1983.
This publication focuses on the actual conditions of the patent cartels before WWII, especially in the enterprises of Germany and America.
[6] Tomita, Tetsuo, *Shijō kyōsō kara mita chiteki shoyūken* (Intellectual property rights from the perspective of market competition), Daiyamondosha, 1993.

The data referred to in this Chapter, camc from the following materials in the possession of the Strand Annex Building which contains American National Public Documents. Box STD-2 (5), 'Patents' (File # 16), March 1948–July 1948, (Japanese National Diet Library 580. 1, 621, 631).

20 The Public Health Policy of GHQ: The Quarantine Program and the Influence of DDT

Hazime Mizoguchi

The Establishment of the Public Health and Welfare Bureau

The total number of Japanese overseas, including soldiers, civilian employees in the service of the military and other civilians amounted to approximately 6.6 million on 15 August 1945, the day that the Japanese surrendered.[1] The Japanese Government commenced repatriation in accordance with Article 9 of the Potsdam Declaration, which stated: 'The Japanese military forces, after being completely disarmed, shall be permitted to return to their homes with the opportunity to lead peaceful and productive lives.' However, both the Japanese Government and the Allied Powers shared grave concerns about the deterioration of social conditions, the scarcity of medicine, medical equipment and facilities, and the repatriation of so many people from areas infested with infectious diseases.

On 22 September 1945, the General Headquarters, Supreme Commander for the Allied Powers (GHQ/SCAP, or GHQ) issued a nine-point memorandum about 'public health measures,' which opened with the statement: 'The Supreme Commander for the Allied Powers directs that the Imperial Japanese Government take the following action...' This directive, known as SCAPIN 48, established the foundations for the direction that public health policy in Japan would take in the postwar years.[2] The memorandum would later be recognized for having provided the 'engine of public health administration.'[3]

Item six of SCAPIN 48 directs the Japanese Government to: 'Inaugurate port quarantine control in cooperation with United States Naval Forces' and orders that port quarantine 'be established by Japanese civilian control.' In response, the Japanese government implemented various measures, directing that: 'The Ministry of Health should exercise unified control over marine-port quarantine of returning servicemen, civilians attached to the military and other civilians repatriated from overseas. The former Army and Navy quarantine stations and their personnel must be absorbed into the Ministry. Stations under the supervision of the Ministry of Transport that require expansion for quarantining repatriates should be expanded under the direction of the Ministry of Health, with whom the Ministry of Transport should cooperate.'[4] The Hygiene Bureau of the Ministry of Health established the Extraordinary Quarantine Department to address the situation.

The Public Health and Welfare Section (PHW), was created within the Special Staff Section of GHQ/SCAP, a non-military branch, to oversee public hygiene and social welfare. Colonel Crawford F. Sams (later promoted to Brigadier General) had been continuously involved in the creation of PHW and was assigned to lead it once it was established. As early as May 1945, he was appointed to the post of Chief, Health, Education and Welfare Division, Military Government Section, US Armed Forces, Pacific. In August, the Public Health and Welfare Section was established within the Military Government Section of the US Armed Forces Pacific (USAFPAC). GHQ/SCAP's PHW was established on 2 October 1945, with the view to providing the foundations for the implementation of the Occupation Forces' policy. PHW continued operating until 28 April 1952, the day when GHQ/SCAP ceased to exist.[5]

PHW's public health policies under the leadership of Sams were wide-ranging. They implemented vaccine programs, amended the Tuberculosis Prevention Law, improved nutrition levels, reorganized public health centers, refined various statistics of infectious diseases, deaths and births, reformed hospitals, structured a system of education as well as national examinations of doctors, nurses and midwives, rearranged the Japanese Red Cross, and instituted schemes for technical instructions and authorizing the manufacture of medical supplies. Among these multi-faceted reforms, Sams and his group made it their primary objective to improve the quality of public health activities and establish the system of health administration. Based on the enormous power of the Occupation Forces, they made efforts to ameliorate the inferior health conditions of Japan at the time, while placing a high priority on protecting the health of Occupation personnel. Through these efforts, the Occupation authorities attempted to decrease the level of social anxiety in Japan and thereby secure Japanese cooperation in the implementation of the Occupation policies.[6]

In 1946, a cholera epidemic swept the nation, with typhus fever, smallpox, Japanese encephalitis and other infectious diseases being prevalent in many places. As a counter measure, the Occupation health officials implemented a program to sprinkle DDT on the heads of the Japanese at repatriation ports, stations, streets, workplaces, schools and hotel accommodations. Many Japanese felt that this was the most humiliating insult that they had to suffer from the Occupation Forces.[7] The smell of DDT still reminds some Japanese of the Occupation years.[8]

The next section will examine some of the incidents related to DDT spraying, the measures taken to prevent venereal disease among Occupation soldiers, and the reaction of the Japanese Ministry of Health to the unfolding situations.

Repatriation Quarantine and DDT

A typhus epidemic swept the city of Osaka for four months from 27 December 1945 to 25 April 1946, during which some 4,000 suffered from the disease, with 387 fatalities. The *Asahi Shimbun* reported on 16 February 1946: 'All inhabitants of Osaka, Sakai and Fuse must be sprinkled with DDT in accordance with the special directive of the MacArthur headquarters. If they do not do so, their lives cannot be guaranteed.'[9]

The US armed forces provided the medical supplies required for DDT spraying. In March 11,735 pounds (approximately 5,300 kg) of absorbent cotton, 6,350 10ml hypodermic syringes, 25,000 20ml anti-typhus vaccination bottles, 467,500 pounds (approximately 212,000 kg) of DDT, and 5,000 DDT spraying devices were unloaded at Uraga Port.[10] In the same month, DDT spraying teams and vaccination teams from the Tokyo Metropolitan Government visited the Shinagawa, Kamata and Setagaya areas in an attempt to stamp out typhus and smallpox.[11]

In April 1946, cholera patients were identified for the first time on a repatriation ship. Later, cholera-contaminated ships arrived in repatriation ports one after another. On 7 March 1946, GHQ announced a program to sprinkle DDT on all Japanese people. SCAPIN 927 dated 7 May 1946 and entitled 'Repatriation' includes an appendix, entitled 'Medical and Sanitary Procedures,' concerning disinfestations 'by DDT of all repatriates arriving in Japan.' It instructs that: 'Disinfestation will include the clothing and baggage of such persons and other articles susceptible to infestation...Disinfestation with DDT will be accomplished monthly on all vessels operating on shuttles where less than a month is required for a round trip. Vessels requiring more than a month for a round trip will be disinfected with DDT prior to departure from Japan on each voyage.'

GHQ conducted thorough DDT sterilization, typhus vaccination, and excrement and urine treatment at ports of departure and arrival as well as on board ships. Further, GHQ also assembled and isolated cholera-contaminated ships at the ports of Uraga and Sasebo. At one time, more than seventy thousand repatriates were isolated on board a variety of ships. Officials imposed quarantine on all repatriation ships arriving in a port. Only when safety was assured were the ships authorized to enter port and the repatriates allowed to land. After landing, each repatriate was subjected to close examination. As a result of these efforts, the number of genuine typhus patients in Tokyo declined from 9,864 in 1946 to 217 in 1947. Nationally, the number radically decreased from 32,000 to 2,500.[12]

From the beginning of repatriation, quarantine facilities investigated the physical conditions, diseases, nutrition and body strength of each repatriate. Though only a limited number of studies on DDT were conducted in Japan at the time, it is noteworthy that some quarantine practitioners attempted to demonstrate that DDT was not omnipotent. For example, Fujio Fujiki who was engaged in quarantine work at the port of Ujina in Hiroshima delivered a lecture at a study group meeting held at the Kokuritsu Ryōyōsho, Kurihama Byōin (National Kurihama Recuperation Clinic) on 28 and 29 January 1947 under the auspices of the Quarantine Bureau of Hikiage Engoin (Repatriation Assistance Agency). His talk was entitled 'The test on the effects of DDT spraying on vaccination and its activation and DDT's disinfecting effects on various infectious pathogenic germs.' The results of Fujiki's study were published in *Igaku to seibutsugaku* (Medicine and biology; volume 10, number 3, 1947). In the introduction to the article, he suggested that DDT has pesticide effects on insects but not on germs and bacteria and warned against excessive confidence in, and complacent attitudes towards DDT's disinfecting power. In his experiments, he sprinkled DDT on rabbits and demonstrated that DDT had no effects on the activation of vaccination or pathogenic bacteria in intestines.

DDT Spraying and its Context

DDT is a kind of organic chlorine insecticide first synthesized by O. Zeidler, an Austrian, in 1873. P. Müller and his group at J. R. Geigy A.G. in Basel, Switzerland, discovered its pesticide effects in 1938. P. Läuger, also of Geigy, reported these findings in detail in January 1944.[13] Müller received a Nobel Prize in Medicine in 1948 for 'his discovery of the high efficiency of DDT as a contact poison against several arthropods.'

DDT is a white crystalline solid prepared by the reaction of monochloral to chlorobenzene. Mass production facilities for DDT were established in the United States during World War II, while trial manufacturing and experimentation for practical use of DDT began in Japan in 1944. On 11 September 1945, immediately after Japan's defeat, the Japanese Government requested permission from GHQ/SCAP to manufacture DDT in Japan. However, the request was rejected on the grounds that it was difficult to import raw materials (SCAPIN 106: dated 8 October 1945). Hence, DDT was imported to Japan under the direction of GHQ/SCAP. Meanwhile, Nihon Soda Company's Nihongi Factory commenced research in 1945 and first used homemade DDT in a vegetable field in Nariwa town in Okayama Prefecture.[14] On 4 May 1946

GHQ lifted the ban on Japanese production of DDT (SCAPIN 922, dated 4 May 1946), after which it was openly produced.

On the basis of SCAPIN 1261, the Ministry of Health issued a directive dated 23 December 1946 entitled 'On the Distribution of DDT and the Typhus Vaccine.' Following the directive, the Japanese government purchased all of the DDT produced in Japan and distributed it across the nation through Nihon Yakugyō Company. In the same year, the Agricultural Experiment Site of the Ministry of Agriculture processed DDT into ester emulsion and examined its effects on the larvae of *nikameichū* (rice stem borer). In 1947, Nihon Soda began full-scale production. Around the same time, Nichizui Bōeki entered into a patent contract with Geigy.[15] In May 1948, GHQ allowed companies to sell DDT freely. In September 1948, Kureha Chemical Industry started production of DDT concentration.

Immediately after the Allied Powers occupied Japan, large quantities of DDT were sprayed from airplanes, well before it was used for direct spraying on human bodies. For example, on 30 August 1945, the US Navy's Third Fleet notified the Japanese Navy's Deputy Chief of Staff that disinfectants would be sprayed from the sky, twenty-four hours prior to the Allies' landing in Japan, to eradicate mosquitoes and other harmful insects. On 6 September, the Occupation Forces notified the Tachikawa City Office of Tokyo that gas-form DDT would be sprayed from airplanes.[16] In the middle of September, the Americans brought in massive amounts of powdered DDT and commenced DDT disinfection, with large quantities for aerial spraying. *The Bulletin of the US Army Medical Department* ran at least two commentaries, one entitled 'Use of DDT as a mosquito larvicide' and the other 'Use of DDT as insecticide to kill adult mosquitoes' which were translated into Japanese. They suggested: 'To prepare a 5 percent (weight/volume) solution, DDT (larvicide, DDT, powder, dissolving) is dissolved at the rate of 7 ounces DDT to 1 gallon of oil' and 'The most practicable method for large-scale outdoor distribution of DDT is by airplane...[F]or larger pay loads of insecticide, B-25 and C-47 aircraft equipped with spray apparatus has been used extensively to obtain wide and efficient distribution of DDT-oil solutions for immediate kill.'

The *Asahi Shimbun* reported on 28 June 1946, under the headline 'American Military Planes Spraying DDT from the Sky' that they were not large planes like the B-25 or C-47, but small, like training planes. The article was accompanied by a photograph of planes spraying DDT which formed white lines in the sky.

Since the end of the 1940s, it has been reported that DDT does not fully dissolve upon entering the human body, but rather tends to linger in it. Furthermore, it was discovered that some of the insects being targeted by the spraying were developing resistance to DDT, and since it also accumulates in their bodies, it was becoming condensed through the food chain. The Overseas News and Literature section of *Kōshū eiseigaku zasshi* (Journal of Public Health; vol. 6, no. 3, 1949) ran an article entitled 'The Danger of the Use of DDT.' It referred to a study in which forty-six cases of DDT poisoning were detected among American children and reported that the use of DDT had been prohibited on cattle farms in the US from 1 April 1949. The October 1949 issue of *Kagaku Asahi* (Science Asahi) published an article titled 'Prohibiting DDT from Dairy Farming,' which reported that the US Department of Agriculture banned the use of DDT – used to get rid of flies – from dairy farms and processing firms. However, it was 1971 before DDT was banned as an agricultural chemical and insecticide in Japan, and 1981 before its manufacture, distribution and use were totally prohibited.[17]

Countermeasures against Venereal Disease

As stated earlier, GHQ/SCAP was primarily concerned about the health and hygiene of Occupation soldiers. Item nine of the previously mentioned directive SCAPIN 48 instructed the Japanese Government to: 'Place special emphasis on adequate measures for the control of all venereal diseases occurring in indigenous Japanese personnel. This will be accomplished by using existing Japanese agencies.' The Japanese Government's response to the memorandum declared that 'women who are engaged in this special work should be medically inspected with increased intensity. We will let Japanese doctors establish clinics which primarily deal with venereal disease.'

However, in a memorandum dated 16 October 1945, Colonel H. W. Allen, Aide-de-camp, expressed GHQ's dissatisfaction with these Japanese measures for the prevention of venereal diseases and ordered the Japanese government to designate syphilis, gonorrhea, and *nansei gekan* (chancroid) as infectious diseases, to report the name, age, sex and address of each patient, to implement rigorously the existing regulations for the prevention of venereal diseases, and to inspect and treat everyone who was potentially at risk of catching and spreading these diseases.

These were landmark instructions; doctors were now required to report to authorities the cases of venereal disease amongst ordinary citizens, not only the prostitutes who were considered most likely to spread sexually transmitted diseases.

The background of these countermeasures is informative. Upon hearing the news of 'enemy forces' landing on and occupying Japan, the Japanese authorities became extremely anxious. The prevailing sentiment is reflected in an official notice of 18 August 1945, which was issued by the head of the Public Security Bureau of the Ministry of the Interior and distributed to its regional heads. Entitled 'Rules of Conduct during the

Forthcoming Occupation,' the notice suggests that 'women should not walk alone. They should refrain from wearing slovenly clothes.' On the same day, the Bureau head sent another notice to the police forces around the nation, spelling out some specific plans for building 'breakwaters of sex.'[18] In Tokyo, the Superintendent-General of the Metropolitan Police took the initiative of establishing *Tokushu Ian Shisetsu Kyōkai* (Recreation and Amusement Association or RAA) with the first one being set up in the Ōmori area. Newspapers ran advertisements recruiting 'comfort women.'

Despite these measures, the infection rate of sexually transmitted diseases among Occupation soldiers did not decline, unlike other diseases. In January 1946, the prostitution system was officially abolished, with RAA being closed on 27 March. In reality, however, the Japanese government designated 'red-light districts' and acquiesced to prostitution in these areas. Some of the 55,000 women who had worked at RAA moved to these quarters. On 13 December, the *Sun Shashin Shimbun*, a photo newspaper, carried a picture of the Sankyō Shinagawa firm where neo-salvarsan, a supposed wonder drug for syphilis, was mass-produced.

Reform of the Administrative Organizations

On 11 May 1946, the Occupation authorities issued a memorandum entitled 'Reorganization of Government Public Health and Welfare Activities' which states: 'In order to cope adequately with the emergency health and welfare situation,...the Imperial Japanese Government will immediately reorganize the administration of health and welfare activities to provide for the following administration of functions.' The Ministry of Health was instructed to establish (1) 'a Bureau of Health, the responsibility of which will be public health (maternity, child, and adult hygiene), health education, vital statistics activities and nutritional activities,' (2) 'a Bureau of Medical Treatment, the responsibility of which will be general affairs (medical relief programs), administration of hospitals, administration of sanatoria, medical affairs, pharmaceutical affairs (distribution), drug production (including biologicals) and pharmaceutical standardization,' (3) 'a Bureau of Preventive Medicine, ...the responsibility of which will be those concerned with sanitary engineering, communicable diseases and chronic infectious diseases,' and (4) 'a Bureau of Social Affairs, the responsibility of which will be those in connection with public assistance, public welfare and procurement and disposition of materials necessary to implement such functions.' The memorandum also states: 'The Imperial Japanese Government will cause to be established in prefectural governments a Bureau of Health and a Bureau of Welfare whose functions will include those outlined for the Ministry of Health.'

These organizational reforms were attributable to the Occupation authorities having concerns about Japan's public health administration. They were 'surprised that science and technology did not play an important role in administrative work and felt that it should form the core of public health administration.'[19]

The Bureaus of Public Health, Hygiene and Extraordinary Infectious Disease Prevention, which had been created in October 1945, were merged into the newly established Bureau of Hygiene. The Bureau of Extraordinary Infectious Disease Prevention, composed of the Division of Infectious Disease Prevention and the Division of Quarantine was abolished, with the functions of the former handed over to the Bureau of Hygiene and those of the latter taken over by the newly established Repatriation Assistance Agency. In accordance with the SCAP memorandum, the Ministry of Health and Welfare amended the ministerial rules and regulations on 5 November 1946 to establish the Bureaus of Health, Medical Treatment, and Preventive Medicine. Initially the Bureau of Health was made up of the Division of Investigation (whose functions included planning for the improvement of public health, surveys of hygiene and sanitation, matters related to national parks, hot and cold springs and public health agencies), the Division of Public Health (health clinics, health nurses, the diffusion and improvement of knowledge of health among the people, eugenics, the sanitation of public baths and theaters), and the Division of Nutrition (nutrition, food hygiene, abattoirs, and the National Institute of Nutrition). Later in August 1947, the Division of Health Statistics was established as the fourth division.

In the Bureau of Medical Treatment, there were five divisions: the Division of Medical Treatment (matters related to medical practitioners and the enforcement of the National Medical Law), the Division of Medicine (the Law on Medicine, government-regulated companies, the distribution, export and pricing of medicine, policing of poisonous and powerful medicine, and the production of health materials), the Division of Pharmaceutical Manufacture (licensing the production and manufacture of pharmaceutical goods, the Japanese Pharmacopoeia, the cultivation of medicinal plants, and non-medical drugs), and the Division of Hospitals (matters related to national hospitals). In February 1947, the Division of Narcotics was set up. Further, in August 1947, the Division of Services (goods, building and repairs of national hospitals and sanitaria) was established.

The Bureau of Preventive Medicine was composed of three divisions: the Division of Preventive Medicine (tuberculosis, leprosy, trachoma, parasite diseases, myopia, cancer and mental illnesses), the Division of Infectious Disease Prevention (acute infectious diseases and venereal diseases), and the Division of Quarantine (quarantine activities at marine and air ports). In June 1947, the Division of Licensing (matters related to bacterio-therapy and the Institute of Preventive Hygiene) was established.

In December 1947, the Division of Health Facilities (tap water and sewage works, garbage hygiene, building sanitation, the extermination of rats and insects) was created.

Ryūtarō Azuma (graduate of the School of Medicine, Tokyo Imperial University in 1917) was appointed to the post of Director-General of the Bureau of Medical Treatment), Kikuo Hamano (graduate of the School of Medicine of Keiō University in 1920) to Director-General of the Bureau of Preventative Medicine) and Yukiharu Miki (graduate of Okayama Medical College in 1929) to Director-General of the Bureau of Health. Thus the directorates of the three bureaus of the Ministry of Health were all occupied by technical officials, an arrangement which Minoru Katsumata (graduate of the School of Medicine), Kōsei Gikan (Technical Superintendent) of the Health Ministry (graduate of the School of Medicine of Tokyo Imperial University in 1919) and Sams all considered to be desirable.[20]

In assessing the postwar public health policies, Nobuo Tatebayashi, then head of the Department of Infectious Disease Prevention, observed in a round-table discussion: 'My frank impression was that the highest priority was the hygiene of Occupation soldiers rather than the Japanese people, particularly in the initial phase of the Occupation...The Occupation authorities set to work in earnest because the health of their soldiers was at stake.'[21]

Elizabeth Fee, then Associate Professor of Health Policy, School of Public Health and Hygiene, Johns Hopkins University, attributed the prolongation of the average life expectancy of the Japanese in the first five years after the war to the importance attached to public health. She also maintained that Japan's disarmament made it possible for large sums of funds to be allocated for the budgets for public hygiene.[22]

Thus, the Occupation mounted its public health policies in order to carry out the overall Occupation programs by removing the social anxiety of the Japanese public about the spread of infectious diseases and preventing Occupation soldiers from suffering from venereal diseases.

The number of people who died of typhus, cholera, Japanese encephalitis and tuberculosis decreased dramatically in Japan as a result of insecticide effects of DDT as well as the widespread use of penicillin and vaccination. GHQ/SCAP-PHW was most successful in implementing a grandiose scheme of focusing on public health and hygiene as the basis of carrying out the Occupation policies. In comparing bombing and tuberculosis on the same dimension, Sams argued that 'the number of Japanese who died from tuberculosis among the civilian population that year [1945] was greater than the number who died from all of the bombing, the fire raid, and the two atom bombs,' a statement reflecting the perceptions of the Occupation policy makers at the time. Yet, there is little doubt that PHW activities contributed greatly to the fact that both the rate of the annual growth of the average height and the overall life expectancy of the Japanese exceeded those of Americans in postwar years.

Notes

1 Kōseishō Gojūnen-shi Henshū Iinkai (Editorial Committee for the Fifty-year History of the Ministry of Health and Welfare), 1988 [3] pp. 698–704.
2 Shimizu, 1983 [5] pp. 605–35.
3 Nagatoya, 1961 [6] pp. 53–7.
4 'Rengōkoku saikō shireibu yori hoken eisei...(Summary of the measures taken...),' 1946 [2] pp. 17–22.
5 Takemae, 1983 [10]
6 Sams, 1986 [8] pp. 133–205.
7 Kōdansha, 1989 [18] pp. 226–9.
8 Ishikawa, 1989 [22] p. 29.
9 Nihon Kagakushi Gakkai (History of Science Society of Japan), 1967 [4] pp. 317–41.
10 Hikiage Engo-in Ken'eki-kyoku (Repatriation Assistance Agency, Quarantine Bureau), 1949 [1], p. 34.
11 Mainichi Shimbun, 1980 [19] pp. 228–9.
12 Kōdansha, 1989 [18] p. 228.
13 Ishii, 1946 [24] pp. 333–42.
14 Matsubara, 1984 [25]
15 Uemura, et al., 1988 [26] pp. 156–9.
16 Sams, 1986 [8] pp. 18–19.
17 Uemura, et. al., 1988 [26] pp. 156–9.
18 Kōdansha, 1989 [18] p. 22.
19 Katsumata Minoru Sensei Tsuitōroku Kankōkai, 1970 [28] p. 242.
20 Katsumata Minoru Sensei Tsuitōroku Kankōkai, 1970 [28] pp. 203–41.
21 Katsumata Minoru Sensei Tsuitōroku Kankōkai, 1970 [28] pp. 203–41
22 Fee, 1989 [27] pp. 1–7.

Bibliographic Guide

As data for repatriation and quarantine

[1] Hikiage Engo-in Ken'eki-kyoku (Repatriation Assistance Agency, Quarantine Bureau), *Hikiage ken'eki-shi dai ichibu* (History of repatriation and quarantine, Part 1), 1949.

On overall public hygiene policies during the Occupation

[2] 'Rengōkoku saikō shireibu yori hoken eisei kankei no shirei oboegaki ni

taishi toritaru sochi no gaiyō (Summary of measures taken according to the GHQ memoranda concerning health and hygiene),' in *Nihon kōshū hoken kyōkai zasshi* (Journal of the Japan Public Health Association), vol. 21, 1946, pp. 17–22.

[3] Kōsei Mondai Kenkyū-kai (Welfare Research Foundation), *Kōseishō gojūnen-shi* (Fifty year history of the Ministry of Health and Welfare), 2 volumes, 1988.

[4] Nihon Kagakushi Gakkai (History of Science Society of Japan), *Nihon kagaku gijutsu-shi taikei 25, Igaku 2* (A compendium of the Japanese history of science and technology in Japan 25, Medicine 2), Daiichi Hōki Shuppan, 1967.

[5] Katsuyoshi Shimizu, *Zoku kōshū eisei no hattatsu* (Development of public hygiene: A sequel), Nihon Kōshū Eisei Kyōkai (Japan Public Health Association), 1983.

[6] Yōji Nagatoya, 'Sengo Amerika no Nihon igaku ni oyoboshita eikyō (jō) – Kōshū eisei to igaku kyōiku o chūshin to shita shiron (American influence on Japanese medicine after the war, the first volume: An essay focused on public hygiene and medical education),' in *Nihon iji shinpō* (Japanese medical affairs bulletin), no. 1963, 1961, pp.53–7.

[7] Yōji Nagatoya, 'Senryōka eisei gyōsei-shi memo (Memorandum on the history of the administration of public hygiene during the Occupation),' in *Igakushi kenkyū* (Studies of medical history), no. 20, 1966, pp. 1047–9.

[8] C. F. Sams, *DDT kakumei: Senryōki no iryō fukushi seisaku o kaisō suru* (The DDT revolution: Reflections on medical welfare policy during the Occupation), Eiji Takemae (tr. and ed.), Iwanami Shoten, 1986.

[9] Akiko Sugiyama, *Senryōka no kōshū eisei – Shoki seisaku o megutte* (Public hygiene during the Occupation: Early policies), presented at Senryōshi Kenkyūkai (Study Group on a History of Occupation), 18 April 1992.

On organization and personnel affairs of GHQ/SCAP-PHW

[10] Eiji Takemae, *GHQ,* Iwanami Shoten, 1983.

[11] Shisō no Kagaku Kenkyūkai (Study Group on the Science of Thought) (ed.), *Kyōdō kenkyū, Nihon senryōgun sono hikari to kage* (Joint research, the Occupation Forces in Japan, their bright and dark side), Tokuma Shoten, 1987.

[12] Eiji Takemae and Yukuo Sasamoto, 'Senryōka no iryōto fukushi (2), GHQ/PHW no soshiki to jinji (Medical treatment and welfare (2), Organization and personnel of GHQ/PHW),' in *Tokyo Keizai Daigaku kaishi* (Journal of the Tokyo University of Economics), no. 156, 1988, pp. 267–85.
This contains PHW's organizational chart, range of duties of each division, names of people in charge in each department (information was taken from the GHQ phone directory) from October 1945 to January 1951.

GHQ/SCAP-PHW related activities during the same period were arranged in chronological tables by Eiji Takemae (Tokyo Keizai University) and Yukuo Sasamoto (a member of the Study Society on Occupation History). One table is on mainland Japan, another one is on Okinawa and the third one is arranged according to social background.

[13] Eiji Takemae and Yukuo Sasamoto, 'Senryōka no iryō fukushi nenpyō (1) zenshi (Chronological table of medical welfare during the Occupation (1): The antecedent phase),' in *Tokyo Keizai Daigaku Kaishi* (Journal of the Tokyo University of Economics), no. 147, 1986, pp. 259–70.

[14] Eiji Takemae and Yukuo Sasamoto, 'Senryōka no iryō fukushi nenpyō (2) (1945. 9–1946. 12) (Chronological table of medical welfare during the Occupation (2) (1945. 9–1946. 12),' in *Tokyo Keizai Daigaku kaishi* (Journal of the Tokyo University of Economics), no. 157, 1988, pp. 91–123.

[15] Eiji Takemae and Yukuo Sasamoto, 'Senryōka no iryō fukushi nenpyō (3) (1947. 1–1948. 12) (Chronological table of medical welfare during the Occupation (3) (1947. 19–1948. 12),' in *Tokyo Keizai Daigaku kaishi* (Journal of the Tokyo University of Economics), no. 159, 1989, pp. 135–74.

[16] Eiji Takemae and Yukuo Sasamoto, 'Senryōka no iryō fukushi nenpyō (4) (1949. 1–1952. 4) (Chronological table of medical welfare during the Occupation (4) (1949. 9–1952. 4),' in *Tokyo Keizai Daigaku kaishi* (Journal of the Tokyo University of Economics), no. 157, 1989, pp. 131–62.

On the explanations of the shipping lists which were drawn out when GHQ/SCAP-PHW documents were shipped to the United States

[17] Masayuki Miura, 'Sengo Nihon no kōshū eisei kaikaku to PHW – shokatsudo o gaikan suru shippingu risuto (Reform of public hygiene in postwar Japan and PHW: Shipping lists provide a general view of various activities),' in *Ritsumeikan Daigaku Jinbunkagaku Kenkyūjo-kiyō* (Journal of the Humanities Institute at Ritsumeikan University), separate volume no. 5, 1985, pp. 123–48.

The following explain incidents concerning the spraying of DDT during the Occupation

[18] Kōdansha, *Shōwa niman-nichi no zenkiroku 7, haikyo kara no shuppatsu, Shōwa 20nen–21nen* (Complete Records of 20,000 days in Shōwa, vol. 7: Departure from ruin; 1945–1946), Kōdansha, 1989.

[19] Mainichi Shimbun, *Ichioku-nin no Shōwashi, Nihon senryō 1, kōfuku/shinchū/hikiage* (One hundred million people's history of Shōwa, Japan during the Occupation 1, surrender, occupation, repatriation), Mainichi Shimbunsha, 1980.

[20] Toshio Sumimoto, *Senryō hiroku* (Secret records of the Occupation), Mainichi Shimbunsha, 1952 (Chūkō Bunko, 1988). Refer to 'DDT to onna (DDT and women),' pp. 129–42 (Bunko/paperback version).

[21] Akihide Mori, 'DDT to penishirin (DDT and penicillin),' in *Nihon fūzoku jiten, Amerikan karuchaa 1 '45–50's* (Encyclopedia of Japanese manners and customs, American culture 1, 45–50's), Sanseidō, 1981, pp. 18–20.

[22] Hiroyoshi Ishikawa, *Yokubō no sengoshi* (Postwar history of desire), Kōsaidō, 1989.

[23] Hazime Mizoguchi, 'Seiyakugyō shinten ni kōken shita GHQ no "DDT kakumei" (GHQ's DDT revolution contributed to the development of the pharmaceutical industry),' in *Ekonomisuto* (Economist), 22 May 1990, pp. 92–3.

DDT's special qualities and its rise and fall

[24] Shintarō Ishii, 'DDT no sayō kijo (Function and operation of DDT)', in *Igaku no ayumi* (Path of development of medicine), vol. 1, 1946, pp. 333–42.

[25] Hiromichi Matsubara, *Nihon nōyakugakushi nenpyō* (Chronological table of Japanese agricultural chemicals), Gakkai Shuppan Sentaa, 1984.

[26] Shinsaku Uemura, Hiroshi Kawamura, Machiko Tsuji, Shigeyuki Tomita and Shizuo Maeda, *Nōyaku dokusei no jiten* (Encyclopedia of toxicity of agricultural chemical), Sanseidō, 1988.

How public hygiene was understood

[27] Elizabeth Fee, 'Rekishiteki kanten kara mita atarashii kōshū eisei no dōkō (Historical view of the new trends in public hygiene),' in Satoru Sugita (tr.), *Kagaku igaku shiryō kenkyū* (Studies of data on scientific medicine), no. 179, 1989, pp. 1–7.

Memoirs from the administration

[28] Katsumata Minoru Sensei Tsuitōroku Kankōkai (Editorial Committee on Professor Minoru Katsumata's Work in His Memory), *Kindai kōshū eisei no chichi, Katsumata Minoru* (Father of modern public hygiene, Minoru Katsumata), 1970, p.242.

21 The Population Problem and the Birth Control Program During the Occupation

Hazime Mizoguchi

Jurisdiction over the population census moved from the Statistics Bureau of the Cabinet to the Ministry of Health and Welfare in October 1947. A provisional state of the nation survey was then conducted which indicated approximately 2.68 million births and 1.14 million deaths for the previous year, a natural increase of 1.54 million. In 1948, the natural increase in population was 1.43 million and in 1949 it was 1.3 million. Thus a steady increase in the population of Japan was recorded.[1] During this same period as many as 6.6 million Japanese, including soldiers and military personnel, were repatriated from overseas. Thus, in the five years following World War II the population of Japan grew in excess of fifteen percent – from 72 million to 83.2 million.

Concurrent with this population 'explosion' and its associated problems, there were many pregnancies amongst women who had given comfort to members of the Occupation Forces which raised the question of 'mixed-blood' babies. 'Backyard' abortions were rife, commonly conducted by injecting saline solutions into the walls of the abdomen to induce labor. This procedure frequently resulted in the perforation of the womb, bacterial infection and death. Other issues that came to the fore were birth control measures coupled with questions about the welfare of mothers and the 'emancipation' of women. Eugenic questions concerning protecting the 'quality' of the population also arose. This chapter traces the debates and measures taken to address these issues during the Occupation. First, though, I will present a brief outline of historical and theoretical arguments surrounding the population question.

The background to the 'Ogino Method of Contraception' and the catch-cry 'Give Birth and Populate for the Sake of the Nation'

Birth control has been practiced for a wide variety of reasons since the earliest days of human society. The English economist T. R. Malthus advocated its social implementation in his work *A Discourse on the Principles of Population* in 1798. Malthus considered the relationship between population growth and food supplies to be inherently problematic, arguing that population increases geometrically while subsistence production can only increase arithmetically. In other words, left unchecked population growth inevitably outstrips the capacity to sustain it. He concluded that excess population therefore produces adversity and vice. The second edition of his book recommended marrying later in life and chastity after marriage as measures to be taken to redress this issue.

Another Englishman, F. Place, disagreed with Malthus in his work *Demonstration of Population Principles* published in 1822. Place argued that in reality marriage later in life and chastity after marriage are not possible. He therefore advocated early marriage and birth control instead. This position came to be known as 'Neo-Malthusianism.' But enthusiasm in England for Neo-Malthusianism died down in the latter half of the 1870s as the birth rate declined.

In the mid 1920s, while the Japanese Government was becoming increasingly concerned about the deteriorating state of the economy, they were also confronted with poor harvests and the risk of food shortages in the Tōhoku and Hokkaidō regions. There were also an increasing number of young women turning to prostitution. Many of them suffered from sexually transmitted diseases and difficulty in becoming pregnant. Furthermore, abortions and infanticide – practices prevalent in impoverished areas – were negative variables in the population equation. Many voices were raised about the effects of overpopulation.

Neo-Malthusianism had been introduced to Japan by Sadao Oguri and Kan'ichirō Karai's *Shakai kairyō jitsuron* (Discourse on Social Reform Realities (1903)), but the theory did not take root until the arrival of M. Sanger in Japan in 1922. The series of lectures she delivered around the country was widely reported in the press. This proved to be an impetus for the serious advocacy of birth control. Soon thereafter the journal *Sanji chōsei hyoron* (Birth control review) was launched with biologist Senji Yamamoto as editor-in-chief. The 'Ogino Method of Contraception' (the rhythm method), first proposed by the gynecologist Kyūsaku Ogino in 1923, was introduced to the public through the women's magazine *Shufu no tomo* (The Housewife's companion) in 1927.

One response to this 'birth control movement' was the creation of the Minzoku Eisei Gakkai (National Hygiene Society) in 1930. Sen Nagai was appointed Managing Director and members of the medical profession such as Yoshio Koya assumed Directors' positions. The Society's mission was to provide education about birth control. To this end they launched the magazine *Minzoku eisei* (National health). Their concerns, though, were with improving the 'quality' rather than the 'quantity' of the population as can be seen in the following quotation from the organization's prospectus: 'At the present time the Sanger philosophy, based in Neo-Malthusianism, is sweeping the world. It is ceaselessly eating away at the biological power of civilized nations.'

The government's policies on population also reveal a shift in emphasis from concerns about 'quality' to 'quantity' over the ten-year period beginning in the mid-1920s. In 1930 the government attempted to establish a permanent commission of inquiry into the question of the population and food supplies. It was unsuccessful on this occasion, but in 1933 Jinkō Mondai Kenkyūsho (Population Question Research Foundation) was established. Amongst the options considered for addressing the problems of overpopulation were: reducing the standard of living, popular limits on the number of births, encouraging emigration and securing new territories.[2] It appears that the option of securing new territories was chosen. The military soon commenced their invasion of the Chinese mainland, heralded by the Manchuria Incident. The slogan 'Give Birth and Populate for the Sake of the Nation' became the catch-cry of the day – to ensure the human resources needed to maintain the war effort.

In January 1933 Adolf Hitler became the Chancellor of Germany. In July 1933 the 'Hereditary Disorder Descendant Prevention Law' was proclaimed. The Bill had been drafted by the Prussian Health Committee in 1932 to define policies for the sterilization of people with hereditary disorders. It became effective from 1 January 1934. The First Article states that 'Persons possessing congenital disorders shall be surgically sterilized. However this is to be restricted to cases in which it can be expected on medical grounds that there is a high likelihood that any children produced will be afflicted by serious physical or mental congenital disorders.'[3] The Law defines hereditary disorders as: congenital mental retardation, split personalities, manic-depressive psychosis, Huntington's Chorea, epilepsy, congenital blindness, deafness and alcoholism. Thus we can see that in Nazi Germany the policy emphasis was on the 'quality' of the population.

In contrast, the Ministry for Health and Welfare was founded in 1938 to encourage population growth – i.e., an emphasis on quantity. In November 1938 the Nation's Health Research Council was formed within the Ministry and commenced research into intellectual disabilities and mental illness. In 1940 the Ministry inaugurated an award for families with 10 or more children entitled 'Excellent Child Treasure Corps.'[4] Through these two measures the Ministry for Health and Welfare sought policies for simultaneously improving both the quality and the quantity of the population. These policies were underpinned by the 'National Eugenic Law' (proclaimed in 1940 and effective from July 1941) and the 'Outline of the Establishment of the Population Policy' (adopted by the Cabinet in January 1941).

With the backing of the Nihou Minzoku Eisei Kyōkai (Japanese National Health Society, est. 1935) the 'National Eugenic Law' aimed to improve the population in both qualitative and quantitative terms, as we can see from its First Article: 'This law aims to inhibit any increase in the number of people possessing hereditary disorders while planning for an increase in the numbers of those possessing a full and healthy nature and by so doing bring about an improvement in the constitution of the people.'[5] In more concrete terms, it aimed at increasing the number of healthy and robust male children. Under this law and in line with the policy of improving the population stock, four hundred and thirty-five people (one hundred and ninety-two males, two hundred and forty-three females) were sterilized by surgical operation from 1941 to 1945.[6] While the similarities between the Nazi and the Japanese policies and practices are obvious, it is important to note that the Japanese Government was also highly interested in the eugenics movement in the US at the time. Whether these practices led to any 'improvement in the stock of the national population' is quite doubtful.

The essence of the 'Outline of the Establishment of the Population Policy' was that

> Building the Greater East Asian Co-Prosperity Sphere and planning for its permanent healthy development is the mission of the Empire. To this end we establish a population policy to rapidly increase the population of our country and further to provide continued increased development and rapid improvement in its nature...As population increase ensures perpetual development the aim is to have a policy with the key concept of increasing the birth rate and partnering this with a decrease in the death rate...[The policy] aims in the next ten years to reduce the marriageable age, possibly by three years, and achieve an average birthrate of five children per married couple.'[7]

These things could be thought of as bringing together into one the sentiment expressed by Minoru Tachi of the Ethnic Research Institute (Ministry of Health and Welfare) that 'the position we are in is such that if the population of the Empire were 200 million or even 300 million it would still not be enough so by a strong and appropriate population policy we have to produce this degree of vigor'[8] and the path being followed at the time. On this point, while it is in line with the German, Italian and French situations where 'a qualitative population policy greatly enhanced the quantitative policy' Tachi stated: 'I believe there is no need for me to speak at length about the fact that eugenic policies are possible given the precondition of there being many births.[9]

Hence the slogan 'Give Birth and Populate for the Sake of the Nation' neatly encapsulates the Japanese Government's population policy up to the end of World War II, i.e., to increase the overall population. The 'National Eugenic Law,' meanwhile sought to improve the 'quality' of the population. The processes through which these policies were discontinued, however, were complex and not necessarily efficacious.

The Population Problem Immediately after World War II

As previously mentioned the rapid increase seen in the population immediately following the end of WWII caused many problems. First and foremost was the deteriorating situation of food supplies. At a Cabinet meeting on 12 October 1945 – less than two months after the hostilities ended – Mr. Kenzō Matsumura, the Minister for Agriculture and Forestry, announced a shortage of rice and acknowledged that a food crisis might eventuate in June and July 1946. At the command of GHQ, a survey was conducted in December into the nutritional intake of the citizens of the country.

An article entitled 'Open Letter from Mrs. Shizue Katō: Start a Movement Limiting Births' appeared in the *Jiyū Shimbun* on 15 January 1946. Mrs. Kato said 'I believe we Japanese must plan our own national development and that Japanese women can emancipate themselves as a result of a correct appreciation of birth control. With the co-operation of men, we must hold out hope for the future of the Japanese race.' This comment appears to be the first comment made in postwar Japan that advocated birth control. But in fact its implications go further; it should perhaps be interpreted also as an expression of a nascent 'women's liberation movement.' Amongst other reasons, this can be sensed from the fact that, in October 1945, soon after the establishment of GHQ, a second generation Japanese-American visited Mrs. Kato and requested that she express her opinion relating to women privately to GHQ. At the end of January 1946 the Ministry of Health and Welfare called seventeen academics and others with knowledge and experience in the field of population studies to a discussion forum on the questions of population and birth control.

The *Asahi Shimbun* of 10 February 1946 published an interview with Colonel Crawford F. Sams, the Chief of the GHQ Public Health and Welfare Section (PHW), entitled 'Food Self-sufficiency Not Possible: Col. Sams Throws Light on the Japanese Population Problem.' In this interview Sams suggested three options for solving the problems of overpopulation: 1) making the importation of foodstuffs possible through developing industry; 2) overseas migration; and 3) encouraging limitations on births in order to arrest the population increase. He pointed out that determining which of these options to adopt was a matter for the Japanese people themselves. In his memoirs Sams commented 'No matter how much birth control was preached or promoted, unless the economic and social structures made it an attractive option it would not be very effective.'[10]

In April 1946 a 'Discussion on Birth Control' was held in the Japanese Medical House under the auspices of Nippon Igaku Hakushi-kai (the Japanese Medical Doctors Association). Six of the organizers plus Kōtarō Tanaka (Head of the School Education Bureau of the Ministry of Education), Minoru Tachi (Ministry of Health and Welfare Population Ethnicity Dept), Junji Kaneko (Municipality of Tokyo Technical Expert) and Yoshio Imamaki (Managing Director of the Japanese Medical Doctors Association) attended. The majority opinion at this meeting was that 'an immediate increase in population is not desirable' because a population increase would exacerbate food supply problems, employment problems would arise, and population increases were a factor in the cause of wars. Furthermore the discussants voiced an opinion in favor of women's liberation and the protection of infants, although recognizing that these would not contribute to solving the population problem.[11] It was about this time (June 1946) that L. V. Phelps from GHQ/PHW, Mitsuo Seki from the Ministry of Health and Welfare and others began to advocate the development of a proper understanding of population dynamics.

In November 1946 the Nutrition Section was established within the Ministry of Health and Welfare and the first national survey on nutrition was conducted. It recommended that statistical evidence be compiled on the population.

On 14 March 1947 the Japanese Government received an order from GHQ 'in relation to the matter of a proposal concerning vital census statistics' that called for data to be collected on deaths, births, still births, marriages and divorces. In response to this order a 'Population Statistics Survey Preparatory Outline' was provided on 4 April. The survey commenced on 1 October. Through the 'Notification of Stillbirths' provision of the census we learn that there was a reduction in the infant mortality rate during this period. We can surmise from this that the policies and practices implemented immediately after the war for the improvement of public health contributed to this reduction, and also that this reduction, in turn, contributed to the growing population.

Formation of the Eugenics Protection Law

A number of groups interested in birth control and the population question were created in 1947 and 1948. These included the Birth Control Research Group (March 1947), the Childbirth Regulation League of Japan (April 1947), the Population Problem Research Association (April 1948) and the Population Institute of Japan (November 1948). It was at this time that we saw the development and marketing of contraceptive devices such as the Ōta Ring. We also saw a re-evaluation of the Ogino Method of contraception.

In 1947 Masako Fukuda, Shizue Katō, Tenrei Ōta and other Members of the Diet from the Japan Socialist Party prepared the first draft of the Eugenics Protection Bill. The draft proposal was arranged in seven chapters: 1 – General Provisions, 2 – Voluntary Sterilization, 3 – Forced Sterilization, 4 – Notification and Probation, 5 – Temporary Contraception, 6 – Pregnancy Termination, and 7 – Punitive Provisions.

The objective of the Bill was outlined in Article 1 which stated 'This law protects the life and health of the mother while preventing an increase in the number of inferior descendants. By this means it aims to improve the quality of the population.' Article 2 prescribes the application of radiation to the genitals as a method of sterilization and states: 'Radiation application refers to the application of X-ray or radium radiation to the genitals to render them permanently sterile and indicates a castration level radiation.'[12]

The Eugenics Protection Law was later drawn up based upon this draft Bill, principally aimed at affording protection for the mother while promoting a policy of population control. However, the law was slightly different to the draft Bill. For example, the revised Article 1 states 'This law aims to prevent the birth of eugenically inferior descendents while protecting the health and welfare of the mother,' while Article 2 discusses eugenic (surgical) sterilization and induced abortion but does not mention the use of radiation for sterilization.

However, the Head of the Ministry of Health and Welfare's Population Problem Research Institute, Fuminori Okazaki, reportedly did not believe that birth control was an effective means of population control (August 1946). In June 1948 the reasons for the Eugenics Protection Law proposal were explained to the House of Representatives Health and Welfare Committee meeting. Amongst the reasons given were the prevalence of 'back-room' abortions, the strong negative feelings in the community about 'mixed-blood' children and a general opinion in favor of the termination of fetuses with congenital deformities. The Eugenics Protection Law was announced on 13 July 1948 and went into effect on 11 September 1948. In Okinawa a very similar Eugenics Protection Law was approved by the Legislative Council but was later abolished by a civil administration edict in 1956. In contrast, the 1948 Law in the main islands of Japan has not been repealed or abolished to date, although, as we will discuss shortly, it has been repeatedly revised.

The term 'birth control' did not feature in the draft Bill but was introduced in the first amendment (Article 20, dated June 1949). Initially the Law emphasized the mother's welfare and did not touch upon the question of population (notification of the Vice-Minister for Health and Welfare dated 25 June 1949). However, following the Cabinet's consideration of the matter in 1951 the question of birth control was addressed in earnest. Later on the problematic, so-called 'economic provisions' section, which defined 'Cases where the physical or economic implications involved in the continuation of pregnancy or childbirth could be thought to have a serious effect on the mother's health' as grounds for abortion (Article 14 Clause 3), was changed to 'Cases of pregnancy where there are already several children and childbirth could be thought to have a serious effect on the mother's health.'

When the first amendments to the Eugenics Protection Law were being debated there was a discussion taking place in the national media about the size of the Japanese population. The 20 April 1949 edition of the *Nippon Times* reported that W. S. Thompson, an advisor to the Natural Resources Section of GHQ, had said in March 1949 that 'the increase in the Japanese population is threatening the achievement of plans to attain economic self-reliance. Providing for a Japanese population increase at the rate of 1.5 million annually poses difficulties in terms of building up industry and the promotion of trade. There is no way this can be addressed other than by regulating the number of pregnancies.' However, Catholic groups opposed any form of birth control and the SCAP, General Douglas MacArthur, who was a Presidential hopeful and therefore sensitive to public opinion, told Col. Sams that PHW should refrain from any further comments on regulating the number of pregnancies. This effectively ended GHQ's official contributions to the population and birth control debates.

It was at about this time (August 1949) that the National Public Opinion Survey Institute planned a 'Survey of Public Opinion Relating to the Population Question' which was conducted in September and October. The survey revealed that only three percent of the respondents actually knew the country's population but eighty-seven percent felt that Japan was overpopulated. Furthermore sixty-two percent of the respondents were unclear on details of birth control.[13] The survey results make it clear that while there were widespread opinions and debate about the size of the population – seemingly prompted primarily by concerns about food supplies – the debate was significantly lacking in accurate information about the actual size of the population or the birth control methods that had been proposed to address the problems.

The Public Health Bureau of the Ministry of Health and Welfare approached the National Public Opinion Survey Institute and organized a survey that was conducted in December 1951. In March 1952 the results were published in a report entitled a 'Public Opinion Survey in Relation to Birth Control.'[14] The 'preamble' to the report stated that 'When we consider that overpopulation hinders the growth of trade and industry it is perfectly natural to seek to popularize birth control.' The level of knowledge people had about birth control and its methods, their attitudes towards birth control, their experiences and opinions about abortion and the opinions and hopes they wanted to express to the government concerning birth control were all surveyed.

In 1949 the Ministry of Health and Welfare surveyed twenty thousand households centered around the Tokyo Metropolitan area on contraception and the practical realities surrounding contraception laws. This survey found that approximately thirty-three percent of public servants were practicing contraception but that only about twenty percent of factory workers, farm laborers and those engaged in commerce and trade did so. The means of contraception used were the rhythm method, condoms, suppositories, pessaries and birth control tools, but knowledge about the range of methods available was limited. The report concluded that 'there is a lack of accurate knowledge about birth control' and that unless developmental guidance is provided 'we could end up regressing to the abortion culture of the Tokugawa period.' Whilst it stated that 'birth control itself is one of the very important current population issues,' it also reported that a belief that birth control alone would solve the population problem represented a 'superficial understanding' of the situation at hand. It was necessary, it said, to move 'in the direction of stabilizing the population and to work towards ensuring its soundness and health.'[15]

Birth Control Education and Trends after the Enactment of the Eugenics Protection Law

Allow me here to speak briefly on the question of trends that emerged after the enactment of the Eugenics Protection Law. In April 1949 the Japanese Medical Association for the Protection of Mothers was established at the Japanese Obstetrics and Gynecology Symposium that was held in Sendai. Its president was Yasaburō Taniguchi, who had pushed for the Eugenics Protection Law draft legislation as a member of the House of Councilors, and its advisory posts were held by the Chairman of the Japanese Doctors' Association, the Head of the Medical Affairs Bureau at the Ministry of Health and Welfare and the Head of the Public Health Bureau. Minoru Tachi, Director General of the Population Problem Research Institute, and Mitsuo Seki, Technical Officer from the Ministry of Health and Welfare, also contributed to the organization. The Association's stated objectives were 'the advancement of scientific research regarding eugenics protection and emphasizing a general promotion of the welfare of mothers.'[16]

The first amendment to the Eugenics Protection Law in 1949, focused on abortions provided on the grounds of economic necessity with the 'economic provisions' referred to earlier being added to the amendment. This was also the year in which the baby boom came to a close. At this time there was a surge in educational publications by government officials from the Ministry of Health and Welfare and Members of the Diet who had been involved with the Eugenics Protection Law. A survey was conducted by the *Mainichi Shimbun* in 1950 on the status of birth control and in October 1951 a law concerning the provision of birth control information and guidance was passed by Cabinet. There were further amendments to the Eugenics Protection Law in 1952 and 1955. The procedures involved in obtaining an abortion were simplified such that the only prerequisite was the judgment of a nominated doctor.

In more recent times, at the House of Councilors Budgetary Committee meeting on 15 March 1982, Masakuni Murakami of the Liberal Democratic Party submitted a proposal to the Minister for Health and Welfare, Motoharu Morishita, to abolish the 'economic provisions' of the Eugenics Protection Law. Since approximately ninety-five percent of abortions at the time used the 'economic provision,' it can be assumed that the objective of removing the provision was to reduce the number of abortions. The government began redrafting the law, but in the end the bill was never introduced, perhaps because the public debate in the media became so heated and polarized.

For example, Kenji Suzuki, an *NHK* announcer, said 'If you become pregnant as a result of having sought pleasure and then kill the living fetus, and you are then able to put that behind you without being affected by it and proceed to live the rest of your life without any sign of remorse, you are virtually a prostitute.'[17] The response from women was vehement, and sometimes even vitriolic.

Mariko Shibuya, a lawyer, said 'This is too stupid for words. No matter how many precautions you take there is always the danger of an unwanted pregnancy. They just don't understand that.'

Teruko Yoshitake, a public opinion analyst, reportedly said 'Mr. Suzuki, in his own inimitable theory of women, claims that "women should be submissive and obedient". Then he comes up with "Have the child regardless of the circumstances!" The man is cracked!'

Women who had had abortions made their feelings clear, saying 'At that time I was truly in pain both physically and mentally. But Mr. Suzuki is saying that we should bear that pain for the rest of our lives, is he not? I think he is being quite unreasonable.'

Mr. Suzuki's response was: 'All I can say is let those people who want to say that say it. I have no intention of accepting abortion.'[18]

Since about the middle of the 1970s, approximately half-a-million abortions are reported in Japan annually.

Conclusion

The movement against the proposed amendment of the Eugenics Law in 1982 involved not only physically disabled people and their organizations but also those who had participated in the first wave of the women's liberation movement which swept the nation from the beginning to the middle of the 1970s. During this period a variety of debates over feminism arose, with a considerable focus on abortions as well as the medical technology of reproduction. Eventually, the controversial pro-eugenics articles were removed from the Eugenic Protection law when it was revised in April 1996, becoming the Law for the Protection of the Mother's Body.

In the latter half of the 1990s, the birth rate declined dramatically, falling below 1.5 children per household. In the meantime, the number of women not wanting to marry increased. As a consequence, on the one hand, concern has mounted about the likelihood that the present youth labor force will be incapable of coping with the aging society in the near future. Some leaders openly expressed the view that the birth rate should be increased by all means available. Some municipal governments started to provide bounties to parents with new babies. On the other hand, others argue that it is unnecessary to fill the probable population shortage with increased birth rates. They claim that the population problem can be resolved through increasing the number of foreign immigrants, which has the added bonus of making Japan a more diverse society.

Notes

1 Ministry of Health and Welfare, 1992 [1]
2 Suzuki, 1983 [40] pp. 143–57.
3 Frercks, 1942 [11] pp. 72–81.
4 Kōsei Mondai Kenkyū-kai (Welfare Research Foundation), 1988 [5] pp. 211–13.
5 *Nihon ishikai zasshi* (Journal of the Japan Medical Association), 1940 [13] pp. 16–27; Information Bureau (ed.), 1941 [14] pp. 2–11.
6 This figure was quoted by Nen Yokota of the Health Bureau of the Ministry of Health and Welfare to the GHQ Public Health and Welfare Section; in GHQ/SCAP Records, December 1946 [43]
7 Kōsei Mondai Kenkyū-kai (Social Welfare Research Foundation), 1988 [6]
8 Tachi, 1943 [2] pp. 118–19.
9 Tachi, 1943 [2] p. 134.
10 Sams, 1986 [23] p. 386.
11 Andō, et. al., 1946 [16] no. 82, pp. 120–2, no. 83, pp. 143–5, no. 88, pp. 224–5, no. 89, pp. 240–2.
12 'Shakai-tō no yūsei hogo hōan zenbō (The Japan Socialist Party's full draft of the Eugenics Protection Law),' 1947 [30] p. 579.
13 *Jinkō mondai ni kansuru yoron chōsa* (Opinion surveys on population issues), 1950 [55]
14 *Jutai chōsetsu ni kansuru yoron chōsa* (Opinion surveys on birth control), 1952 [56]

15 Shinozaki, 1949 [29] pp. 29–35.
16 Nihon Bosei Hogoi Kyōkai (Japanese Medical Association for the Protection of Mothers), 1970 [35] p. 215.
17 Nihon Kyōbunsha, 1982 [48] pp. 29–32.
18 'Suzuki-san no mondai hatsugen...(Controversial comments by Mr. Suzuki...),' 1983 [49] pp. 54–5.

Bibliographic Guide

Population Statistics, Population Problems

[1] Ministry of Health and Welfare, *Boshi eisei no omonaru tōkei* (Statistics on mother and child ante-natal health), 1992.
[2] Minoru Tachi, *Jinkō mondai setsuwa: Asahi jōshiki kōza* (Treatise on the population question: the Asahi common sense lecture), Bon'yōsha, 1943.
[3] Minoru Tachi, 'Jinkō mondai kara mita sanji seigen (Birth control viewed from the perspective of the population problem),' in Kakuichi Andō (ed.), *Sanji seigen no kenkyū* (Birth control research), Nihon Rinshōsha, 1947, pp. 27–154.
[4] Ministry of Health and Welfare, Population Research Institute, *Sanji seigen mondai no jinkō seisaku-teki kōsatsu* (Study of population policy in regard to the birth control question), July 1949.
[5] Kōsei Mondai Kenkyū-kai (Welfare Research Foundation), *Kōseishō gojūnen-shi: Kijutsu-hen* (Fifty year history of the Ministry of Health and Welfare: Descriptive accounts), 1988.
[6] Kōsei Mondai Kenkyū-kai (Social Welfare Research Foundation), *Kōseishō gojūnen-shi:Shiryō-hen* (Fifty year history of the Ministry of Health and Welfare: Data), 1988.

Prior History of the Eugenics Protection Law

[7] Bonsen Takahashi, *Datai mabiki no kenkyū* (Abortion research), Daiichi Shobō, 1926, republished by Chūō Shakai Jigyō Kyōkai, 1981.
[8] Margaret Sanger, *Sanji chōsei-ron* (A treatise on birth control), Toshisada Oku (tr.), Seika Shoin, 1921.
[9] Senji Yamamoto (chief ed.), *Sanji chōsei hyōron* (Review on birth control), no. 1–8, 1925, republished by Fuji Shuppan, 1983.
[10] Senji Yamamoto (chief ed.), *Sei to shakai* (Sex and society), no. 9, 1925–26, republished by Fuji Shuppan, 1983.
[11] Rudolf Frercks, *Nachisu no yūsei seisaku* (The Nazi eugenics policy), Fumio Hashimoto (tr.), Risōsha, 1942.
[12] RyūsaburōShikiba, *Sangar fujin-den to sanji chōsetsu tenbō* (Mrs. Sanger's biography and the future of birth control), Daigensha, 1947.

[13] *Nihon ishikai zasshi* (Journal of the Japan Medical Association), vol. 15, no. 12, 1940, pp. 16–27.
[14] Information Bureau (ed.), *Shūhō* (Weekly report), no. 244, 1941, pp. 2–11.

On the Debate About the Establishment of the Eugenics Protection Law

[15] Naotarō Kuji, 'Sanji seigen-setsu no taitō to kokumin yūsei-hō no saikentō (The rise of the birth control argument and a re-evaluation of the national eugenics law),' in *Sanka to fujinka* (Obstetrics and gynecology), April 1946, pp. 7–11.
[16] Kakuichi Andō, Sa Yamazaki, Nen Yokota, and Minoru Tachi, 'Zadankai: Sanji seigen ni tsuite (Roundtable discussion on birth control),' in *Nihon rinshō* (Japanese clinical medicine), 1946, nos. 82, 83, 88, and 89.
[17] Tenrei Ōta, 'Sanji seigen no hantai ni kotaeru (In response to birth control critics),' in *Kōshū eisei-gaku zasshi* (Journal of public health and hygiene), 1947, vol. 2, pp. 215–9.
[18] Yasaburō Taniguchi, 'Yūsei hogohō kaisetsu (Explanations of the Eugenics Protection Law),' in *Nihon ishikai zasshi* (Journal of the Japan Medical Association), 1948, vol. 22, pp. 215–9.
[19] Seizaburō Hata, 'Jinkō ninshin chūzetsu ni tsuite (Induced pregnancy terminations),' in *Nihon ishikai zasshi* (Journal of the Japan Medical Association), 1948, vol. 22, pp. 248–53.
[20] Yūkichi Abe and Kōji Kondō, 'Jinkō ninshin chūzetsu mondō (Questions and answers on induced pregnancy terminations),' in *Nihon ishikai zasshi* (Journal of the Japan Medical Association), 1948, vol. 22, pp. 263–6.
[21] Yūkichi Abe, 'Yūsei hogohō ni tsuite (On the Eugenics Protection Law),' in *Nihon iji shinpō* (Japan medical affairs bulletin), 1948, no. 1270, pp. 1005–6.
[22] Yasaburō Taniguchi, 'Yūsei hogohō ni okeru shitei ishi no sentei hōhō ni tsuite (On the nominated doctor selection method under the Eugenics Protection Law),' in *Nihon iji Shinpō* (Japan medical affairs bulletin), 1948, no. 1272, pp. 1076–7.
[23] C. F. Sams, *DDT kakumei: Senryō-ki no iryō fukushi seisaku o kaisō suru* (The DDT revolution: Reflections on medical welfare policy during the Occupation), Eiji Takemae (tr.), Iwanami Shoten, 1986.

Interpretations on the Situation of Contraception and Birth Control

[24] Nagamune Aida, 'Shizan to jinkō ninshin chūzetsu (Stillbirth and induced pregnancy termination),' in *Kōsei jihō* (Health and welfare bulletin), July 1949, pp. 24–6.

[25] Yutaka Morishima, 'Jutai chōsei no hōhō to hihan (Birth control methods and criticisms),' in *Sanka to fujinka* (Obstetrics and gynecology), August 1949, pp. 330–40.
[26] Nobuo Shinozaki, 'Gendai seiseikatsu no jittai (Present day sex lives),' in *Kōsei jihō* (Health and welfare bulletin), December 1949, pp. 6–14.
[27] Fuminori Okazaki, 'Shussan chōsei wa ikani okonau bekika (How should birth control be handled?),' in *Kōsei jihō* (Health and welfare bulletin), December 1949, pp. 16–17.
[28] Ministry of Health and Welfare, Population Research Institute, *Shogaikoku ni okeru sanji seigen no fukyū jōkyō* (The situation concerning the popularization of birth control in several overseas countries), 1949.
[29] Nobuo Shinozaki, 'Sanji seigen no jitsujō o saguru (Exploring the realities of birth control),' in *Kōsei jihō* (Health and welfare bulletin), May 1949, pp. 29–35.

Interpretations of the Eugenics Protection Law and the Circumstances Surrounding its Establishment

[30] 'Shakai-tō no yūsei hogo hōan zenbō (The Japan Socialist Party's full draft of the Eugenics Protection Law),' in *Nihon iji shinpō* (Japan medical affairs bulletin), no. 233, 1947, p. 579.
[31] Kakuichi Andō (ed.), *Sanji seigen no kenkyū* (Studies of birth control), Nihon Rinshōsha, 1947.
[32] Nobuo Shinozaki, *Sanji chōsei to fūfu sei-seikatsu no jittai hōkoku* (Report on birth control and the sex life of married couples), Muramatsu Shoten, 1949.
[33] Tenrei Ōta, *Datai kinshi to yūsei hogo-hō* (Abortion prohibition and the Eugenics Protection Law), Keieisha Kagaku Kyōkai, 1967.
[34] Tenrei Ōta, *Sei no kenri: Datai kaihō no rekishi* (Sexual rights: a history of abortion liberation), San'ichi Shobō, 1970.
[35] Nihon Bosei Hogoi Kyōkai (Japanese Medical Association for the Protection of Mothers), *Nijūshūnen kinenshi* (Twentieth anniversary commemorative publication), 1970.
[36] Harumasa Takemura and Jirō Harada (ed.), *Bosei hogo* (Mother's welfare), Shin Nippon Igaku Shuppan, 1972.
[37] Tenrei Ōta, *Nihon sanji chōsei hyakunen-shi* (One-hundred years of birth control in Japan), Ningen no Kagakusha, 1976.
[38] Shakai Hyōronsha Henshū-bu, *Josei no sei to chūzetsu: Yūsei hogohō no haikei* (Female sexuality and abortion: Background to the Eugenics Protection Law), Shakai Hyōronsha, 1983.
[39] Shūkan Books Tokubetsu Shuzai-han (Special Investigation Team of the Weekly Books magazine) (ed.), *Naze ima yūsei hogohō 'kaisei' ka* (Why

the proposed 'revision' of the Eugenics Protection Law Now?), Gendai Shorin, 1983.
[40] Zenji Suzuki, *Gendai no yūseigaku* (Contemporary eugenics), Sankyō Shuppan, 1983.
[41] Michiko Ōbayashi, *Josan-fu no sengo-shi* (Postwar history of midwives), Keisō Shobō, 1989.
[42] GHQ/SCAP Records, 'Eugenic Protection Law,' May 1948–November 1949, no. 9322, PHW, 1176–1179.
[43] GHQ/SCAP Records, 'Eugenics,' December 1946, Box no. 9347, PHW 2003.
[44] GHQ/SCAP Records, 'Genetics Program,' NHI and ABCC, May 1947–November 1948, Box no. 9354, PHW 1726–1727.
[45] GHQ/SCAP Records, 'Eugenics,' November 1946–October 1950, Box no. 9425, PHW 4821–4825.
[46] GHQ/SCAP Records, 'Birth control,' March 1949–May 1951, Box no. 9468, PHW 2480–2483.

On the Revision of the Eugenics Protection Law

[47] 'Yūsei hogohō kaisei wa zeka hika? Jinkō ninshin chūzetsu (Pros and cons of the planned amendment of the Eugenics Protection Law: Induced pregnancy terminations),' in *Gendai sei kyōiku kenkyū* (Contemporary studies of sex education), December 1982.
[48] Nihon Kyōbunsha (ed.), *Taiji wa ningen dewa nainoka: Yūsei hogohō no gimonten* (Aren't fetuses human beings? Problems of the Eugenics Protection Law), Nihon Kyōbunsha, 1982.
[49] 'Suzuki-san no mondai hatsugen: "Chūzetsu suru onna wa baishunfu da!" no ōsōdō (Controversial comments by Mr Suzuki: Turmoil over 'women who have abortions are prostitutes!),' in *Josei jishin* (Women's weekly), 7 April 1983, pp. 54–5.
[50] 'Umu, umanai, umenai: Yūsei hogohō to yūsei shisō o kangaeru (Giving birth, not giving birth, and being unable to give birth: Reflections on the Eugenics Protection Law and the philosophy of eugenics),' in *Agora*, no. 28, BOC Shuppan-kyoku, 1983.
[51] Fujin Kyōdō Hōritsu Jimusho (Female Collective Law Office) (ed.), *Ima naze yūsei hogohō kaiaku ka?* (Why is the Eugenics Protection Law amended in the wrong direction?), Rōdō Kyōiku Sentaa (Labor Education Center), 1983.
[52] Nihon Kazoku Keikaku Renmei (Japan Family Planning League) (ed.), *Chūzetsu kinshi e no hanmon: Kanashimi o sabake masuka* (Refutation of the prohibition of abortions: Can we adjudge one's grief?), Ningen no Kagakusha, 1983.

[53] Tenrei Ōta, *Chūzetsu wa satsujin dewa nai* (Abortions are not murders), Ningen no Kagakusha, 1983.
[54] Nihon Kazoku Keikaku Kyōkai (Japan Family Planning Association) (ed.), *Gendai no kazoku keikaku* (Contemporary family planning), Nihon Kazoku Keikaku Kyōkai, 1984.

Opinion Surveys

[55] Kokuritsu Yoron Chōsasho (National Institute of Opinion Surveys), *Jinkō monai ni kansuru yoron chōsa* (Opinion surveys on population issues), 1950.
[56] Kokuritsu Yoron Chōsasho (National Institute of Opinion Surveys), *Jutai chōsetsu ni kansuru yoron chōsa* (Opinion surveys on birth control), 1952.

22 The Development of Japanese Style Quality Control

Shigeru Nakayama

During the 1980s, Japanese postwar achievements in science and technology were called a miracle in the West. Western observation teams traveled to Japan to discover the secret of the miracle. They finally agreed that quality control (QC) was the key. Soon, numerous books were being written about Japanese QC in European languages.[1] But quality control is not a unique Japanese creation.

The Prehistory of Quality Control

The US's prewar mass-production system, built on an unskilled labor-force, faced product quality problems that Europe had not experienced. Statistical Quality Control was developed as a solution; introducing random sample inspections to identify defective products before the final production stages. These techniques were further developed during WWII into Total Quality Control (TQC, quality control that applies to the whole corporation as a system) for munitions production. Munitions required strict quality control to ensure proper performance under the severe conditions of warfare. To meet military specifications, special statistical technology was developed.[2]

In Japan the Mathematical Statistics Society was founded in 1941 with Toshio Kitagawa, Motosaburō Masuyama and Heihachi Bandō as prominent members. Their aim was to apply their methods to the industrial sector. In 1944 the Institute of Statistical Mathematics was founded with the support of the military to adapt QC methods to military-oriented production. Masuyama coined a new word 'Suikei' (estimate, stochastics) for random sample inspection to distinguish it from ordinary statistics. But statistical quality control was not sufficiently understood by the Japanese manufacturing industry at that time.[3]

In December 1945, Nihon Kikaku Kyōkai (Japanese Standards Association: JSA) was established. It began disseminating manufacturing standards by publishing the magazine *Kikaku to Hyōjun* (Standards), which was later renamed *Hyōjunka to Hinshitsu Kanri* (Standardization and Quality Control).

In May 1946 Ken'ichi Koyanagi organized a professional union in the private sector called Nikka Giren (Japanese Union of Scientists and Engineers, JUSE) with seven fellow engineers. This group was always located at the center of the movement advocating quality control. In 1947, Nikka Giren (JUSE) participated in the anti-establishment Kenkyū Fukkō Kaigi (Research Recovery Conference) that Minka organized and acted as its secretariat for a period. But soon thereafter, JUSE clarified its role as a technocratic professional association. In 1948 they established the Quality Control Research Group (QCRG). Its members included university professors of statistics, as well as bureaucrats and engineers from the private sector. They were donated literature on quality control, such as Walter A. Shewhart, *Economic Control of Quality of Manufactured Products* (Van Nostrand, 1931) by a Bell System engineer, who was working for GHQ and recognized the urgency of employing quality control methods for increasing productivity.[4]

In September 1949, JUSE began a regular seminar course in basic quality control. In May 1949, through another society, Nihon Nōritsu Kyōkai (Japan Efficiency Society), Eisaburō Nishibori began to provide consultancy services to corporations on QC.

Demand for Quality Control by GHQ

GHQ began to introduce SQC (Statistical Quality Control) into Japan in 1946. The Occupation Forces were troubled by failures in the telephone system. They identified the cause as poor and uneven quality of the communication instruments. Accordingly they advised the Japanese telecommunications industry to adopt new quality control methods and provided guidance. John W. O'Brien, the head of ESS/ST, gave a talk to the Japan Standards Association, warning that unless quality control is maintained the Japanese could not win in the international competition for exports in peacetime.[5] In fact, it was a widely-held belief in the West that Japanese exports were poor quality at cheap prices.

In July 1949, the Industrial Standardization Act was enacted as part of a government effort to encourage rationalization of corporations. In 1946–49, however, the major overseas markets for the Japanese industrial sector were in Asia, where prewar export networks were revived. According to ESS/ST, Asian importers did not demand high quality from Japanese goods[6] and the Japanese had therefore not engaged with quality control issues very enthusiastically.

In the latter phase of the Occupation, Bowen Dees, deputy director of ESS/ST, and other staff members tried to introduce quality methods to contribute to the economic recovery of Japan.[7] In April 1950, with the guidance of invited American engineers, Heith and McNaught, quality control had considerable affect on heat control in the iron and steel industry and on open-hearth furnace work. Thus, ESS/ST played a vanguard role in the full-scale introduction of quality control practices in Japan.[8]

Deming's Contribution

Japanese quality control is generally associated with the name of a US management scholar, E. Edwards Deming. Some people question why this is so since he was not well known in the United States.

Dr. Deming, Professor of Statistics at New York University, arrived in Japan as a member of the US Statistics Mission. He gave talks at the Institute of Statistical Mathematics and JUSE. In 1948, with the guidance of Sarason from GHQ/CCS, quality control was put into practice at the NEC Factory at Tamagawa. This was a faithful replication of a US system of quality control that had some favorable effect, but its influence was limited to the communications technology sector.[9]

Then, when the Korean War broke out on 22 June 1950, Deming happened to revisit Japan and conducted a seminar on quality control. Deming had experience in supervising the introduction of quality control in the US munitions industry during WWII. In order to meet the specifications for special procurements by the US Army,[10] and to meet international interchangeability requirements, especially for weaponry, Japanese enterprises now zealously embraced quality control. Until the Korean War, quality control was primarily a research topic for scholars and scientists, with only a few enterprises having tried it on the shop floor. After the Korean War, though, quality control became an integral and indispensable part of Japanese manufacturing.

In the 1950s and '60s, quality control practices in the West were employed only in the munitions industry.[11] Consumer industries in the US focused on raising productivity and did not seriously consider quality control. Thus, Deming's advocacy of quality control was generally not well received. For most US factory managers, quality and productivity were considered to be incompatible; in other words, to raise productivity, quality had to be sacrificed.

For Japanese industry, though, after adapting to the quality control requirements of the special procurements for the Korean War, the market orientation of its export industry switched from being Asia-centered to a focus on the more advanced countries such as the US, which demanded higher quality goods. To fulfill these demands, Japanese industry seriously and thoroughly adopted QC practices, even ahead of the Western countries, eventually resulting in increasing exports to the Western market. It is an irony of history that the quality of Japanese goods surpassed others, triggered by the demands of the US military.

The Korean War created a temporary market of urgent demand. By February 1951, it appeared that the Korean War would remain a local war (i.e., not escalate to another World War); then, the development of the war-oriented business sector – which had started to boom – stagnated. However, in April 1951, W. F. Marquat, the Director of ESS, proposed to Washington D.C. to incorporate Japanese industry into the US Far East economic policy 'on the basis of the quality of products and cheaper price.' Responding to this policy change, the Japanese business community anticipated that after the San Francisco Peace Treaty their primary export market would shift from Korean War special procurements to a new phase of procurement for an 'ever-lasting' Cold War, hopefully leading to the recovery of Japanese industry. The Bōei Iinkai (the Committee on Defence Industry) of the Keidanren (Keizai Dantai Rengōkai, Japan Federation of Economic Organizations) advocated preparing for the production of finished and assembled weapons that would be procured by the US military for the Cold War under the US-Japan Security Treaty.[12] Thus there was an increased sensitivity to demands for systematic and rigorous quality control.

Deming went home after presenting a basic QC course from 10–17 July 1950. JUSE published his lecture notes under the title *Hinshitsu Kanri Kōgiroku* (Lecture on Quality Control) which sold widely.

Encouraged by the good sales figures, JUSE invited Deming again from July to October 1951. This time top managers attended his lectures on quality control. Deming was somewhat bitter from the experience that US managers did not accept his ideas about quality control, so he had insisted on the necessity of selling his ideas to the top Japanese management.

In September 1951 JUSE created the Deming Prize at the first Quality Control Meeting in Osaka with the proceeds from his book sales. The main Deming Prize was given to individuals who contributed to the theory and dissemination movement of quality control. But more significantly, JUSE established the Deming Execution Prize to honor the corporations that enthusiastically introduced quality control practices in the hope that it would change the character of the workplace. The inaugural execution prizes were awarded to four corporations; Shōwa Denkō, Yawata Seitetsu, Fuji Seitetsu, and Tanabe Seiyaku. The execution prize was quite significant; externally it improved the corporate image and advertised the good quality of their products while internally the QC practices strengthened the corporate identity, provided targets for workers and became an integral function of labor management. With all corporations seriously competing for the Deming award, it contributed greatly to an improvement in the quality of Japanese products.

The Korean War certainly contributed to the recovery of Japanese industry, which had been in a slump throughout the postwar period. Quality control was the technological factor in this recovery. Deming found many enthusiastic supporters in Japan. In fact, in the Japanese industrial sector, Deming is often regarded as the savior of the Japanese economy. While the scarcity of exchange currency limited the importation of foreign machineries, quality control was vigorously welcomed as a means of achieving rationalization targets without financial cost.[13] Thus, although introduced to Japan somewhat accidentally, Deming became quite famous in Japanese industry even whilst being neglected in the Western industrial sector.

Total Quality Control for the Whole Corporation

In the early half of the 1950s, statistical quality control had further developed but at the same time self-reflection revealed some issues that can be summarized in the following three points:[14]

1 over dependence on statistical methods;
2 advanced design but overly rigid schematization;
3 slow understanding by top management.

These issues appear from the perspective of mid-level engineers who were the main promoters of quality control. The third item was based on Deming's reflections that in the US engineers understood the significance of quality control but the management did not accept it.[15]

In 1954, J. M. Juran was invited to Japan by JUSE. He argued that it was necessary for quality control to be at the center of management practices rather than a peripheral or technical matter. Following his advise, the Company Wide Quality Control (CWQC, or TQC) movement began in the late 1950s. Its aims were to involve everybody from the company president to shop floor labor in the study of statistical methods and the implementation of quality control practices. Technically speaking, TQC is not radically different from the seven tools of SQC (Statistical Quality Control) but while SQC methods check for quality at the final stage of production, TQC checks every stage of the production process. Hence, TQC is not just a technique in manufacturing, but a different way of thinking about manufacturing.

In April 1962, JUSE launched the magazine *Genba to QC* (Workplace and quality control), which became the primary publication to promote the 'QC circle' organization of the workplace. A QC circle is a group consisting of five to ten workers at the shop floor level for discussing and proposing improvements (Kaizen) to the manufacturing processes in their particular work area. Top management then reviews the proposals to decide which to adopt and acts to implement them. The ideas born at the shop floor level do not generally lead to major technological innovations, but many minor labor-saving and cost-effective ideas are generated. This system was not original or unique; it had its roots at the Ford Motor Company, but was limited there to the engineers, while the Japanese approach involved all of the workers in a factory – both production and support staff. Thus, they succeeded in manufacturing less expensive, higher quality, safer and more durable products. The advocates of the QC movement congratulated themselves, claiming that they had inaugurated the special character of postwar Japanese industrial technology. This movement was initiated by private sector mid-level engineers but has now been incorporated into the Japanese Government's campaign for upgrading productivity.

Although initiated by mid-level engineers, the QC movement soon involved top-level management and became essentially a top-down movement within each corporation. It was neither instigated, guided nor supported by the Government. Its main promoters were JUSE and Nihon Nōritsu Kyōkai (The Japan Management Association). In 1960, celebrating the tenth anniversary of the publication *of Hinshitsu Kanri* (Quality Control), JUSE designated November to be Hinshitsu Kanri Gekkan (Quality Control Month); various events are organized nation-wide under the sponsorship of corporations and the auspice of MITI.[16] In March, local QC conferences are held for specialist QC coaches and mid-level managers. October has been designated Hyōjunka Sokushin Gekkan (Standardization Promotion Month). Moreover, QC circle conferences are held more than eighty times each year, with symposia and on-site observation tours of QC circles. In 1976, there were more than 60,000 participants at these events.[17]

Criticism and International Development

Labor unions complain that TQC is nothing more than an ingenious labor management technique and that QC circles are unpaid overtime. But the anti-rationalization movements of labor unions primarily target traditional demands for wage increases, shorter working hours and resistance to worker redundancies. Mainstream labor leaders are neither concerned with the quality of production in the workplace nor opposed to the quality control movement. For example, Satoshi Kamata, a critical writer with personal experience of laboring in a car factory, wrote a criticism of overwork in the workplace.[18] But he dealt with general labor conditions without mentioning the quality control movement at all. During the 1970s, the quality of work was an issue raised internationally but Japanese labor unions did not take it up very seriously. These criticisms were all rather superficial and had little real influence. 'QC movement' in which 'bottom-to-top' suggestions are seemingly democratic was encouraged by management as a clever labor management measure to reduce alienation in the workplace. In fact, one leader of a Japanese labor union cynically claimed that the modern workplace was so boring that even QC circles were welcome.

The promoters of the quality control movement in Japan wanted their success to extend to Japanese industry's subcontractors in neighboring countries. In 1968 the Overseas Technology Cooperation Agency (OTCA; later JICA – the Japan International Cooperation Agency) started an internship program that provided training in industrial standards and quality control. In August 1970, they organized the first international convention on QC circles, held in Taiwan.[19] In the same year, quality control was adopted as Korean government policy and successfully adopted by Korean industry.[20]

In fact, the quality control movement was, in part, an old-fashioned brainstorming movement to raise labor morale without cost, thereby raising technical standards on the shop floor. With progress in automation, labor alienation became a significant problem common for the advanced countries in the late 1970s. For instance, young workers in the US deserted the Ford Motor Company factories because of the boring work. Management was thus faced with a hitherto unrecognized labor-generated problem in the quality of work. Accordingly, production systems were altered, in some instances changing the work assignment every hour in an effort to overcome boredom; in other cases the traditional production line was modified such that the workers moved along the line with the product on the conveyor in an effort to develop a sense of achievement or craftsmanship. At that time TQC attracted world-wide attention as a Japanese style labor management approach for preventing the alienation of workers.

In the 1980s, TQC (and Deming) was reevaluated in the USA. It was adopted by the Ford Motor Company's factories in 1983. In the latter half of the 1980s, its influence reached Europe. TQC was, however, frequently introduced as a temporary fashion or adopted by managers and management scholars as simply another method of labor management. In Japan also, the factor of special Japanese labor management to enhance morale in the workplace has gradually faded away (cf. vol. 5, ch. 5).[21]

Academic Development

Japanese quality control began in the postwar era as a movement of engineers and developed into a corporate labor management approach. In contrast, its academic development was relatively late. In order to theorize and systematize as an academic discipline and adopt it at the university teaching level, Nihon Hinshitsu Kanri Gakkai (Japanese Society for Quality Control) was founded in 1970 and published its periodical *Hinshitsu* (Quality). They have adopted the standpoint of consumers rather than corporations. Academically, quality control was located in the discipline of Keiei Kōgaku (management technology). Keiei Kōgaku must be located in the business schools but in practice it is quite technical and hence generally taught by engineers with engineering minds at engineering schools.

Notes

1 Sasaki and Hutchings, 1984 [14] and Mann 1988 [21] p. 46, for instance, describe the state of affairs at the time.
2 Hutchings, 1984 [14] p. 45.
3 Nihon Kagakushi Gakkai, 1964 [18] p. 446.
4 E. Edwards Deming, *Out of Crisis*, 1982, MIT, Center for Advanced Engineering Study, pp. 2–3.
5 'Japanese Standards Agency Holds Vital Industrial Role,' in *Nippon Times*, 29 December 1946, p. 4.
6 Ichikawa, 1987 [19] p. 14.
7 Ichikawa, 1987 [20]
8 Nihon Kagakushi Gakkai, 1964 [18] p. 262.
9 Ishikawa, 1981 [10] p. 20.
10 Keizai Dantai Rengōkai, Keizai Kyōryoku Kondankai, 1953 [17] p. 340.
11 Hutchings, 1984 [14] p. 45.
12 Keidanren Jimukyoku (Secretariat of the Keidanren), 1952 [16] p. 22.
13 Nishimura, 1951 [15] p. 257.
14 Ishikawa, 1981 [10] pp. 25–6.
15 Mann, 1989 [21]
16 Ishikawa, 1981 [10] p. 5.
17 Ishikawa, *Quality Control Circles at Work*, Asia Productivity Organization, 1984, pp. 3–4.
18 Kei Kamata, *Jidōsha zetsubō kōjō* (Desperate car factory), 1984. Also Kei Kamata, *Jidōsha ōkoku no kurayami* (Gloom of automobile empire), 1984.
19 Mizuno, 1985 [12] p. 367.
20 Boku Uki, *Kankoku no gijutsu hatten* (Technological development in South Korea), 1988, p. 100.
21 Ikurō Kusaba, 'Nihon no Hinshitsu Kanri (Japanese Quality Control),' in *Hinshitsu*, July 1992 [2] vol. 22, no. 3, p. 260.

Bibliographical Guide

Journals of the movements for disseminating quality control

[1] *Hinshitsu Kanri*.
Monthly journal published since March 1950 by JUSE, (Japanese Union of Scientists and Engineers), a professional union of scientists and engineers organized in 1946. The opening address was written by Ichirō Ishikawa, the President of both Keidanren and JUSE.

[2] *Hinshitsu*.
Bulletin of the Japanese Society of Quality Control. While the Quality

Control movement began in Japan in the 1950s as a practical approach, its academic journal appeared rather late in 1971. It includes some historical descriptions.

[3] *Reports of Statistical Application Research*, JUSE.
English journal of JUSE, published since March 1951.

[4] Nikka Giren (JUSE), *Genba to QC* (Shop Floor and QC, monthly).
Published for shop floor workers from April 1962. From 1973 entitled: *FQC* (in Japanese). From January 1988: *QC saakuru* (QC circle).

[5] Nikka Giren (ed.), *Nihon Kagaku Gijutsu Renmei jūgoshūnen kinen* (The first fifteen years of JUSE), May 1961.

[6] Nihon Kikaku Kyōkai (Japanese Standards Association), *Kikaku to kijun* (Standard), later renamed *Hyōjunka to hinshitsu kanri* (Standardization and quality control).

[7] Nikka Giren (tr. & ed.), *Demingu hakase kōgiroku: Tōkeiteki hinshitsu kanri no kiso riron to ōyō* (Lecture notes of Dr. Deming: Basic theory and the applications of statistical quality control).

[8] Nikka Giren (tr. & ed.), *J. M. Juran jutsu; Hinshitsu kanri seikōhō* (Juran talked about how to succeed in quality control), 1956.

[9] Kōzō Koura, 'Hinshitsu kanri nenpyō (Chronological table of the quality control movement)', in *Hinshitsu kanri* (Quality control), vol. 41, 1990, pp. 391–5.

While countless books on 'How to control quality' were published, some of them include historical descriptions of the promoters of the movement

[10] Kaoru Ishikawa, *Nihonteki hinshitsu kanri; TQC to wa nanika* (Japanese style quality control: What is TQC?), 1981.
The author is one of the advocates of quality control. Starting with his involvement in the movement, a chronological table and bibliography of quality control is attached at the end.

[11] Kōgyō Gijutsu-in Hyōjunbu (Agency of Industrial Science and Technology, Standard Section), *Wagakuni no kōgyō hyōjunka; nijūnen no ayumi* (Japanese industrial standardization: A twenty year history), 1969.
Follows the development of industrial standards. In Chapter 8, quality control is illustrated in a more cool-headed manner than the JUSE approach.

[12] Shigeru Mizuno, *Kaisha sōgō hinshitsu kanri*; *TQC no dōnyū to tenkai* (Company-wide quality control; introduction and development of TQC), 1985. Particularly see chapter 15 'Hinshitsu kanri no rekishi to tenbō (History and future scope of quality control).'

[13] Masao Kogure, *Nihon no TQC; sono saiginmi to shintenkai* (Japanese TQC; its reevaluation and new developments), 1988.

Numerous English texts were written to introduce Japanese quality control. In the following, a historical account is given

[14] David Hutchings, 'The Historical Development of Quality Control in the West,' in Naoto Sasaki and David Hutchings, (eds.), *The Japanese Approach to Product Quality: Its Applicability to the West*, Pergamon, 1984, p. 45.

Articles related to quality control appeared in *Keizai rengō*, the bulletin of the Keidanren

[15] Keizō Nishimura, 'Keizai no gōrika to Demingu hakase (Rationalization of economy and Dr. Deming),' in *Keizai rengō* (Economic federations), September 1951, p. 257.
The author is the President of Furukawa Denki Kōgyō (Furukawa Electric Co.). He was an ardent adherent of statistical quality control and market research.

[16] Keidanren Jimukyoku (Secretariat of the Keidanren), 'Shintokuju no igi to kenzai kyōryoku no kōsō ni tsuite (On the significance of new special procurement and the idea of economic cooperation),' in *Keizai rengō* (Economic federations), August 1952, p. 22.

[17] Keizai Dantai Rengōkai, Keizai Kyōryoku Kondankai (Economic Cooperation Conversazione), 'Tokuju keiyaku jōken tō ni kansuru yōbō iken (Request and opinion about the conditions of special procurement contracts),' in *Keizai rengō* (Economic federations), no. 7, 1953, p. 340.

Others

[18] Nihon Kagakushi Gakkai (History of Science Society of Japan) (ed.), *Nihon kagaku gijutsu-shi taikei 5; Tsūshi 5* (A compendium of the Japanese history of science and technology, vol.5, Historical overview 5), chapter 10, 1964, and *Nihon kagaku gijutsu-shi taikei, Sūri kagaku* (Mathematical sciences), chapter 13, 1969, Daiichi Hōki Shuppan.

[19] Hiroshi Ichikawa, 'GHQ kagaku gijutsuka no seisaku to katsudō ni tsuite (On the policy and activity of ESS/ST),' in *Osaka Shiritsu Daigaku Ronsō* (Journal of Osaka Municipal University), no. 54, 1987.

[20] Hiroshi Ichikawa, 'Chōsen Sensō izen ni okeru hinshitsu kanri to kanri gijutsu no igi ni tsuite; GHQ no tai-nichi sangyō seisaku no shiten kara (On the significance of quality control and management technology before the Korean War; from the perspective of GHQ's industrial policy for Japan),' in *Osaka Shiritsu Daigaku Keiei Kenkyūkai Keieigaku kenkyū* (Journal of management studies, Osaka Municipal University), vol. 38, no. 5, 1987.

A path breaking paper evaluating the effort made by ESS/ST to introduce quality control in Japan.

[21] Since the 1980s Deming has been associated with the success of Japanese quality control. As in Japan, several books appeared in the US to advocate the 'Deming philosophy.' The following books discuss the Japanese QC Circle movement.

- Mary Walton, *Deming Management at Work*, Perigee books, 1991.
- William W. Scherkenbach, *The Deming Route to Quality and Productivity*, CEEP Press Books, 1986.
- Nancy R. Mann, *The Key to Excellence: the Deming Philosophy*, Study of Prestwick Book, 1989
- M. P. Cumbers, (ed.), *Quality Management and Product Liability including the Lesson from Japan*, 1981.

23 Industrial Safety

Seikan Ishigai

Introduction

Before one can develop an understanding of the place of industrial safety during the Occupation of Japan, it is necessary to understand the general principles and the status of industrial safety in Japan before the end of World War II. Section 1 of this chapter provides an outline of the general principles of industrial safety and section 2 reviews the pre-Occupation history of industrial safety in Japan. Section 3 then analyzes the developments in industrial safety practices during the Occupation.

The General Principles of Industrial Safety

Industry is an important aspect of human civilization. In short, its purpose is to supply the goods and services necessary for human life. Provided the outcome is the same, the lower the cost the better. However, as every person instinctively tries to avoid death and injuries, in this era of well-established human rights, some people maintain that 'a human life is more precious than the very earth itself.' Safety, therefore, is paramount. But safety generally incurs costs.

There is thus a fundamental conflict of interests involved. What is considered to be an acceptable compromise on this issue varies according to the time, place and an individual's position in the power relations of any given situation. For example, in certain times and places the belief that 'human rights cease at death' has been so widespread that 'the protection of the law did not extend to the dead.' Under such circumstances there was no financial compensation provided to the survivors of those who die as a result of workplace accidents – and thus no financial penalties for the employer. In contrast, today the large sums paid for both compensation and punitive damages in product and employer liability trials in the USA sometimes create problems for manufacturers and other industrial employers.[1]

Generally, workplace safety is in the interests of industry itself and ensuring a bare minimum level of safety is an absolute necessity. This can be illustrated by looking at an extreme example: even in a system of slave labor with total disregard for human rights, the workers must be looked after to some extent since it is not possible to maintain the supply of slaves if each slave is worn out in a short period of time.

Many branches of technology have been abandoned because they could not be made safe,[2] such as very deep coal mining in Japan, the aceto-aldehyde production methods that caused the Minamata disease, and many others. Of course, we must recognize that there is no absolute safety anywhere; there will always be accidental deaths – even in the home. It follows then that spending money does not necessarily mean an improvement in safety.

The history of industrial society is so full of industrial disasters and environmental pollution that one might justifiably say that industry is inherently dangerous. But any written history that does not also record the continuous efforts at improving industrial safety would be terribly incomplete. The development of new technology is always accompanied by unintended – and often unforeseeable – dangers. Fortunately there are almost always forewarnings before those hidden dangers manifest themselves as unmitigated disasters.

Unfortunately, because the persons responsible for implementing technological changes must always consider a variety of factors such as budgetary concerns, production schedules and a multitude of other things, these warnings are often misinterpreted or even blatantly ignored. To prevent such imprudence, it is therefore necessary that neutral safety specialists be involved in any such projects; people who have both specialized knowledge and relevant experience and for whom safety – rather than budgets or productivity – is the primary concern. In order for such safety specialists to exist and to have the authority necessary to effect changes to work practices though, certain social measures must be in place. To date, these social measures have not been adequately instituted in Japan.[3] There have, however, been developments towards instituting them – albeit sometimes very slow and with occasional setbacks. The aim of this chapter is to trace these developments.

Developments in Industrial Safety Prior to the End of WWII

The industrial disasters recorded in Japan prior to 1850 occurred in mines and metal refineries. Since the Meiji period, the railway construction industry, light industries and the transport industry join this list. As in all developed nations, industrial development in Japan was accompanied by industrial accidents and environmental pollution. The more developed nations had learned from their experiences while solving a variety of problems. With Japan's industrial development being guided by the government's devotion to catching up with more developed nations – and quickly – the idea of 'high bureaucrats and low commoners was hardly challenged in Japanese society. Certain elements reminiscent of traditional societies were preserved in the process of transition. Thus, the practice of 'local empire building' where bureaucrats attempt to acquire more authority and larger budgets, and therefore vigorously defend any reorganization that would reduce their 'empire' – a principle characteristic of feudal Japanese society – was fully translated into the principle of *tatewari gyōsei* (vertically partitioned administration in which ministries compete with each other to optimize their spheres of regulation and influence without inter-ministerial communication and coordination, while partitioning their activities along ministerial lines). Jurisdictional sectionalism of this kind can still be found in various government ministries today.

The construction of a Western-style navy, which began under the direct management system of the Bakufu government, and the construction of steel warship building facilities (including steel mills) were both successfully continued by the new Meiji government. The successful experience of these enterprises was taken over by the nationalized enterprise of the Meiji government as early as 1868 to assume the management of the Ikuno mine and other mines under Bakufu management. These experiences also influenced the decision of the government to commence the construction of numerous government-managed factories; the Ministry of Engineering was established in 1870 for this purpose (jurisdiction over mines, steel manufacturing, railways, lighthouses and telegraphs was transferred from the Civil Sector Ministry). The Ministry of Home Affairs was established in 1873 and was given responsibility for non-government-managed general industry as well as regional administration and the police. When the Ministry of Agriculture and Commercial Affairs was established in 1881 it assumed responsibility for industrial administration and the Ministry of Home Affairs' portfolio was reduced to local administration and police. With the collapse of the feudal clan structure of the Bakufu era, the unemployed descendants of samurai and poor peasants joined the labor pool that was required for this government-driven industrial development program.

As Japan's modernization and industrialization progressed, with ups and downs along the way, the emergence of tragic groups of workers similar to those in various developed nations became a social issue. In response, pioneers from the government and private industries started The Factory Law Enactment Movement. Industrialists, however, strongly opposed the enactment of the Factory Law and heated conflict arose between the Ministry of Home Affairs and the Ministry of Agriculture and Commercial Affairs over jurisdiction of the law. The law finally passed the Diet in 1911 and went into effect in 1916 under the Ministry of Agriculture and Commercial Affairs. Paralleling this development, the Mine Law was enacted, creating the role of industrial safety specialists. This resulted in the appointment of inspectors to the Factory Section and the Mine Office of the Ministry of Agriculture and Commercial Affairs, and factory inspectors in eight different prefectures of the Police Section. Unlike the situation in Europe and America where industrial safety inspections were demanded and implemented by neutral third parties, the Japanese safety specialists acted on behalf of and in the interests of the government bodies.

The flood of unemployed people that resulted from the depression and disarmament following World War I made unification of labor administrations an urgent topic for consideration. The Katō Cabinet planned to establish the Social Agency (Shakaiin) to promote the unification of labor administration and to liaise with the International Labor Organization (ILO). A dispute over who would control the Agency raged between the Ministry of Home Affairs and the Ministry of Agriculture and Commercial Affairs. In 1922, almost all of the labor administration of the various ministries were united in the Social Bureau (Shakaikyoku) that was established as an external office of the Ministry of Home Affairs. However, matters relating to 'the protection of life and hygiene' for mine workers remained the responsibility of the Ministry of Agriculture and Commercial Affairs. This disintegration in the administration of industrial safety survived the reforms introduced by GHQ during the Occupation.

Several government ministries such as the Ministry of Commerce and Industry and the Ministry of Communications, had, for example, bureaus for industrial safety to cover domains that were not covered by the Ministry of Labor. The division of labor administration along ministerial lines still persists today and is often referred to as a 'vertically partitioned administration system.' In Western nations, the structure of the industrial safety system evolved such that private insurance companies demanded safety certification issued by private ('third party') safety inspection companies. In Japan, though, the entire process operates as a part of the control administration of the central government and some prefectural offices and thus serves to reinforce the authority of particular government offices. Fire protection and fire fighting function in the same way.

As the war between Japan and China progressed towards World War II, a shortage of labor developed and control reinforcement was emphasized. In 1937, the labor unions were ordered to disband and in 1938 the Ministry of Public Welfare was established (industrial safety and hygiene were controlled by a labor office, separate from the Ministry) and in 1940 the Japan Industry Serve-the-Country Association (Dainihon Sangyō Hōkokukai) was established for the national integration of labor. In 1943, the Munitions Ministry (Gunjushō) was established to control eight different categories of employees in munitions-related fields such as aviation, machinery and steel manufacturing. Responsibility for labor administration was split between the Ministry of Public Welfare and the Munitions Ministry until the end of the War.

As the War became more entrenched the number of workplace related disasters increased dramatically. There are no official statistics recording the number or type of workplace accidents between 1940 and 1957 but there can be little doubt that human lives were lost futilely because of an attitude that 'Soldiers are dying everyday on the front: it should be honorable for workmen and volunteer members to die in a factory.'[4] From 1928, however, a National Safety Week has been in place every year even during the war as an attempt to maintain a bare minimum level of safety.

Industrial Safety During the Occupation

There are no official statistics on the workplace accidents that occurred in the first half of the Occupation and, thus it is not easy to develop an accurate picture of the industrial safety situation at the time.[5]

The Early Occupation Years

From August 1945 to September 1947 (from the end of the war to the establishment of the Ministry of Labor), the Occupation of Japan was formally overseen by the General Headquarters of the Supreme Commander for the Allied Powers (GHQ). While the Potsdam Declaration – which declared freedom of speech, religion, thought and a respect for basic human rights – and the Far Eastern Commission – constituted by eleven different countries – had some degree of influence in determining the nature of the Occupation of Japan, in fact it was essentially the US and the US alone that occupied Japan. With the exception of Okinawa, where direct military governance was implemented, the Occupation Forces employed an indirect form of rule via the Japanese government.[6] However freedom of expression was directly curtailed via the press code (cf. ch. 24 and 27).[7]

The Occupation Forces required bases and housing, offices and the provision of services for six hundred thousand troops.[8] The cost of supplying these facilities and maintaining the railways, ships, communication and electricity services for the Occupation Forces' usage was borne totally by the Japanese government and became an important industry at a time when other industries were in chaos.[9]

At the time, domestic coal and hydroelectric power were the only energy resources available. Although the hydroelectric generating facilities were not damaged by air raids, thirty percent of steam power plants were destroyed and the remainder of the plants were rendered inefficient by a shortage of fuel and the poor quality of the fuel that was available. In the cities, power failures were a constant problem because of the destruction of the power line distribution network by air raids. However, nationally there was surplus capacity to generate power (cf. ch. 17). Indeed other ways of utilizing power, such as electric salt manufacturing, were seriously considered. Thus, at a Cabinet meeting in December 1946, the Yoshida Cabinet decided on prioritized production, emphasizing the role to be played in an economic recovery by coal and steel.

At the beginning of this period the Occupation Forces pursued policies for the democratization of Japan in accordance with the idealism of the American 'New Deal.' Thus, labor unions and the Industrial Safety Movement were revived and in 1946 the nineteenth National Safety Week was held as an initiative of the Ministry of Public Welfare. In 1947 various epoch-making regulations were effected, including the Labor Standards Law and the Work Safety and Hygiene Standards.[10] The Ministry of Labor was also established at this time, acquiring responsibility for the administration of industrial safety from the Ministry of Public Welfare. Approximately three hundred and twenty-four Labor Standard Management Offices were set up around the country soon thereafter.

During this period there were some bizarre accidents recorded. For example on 7 October 1945, the Muroto Maru, a passenger boat, struck a mine and sank with a death toll of 236; on 12 November 1945 an explosion in the Futamata Tunnel resulted in 147 dead and 149 injured. There were also a number of railway accidents – e.g., on 25 February 1947, the Hachikō line derailment left 184 dead and 497 injured – and several coal mine incidents.

The Middle Occupation Period: September 1947 to June 1950

(From the establishment of the Ministry of Labor to the outbreak of the Korean War).

In 1947, the February First nationwide general strikes were banned by GHQ. The Potsdam Declaration was effectively annulled in stages by the Red Purge in 1949 and the explosive declarations by US Senator Joseph McCarthy as anti-communist sentiment was growing. But the Japanese people, who had only recently seen democracy sprout forth and spread its roots at a tremendous rate defended the Potsdam Declaration. Their democracy, however, was still in its infancy and the authority of the Occupation Forces still strong. The power of the bureaucrats increased exponentially as Occupation policy changed: demilitarization was de-emphasized, reparations were cancelled, the reconstruction of industry was picking up steam and the conservatives exploited the commodities that had been stockpiled by the Army and the Navy as war reserves.[11]

In this environment, advocates for the unification of safety administration had very little influence in either the financial, political, labor or academic spheres. Although the Labor Standards Law was intended to unify labor administration, in December 1948 responsibility for the administration of mining safety was transferred from the Ministry of Labor to the Ministry of Commerce and Industry as a result of a Cabinet decision under pressure accompanying the policy to adopt priority production. In every ministry the principle of defending sectionalism was revived.

In parallel with this trend of weakening work safety regulations and practices, the first article of the Factory Standardization Law – 'The Purpose of the Law' – which was proclaimed in June 1949, did not mention the term 'safety' at all, although reference was made to the 'promotion of public welfare.' However, the second article of this law specifies 'definition,' 'safety levels,' 'examination, analysis, judgment, inspection and the means for approval and measurement' as criteria for standardization. Basic matters in the certification, for example, 'the kind of procedure that should be used to identify needs and who should carry out this identification' were absent until the Japan Institute of Standards (JIS) Standard Z9900 was translated from the International Standard Organization (ISO) and was adopted in 1990. It is indicative of Japan's safety system that responsibility for safety administration lies with individual ministries to supervise in accordance with compulsory controlling regulations, without unification.

Nevertheless, the reforms implemented during the Occupation – establishing principles and standards for operating safety administration with a basis in law, and making them public – represented huge progress. After the Labor Standards Law was proclaimed, the Fire-fighting Law (1948) and the Mining Security Law (1949) were enacted. Other laws such as the Gas Enterprise Law, the Electricity Enterprise Law and the Electric Products Regulation Law all came about after the Occupation. Until they were enacted, however, these matters were governed by ministerial ordinance from the different ministries concerned.

In February 1949 the Ministry of Labor announced its intent to strengthen and reinforce the Labor Criteria Supervision in response to an increase in slavery-like human trade involving juveniles in the Tōhoku region and the north of the Kantō district. In January 1950 the postwar reforms began to be effective. For example, the Ministry of Labor publicly identified enterprises with high accident rates (in the first listing there were 643 enterprises named), promoted special safety management guidance and established several associations for promoting industrial safety.

The first huge accident involving a manufacturer in postwar Japan was a gas explosion at the Shōwa Denkō factory in Kawasaki on 24 June 1949 (17 dead and 63 injured). There was also a rear-end train collision at Hanazono station on the Kintetsu Nara line on 31 March 1948 (49 dead and 272 injured) and other railway accidents such as the Shimoyama incident, the Mitaka incident and the Matsukawa incident which occurred in July and August 1949. And, as usual, there were more coal mine disasters.

The Final Occupation Period: June 1950 to April 1952

(From the outbreak of the Korean War to the dissolution of GHQ – via the San Francisco Peace Treaty and the Japan-America Security Treaty).

With the outbreak of the Korean War, Japan became a forward base for the United Nations. The US was the primary force involved. Immediately prior to this, the Occupation Forces had numbered 100,000, this being reduced to 20,000 because of the Korean War. In July 1950 GHQ ordered Japan to establish a reserve police party of some 75,000[12] to replace the 80,000 troops now gone.

Shortly before the outbreak of the Korean War, Japan's economy had been in a state of 'stable panic' thanks to what is referred to as the 'Dodge Line.' But the United Nations' forces purchased vast quantities of local 'special procurements' from Japan for the Korean War mobilization, thereby stimulating Japan's economy; exports increased rapidly. In 1951 the mining sector's output virtually returned to its prewar level (1934–1936 average), forming the basis for future development.

At the same time, the industrial safety movement began to flourish, accompanied and aided by the use of statistics to quantify levels of safety. Statistical methods adopted included, for example, the Frequency Rate (the Disaster Occurrence Rate defined by the ILO; it measures the number of dead and injured per 1,000,000 working hours), the Intensity Rate (measuring the number of working days lost due to disabilities per 1000 working hours. Deaths on a per capita basis were calculated as occurring every 7500 days).

Around 1950 the Frequency Rate in many factories exceeded 100. Six steel companies with an average Frequency Rate of 65.35 organized a Safety Liaison Conference in 1950. Five years later the average rate for these six companies had decreased by ten percent.[13] When a safety system was organized by Hitachi Shipbuilding, the average Frequency Rate over five work sites decreased rapidly; from 157 in 1950 to 125 in 1951, 57 in 1952 and 11 in 1953.[14]

The railways had recorded approximately 500 work related deaths per annum until about 1948, with the number of injured numbering around 25,000. In 1953 the death toll was less than 200 and in 1963 it was under 100. The Frequency Rate also decreased rapidly: from 12.14 in 1951 to 4.67 in 1960 and 1.63 in 1965.[15]

In 1950 the Mining Law was proclaimed and in 1951 the Road Vehicle Transportation Law and the High Pressure Gas Regulation Law were affected.

Summary of the Occupation Period

For those Japanese citizens who had survived the rain of incendiary bombs and other wartime horrors the end of the war found them just one step away from death's door. Soon after the war, unemployed people flooded into the cities. Those 'fortunate' enough to be employed had to work under conditions where the Frequency Rate for workplace accidents was in excess of 100; in fact, the life expectancy of a worker was about three to four years. Under these circumstances an awareness of human rights and democracy grew and, as industry started to get back onto its feet, industrial safety became an item on the social agenda. Compared to the prewar years, this period saw tremendous progress made towards establishing a system where safety was regulated according to specified standards. However safety administration remained locked in a system of bureaucratic leadership with a vertical-orientation and responsibility dissipated across numerous government ministries. It would be 1994 before the Japanese industrial safety certification system approximated the long-established system in other developed nations where safety-certification was awarded by independent third-party mechanisms.

Notes

1 Yasuda Sōgō Kenkyūsho (Yasuda Research Institute), 1989 [4] p. 117.
2 Ishigai, 1985 [2] p. 11.
3 Ishigai, 1985 [2]; Ishigai, 1977 [3]; Temporary Administration Investigation Committee Secretariat, 1983 [6]
4 Chūō Rōdō Saigai Bōshi Kyōkai (Central Work Disaster Prevention Association), 1971 [8] p. 245.
5 Takemae, 1980 [10]; Sato, 1991 [9]
6 Yui, 1989 [11] p. 207; Masamura, 1990 [12] p. 45.
7 Satō, 1991 [9] pp. 71–82; Kiyoko Horiba, 'Puresu kōdo ka no genbaku hōdō (The atomic bomb reports under the press code),' in *Tosho*, no. 494, 1990, p. 8.
8 Satō, 1991 [9] p. 69.
9 The 1946 national budget totaled 110 billion yen, of which 68.2 billion yen was spent on postwar management, including the costs associated with stationing the Occupation Forces in Japan. This was one of the causes of inflation at the time. Satō, 1991 [9]
10 The original plan of the Labor Standards Law was drawn up primarily by Hogo-ka (the Public Assistance Division) of the Ministry of Health and Welfare, not by the US Occupation Forces. In 1950, Dr D. E. Walter came to Japan as a GHQ safety adviser. Before leaving the country he submitted a report to GHQ recommending: 'In order to avoid duplications of laws and competition between ministries, Safety Hygiene Supervisors should be separated from other ministries and be put under the control of the Labor Standards Office in the Ministry of Labor.' But this was too late to be committed to writing. However, we should note

that GHQ indicated that a Safety and Hygiene Department should not be established in the ministries that were primarily concerned with industrial productivity. Japan's vertical safety administration was not affected. The typical industrialists' views on government regulation of workplace safety is summed up in the following analogy: 'I wish to be able to drive a car freely, so I would like you to leave the questions of acceleration and braking to the driver.' The history of industrialization clearly demonstrates, though, that when workplace safety is left to the prerogative of the 'driver,' workers invariably suffer. Japan is certainly no exception. The wartime experience indicates that under certain circumstances industrialist 'drivers' will drive even more recklessly – that is, under circumstances when they can get away with it. Yet, when priority production was demanded, this analogy was again presented to support a case for a 'hands-off' government policy, demonstrating the level of (im)maturity existing in Japan's democracy at the time. Chūō Rōdō Saigai Bōshi Kyōkai (Central Work Disaster Prevention Association), 1971 [8] pp. 250–83.

11 The stockpiled commodities were intended to sustain the war effort through a battle of two and a half years duration. At the time, these commodities were commonly referred to as 'secret hoarded commodities.' Any clarification or disclosures of them was suppressed and efforts to expose these commodities by the democracy movement were weak. Satō, 1991 [9] pp. 92–3.

12 Satō, 1991 [9] p. 70.

13 Chūō Rōdō Saigai Bōshi Kyōkai (Central Work Disaster Prevention Association), 1971 [8] pp. 298–301.

14 Chūō Rōdō Saigai Bōshi Kyōkai [8] pp. 301–5.

15 Chūō Rōdō Saigai Bōshi Kyōkai [8] pp. 312–14.

Bibliographic Guide

The General Principles of Industrial Safety

[1] Kenzō Yano, *Kōjō saigai fujo-ron* (Factory disaster support theory), Sanseido, 1931.
Explains the principles of industrial safety, and the Factory Law (Chapter 8). Selections from the Factory Law are included in an appendix. 'The Responsibilities of the Factory Owner' (in the prefatory remarks to Chapter 3) clarifies the changes to laws in response to industrial disasters since Japan's industrial revolution. 'Grounds for Responsibility' presents industrialists' arguments against the laws and discusses irrational points.

[2] Seikan Ishigai, 'Anzen gijutsu no shohōsoku to kihon mondai (Various laws and basic problems concerning safety techniques),' in Nihon Kikai

Gakkai (Japan Machinery Society) (ed.), *Gijutsu no kokoro* (The heart of technology), vol. 2, Maruzen, 1985.
[3] Seikan Ishigai, *Kōgaku gairon* (Zōhoban) (Introduction to technology, [Enlarged edition]), Corona-sha, 1977.
Presents the background to industrial safety principles; deals with the balance between cost and safety; discusses the fact that although safety and financial interests often work against each other, society has to find ways to make them compatible. Also discusses the causes involved, centering on the question of human nature.
[4] Yasuda Sōgō Kenkyūsho (Yasuda Research Institute), *Seizōbutsu sekinin* (Products liability), Yūhikaku, 1989.
Begins with the laws of product liability and specifically explains contemporary business insurance in the global economy. A variety of judicial precedents are appended.
[5] Shinsaku Onodera, *Kensa: Seizōbutsu no sōgō-teki gijutsu tenkai* (Inspection: General technological development of products inspection), Nihon Kikaku Kyōkai (Japan Standards Association), 1989.
Onodera has a wealth of experience both in workplaces and at international conferences. He is a cosmopolitan figure who is well acquainted with the standards of Third Party Inspection Institutions such as ASME, TUEV and Lloyd's in various countries around the world. He is also familiar with the criteria and the operation of these institutions. He provides an academic discussion of what product inspection should entail. In 1983, the Second Temporary Administration Investigation Committee presented a written report of the Rinchō' final proposal [6] to the government and recommended that the establishment of an international authoritative institution (from here on referred to as the Third Party Inspection Mechanism) which would carry out third party-like inspections independent from the administration institution, was necessary. Since then the level of awareness of its importance has not been heightened and each ministry retains its inflexible attitude to vested interests. The search for a means of changing their attitude in order to establish such an institution continues. Onodera suggests that clarification of the correlation between industrial safety, public welfare and product inspection is necessary. The book is a valuable academic inquiry originating from the workplace perspective.
[6] Temporary Administration Investigation Committee Secretariat, *Rinchō saishū teigen* (Final proposal of the Temporary Administration Investigation Committee), Gyōsei Kanri Kenkyū Sentaa (Administration Management Research Center), 1983.

Developments in Industrial Safety Prior to the End of World War II

[7] Yutaka Miura, *Rōdō to kenkō no rekishi* (A history of work and health), Vol.1

– 4, Rōdō Kagaku Kenkyūsho (Research Institute of Labor Science), 1978.

[8] Chūō Rōdō Saigai Bōshi Kyōkai (Central Work Disaster Prevention Association), *Nihon no rōdō anzen eisei undō: Gojūnen no kaiko to tenbō* (Japan's safety and hygiene movement: Reminiscing on fifty years and future prospects), 1971.
The chapter entitled 'Fukoku kyōhei, shokusan kōgyō (The national prosperity and military strength, increased industrial enterprise)' in the part on 'Zenshi: Kōjō-hō ga dekiru made (Pre-history: The process of establishing the factory law)' concludes as follows: 'The youthful bureaucrats in the Meiji government carried out policies with diligence, organization and an enterprising spirit not only in economic policies, but also in the fields of education, culture and administration. The details of the establishment of the Factory Law subsequently prove this fact. This kind of "national leadership style" for initiating matters was to be seen in other areas later on, but it was already shaped by means of the capitalization which proceeded from the higher echelons.'

Industrial Safety during the Occupation

[9] Kōji Satō, *Bei gunsei kanri to heiwa undō* (The US military administration and peace movements), Seseragi Shuppan, 1991.
The first edition of this work, produced in 1986, was not released for sale, but the enlarged third edition was in 1991. The original version of the book was entitled *Tachiagaru kichi Keihanshin* (Military bases that stand up in the Kyoto-Osaka-Kobe area) and published in April 1954. The process by which this book was published as a not-for-sale item in 1986 is described in the 'Foreword' of the enlarged edition. This work provides a valuable lesson in historical research during a period when publications were suppressed. Recently publicized Japanese and US substantiated research papers on documents relating to the Occupation are valuable, but the book is also a valuable item of literature and should be read as such.

[10] Eiji Takemae, *Senryō sengo-shi: Nihon kanri seisaku no zen'yō* (A history of the Occupation after the War: A total picture of the management policies for Japan), 1980.
This is the first book in a series of research papers on the postwar history by the author.

[11] Daisaburō Yui, *Mikan no senryō kaikaku* (Unfinished Occupation reforms), University of Tokyo Press, 1989.
Concerned with the international research organization, Taiheiyō Mondai Chōsakai (Institute of Pacific Relations: IPR) activities, primarily in the US. Investigates the processes that were left unfinished as a result of McCarthy's stormy 'Red Hunting.' In the early stages of the period the results of the research influenced the military policies of the Occupation Forces.

[12] Kimihiro Masamura, *Sengo-shi* (Post-war history), vol. 1, Chikuma Bunko, 1990.
Describes Japan's development in the context of trends in the world economy and includes a brief chronological table.

Part IV

Scientists and Engineers in the Postwar Democracy

24 Reporting on the Atomic Bomb and the Press Code

Yukuo Sasamoto

From the Dropping of the Atomic Bomb Until Defeat in the War – The Atomic Attack Propaganda Counter-strategy

One cannot ignore the strict military and governmental controls on reporting and 'free speech' when examining the way the atomic bomb was reported during the war. Reports relating to air raids were subject to censorship. According to the Department of Home Affairs' 'Essentials of the Reporting Measures in Relation to Foreign Attacks,' reporting on damage to such things as telegraphs, telephones, railways, roads, bridges, harbors, electricity, gas and water supplies was prohibited. Restrictions included: 'a) locating damage any more specifically than the town or village name; b) reporting any details relating to the loss of life (human or livestock) including the numbers of dead and injured; and c) reporting details of the damage sustained by houses and other structures. In other words, reporting on almost any details of air raid damage was prohibited.[1] To this end the 10 March 1945 blanket air raids on Tokyo were reported in the *Asahi Shimbun* of 11 March as follows: 'The Imperial Headquarters reports (10 March, 1945, noon) that from about midnight for two hours and 40 minutes a force comprised of 130 mainly B29s attacked the Imperial capital with indiscriminate bombing causing fires across the city. The Imperial Household Department Equerry Dormitory put fires out at 2:35 and again at about 8 o'clock. At present, the achievements of the military are: 15 planes shot down and approximately 50 damaged.' The fact that 100,000 people lost their lives in this air raid was not reported.

At 8:15 am on 6 August 1945 an American B29 bomber, the 'Enola Gay', dropped a uranium type atomic bomb on the city of Hiroshima. After the atomic attack, the United States President, Harry S. Truman, made a public announcement:

> Sixteen hours ago an American airplane dropped one bomb on Hiroshima, an important Japanese army base...It is an atomic bomb. It is a harnessing of the basic power of the universe...Let there be no mistake, we shall

> completely destroy their capacity to make war. It was to spare the Japanese from utter destruction that the ultimatum of July 26 was issued at Potsdam. Their leaders promptly rejected that ultimatum. If they do not now accept our terms, they may expect a rain of ruin from the air, the like of which has never been seen on this earth.[2]

This announcement was made by short-wave radio transmission to the whole world and was monitored by the Japanese. The monitoring of overseas transmissions was one of the wartime information gathering strategies the military and government employed.[3]

The first newspaper report concerning the atomic attack (*Asahi Shimbun*, 7 August) said: 'Hiroshima Explodes in Fire. On 6 August at approximately 7:50 am two B29 aircraft invaded the skies over Hiroshima carrying an incineration bomb, delivering some damage to that city and its environs (Osaka).' It was some 31 hours after the attack on Hiroshima – on 7 August at 3:30 in the afternoon – that the Imperial Headquarters announced it to the public. That report stated: 'Imperial Headquarters Announcement (3:30 pm 7 August 1945). 1: On 6 August Hiroshima was subjected to considerable damage as a result of an attack by a small number of enemy B29s. 2: The enemy used a new type of bomb in the attack. Preliminary investigations are currently being conducted to ascertain details.' (*Asahi Shimbun*, 8 August). These reports, appearing immediately after the event, were typical of the reporting on air raids in general.

The fact that the Imperial Headquarters Report used the term 'new type of bomb' is worthy of attention. US President Truman had already announced to the whole world that it was an 'atomic bomb.' It would have been reasonable for the Japanese Imperial headquarters to also have used this term, but it appears that the debate over this issue between the military and the government was decided in the negative.

On 7 August, the day after having suffered the atomic attack, the Japanese set up an Atomic Bomb Committee comprised of the Army and Naval forces plus the Government. On this same day the division heads of the Information Bureau met to discuss how to maximize the propaganda value of the atomic bomb attack.' Their discussions were held centering on the question of counter-measures that might be employed in terms of propaganda reporting relating to the atomic attack. According to Tatsuo Hisatomi, the Deputy Chief of the Information Bureau, the following decisions were taken at this meeting: '1: they would appeal to the public opinion of the world through a world-wide publicity campaign that reported the inhumanity of the atomic weapons deployed; 2: announce domestically that it was an atomic bomb, and 3: deliver a new awareness of the achievements of the war to the populace. To this end they set out to "investigate the true facts" while instigating "instant reporting of facts."'

Although the Foreign Ministry supported this policy, it met opposition from the military and the Home Ministry. The military's stated reasons for opposing it were '1: The enemy announced that it was an atomic bomb but this could have been a fabricated propaganda ploy. We cannot immediately conclude that it was an atomic bomb without first seeing proof of sufficiently scientific investigations; and 2: On the grounds of war morale, we are opposed to risking such a strong shock to the psychology of the people as might result from an announcement of such gravity.' In the end the government decided not to use the term 'atomic bomb' in the newspapers or radio broadcasts and the people were not informed about the atomic bomb until after the war. The press simply reported a 'new type of bomb' following the lead of the announcement from Imperial Headquarters.[4]

It is important to note that this debate between the military and the government was limited to the issues surrounding the second point of the Information Bureau Chief's conference decision. Hisatomi's account reveals a total absence of debate concerning the implementation of an external propaganda counter-strategy (the first point). This counter-strategy was implemented unadulterated; the government set out on a propaganda campaign whose aim was to appeal to the international community about the inhumanity of the atomic bomb. At the same time, the government prohibited the use of the term 'atomic bomb' by domestic newspapers and broadcasters. In other words, the Japanese Imperial Government maintained strict censorship regulations prohibiting the dissemination of any knowledge about the atomic bomb to the Japanese public whilst simultaneously conducting an international publicity campaign to ensure the widest possible dissemination of this information to the international community. This is an important point to remember when examining the Japanese reporting of the atomic bomb.

The Information Bureau's policy of 'investigating the true facts' has already been discussed in chapter 2 ('Atomic Bomb Damage Investigations'). The crucial point here is that the atomic bomb investigations carried out by the Japanese were directed by a national policy intended to maximize the propaganda value of the atomic bomb damage.

Characteristics of the Newspaper Reporting on the 'New Type of Bomb'

Newspaper reports about the 'new type of weapon' following the Imperial Headquarters' announcement of 7 August, display several distinctive characteristics. One of these is the reporting of the instructions issued by the Air Defense General Headquarters of the Ministry of Home Affairs regarding defensive strategies to be used against the new type of weapon. This can be seen in newspaper articles such as 'There is a danger of burns.

Be certain to take refuge in an air raid shelter. This is most important' (*Asahi Shimbun*, 9 August). This reporting of defense strategies was aimed at keeping up the people's will to fight and persevere with the war and continued until the Japanese surrendered on 15 August 1945.

Another discernible characteristic of the reporting is the repeated emphasis on the inhumanity of the 'new type of weapon' in accordance with the propaganda policy of the Information Bureau. We see this in headlines such as 'The New Type of Inhumane Bomb Floats Down by Parachute and Explodes in Mid-air, Having No Regard for Humanity' which appeared immediately after the attack on Hiroshima (*Asahi Shimbun*, 8 August).

On 10 August the government announced to the Allied Powers, through the Swiss and Swedish governments, that it accepted the Potsdam Declaration with conditions. At the same time Radio Tokyo and the Dōmei News Agency broadcast the declaration of acceptance overseas. They did not, however, report it domestically.[5] The American and British presses reported the Imperial Government's surrender in a big way, with headlines such as 'Japan Offer to Surrender; U.S. may let Emperor Remain; Master Reconversion Plan Set' (*The New York Times*, 11 August). They reported that people the world over were delighted with Japan's defeat. Certainly, the conditional acceptance of the Potsdam Declaration should, for all intents and purposes, have been reported to the Japanese people.

Also on 10 August, with the Swiss Government acting as an intermediary, the Japanese Imperial Government submitted a letter of protest to the US Government claiming the 'new type of bomb' was a cruel and inhumane weapon that showed no regard for international law. The same message was sent to the International Committee of the Red Cross.[6] In other words, at the same time as announcing the conditions for their acceptance of the Potsdam Declaration, the government was protesting the inhumanity of the atomic bomb.

It is not clear whether or not, hitherto, protests about the atomic bomb had been broadcast overseas, but records in Melbourne, Australia indicate that when, at midnight on the morning of 11 August, Radio Tokyo broadcast the conditional acceptance of the Potsdam Declaration to North America, it

> criticized the dropping of the new type of bomb on Hiroshima as "a crime which should not be perpetrated on humanity" and reported that through the intermediary of the Swiss government, the Japanese government submitted a letter of complaint stating: "the new type of bomb caused a very great number of civilian casualties and deaths in a provincial city which had no military significance".[7]

It is clear from this record that the Japanese announcement of the acceptance of the Potsdam Declaration and the protest over the atomic bomb took place as one and the same exercise – a single entity as it were. The propaganda policy of appealing to the world about the inhumanity of the atomic bomb was very closely aligned to the Japanese surrender. The government did not subsequently lodge a formal complaint over the atomic bomb. It can be surmised that the government did not use the term 'atomic bomb' in its protest out of consideration for the United States after the Occupation. The only record of this protest in the US Government records is a reference to the receipt of a protest memorandum made in a Department of State reply on 24 October 1945, after defeat in the war. We do not know what sort of counter-strategy was devised.[8]

Domestic newspapers reported this protest thus: 'Imperial Japan Submits a Protest to the American Government Over Cruel New Type of Bomb which Ignores International Law' (*Asahi Shimbun*, 11 August). Following that the newspaper reporting on the inhumanity of the 'new type bomb' continued: 'Atomic Bomb "Deals a Backhander" to Civilization and Humanity, Japanese Government Soon to Announce Censure, talking with Mr Kiyose' (*Asahi Shimbun*, 14 August). There are other similar articles, but in this one the term 'atomic bomb' is used. However, even the reports that painted the atomic bomb as an inhumane bomb do not reveal the force it unleashed or the damage wrought. To put it more simply, there were no articles about the casualties of the atomic bomb.

On this point it appears that the prohibitions on reporting the human and livestock casualties of air raids – the numbers dead and injured – were respected. An exception to this is the wording of the protest to the United States mentioned earlier: '...damage was widespread, people in the above mentioned area, irrespective of whether they were combatants or otherwise and without distinction between male and female, young and old, were killed indiscriminately by the wind blast and the radiation rays.'

There is something of great interest relating to the atomic bomb reporting prior to the end of the war. The *Yomiuri* report of 13 August incorporated an atomic bomb campaign feature over two pages. It stated '"Atomic Bomb" Enemy of the Entire Human Race: British Magazine Condemns the Potsdam Decision,' 'The Atomic Bomb Uses Astronomical Explosive Force. The Atom's Destructive Energy is a Tyrant Which Invites Destruction of the Human Race' and '"Can it be an Atomic Bomb?" We ask Mr Osamu Yazaki about the Supremacy-Craving American Research. Certain Doubts about the Hiroshima Bombing.' Clearly this article violated the government's prohibition on the use of the term 'atomic bomb.' The government ordered a ban on the distribution of that day's *Yomiuri* report, but by the time the order was given all the papers had already been delivered.[9] As we will see, this *Yomiuri* report was the forerunner of other features reporting on the atomic bomb and similar matters.

From the end of the war to the beginning of the Occupation 'the gathering of information and reporting on it were conducted "freely"'

On 15 August 1945 Emperor Hirohito made a recorded broadcast about the termination of the war. The *Asahi Shimbun* of 15 August reported 'An Imperial Rescript Proclamation on the Imperial Will Concerning the Horrors of the New Bomb. The Empire Accepts the Four Country Declaration Respectfully Paves the Way for Peace for Eternity.' The Imperial Edict explained that Emperor Hirohito's decision to terminate the war was determined by the two explosions of the 'new type of bomb.'

It is surprising how little is known about the special guidance the Information Bureau gave regarding the production of the 15 August paper. I will touch upon this briefly in relation to the post-surrender reporting of the atomic bomb. At 5 pm on 14 August the Information Bureau sent each newspaper house a copy of the 'Public Opinion Guidance Policy in Accordance with the Great East Asian War Termination Negotiations.' This policy instructed them to not venture into the area of the negotiations and ordered them to work to: maintain the national polity; promote 'Harmony and Unity Between Sovereign and Subject,' and; emphasize the difficulties faced by the nation and the fact that the responsibility for these difficulties was shared by all of the people of the nation. Furthermore, it informed that arguments in favor of a continuation of the war or criticism of leaders of the army or the government were subject to strict controls.[10]

Newspapers reporting on the Imperial Proclamation about the termination of the war were ordered to be distributed only after Emperor Hirohito's broadcast, although they were printed and ready to go before that. The instruction from the Department of Home Affairs Police Bureau notifying the Iwate Prefectural Police Chief of this order said: 'An announcement about the Imperial proclamation will be made today. There will be a major radio broadcast at noon tomorrow. It is intended not to bring this matter to the attention of readers before that. Deliver the newspaper carrying the article about this tomorrow, 15 August after 1 pm. These instructions [missing] should be sent to the Chief of the Special Secret Police, the newspaper chief-of-staff and the Censorship Director. 10 pm receipt.'[11] Everyone at the newspaper company, from the editors to the printing staff, knew of the surrender. But this news was not delivered to the people ahead of the noon broadcast. This fact very clearly demonstrates one of the characteristics of the reporting of Japan's defeat in the war, i.e., that the release of information was always controlled in the interest of power politics – or to put it another way: that the people's 'right to know' was always subordinate to the Government's need to control.

Furthermore, on 16 August the Information Bureau placed new controls on all newspaper reporting, with the public opinion guidance policy entitled 'An Outline of the Information Bureau Liaison Conference.' This policy statement provided more detailed guidance than the earlier directive. It stated:

> 1 On the policy of guiding public opinion involving the termination of the Greater East Asian War…;
> 2 It will be necessary to censor all foreign telegrams in accordance with the policy of adopting prudence in relation to them…;
> 3 As for articles relating to the resignation of all Cabinet members, everything other than those announcements emanating from this office, or those which carry our approval, will be subject to censorship…;
> 4 As for the handling of future articles, objections will not be raised on minor matters but with regards to handling matters where rigorous attention is required, take heed that very firm prohibitive measures will be put in place.'[12]

As for the post-declaration reporting of the atomic bomb, Tatsuo Hisatomi, said 'After the war, the Information Bureau, in collaboration with the Department of Home Affairs, put all its effort into drafting announcements which described the reality of the damage in Hiroshima and Nagasaki, in as much detail as possible and as quickly as possible. The Dept. of Home Affairs authorities requested that this be put on hold until the swearing in of the new Cabinet Ministers. That policy was partially implemented, but in the end the press agency was granted freedom to gather information and report on the atomic bomb damage.'[13] In accordance with the propaganda objectives of the Information Bureau, an 'instant reporting of facts' immediately followed the atomic blasts. We should note, however, that after the surrender, 'the reality of the destruction in Hiroshima and Nagasaki' was reported. This point indicates that the atomic bomb propaganda counter-strategy changed after the war. The policy differentiation between 'domestic and overseas announcements,' however, remained unchanged. The fact that the 'press agencies were allowed to freely gather information and report on it' in reality meant that the responsibilities of the Information Bureau could now be laid at the feet of the press agencies.

At the same time a strengthening of newspaper censorship immediately following the war is indicated by hints about banning publications. The Imperial Government's policy for reporting on the atomic bomb was ambiguous, allowing the 'free' collection of information on the damage caused by the atomic bomb while at the same time issuing a directive calling for the delayed distribution of the 15 August edition of the newspaper. I must stress again, however, that the right to 'freely' collect information and report on it was granted for the purposes of pursuing the international propaganda campaign rather than for domestic dissemination. This point has been left unclear in discussions about postwar atomic bomb reporting, thereby strengthening the idea that during the period of the power vacuum – before the reporting and discussion controls of the Occupation Forces were instigated – reporting on the atomic bomb was literally 'free'.[14]

An examination of the 'external' propaganda more bluntly reveals the message behind the Information Bureau's allowing for the 'free' collection and reporting of news about the atomic bomb. However, the records of the Information Bureau, the Army and Government do not reveal any 'external' propaganda other than the protests directed towards the US Government relating to the 'new type of weapon.' To rectify this, I will now examine records of Japanese external propaganda intercepted in places such as the United States and Australia.[15]

Immediately after the atomic bomb attack on Hiroshima the Japanese, via Radio Tokyo and Dōmei News Agency, instigated an information war.[16] We have already seen one facet of this in the item on 10 August, but in summary, the atomic bomb information war I would like to examine here relates to exposing the effect on, and damage to, the human body brought about by the atomic bomb. The following article can be cited as typical of the reporting on this issue:

> Japanese broadcasts said today that the atomic bomb dropped on Hiroshima on 6 August impregnated the earth with 'radioactivity' or ultra-violet rays that fatally burned 30,000 victims during the first two weeks following its explosion. The death toll at Hiroshima and Nagasaki, the other Japanese city blasted by one of our atomic bombs, was still rising, according to the enemy broadcasts recorded here by the United Press...Now Hiroshima is peopled by 'ghost people', the living doomed to die of radiation burns, Tokyo asserted ...One broadcast, a Domei dispatch, asserted that 'radioactivity caused by the fission of uranium used in atomic bombs is taking a toll of mounting deaths and causing reconstruction workers in Hiroshima to suffer from various sicknesses and ill-health'. (*The New York Times*, 25 August 1945)

This sort of propaganda appears in newspapers in the US from 22–25 August.

We can learn several things from these reports. First, we observe a departure from the Dept. of Home Affairs prohibition on reporting the 'toll of people and animals and the numbers of dead and wounded.' Second, references to the after effects of the atomic bomb's radioactivity in terms of the injuries inflicted on the human body are particularly noteworthy. They reveal that the reporting was not limited to the physical damage sustained, but also included authoritative reporting of the results of the

Army and Government investigations carried out immediately after the atomic attack on Hiroshima. In the protest of 10 August the damage situation was utilized in order to express the nature of the indiscriminate attack using the 'new type of bomb,' but we can see in this overseas reporting, figures which considerably exceeded the real figure. This very thorough exposing and reporting of damage is an example of where the atomic reality investigation counter-strategy, which was based on Imperial Japan's national policy of atomic damage utilization, and the counter-strategy of instantly announcing the atomic damage domestically and abroad, dovetailed almost perfectly.

In response to the Japanese information war, Robert Oppenheimer, a leading scientist with the Manhattan Project, denied the Japanese reports, claiming that 'Based on all our experimental work and study and on the results of the (16 July) test in New Mexico...there is every reason to believe that there was no appreciable radioactivity on the ground at Hiroshima and what little there was decayed very rapidly' (*The New York Times*, 25 August). Furthermore, Manhattan Project Commander, Major General Leslie Groves, not only denied the Japanese reports by saying 'Studies by scientists in this country do not bear out the death reports...While the people of the United States would be committing suicide if work on atomic energy were not continued...I think our best answer to anyone who doubts this is that we did not start the war, and if they don't like the way we ended it, they should remember who started it' (*The New York Times*, 31 August). He also appealed to the people of the US to accept the legitimacy of the atomic bomb. In other words, the Japanese reporting of the atomic bombs touched a vulnerable spot in the US.[17]

Let's now examine the domestic reporting of the atomic bomb. Typical of the newspaper reporting in Japan about the damage caused by the atomic bomb are articles such as 'Neutral Country's National Newspaper (*Zurich Special Dispatch* of 8 August) Reports: Seen from the Air, 10 Square Kilometers of Hiroshima were Destroyed Immediately After the Bomb was Dropped' (*Asahi Shimbun*, 16 August 1945). Photographs of the devastated areas of Hiroshima and Nagasaki began to be published and reports with a similar content to those broadcast to the US and elsewhere started to appear. For example 'The Horror of the Century – the Atomic Bomb, The Cruelty Further Exposed, Agonizing Death for the Wounded, For the Next 70 Years No Living Thing Will Inhabit the Area' (*Mainichi Shimbun*, 23 August). 'We See the Cruelty of the Area Devastated by the Atomic Bomb – 90% of Homes Destroyed. The Slaughter Leaves Not Even a Mosquito Alive: 51,000 Dead in Hiroshima. In Nagasaki 22,000 are Roasted to Death in Tunnels and Air Raid Shelters, Lost or Unaccounted for' (*Asahi Shimbun*, 23 August). 'The "Demon" Possessing Hiroshima: the Death Toll Doubles in a Fortnight (Osaka)' (*Asahi Shimbun*, 25 August).

This genre of article died down after 25 August, and at the end of August, as seen below, there was a change in the reporting of atomic bomb damage. The article 'The Horror of the Atomic Bomb Shakes Even Medical Science, Irradiation Eats into the Flesh, No Assistance Measures, Actress with Abrasions Finally Succumbs to Death' (*Asahi Shimbun*, 29 August) have characteristics not found in earlier damage reporting.

The actress referred to in this particular article was Midori Naka, a starring actress from the Sakura Troupe of the Japan Traveling Performance Association. After being a victim of the bomb attack she returned to Tokyo and was admitted to the hospital attached to Tokyo Imperial University Medical Department. Although she had no visible injuries she passed away suddenly at the hospital on 24 August, nineteen days after being exposed to the bomb. An autopsy was carried out one hour later at the request of Surgical Professor Masao Tsuzuki of the Medical Department of Tokyo Imperial University. (I touched on the relationship between Masao Tsuzuki and the atomic bomb research in ch. 2.) An article based on discussions with Tsuzuki stated that although 'she sustained only slight abrasions and no burns or broken bones…the results of the autopsy indicated a striking change had taken place in her internal organs. Namely the bone marrow which produces blood, the liver, the spleen, the kidneys, the lymph glands and others were conspicuously damaged. It was determined that these conformed exactly to the worst symptoms resulting from the powerful use of X rays and radium rays. Thus, while hitherto the effects of the atomic bomb were thought to have been of two kinds, the damage and injuries attributable to the atomic bomb blast and burns attributable to heat rays, now there was proof of a third type of harmful effect – the effect of "residual radioactivity."'

This article also quotes from research on the effects of X ray irradiation on living creatures which Tsuzuki had reported in an address to the American Roentgen Academy in 1926. This article also states that 'from here on in those in this field of endeavor must do their utmost' to research the effects of the irradiation of materials made radioactive by the atomic bomb. The gist of the article is that research relating to atomic bomb injuries and the atomic bomb victims and their treatment is a great duty the Japanese medical world has to shoulder. Another article entitled 'Two Weeks On, Hair Falls Out in Handfuls "Something Has to be Done" – Moves Toward a Thorough Inquiry by Our Medical Fraternity' (*Mainichi Shimbun*, 31 August) reports the same message.

These articles have a very different flavor to those that were published before about 25 August. They include many angles that were previously left unreported, such as: the citing of the personal story of a bomb victim, who was a well known actor; detailed descriptions of the damage caused; the identification of radioactivity as a cause of death; the instituting of treatment for victims, and; the announcement of a policy for investigative research into the cause of death.

At the same time articles started to appear based on personal experiences of the atomic bomb horrors. Amongst these we find the article by Yōko Ōta entitled 'Light like that from the Bottom of the Sea – Exposed to the Atomic Bomb Attack (Hiroshima)' (*Asahi Shimbun*, 30 August). This is a well known first hand account of the damage wrought by the atomic bomb. Ōta says 'The damage wrought upon Hiroshima was, ultimately, profound. However, if those scenes are found to be offensive that is the offensiveness of the other party. Hiroshima City was not offensive. Rather I would like to think that the beauty of the sacrifice of the victims served to decorate the end of the war.'

It was exceptional for an air-raid victim to be able to publish their personal experience of a wartime disaster in the newspaper. In fact, as we have seen, such an account would have been subject to censorship during the war. If we look at the fact of Ōta's publication in isolation from any surrounding context, we might conclude that at that time there was a new freedom in Japan for reporting on the atomic bomb. But as we have seen, the Japanese were conducting damage exposure investigations while reporting both domestically and overseas on the effects of the atomic bomb's radioactivity on the human body. Had Yōko Ōta's account gone beyond the parameters of reporting established by the army and government's published research results, it most likely would not have been cleared for publication. But this account neither exceeded nor contradicted Imperial Japan's national policy of extracting propaganda value from the damage caused by the atomic bomb.

These reports were being published at the time when the Supreme Commander of the Allied Powers, Douglas MacArthur, had landed in Atsugi (28 August) and the Occupation Forces were starting to be stationed on the Japanese home islands. I argued in chapter 2 that the Japanese atomic bomb investigations were carried out in accordance with a national policy for getting mileage out of the damage caused by the bomb. After the stationing of the Occupation Forces in the country, this same national policy was pursued by conducting investigations into the atomic bomb damage in co-operation with the Occupation Forces. From this point of view, you could say that all of the newspaper reporting contained an element of propaganda that served to camouflage the extent of the co-operation provided to the Occupation Forces' investigations.

We should also note reports such as 'Chamberlains Dispatched to Hiroshima and Nagasaki' (*Mainichi Shimbun*, 31 August) which report that, in the lead up to the signing of the surrender documents aboard the USS Missouri on 2 September, Emperor Hirohito dispatched two chamberlains to the cities of Hiroshima and Nagasaki to visit the victims.

It is conceivable that the chamberlains were dispatched to display the Emperor's prestige and authority.[18]

Apart from atomic bomb casualty reports, the Information Bureau allowed 'free' reporting on subject matters relating to inside information on the development of the atomic bomb, the theory relating to the atomic bomb and the international management of atomic weapons. Newspapers on 15 August, the day of surrender, reported 'Full Mobilization of US and British Intelligence, Praise for the Success of Truman (Zurich Special Telegram sent on 13th)' (*Mainichi Shimbun*, 15 August), 'Talking with Dr Nishina about the Atomic Bomb, Tremendous Energy Rivaling 20,000 Tons of Gunpowder' (*Mainichi Shimbun*, 15 August) and 'Mutual Self-destruction, British Tabloid Advocates International Management' (Zurich Special Telegram sent on 13 August) (*Mainichi Shimbun*, 15 August). One of the two pages of that edition of the *Mainichi Shimbun* was devoted to a big atomic bomb campaign. The same was the case for the *Asahi Shimbun* and others. Yoshio Nishina's participation in the atomic bomb investigations conducted in Hiroshima and Nagasaki immediately after the atomic attacks, meant that he was prohibited from speaking about the theory of the atomic bomb without receiving prior permission from the army and the government.

The radio reported on the atomic bomb at 7:21 am and in the 3 pm and 5 pm news broadcasts of 15 August. The morning broadcast was made with the cognizance of Yoshio Nishina and carried the same content as the newspaper. The 3 pm broadcast reported on the measures taken by the Hiroshima Prefectural authority which gave careful thought to the future. The 5 pm broadcast reported on the telegram sent by the 'Lisbon Alliance,' which relayed US President Truman's announcement from the *New York Herald Tribune* of 7 August.[19]

The article 'Re-examination of Special Education for Gifted Children – Towards the Establishment of a State Built on Science' (*Yomiuri Houchi Shimbun*, 20 August) states 'We were beaten by the enemy's science. This fact is proven by the single atomic bomb dropped on Hiroshima. Upon his appointment to office Mr Maeda, the new Minister for Education, spoke about how keen he was to see the new Japanese state built on science and his desire to see Japanese culture restored with a broad base that incorporated science.' While conducting an information war with the US based on the exposure of the damage resulting from the atomic bomb blast, domestic announcements portrayed the atomic bomb as a product of science in an effort to pave the way for the defeated Japanese nation to establish itself as a state built on science. The contradictory reporting that both condemned American 'inhumanity' for deploying the bomb and applauded the sophisticated science that created it implicitly made concessions to the enemy as it welcomed the Occupation Forces.

The Occupation Forces' Free Speech Policy and the Continuation of the Information Bureau's Structure and Discord

The Allied Occupation Forces stationed in Japan recognized the Potsdam Declaration to be the mainstay of Occupation policies. Article 10 of the Potsdam Declaration declared that the Japanese government should remove all obstacles to the revival and strengthening of democratic tendencies, stating 'freedom of speech, religion and of thought as well as respect for fundamental human rights shall be established.'[20] Establishing the right of free speech was an important pillar in the Allied Forces' policy for the democratization of Japan. Meanwhile the Occupation Forces censored and restricted speech in the occupied land in accordance with normal military occupation practice.

On 2 September, aboard the battleship USS Missouri, the Imperial Government of Japan, through a representative of the General Staff Headquarters, signed the surrender documents. The first control measure was adopted by the Supreme Commander of the Allied Powers (SCAP) on 3 September. A directive was issued to NHK prohibiting foreign language broadcasts from its program. This directive became effective on 4 September. On 10 September SCAP ordered the Dōmei News Agency's overseas short-wave broadcasts to be shutdown and on 14 September overseas Morse broadcasts were subjected to a 24 hour ban.[21] These actions reveal how the US, immediately after the commencement of the Occupation, put a halt to the international propaganda campaign conducted by Japan from immediately after the bomb was dropped on Hiroshima.

On 8 September Major General Elliot R. Thorpe called Taketora Ogata, Director of the Information Bureau, and Shigeyoshi Matsumae, Director of the Communications Congress, to the United States Pacific Army's Forward Headquarters in Yokohama and instructed them to censor newspapers, the radio and postal services on the pretext that this was necessary for maintaining public peace.[22] Following this, on 10 September, SCAP sent 'A Memorandum Concerning Freedom of Speech and the Press' to the Imperial Government. It ordered the Government to keep restrictions on the right to freedom of speech to a minimum while at the same time directing that 'Subjects which cannot be discussed include Allied troop movements which have not been officially released, false or destructive criticism of the Allied Powers, and rumors...The Supreme Commander will suspend any publication or radio station which published information that fails to adhere to the truth or disturbs public tranquility.'[23] SCAP formally ordered the Japanese to enforce restrictions on and censorship of public debate. The directive was relayed to the relevant Japanese authorities via the Information Bureau.

As we have seen, after surrendering the government maintained tight restrictions on reporting. These were not relaxed when the Occupation commenced. Editorial guidance for articles was provided by way of internal directions in addition to the restrictions on reporting and censorship guidelines ('Censorship Weekly Bulletin' of the Home Affairs Department's Police Bureau Censorship Division and the Information Bureau from 10 to 16 September 1945).[24] The basis of the government's restrictions was to control conduct and speech that ran contrary to, or was disrespectful of, government policy. However, the policy of allowing 'free' news reporting about the atomic bomb damage was maintained. The following examination clearly demonstrates that the government made use of the atomic bomb question.

During the war the US had cracked the code used by the Japanese. Thus, after the commencement of the Occupation they were able to read communications between Tokyo and Japan's overseas legations. According to these reports, on 13 September Foreign Minister Mamoru Shigemitsu sent an official telegram to the Japanese legations in Switzerland, Sweden and Portugal that emphasized the wide reporting of the damage caused by the atomic bombs in Hiroshima and Nagasaki in domestic newspapers as well as the informal talks between the American Atomic Bomb Investigation Team and its leader, Brigadier General Thomas F. Farrell. Furthermore, the Dōmei News Agency transmitted all this reporting overseas in detail. He wanted to know to what extent these reports had received coverage in the countries mentioned. Shigemitsu pointed out the 'noise' the United States was making in relation to cruel treatment meted out to the POWs held by the Japanese and 'thought that Japan "should make every effort to utilize the matter of the atomic bomb in our propaganda." The Foreign Minister sought advice from the overseas Japanese legations about the desirability of such a campaign.' This was sent in a coded official telegram.

The US official who cracked the code guessed that the reasons Japan emphasized the atomic bomb were 'to explain Japan's defeat' and 'to offset Japan's reputation in relation to its treatment of Allied POWs.'[25] This guesswork struck the very heart of the Japanese national policy of getting maximum advantage from the atomic bomb damage. Article 10 of the Potsdam Declaration states: 'stern justice shall be meted out to all war criminals, including those who have visited cruelties upon our POWs.' It is not inconceivable that the Japanese were seeking some kind of concessions from the Allies.

At this time, disputes between the Japanese and the Allies about the question of domestic atomic bomb reporting were surfacing. On 14 September the CCD/PPB (Civil Censorship Division/News, Films and Broadcasting Department) of the CIS (Civil Intelligence Section) of the G-2 (Staff No 2 Div), which comes under the command of SCAP, ordered the Dōmei News Agency to stop operations for 24 hours. There were 6 reasons for the order being issued. One of these concerned a report that stated: 'had it not been for the atomic bomb Japan might have won the war. This bomb is very difficult to deal with as it is just too formidable. It is a weapon only barbarians would be able to bring themselves to use.'[26]

On 18 September SCAP accused the *Asahi Shimbun* of infringing the directive of 10 September and sent a memorandum suspending publication for 48 hours. Amongst the reasons for the suspension of publication was a comment by Ichirō Hatoyama, who said: '...whilst America professes that "righteousness is power" the use of atomic bombs and the slaughter of innocent civilians represents a greater transgression of international law than attacking hospital ships or using poisonous gas. It cannot be denied that this is a war crime...' in the article 'The Idea of Forming a New Party (vol. 1)' that appeared in the 15 September edition of the newspaper.[27] These measures by SCAP clearly illustrate the censorship of atomic bomb reporting and the restrictions placed on the right to freedom of speech.

Many reports about the atomic bomb appeared in the newspapers at this time. There were articles on the movements of the Americans such as: 'The Atomic Bomb Tyranny – Will it Last for 70 Years? Brilliant Scientists Investigate the Devastated Area' (*Mainichi Shimbun*, Osaka, 3 September), ' "More Atrocious than Anywhere in Europe" – 21 American Reporters Observe Hiroshima' (*Mainichi Shimbun*, Osaka, 5 September), 'American Investigating Group Visits Hiroshima' (*Asahi Shimbun*, 10 September) and 'The Atomic Bomb Lays Hiroshima to Unimaginable Waste: American Investigators Appalled' (*Asahi Shimbun*, 11 September).

The full title of the American Investigating Group was the 'Manhattan Project Atomic Bomb Investigating Group.' Major General Groves, commander of the Manhattan Project, dispatched the Group, which consisted of 13 members and was captained by Brigadier General Farrell. The Investigating Group's movements were reported in minute detail as can be seen in the following quotation: '...on the next day, the 9th, with Tokyo University Professor Masao Tsuzuki acting in a guiding capacity, after hearing a detailed explanation, the group was split into a medical research group and physics research group. The medical research group observed the patients housed in the temporary accommodation at the No 1 National Elementary School and the Hiroshima Red Cross Hospital. The physics research group followed a separate itinerary and after studying radioactivity etc at the epicentre of the blast flew out to Atsugi the same afternoon...' (*Asahi Shimbun*, 11 September) as well as in the headline: 'The Atomic Bomb Investigating Group Heads to Nagasaki' (*Asahi Shimbun*, Osaka, 13 September). In addition the article 'Oh! Element Splitting Apparatus! Dr Compton Visits the Imperial University' (*Asahi Shimbun*, 16 September) reports that Karl T. Compton of the Manhattan Project (who is introduced as being responsible for the American science mobilization), visited Tokyo Imperial University to carry out research (cf. ch. 1).

Reports that stress the cruelty of the atomic bomb were still being published, as can be seen in the following headlines: 'Unbearable Atomic Bomb Brutality' (including 4 photographs) (*Asahi Shimbun*, Osaka, 4 September), 'Atomic Bomb: 270,000 Dead and Injured' (*Asahi Shimbun*, Osaka, 6 September), 'The Rising Death Toll: The Tragedy of a Month in Hiroshima' (*Asahi Shimbun,* Osaka, 6 September), 'A New Symptom – "The Fourth Death", Death after Restoring Normal White Corpuscles, The Atomic Bomb' (*Asahi Shimbun*, Osaka, 16 September).

Moreover, reporting on the development, principles and international maintenance of atomic bombs continued. The article entitled '9 Years in the Making: 100,000 Employees Parliamentary Report' (*Mainichi Shimbun*, 6 September) examines the report 'About Uranium,' which the government submitted to a plenary session of the House of Representatives. According to this article, 'About Uranium' explains the principles of the atomic bomb and the history of its development. It also states: 'Whilst there is great interest in this research, in our country we are poorly endowed with uranium as a natural resource and as scientists we were urgently mobilized into this area of research. We were not able to establish a counter-strategy to avoid this horror.' Furthermore, there were articles relating to the Manhattan Project such as: 'Photographs – Mighty Factory Facilities (in the US *Time* magazine of 20 August)' (*Mainichi Shimbun*, 6 September), 'Striking at the Atomic Bomb's True Colors (pt. 1); The Continuation of "Linked Functions", the Key to the Release of Atomic Bomb Energy, American Research: Fruition at Last' (*Mainichi Shimbun*, 9 September), '(pt 2) Suddenly a Huge Factory in the Wilderness, Construction Cost $2000 Million Dollars, Disguised Manhattan Project, 120 Thousand Employees' (10 September), 'Appraisal of National Might, Benefits and Dangers – The Two Faces of the Revolutionary Discovery, 10,000 Times the Power of the Latest Explosives' (11 September).

During this period the Japanese continued reporting on the medical or physiological effects being discovered in the atomic bomb investigations. The following articles are examples of this: 'Great Destruction of Bone Marrow, "The Third Fire": Researching Cases where No Burns are Present' (*Mainichi Shimbun*, Osaka, 5 September), 'Atomic Bomb's Anatomical Conclusion, Dr Tsuzuki from Tokyo University, Prospects Held Out for Patients suffering from Moderate Levels of Unpreventable Neutrons' (*Mainichi Shimbun*, Osaka, 6 September), 'Scalpel Taken to Atomic Bomb in Osaka, Radioactivity in the Phosphorus of Bones' and 'Integrated Investigative Facility in Kyoto University' (*Mainichi Shimbun*, Osaka, 14 September).

There was also considerable reporting about the organization of the research: 'Atomic Bomb Committee, Our Science Corps Included Through

National Research Council' (*Asahi Shimbun*, 16 September) and 'Inquiry into Atomic Bomb Disaster, Investigative Team of 200, Education Department Establishes a "Special Committee"' (*Mainichi Shimbun*, Osaka, 16 September). A lengthy quotation from an article that appeared in the *Mainichi Shimbun* reveals the type and depth of the details being reported:

> As for the real situation resulting from the atomic bombs in Hiroshima and Nagasaki and the measures to be taken from here on in, authoritative figures from the academic world have been individually going to the areas concerned and carrying out investigative research. This sort of fragmentary research cannot be expected to bear much in the way of outcomes so a comprehensive and fundamental investigative research effort will be carried out. It will include the best scientists in the land. The National Research Council (NRC), which forms the backbone agency of scientific research, will organize it. The NRC will immediately establish the "Special Committee for the Investigation of the Effects of the Atomic Bomb". This will pick up on the results of the partial research work which has already been done at the site. In order to conduct the radical investigations the following committee structure and selection of office bearers has been organized. Preparations will go ahead under the auspices of the relevant government agencies and a team of approximately 200 committee members, including research officers, will depart on about the 20th under the Committee Chairman Haruo Hayashi. The team members will be divided into 9 groups to conduct research according to their different areas of expertise. The result of the research will be reported immediately as it becomes known. This should work to bring stability to the lives of the people of the prefecture.

Immediately following that article is a list of the Special Committee's chairman, vice-chairman, the chairmen of the nine sub-committees (Physics/Chemistry/Physical Geography, Biology, Machine/Metal/Raw Materials, Electrical Communications/Civil Engineering, Architecture, Medicine, Farming/Fisheries, Forestry and Veterinary Science and Animal Husbandry) and their members.

Soon there followed a series of articles based on informal discussions with Military Doctor Lieutenant General Kenji Ibuka, Principal of the Army Military Medical School. These articles reported on the latest findings about the treatment of the victims of the atomic blast under headlines such as: 'Atomic Bomb Injury Treatment, the Definitive Edition, Rest and Bone Marrow Regeneration' (*Mainichi Shimbun*, Osaka, 19 September), 'Atomic Bomb Injury Sufferers Have Your Blood Tested Every Week for 23 Months, Rest and Nutrition as Treatment' (*Yomiuri Houchi Shimbun*, 19 September). These articles are exceptions inasmuch as the Army Military Medical School indicates treatment strategies for atomic bomb victims. What was behind the Japanese military's thinking, however, was to turn the blame for the military's past history and the responsibility for defeat in the war away from themselves and towards the atomic bombs by indicating treatment strategies for bomb victims. They also had it in mind to camouflage the co-operation being offered to the Occupation Forces in their atomic bomb investigations.

The Press Code

On 19 September SCAP ordered the Japanese Government to instigate a 'Memorandum Concerning a Press Code for Japan' that included rules governing Japanese newspapers and publications. On 21 September a memorandum from SCAP was forwarded to the newspapers. The fact that it had been sent from the Information Bureau and from the Department of Home Affairs to administrators across the length and breadth of the country was also reported in 'Directions on Newspaper Laws Given by Allied Powers Headquarters Applicable to All Publications in Japan' (*Asahi Shimbun*, 23 September). The 10 articles of the press code are:

> 1. News must adhere strictly to the truth; 2. Nothing shall be printed which might, directly or by inference, disturb the public tranquility; 3. There shall be no false or destructive criticism of the Allied Powers; 4. There shall be no destructive criticism of the Allied Forces of the Occupation and nothing that might invite mistrust or resentment of those troops; 5. There shall be no mention or discussion of Allied troop movements unless such movements have been officially released; 6. News stories must be factually written and completely devoid of editorial opinion; 7. News stories shall not be colored to conform to any propaganda line; 8. Minor details of a news story must not be over-emphasized to stress or develop any propaganda line; 9. No news story shall be distorted by the omission of pertinent facts or details; 10. In the make-up of the newspaper no news story shall be given undue prominence for the purpose of establishing or developing any propaganda line.[28]

The 'press code' was followed on 22 September by a 'Memorandum Concerning a Radio Code for Japan.'[29] Combined, the press code and radio code formed the criteria by which the contents of newspapers, publications and broadcasts were censored under the Occupation Forces. Meanwhile, in order to establish freedom of speech, SCAP issued a 'Memorandum Concerning Disassociation of the Press from the Government' [30] on 24 September and a 'Memorandum Concerning the Removal of Restrictions on Political, Civil and Religious Liberties' [31] on 14 October. It ordered the government to totally abolish control on political, civil and religious legislation and to release political prisoners. It directed that the Ideology

Police be totally disbanded and that the Home Secretary and all members of the Special Police be dismissed. It also ordered a memorandum on the abolition of film controls by the Japanese Government on 16 October.[32]

From 19 September until 8 October, when SCAP ordered the five newspaper offices in Tokyo to carry out pre-publication censorship, we find the following sorts of articles relating to the atomic bombs in the newspaper: 'Murderous Light on Me, Torpedo Finds its Mark, Atomic Bomb Research No Success Yet' (*Mainichi Shimbun*, 21 September), 'Atomic Bomb Special Committee' (*Yomiuri Houchi Shimbun*, 22 September), 'American Senate Approves Atomic Bomb Management Plan' (*Asahi Shimbun*, 23 September), 'Professor Mashita and Others Lose Lives in Course of Duty, Kyoto University Atomic Bomb Integrated Research Team Meet Disaster in Hiroshima Typhoon' (*Asahi Shimbun*, 23 September), 'Management of the Atomic Bomb, US Maintains Secrecy, Power for the Welfare of Humanity' (*Asahi Shimbun*, 25 September), 'Ferocious Might: America Succeeds in Producing a New Type of Atomic Bomb Many Times More Powerful than "Hiroshima"' (*Yomiuri Houchi Shimbun*, 26 September), 'Atomic Bomb Maintenance, US President Recommendations to Coming Conference' (*Asahi Shimbun*, 26 September), 'Atomic Bomb Open to Public, US Observation Group's Memo of Opposition' (*Yomiuri Houchi Shimbun*, 3 October), 'Atomic Bomb Defence Completed' (*Yomiuri Houchi Shimbun*, 3 October) and 'US Senate Advised to Establish Atomic Energy Commission' (*Mainichi Shimbun*, 5 October). The advertisement 'The Alliance Atomic Bomb Series' by Takao Takei (Dōmei News Agency) appeared in the 29 September edition of the *Mainichi Shimbun*. These articles were mostly based on overseas dis-patches, which suggests that the Japanese were chasing information about what was happening with the atomic bomb in the postwar United States.

Before and during the war the Information Bureau which, by virtue of regulations governing newspaper and magazine reporting, guidance and page layouts etc, was at the centre of controls on the right to free discussion in Japan, sent notification to all newspaper offices on 1 October 1945: 'Please be aware that in accordance with the recent instructions from the Allied Powers Supreme Headquarters, guidelines put out by the Dept. of Home Affairs and the Information Bureau regarding enforced bans and editing of articles are now cancelled.'[33] This meant that the Japanese controls on free speech were essentially over. I have already referred to the question of the continued existence of the Information Bureau's set up after 15 August. Precisely how the Japanese press controls were taken over by SCAP has never been made clear. We await elucidation on this in the future. Immediately following the commencement of the Occupation SCAP considered censorship measures on individual reporting agencies, such as the Dōmei News Agency and the *Asahi Shimbun* relating to particular problem articles. On 8 October SCAP ordered five Tokyo newspaper offices (the *Asahi*, *Mainichi*, *Yomiuri*, *Tokyo* and *Nihon Sangyo Keizai* newspapers) to undertake pre-publishing censorship. From that time on newspaper houses were obliged to submit all articles to the Civil Censorship Detachment (CCD) to be censored prior to publication. Pre-publishing censorship of books in Tokyo commenced on 29 October.[34]

The CCD, which was in charge of censorship, was incorporated into the CIS (Civil Intelligence Section) of GHQ's G-2 (General Staff 2). It censored telegrams, telephones, the postal service, newspapers, publications, movies, theatre performances and broadcasts. The PPB (Publications [newspapers/books/magazines], Performances [movies/theatre performances] and Broadcasts censorship division), which controlled the presentation media within the CCD, divided Japan into three districts. The first district incorporated Tokyo, Hokkaidō, and the Tōhoku and Kantō regions. The second district incorporated the Kinki, Hokuriku, Shikoku and Chūgoku regions and the third district covered Kyūshū (and including Hiroshima, Shimane and Yamaguchi).[35]

The censorship system was relaxed and revised several times, the main revision taking place on 15 October 1947. All publishers were now subject to post-publishing censorship with the exception of fourteen publishing companies representing both the extreme right and the extreme left of the political spectrum. These fourteen companies remained subject to pre-publishing censorship. On 26 July 1948 pre-publication censorship of newspapers stopped completely and even the extreme papers were subject only to post-publication censorship. Then, on 31 October 1949, all censorship activities of the civil censorship corps ceased and on 10 November all civil censorship bodies were abolished.[36]

Concrete Examples of Censorship

Let us now turn to some concrete examples of censorship relating to the atomic bomb. In June 1946 the Hiroshima poet Sadako Kurihara submitted a proofreading copy of a collection of poetry, 'The Black Egg,' to the CCD Fukuoka Division. Permission was given for the publication of the poetry collection but entire lines such as 'Drawing up to the war…listening to the photographic broadcast of sounds of the battlefield' and 'handshake' were deleted.[37]

In March 1947 a censorship problem arose in relation to *Masako taorezu* (Masako won't succumb) written by a fifteen year old girl from Nagasaki, Masako Ishida. *Masako taorezu* was sent from Nagasaki to the Fukuoka area CCD where the censorship officer in charge considered the atomic bomb experiences it portrayed to be graphic and advised that it be banned from publication for a while. The reason given was that 'this District believes that the novel *Masako taorezu* would disturb public tranquility in Japan and that it implies the bombing was a crime against humanity.'[38]

Monica Braw's 'The Atomic Bomb Suppressed: American Censorship in Japan 1945–1949'[39] details the particulars of the debates within the various sections of GHQ about the publication of *Nagasaki no kane* (The bell tolls for Nagasaki), which was written by Takeshi Nagai, a victim of the atomic bomb. In 1947 its Tokyo publisher sent the book to the censor. It was eventually published in 1949, but the permission was conditional on it being published in combination with *The Manila Tragedy* which reports on the Japanese army's plundering activities in Manila.

What is interesting about the arguments that went on within GHQ are the 1947 evaluations by ESS (Economic and Scientific Section) and PHW (Public Health and Welfare Section), both of whom sought counsel from CCD. While ESS had no problem with the technical content and PHW had no problem with the publication of information on the physical effects that the atomic bomb had on the human body, the CCD censor, nevertheless, advised that its publication be banned because the book was deemed likely to cause enmity towards the United States. The outcome of the debate was a compromise, when Charles Willoughby, Head of G-2, ordered a six month hold be put on publication of this book. He said: 'it described the horrors and the great death toll and thus was likely to disturb public tranquility and create ill will toward the United States.' Both of these reasons are based on the infringement of the Press Code article 'Threatening Public Order.'[40]

A *Yomiuri Shimbun* article about the World Religions' Peace Conference on 18 June 1948 was totally banned.[41] In 1947 the article 'Three Subjects on the Atomic Bomb' was deleted from the general reading magazine *Nihon Hyoron*.[42]

'Nihon News Reel No. 257' is a typical example of the Occupation censorship of atomic bombs and films. This was a documentary type news film produced by the Japan Movie Co. which had been doing the same business since the wartime. The first half of 'No. 257' was about the atomic bomb and the tragedy of Hiroshima. The second half was about the stationing of American Forces in Yokohama. The CIE (Civil Information and Education Section) was called upon to give its opinion. In his response on 24 September 1945, David W. Conde, Chief of the Performance Section, made a most interesting comment. He said 'the latter part of this picture dealing with US Forces in Japan is satisfactory. The first part, and the part that gives the title to the picture, is the atomic bomb. The fact that the Japanese have never shown any pictures of bomb destruction to her cities makes the impact of this picture that much more important. The fact that this picture is titled and begins with the atomic bomb increases the propagated view that the atomic bomb ended the war. This seems to be deliberate and plays down the fact that Japan was actually militarily beaten. The fact that no shots have been shown of the widespread damage suffered by Tokyo and other cities also would seem to confuse the Japanese in the country…'[43] The CCD finally allowed 'No. 257' to be screened, but only on the condition that the first and second halves be reversed.

Conde's comment very aptly identifies the Japanese policy of getting the most mileage out of the atomic bomb damage. During the war Japan's own censorship rules had prohibited reporting on damage caused by air raids. Subsequently, it was government policy to treat the atomic bomb as the main cause of Japan's defeat.

Reporting and Censorship of the Results of the Atomic Bomb Investigations

The Special Committee for the Investigation of the Effects of the Atomic Bomb of the National Research Council produced some 80 medical reports between its establishment in September 1945 and December 1946. These reports were submitted to SCAP. Such reports were either translated into English by the Japanese scientists themselves before submission or they were translated by ATIS (Allied Translators and Interpreters Service) of the GHQ. Specific permission was required prior to the publication of each research paper.

In June 1946 the Head of the Medical Section of the Special Committee, Tokyo Imperial University Professor Masao Tsuzuki, applied to SCAP for permission to publish these reports in a Japanese science magazine.[45] In the final analysis the publication of all the Japanese reports, including the reports in this submission, took place only after the 1953 peace treaty came into effect. The publication requests made by Masao Tsuzuki were all denied by the Occupation Forces. The details of denials were made clear as a result of the investigation made by Braw.[46] This fact touches on something that doesn't stop simply at the matter of censorship. Whereas the examples of censorship examined in the previous chapter dealt with the works of private authors, the results of the atomic bomb investigations emanated from co-operative research. Publicizing them in the country that had suffered the atomic attacks and was occupied by a foreign army brought quite a different argument into the picture, even though the authors of the reports might themselves be individuals as were the authors of the works mentioned above.

Through the Manhattan Project the Americans practiced strict controls on information and were extremely wary of atomic bomb secrets leaking out. Even after the war, information relating to the atomic bomb or atomic energy that aimed to touch on the question of the atomic bomb monopoly was very strictly managed. For example, the 'Smythe Report' – published immediately after the dropping of the bomb on Nagasaki – announced publicly that information that had been published from and about the Manhattan Project had been properly examined. In 1946 the Atomic Energy Act was enacted and rules were put into place for the management of information. The Atomic Energy Commission treated 'all information relating to the manufacture or use of atomic weapons or the manufacture of fissionable substances used to produce energy' as secret information.[47] From the Spring of 1946 the plan to do away with the secrecy surrounding materials relating to the atomic bomb moved ahead in earnest. The first collection of 270 documents relating to the atomic bomb was released at the end of January 1947. Opening this information to the public required not only the approval of the Atomic Energy Commission but also the approval of the Secretary of the Department of the Army.[48]

The question of the publication of the Japanese atomic bomb investigation reports was very closely tied to the US position on the provision of atomic energy information to the public. The ABCC preliminary commission of investigation, which came to Japan at the end of 1946, advised that publication of the Japanese reports should be allowed as it would encourage the continuation of research. Within GHQ not only CCD, but also ESS and PHW had examined this question. In ESS, Harry C. Kelly, Division Chief of ESS/ST (Scientific and Technical Division), was put in charge. Kelly himself was dispatched to the US Atomic Energy Commission where he requested instructions on this matter. But the question proved to be one that was not easily resolved.

Finally, in September 1948, reports on the Japanese investigations into the effects of the atomic bomb were cleared for publication on the condition that the authors had no opportunity to utilize American information designated as secret and that they had never worked in co-operation with any Americans who were able to make use of any secret information. There was to be no special publicity about this new policy because 'ill-informed people would conclude that the Japanese were being allowed to engage in general research in the field of atomic energy in contradiction to the Far Eastern Commission policy on such matters.'[49]

However, such a conclusion appears almost unavoidable, since in December 1948 the Far Eastern Commission prohibited research by the Japanese in the field of atomic energy.[50] The September decision to allow the publication of the Japanese reports certainly contradicted this Far Eastern Commission decision. If we go further back, we see that on 22 September 1945 SCAP had ordered a ban on all atomic energy research being conducted by the Japanese.[51] Thus, on the one hand, the Occupation Forces would allow neither the existence of a major Japanese investigation project on the atomic bomb damage nor its implementation. But, on the other hand, the US needed the results of the Japanese investigations in order to maintain its superior position in the atomic energy field, a position centered on the postwar mass production of atomic bombs. Maintaining this position was in the American national interest. At the same time, as we have seen, Japanese national policy had foreseen that cooperation in the investigations would be in Japan's national interest. It is possible to argue that this perception was proven correct when the cooperation yielded greater access to the research results than might otherwise have been the case.

These contradictions in Occupation policy are most apparent in the fundamental incompatibility between the principle of allowing freedom of speech and the desire to control any debates that could possibly disrupt public order. As Braw points out on this point: 'The question of freedom of speech was not an argument when publication of Japanese manuscripts was recommended. Neither did the recommendations emphasize the need for free medical discussion in order to help the atomic bomb survivors with their incurable diseases.'[52]

The censors' judgment on the movie 'Effects of the Atomic Bomb,' a documentary film completed by a film crew from the US Strategic Bombing Survey as a part of the investigation of the Special Committee for the Investigation of the Effects of the Atomic Bomb also reflects this basic policy conflict. A memo put out by PPB (Publication, Performances and Broadcasting) on 15 January 1947 states: 'However, I do not see how it would be "objectionable from a security or public safety point of view" or desirable for seizure by CIC because: a) Its production is being supervised by a SCAP agency and, b) The film is to be sent to the Army Medical Museum, Washington, D.C.'[53] Atsushi Furukawa surmises from this advice that 'As civil censorship authorities it was probably considered that there was no other judgment that could have been made in connection with a type of military strategy carried out by the highest echelons of the US forces.'[54]

In January 1949 the secrecy status of many of the Japanese manuscripts concerning research into the effects of the atomic bomb was lifted.[55] By this time, though, there had already been many reports about research into the effects of the atomic bomb published in Japanese medical magazines. For example: there were five such articles in 1945, twenty-six in 1946, fourteen in 1947, three in 1948 and another three in 1949. Subsequently, there were seven books published in 1950 and thirteen books in 1951.[56]

An examination of the censor's reports on the proofreading galleys of the medical magazines revealed that in the section 'Ophthalmology Clinical Medical Report' vol. 40 no. 5 (15 July 1946) there was an article by Shigemi Tamura and four others entitled 'Clinical Findings on Eye Disabilities Resulting from the Atomic Bomb.' Yet there was no directive to delete those censored proofreading galleys. There was likewise no directive to remove the censored proofreading galley for the article by Masakazu Ishizawa 'Investigative Research into the Lack of Menstruation in Hiroshima High

School Girls' or that by Kiyomi Ōtsubo 'Hygenics Investigative Report on Atomic Bomb Casualties in Hiroshima and After-effects of the Bomb' found in *Ganka rinshō ihō* (Ophthalmological clinical research bulletin, vol. 24, no. 1, January 1947).[57] We do not know whether these were censored or escaped censorship because of the specialized nature of journals.

On the censored proofreading galley of 'On "Atomic Bomb Injuries", with Particular Reference to Counter-Measures from the Medical Standpoint' by Masao Tsuzuki, which appears in *Sōgō igaku* (Comprehensive medical science, vol. 2, no. 14, October 1945), there is a directive to delete thirteen lines from page three. The Japanese manuscript is very unclear. It is difficult to decipher what the deleted section says, but judging from the English version it seems to have said: 'Having looked at it from a variety of angles, we recognize that there is a release of poisonous gases which accompanies the explosion and it is not difficult to imagine that there were probably some victims who died as a result of these poisons. At this stage we do not have any clue as to the question of whether or not this explosion was deliberately designed to emit poisonous gases. I would like to broach this matter with the Americans if an appropriate opportunity presents itself.' Mr Yamamoto, the censorship officer who censored this article justified deleting this text on the basis 'that it would disturb public tranquility.'[58]

Similarly, eight lines from page four were deleted from 'An Account of a Journey: Atomic Bomb Damage Investigation Bound' by Yoshifumi Uchida which appeared in *Kagaku gahō* (Science graphics, vol. 35, no. 1, 1 January 1946) on the grounds that 'it was untrue.' An entire article by the National Institute of Health (also known as the 'Atomic Bomb Medical Research Institute') that concerned the establishment of 'The Atomic Bomb Effects Research Institute' was removed from the *Nihon iji shinpō* (Japan medical affairs bulletin, no. 1246, 21 February 1948).[59] The accompanying materials provide no reasons for these two articles to have been edited or suppressed.

Atomic Energy Related Articles and Censorship

The first twenty-one lines of Kazuharu Kimura's 'The Research of Atomic Energy and its Future' (in *Kagaku sekai* [World of Science], vol. 21, no. 11, 1 November 1946) were the subject of a directive deeming them un-acceptable. Two lines from Kōji Fushimi's 'On Velocity Betatron' were deleted on the grounds that they 'criticized' the Occupation policies. The parts that were deleted from the Japanese script have been inked out, but the English version is still available: 'The destruction of the cyclotron seems to be the only uncultured measure taken by the US Occupation Forces.'[60] The Quantum Theory Group's 'An Excerpt – Schrödinger's Interpretation of Quantum Mechanics (3)' which is found in *Shizen kagaku* (Natural science, vol. 1, no. 3, 17 February 1947), was expurgated in its entirety.[60]

It is noteworthy that the deletions are surprisingly few in magazine articles concerning atomic energy. Conversely, articles which received permission for publication were surprisingly numerous. *Sengo zasshi mokuji* (List of postwar magazines) from the period of censorship, i.e. between 1945 and 1949, reveals that articles were published about the economical use of atomic energy, the management of atomic energy, US atomic bomb experiments and the principles of the atomic bomb in magazines such as *Kaizō* (Rebuilding), *Sekai* (World), *Nihon hyōron* (The Japan review), *Kokumin no kagaku* (Science for national people), *Ondori tsūshin* (Rooster report), *Shisō no kagaku* (The science of thought), *Chūō Kōrōn* (Central public opinion) and *Sekai hyōron* (The world review).[61]

The tendency was similar in the newspapers. In relation to the atomic damage caused in Hiroshima and Nagasaki 'Propitious (and non-factual) reporting was published to the effect that the survivors of the Hiroshima and Nagasaki atomic bombs were recovering their health.'[62] 'Radioactivity Symptoms Cured: the Denken (Institute of Infectious Diseases) Atomic Bomb Investigation' from the *Asahi Shimbun* of 25 November 1945 is a prime example of this sort of article.

Re-examination of 'The Press Code Myth'

Braw's study is considered to be the first proper research into the topic of the atomic bomb and censorship during the period following the bombing of Hiroshima and Nagasaki. At the start of *Censorship*, Braw states that 'As a result of this censorship those affected by the bombs were unable to explain their personal experiences to others. As a result of this censorship doctors were not permitted to freely discuss "atomic bomb symptoms". This made the treatment of patients by doctors all the more difficult.'[63] Everything that has been said to date in relation to the atomic bomb and the press code under the Occupation Forces is encapsulated in these three short sentences. I would like to call this the 'press code myth.' The 'press code myth' created a false impression about Hiroshima and Nagasaki under the Occupation Forces. Braw is not the only person who was caught-up in this myth.

In researching this paper I discovered afresh how literary works recording the atomic bomb damage, and newspaper and magazine articles reporting on those affected by the bomb were surprisingly unaffected by censorship. Even the fat censorship magazine of proofreading galley proofs stored in the National Diet Library gives the impression that there were few galleys that touched on the atomic bomb damage itself. In other words the query arises as to whether or not the people of the day really wanted to know or be informed about Hiroshima and Nagasaki.

Also, as examined in this paper, some part of the results of medical research conducted under the Occupation, even if conducted as part of the research carried out in co-operation with the Occupation, would have been accepted for publication only after passing through the hands of the censors. Moreover, the magazines they were published in were specialist medical magazines. It wasn't as if there was absolutely no prospect of doctors discussing the 'atomic bomb symptoms' with one another if they attempted to do this. As already mentioned in the chapter 'Investigations into the Atomic Bomb Damage' it can be said that the medical men and doctors of the day, while faithful to the national policy of the government, did not closely align themselves with the cause of the atomic bomb victims. Incidentally, while the Occupation Forces banned the research of atomic energy by the Japanese, they did not prohibit research into treatments for atomic bomb victims. Responsibility for research into the treatment of victims was indisputably left to Japan. I am of the opinion that the 'press code myth' was borne out of the victims' consciousness. The problem with understanding the actual history is that the Japanese people still continue to talk about the 'press code myth' to historians like Braw.

Notes

1 Matsuura, 1984 [22] p. 237.
2 Takeyama, 1994 [39] p. 131–4.
3 Takeyama, 1994 [39] pp. 123–8.
4 Shimomura, 1949 [1] pp. 97–8.
5 Ministry of Foreign Affairs, 1990 [5] pp. 833–8.
6 *Asahi Shimbun*, 11 August 1945.
7 Kitayama, 1988 [3] p. 290.
8 'The Swiss Legation to the Dept of State, Memorandum, Reference: 1–10.' Washington, 11 August 1945, Foreign Relations of the United States Diplomatic Papers, 1945, vol. IV, The British Commonwealth, The Far East, US Government Printing Office 1969, p. 472–3.
9 Sodei, 1972 [17] pp. 211–2.
10 Takakuwa, 1984 [33] pp. 35–6.
11 Awaya, 1980 [2] p. 366.
12 Takakuwa, 1984 [33] pp. 36–7.
13 Shimomura, 1949 [1] p. 99.
14 Imabori, 1959 [14]; Sodei, 1972 [17]; Fukushima, 1984 [21]; Braw, 1988 [23]
15 Kitayama, 1988 [3] p. 289; Braw, 1988 [23] p. 190–2.
16 Sasamoto, 1989 [24] p. 115–8.
17 Sasamoto, 1989 [24] p. 117.
18 Sasamoto, 1989 [24] p. 121.

19 Takeyama, 1994 [39] p. 171–6.
20 Etō, 1986 [4] p. 339.
21 Nippon Hōsō Kyōkai Hōsō Bunka Chōsa Kenkyūsho Jōhō Chōsa-bu (NHK Research Institute on Broadcasting Culture, Broadcast Information Research Division) (ed.), 1987 [36] p. 17.
22 Hōsōhō Seiritsu Katei Kenkyū-kai (Research Group on the Broadcasting Legislation Enactment Process), 1980 [35] pp. 9–17.
23 Hōsōhō Seiritsu Katei Kenkyū-kai (Research Group on the Broadcasting Legislation Enactment Process), 1980 [35] p. 18–19.
24 Awaya, 1980 [2] p. 368–9.
25 Braw, 1988 [23] p. 50.
26 Braw, 1988 [23] p. 130–1.
27 Haruhara, 1984 [9] no. 396, p. 97–8.
28 Hōsōhō Seiritsu Katei Kenkyū-kai (Research Group on the Broadcasting Legislation Enactment Process), 1980 [35] p. 21–3.
29 Hōsōhō Seiritsu Katei Kenkyū-kai (Research Group on the Broadcasting Legislation Enactment Process), 1980 [35] p. 23–7.
30 Hōsōhō Seiritsu Katei Kenkyū-kai (Research Group on the Broadcasting Legislation Enactment Process), 1980 [35] p. 32–4.
31 Takemae, 1994 [38] vol. 2, p. 144–8.
32 Takemae, 1994 [38] vol. 2, p. 219–21.
33 Takakuwa, 1984 [33] p. 49.
34 Furukawa, 1991 [30] p. 287–8.
35 Furukawa, 1991 [30] p. 287–8.
36 Furukawa, 1991 [30] p. 288.
37 Matsuura, 1984 [22] p. 111–3.
38 Braw, 1988 [23] p. 132–3.
39 Braw, 1988 [23] p. 133–43.
40 Braw, 1988 [23] p. 134–5.
41 Braw, 1988 [23] p. 143–4; Takakuwa, 1984 [33] p. 268.
42 Matsuura, 1984 [22] p. 117.
43 Furukawa, 1992 [25] p. 529.
44 Furukawa, 1992 [25] p. 546.
45 Braw, 1988 [23] p. 169.
46 Braw, 1988 [23] p. 167–80.
47 Braw, 1988 [23] p. 163.
48 Braw, 1988 [23] p. 165–6.
49 Braw, 1988 [23] p. 176.
50 Ministry of Foreign Affairs, 1989, [40] p. 231–2.
51 Ministry of Foreign Affairs, 1989, [40] p. 79–86.
52 Braw, 1988 [23] p. 180.
53 Furukawa, 1992 [25] p. 535–6.
54 Furukawa, 1992 [25] p. 536.

55 Braw, 1988 [23] p. 176.
56 Hiroshima Genbaku Iryō-shi Henshū Iinkai (Editorial Committee on the History of Medical Treatment after the Hiroshima Atomic Bomb), 1961 [15] p. 464–7.
57 Okuizumi, 1982 [29] reel no. R43, R156.
58 Okuizumi, 1982 [29] reel no. R230.
59 Okuizumi, 1982 [29] reel no. R84, R135.
60 Okuizumi, 1982 [29] reel no. R84, R223.
61 Tokyo Daigaku Shakai Kagaku Kenkyūsho Sengo Kaikaku Kenkyū-kai (Research Committee on Postwar Reforms, Social Science Research Institute, University of Tokyo), 1976, 1977 [20]
62 Braw, 1988 [23] p. 151.
63 Braw, 1988 [23] pp. i–ii.

Bibliographic Guide

Japanese Press Controls

[1] Kainan Shimomura, *Shūsen-ki* (Notes on the end of the war), Kamakura Shoten, 1949 (Kōdansha Bunko, 1985).
[2] Kentarō Awaya (ed.), *Shiryō Nihon gendaishi* (Data: History of modern Japan), vol. 2, Ōtsuki Shoten, 1980.
[3] Setsurō Kitayama, *Rajio Tokyo: Senji taiseika no Nippon no taigai hōsō – III Haiboku e no michi* (Radio Tokyo: Japanese wartime overseas broadcasts, vol. III: The road to defeat), Tabata Shoten, 1988.
[4] Jun Etō, (editorial supervision), Ken Kurihara and Sumio Hatano (eds.), *Shūsen kōsaku no kiroku* (A record of moves towards the ending of the war), vol. 2, Kōdansha Bunko, 1986.
[5] Ministry of Foreign Affairs (ed.), *Nippon no sentaku: Dai 2-ji sekaitaisen shūsen shiroku* (Japan's choice: A history of the end of World War II), vol. 2, Yamate Shobō Shinsha, 1990.

Censorship Policies during the Occupation

[6] Eisaburō Okuizumi and Atsushi Furukawa, 'Nihon senryō-ki no kyokutō beigun jōhō shūshū katsudō to soshiki (The information gathering activities and organization of the Far East American military in Occupied Japan),' in *Tokyo Keizai Daigaku kaishi* (Journal of the Tokyo University of Economics), nos. 109–10, December 1978, pp. 115–43.
[7] Jūrō Fukushima, 'Senryō-ka ni okeru ken'etsu seisaku to sono jittai (The realities of censorship policies during the Occupation),' in Takahide Nakamura (ed.), *Senryō-ki Nihon no keizai to seiji* (Japanese economics

and politics during the Occupation years), University of Tokyo Press, 1979, pp. 339–82.
[8] Atsushi Furukawa, 'Senryō-gun no chōhō katsudō: "MacArthur Report" yori (Intelligence activities of the Occupation Forces – from the "MacArthur Report"),' in *Tokyo Keizai Daigaku kaishi* (Journal of the Tokyo University of Economics), no. 119, January 1981, pp. 165–75.
[9] Akihiko Haruhara, 'Senryō ken'etsu no ito to jittai (The intent and realities of Occupation censorship)', in *Shimbun kenkyū* (Newspaper research), parts 1–3, nos. 395, 396 and 397, June, July and August 1984.
[10] Jun Etō, '*Tozasareta gengo kūkan: Senryō-gun no ken'etsu to sengo Nihon* (A closed space of speech: the Occupation Forces' censorship and postwar Japan), Bungei Shunjū, 1989, (Bunshun Bunko, 1994).

The Atomic Bombs and Censorship

[11] Takashi Nagai, *Nagasaki no kane* (The bell tolls for Nagasaki), Hibiya Shuppan, 1949 (Included in *Nagai Takashi zenshū* [The complete works of Takashi Nagai], Kodansha, 1971).
[12] H. D. Smythe, *Genshi bakudan no kansei: Smythe hōkoku* (The Completion of the atomic bomb: The Smythe Report), Yoshio Nishina (editorial supervision), Asao Sugimoto, Eizō Tajima, and Eiichi Kawasaki (tr.), Iwanami Shoten, 1951.
[13] Tōichirō Takamatsu, 'Genshi bakudan no hōdō: Sono genkai to seigen (Atomic bomb reporting : The limits and limitations),' in *Tokyo Daigaku Shimbun Kenkyū-sho kiyō* (Journal of the Newspaper Research Institute of the University of Tokyo), no. 3, 1954, pp. 1–26.
[14] Seiji Imabori, *Gen-suibaku jidai – Gendai-shi no shōgen* (The age of atomic and hydrogen bombs – A testimony of modern history), vol. 1, San'ichi Shobō, 1959.
[15] Hiroshima Genbaku Iryōshi Henshū Iinkai (Editorial Committee on a History of the Hiroshima Atomic Bomb Medical Treatment), *Hiroshima genbaku iryōshi* (A history of the atomic bomb medical treatment), Hiroshima Genbaku Shōgai Taisaku Kyōgikai (Counter-measures Committee on atomic bomb damage in Hiroshima), 1961.
[16] Sōzō Matsuura, *Senryō-ka no genron dan'atsu* (Suppression of the freedom of speech during the Occupation), Gendai Jaanarizumu Shuppan-kai, 1969 (enlarged and revised, 1974).
[17] Rinjirō Sodei, 'Genbaku hōdō (Atomic bomb reporting),' in Fumio Etō, Shunsuke Tsurumi, and Akira Yamamoto (eds.), *Kōza komyunikeeshon 5: Jiken to hōdō* (Series on communication studies, volume 5: Incidents and reporting), Kenkyūsha, 1972, pp. 201–25.
[18] Hiroshima-ken (Hiroshima Prefectural Government) (ed.), *Hiroshima kenshi: Genbaku shiryō-hen* (A history of Hiroshima Prefecture: Volume on the atomic bomb), 1972.

[19] Hiroshima-shi Nagasaki-shi Genbaku Saigai-shi Henshū Iinkai (The Hiroshima and Nagasaki Municipal Editorial Committee for Magazines on Atomic Bomb Damages), *Hiroshima Nagasaki no genbaku saigai* (Damages caused by the Hiroshima and Nagasaki atomic bombs), Iwanami Shoten, 1979.

[20] Tokyo Daigaku Shakai Kagaku Kenkyūsho Sengo Kaikaku Kenkyū-kai (Research Committee on Postwar Reforms, Social Science Research Institute, University of Tokyo) (ed.), *Sengo zasshi mokuji* (Index of magazines published after the war), vols. 1 and 2, University of Tokyo Press, 1976, 1977.

[21] Shingo Fukushima, 'Genbaku jōhō no sogai (Neglect of the atomic bomb reporting),' in *Heiwa kenkyū* (Peace research), no. 9, November 1984, pp. 8–28.

[22] Sōzō Matsuura, *Matsuura Sōzō no shigoto 2: Senchū senryō-ka no masukomi* (Sōzō Matsuura's work, vol. 2: The media during the war and under the Occupation), Ōtsuki Shoten, 1984.

[23] Monica Braw, *Ken'etsu 1945–1949: Kinjirareta genbaku hōdō* (Censorship 1945–1949: Forbidden reporting on the atomic bomb), Seiitsu Tachibana (tr.), Jiji Tsūshinsha, 1988.

[24] Yukuo Sasamoto, 'Genbaku shodō chōsa ni okeru Nihon-gun no yakuwari (The role of the Japanese military in the initial atomic bomb investigation),' in *Rekishi to shakai* (History and society), no. 9, 1989.

[25] Atsushi Furukawa, 'Senryō to chōhō: "Genbaku eiga" fairu to kiroku eiga no yukue (The Occupation and intelligence: The files of "Atomic Bomb Movies" and the future of documentary films),' in *Senshū hōgaku ronshū* (Law journal of Senshū University), joint issues, nos. 55 & 56, February 1992, pp. 527–47.

Magazine Censorship

[26] Atsushi Furukawa, 'Zasshi "Kaizō" ni miru senryō-ka no ken'etsu no jittai (1) (The actual censorship situation under the Occupation as evident in the magazine "Kaizō" (1)),' in *Tokyo Keizai Daigaku kaishi* (Journal of the Tokyo University of Economics), joint issue nos. 116–17 (joint issue), September 1980, pp. 133–43.

[27] Atsushi Furukawa, 'Nenpyō: Senryō-ka no shuppan, engei, hōsō ken'etsu (Chronological table: Censorship of publications, entertainment and broadcasting under the Occupation),' in *Tokyo Keizai Daigaku kaishi* (Journal of the Tokyo University of Economics), no. 118, December 1980, pp. 231–341.

[28] Eisaburō Okuizumi, '*Warera no kagaku* oyobi *Warera no Sekai* shishi: Minshushugi Kagakusha Kyōkai henshū-shi ni miru beigun no ken'etsu no jittai (The realities of American military censorship as evident in the

magazines edited by the Association of Democratic Scientists, *Our science* and *Our world*),' in *Toshokan zasshi* (Library magazine), vol. 75, no. 8, August 1981, pp. 461–65.

[29] Eisaburō Okuizumi (ed.), *Maryland Daigaku Karejji Paaku-kō Makkerudin Toshokan Tōa Toshobu Gordon W. Prange bunko kaisetsu: Senryō-gun kenetsu zasshi mokuroku kaidai, Shōwa nijūnen – Shōwa nijūyonen* (Comments on Gordon W. Prange Collection, East Asia Section, McKeldin Library, College Park Campus, Maryland University: An inventory and synopsis of magazines censored by the Occupation Forces – 1945–1949), Yūshōdō Shoten, 1982.

[30] Atsushi Furukawa, 'Senryō to shuppan ken'etsu: Joron (The Occupation and publication censorship: An introduction),' in *Kenpō no tenbō: Kobayashi Naoki sensei koki shukuga* (An outlook on constitutional studies: In celebration of 70-years of achievements by Professor Naoki Kobayashi), Yūhikaku, 1991, pp. 281–97.

[31] Kazumoto Takano, 'Senryō-gun no zasshi ken'etsu to *Minshu hyōron* (Occupation Forces' magazine censorship and *Democratic voice*),' in *Ōhara Shakai Mondai Kenkyūsho zasshi* (The Journal of the Ōhara Research Institute of Social Problems), no. 406, October 1992, pp. 18–42.

Newspaper Censorship

[32] Jūrō Fukushima, 'Senryōki shoki ni okeru shimbun ken'etsu (Newspaper censorship during the early period of the Occupation),' in Shisō no Kagaku Kenkyū-kai (Study Group of the Science of Thought) (ed.), *Nihon senryō-gun: Sono hikari to kage, jō-kan* (The Occupation Forces in Japan: Their bright and dark sides, vol. 1), Gendai-shi Shuppan-kai, 1978, pp. 115–36.

[33] Kōkichi Takakuwa, *MacArthur no shimbun ken'etsu: Keisai kinshi sakujo ni natta shimbun kiji* (MacArthur's newspaper censorship: Newspaper articles that were banned or deleted), Yomiuri Shimbunsha, 1984.

Broadcasting Censorship

[34] Kazu Takeshima, 'Senryō-ki no ken'etsu to hōsō: Rajio kōdo (Censorship and broadcasting during the Occupation: The radio code),' in *Hōsō kenkyū to chōsa* (Research and surveys on broadcasting), vol. 29, no. 12, December 1979, pp. 24–9, 59.

[35] Hōsōhō Seiritsu Katei Kenkyū-kai (Research Group on the Broadcasting Legislation Enactment Process) (ed.), *Shiryō: Senryō-ka no hōsō rippō* (Data: Broadcasting legislation during the Occupation), University of Tokyo Press, 1980.

[36] Nippon Hōsō Kyōkai Hōsō Bunka Chōsa Kenkyūsho Jōhō Chōsa-bu (NHK Research Institute on Broadcasting Culture, Broadcast Information

Research Division) (ed.), *GHQ bunsho ni yoru senryō-ki hōsōshi nenpyō, Shōwa 20-nen 8-gatsu 15-nichi – 12-gatsu 31-nichi* (A chronological table of a history of broadcasting during the Occupation compiled from GHQ documents, 15 August – 31 December 1945), NHK Hōsō Bunka Chōsa Kenkyūsho Jōhō Chōsa-bu, 1987.

[37] Hideki Kōgō, 'Tai-nichi ken'etsu seisaku no keisei katei to hōsō ken'etsu: Bei tōgō sanbō honbu bunsho o chūshin ni (The formation process of the censorship policies for Japanese broadcasting: Primarily based on the documents of the U.S. Joint Intelligence Headquarters)', in *Hōsō kenkyū to chōsa* (Research and survey on broadcasting), vol. 38, no. 1, January 1988, pp.22–31.

[38] Eiji Takemae (supervising editor), *GHQ shirei sōshūsei: SCAPIN* (A complete collection of GHQ directives: SCAPIN), 15 vols., MT Shuppan, 1994.

[39] Akiko Takeyama, *Sensō to hōsō* (War and broadcasting), Shakai Shisōsha, 1994.

Occupation Policies

[40] Ministry of Foreign Affairs, Political Affairs Bureau, Special Data Division, *Nihon senryō oyobi jūyō kanri bunsho-shū – Dai 1-kan: Kihon-hen, Dai 2-kan: Seiji, gunji bunka-hen* (The Occupation of Japan and a collection of important management documents – vol. 1: Basic and vol. 2: Political, military and cultural affairs), 1949. Reproduced in 1989 as *Nihon senryō jūyō bunsho* (Important documents relating to the Occupation of Japan, Nihon Tosho Sentaa.

25 The Association of Democratic Scientists (Minka)

Shigeru Nakayama

Union of Science and Democracy: the Formation of Minka

In January 1946, Japanese scientists formed a nation-wide organization named Minshushugi Kagakusha Kyōkai (The Association of Democratic Scientists, hereafter abbreviated: Minka). Minka is symbolic of the scientists' movement during the period of postwar democracy. It provided a guiding light towards the future during the postwar ideological confusion.

On 19 January approximately 2,000 social and natural scientists gathered for the Inaugural Assembly of Minka at the Auditorium of the Japanese Red Cross in Shiba, Tokyo. Kinnosuke Ogura, a mathematician and historian of mathematics was elected President. Karoku Hosokawa and Hyōe Ōuchi, both social scientists, were elected as Vice-Presidents. The cadre members of the Association were Marxist activists, mainly in the social sciences, who had been active in the prewar Yuibutsuron Kenkyūkai (Society of Materialism Research) and Puroretaria Kagakusha Dōmei (Union of Proletariat Scientists).[1] Compared to the small membership of their prewar underground organization, the new postwar organization encompassed a large membership from all fields of science and technology. Among the promoters were prominent scientists like Hideki Yukawa, Shin'ichirō Tomonaga and Fujio Egami. W. Hicks of CIE (GHQ's Civil Information and Education Section) was invited to the Inaugural Assembly and delivered a congratulatory speech.

The Role Played by Minka

Science and Democracy as Liberators from Militarism

It was not the Japanese but GHQ who rescued the communists and intellectuals that had been imprisoned for political and ideological offences in wartime Japan. Thus, in the early phase of the Occupation, the Japanese Communist Party regarded GHQ as liberators. Conventional scientists and engineers without any particular ideological or political commitment felt that the ultra-nationalism and militarism of wartime

Japan had been absurd. Thus, they too regarded GHQ as liberators. Hence, the idea of democracy that GHQ brought to Japan – including its demilitarization policy – was welcomed as a kind of liberation ideology.

The logic went as follows: the feudalism and ultra-nationalism that had dominated prewar and wartime Japan were irrational and unscientific ideologies. In contrast, democracy is rational and therefore scientific. Therefore, true scientists should be democratic. Science emerged in modern times to overcome the vestiges of feudalism. In reality, though, the practice of science is not thoroughly democratic because irrational and unscientific reactionary forces interfere with the diffusion of science. Here, the following formula holds: ultra-nationalism = feudalism = unscientific = irrationalism is contrasted to democracy = modernity = science = rationalism. This was understood as obvious among the progressive scientists in Minka. The major objective of Minka was to reconstruct Japan as a democratic nation through the creation and diffusion of science.

Laboratory Democracy

Unlike social scientists who deal with the broader society as their object of research, natural scientists and engineers spend most of their time in laboratory workplaces. Hence their quest for democracy focused on the democratization of their laboratories.

From this perspective, the first goal for democratic scientists should be to wipe out feudalism in their laboratories. The social constructivist view of science outlined by J. D. Bernal (among others) was flourishing at the time. The main Japanese advocate, Shōichi Sakata, a distinguished physicist at Nagoya University, and his followers enthusiastically studied Bernal's *The Social Function of Science* (1939) during WWII. After the war they incorporated Bernal's views in their Kyōshitsu Kenshō (Laboratory Constitution), which soon became the model of institutional reform for the Japanese scientific community.

A simple, ideology-free definition of democracy is simply: 'majority rule.' Laboratory democracy in universities and research institutes is the application of this principle by the scientific community. In fact, it was employed to overthrow the old hierarchical systems and to establish departmental management (including recruiting and appointing new members) on the basis of the egalitarian participation of all the members of the laboratory in the decision-making of the Kyōshitsu Kaigi (Laboratory Council), with all members – graduate students and technical assistants as well as professors – having an equal vote. This Laboratory Council system was welcomed and followed nationwide by those who sought a new way for organizing research, especially the younger researchers in the physical sciences. The Kyōshitsu Kaigi, even now, can be found in many universities, especially in the natural sciences, although in a fossilized form.

The system is quite idealistic by any account, but the problem of the conflict between elitism, which is inherent in any competitive scientific community, and the egalitarian participation of democracy, remains unacknowledged and unresolved. Majority decision is also the operating principle of the JSC, which is incompatible with the principle of an efficient technocratic top-down system as instituted by Kagaku Gijutsu Kaigi (Science and Technology Council).

Intellectual Function

Parallel to the claim of laboratory democracy, Minka gatherings advocated opening new research areas that had been effectively prohibited by the prewar chair system (cf. ch. 15, 16). These new fields included elementary particle theory in physics, ecology, molecular biology and the Michurin Lysenco theory in biology. Advocates of the Michurin-Lysenco theory organized a Theoretical Biology Branch within Minka and published the journal *Seibutsu kagaku* (Biological sciences, Iwanami). The bulletin of Minka was named *Minshushugi kagaku* (Democratic science), later changed to *Shakai kagaku* (Social science), while a new magazine *Shizen kagaku* (Natural science) was also created by Minka. Still later, the academic functions of these bulletins moved to specialized journals as the activities of specialist societies resumed.

The Conflict among Special Sections

For humanist intellectuals in the West, science has been historically identified with various (Western) concepts, such as rationalism, secularism, universalism and democracy. Postwar Japanese scientists and intellectuals identified science with rationalism, based on their abhorrence and reaction to the irrational and emotional ultra-nationalism that had dominated Japanese society in the final stages of WWII. For humanistic intellectuals in particular, science is nothing other than rational thinking. For the cadres of Minka, primarily social scientists and materialist philosophers, science also meant the scientific socialism outlined by Friedrich Engels.

On the other hand, for natural scientists science is primarily research that differs in different disciplines and professions. This is a professional view of science. For instance, the elitism embraced by the Physics Section – headed by a theoretical physicist, Shōichi Sakata – reflects a considerably different view of science to the mass mobilization approach adopted by the geology section of Chidanken (Chigaku Dantai Kenkyūkai, The Association for Geological Collaboration Japan) with Shōji Ijiri as their charismatic leader.[2] While the former was central to the laboratory democracy movement, the latter was oriented towards the mass organization of grassroots researchers (cf. ch. 26). Similarly the History Section wanted to organize local historians.

Based on their different views of science, the Physics and Geology Sections had diametrically opposed attitudes towards the construction of the Research Institute of Nuclear Physics affiliated with Tokyo University. The Physics Section, afraid of being seen internationally as 'backwards' in nuclear physics, tried to enlighten people who expressed 'naïve misgivings' about nuclear power by promoting the merits of academic science. The Chidanken, on the other hand, were representatives for the peoples' opposition to nuclear research. After much persuasion by its director Shin'ichirō Tomonaga, the Institute was established in 1955.

Much later, Nihon Kagakusha Kaigi (Council of Japanese Scientists, the successor of Minka) concluded that the antagonism between the different sections was a significant cause of the decline of Minka and abolished the disciplinary Sections, replacing them with local regional branches instead of disciplinary branches.

The interpretation of science and democracy differed not only between the sections, but also between various localities and universities. Nevertheless, Minka, which had its roots in prewar underground activity, attempted to organize all scientists under the direction of its central secretariat, but such military-like organization did not fit the democratic mood of the intellectual and scientific organization in peacetime.

The People's Front

The mass-enlightenment of society was one of the primary objectives of Minka; however, the interpretation and execution of this objective also varied across the different disciplinary Sections. While the elitist community of the Physics Section did not sufficiently understand mass movements, the History Section with a higher proportion of high school teachers and local historians in its membership, the Geology Section (Chidanken), an organizer of field work involving a wide community, including high school students, and the Biology Section which was involved in assisting farmers in their adaptation of the *Yarovizatsiya* (a Soviet agriculture system based on the Michurin-Lysenco theory), have survived longer than the Physics Section, because of their mass-enlightenment activities, although each Section has its own organizational problems.

The membership of Minka expanded from its early scientist-centered orientation to later include students and school teachers. From 1949 the center of union activity moved from company engineers and technicians to university assistants and students. Since then, Minka's role as a labor union has declined, but, according to Tetsu Hiroshige, it was still quite prominent in the student movement in 1950–51.[3] In 1954, Hideomi Tsuge, the Secretary-General of Minka, tried to participate in the World Federation of Scientific Workers as its Japanese local branch, but activists in the union movement were caught-up in the Red Purge. However, by then Minka's international function had been reduced to issuing ideological and political statements. It could no longer claim to be the union representing professional scientific workers.[4]

In 1952 Tadashi Ishimoda of the History Section of Minka proposed a concept of 'Kokuminteki Kagaku (Japanese People's Science).' The concept was adopted in 1953 by Shōji Ijiri as the leading principle of the Chidanken. But it was not accepted by all of the Minka scientists. For most Japanese scientists, words like 'Kokuminteki (literally: National People's),' or 'Minzoku (Race, or Ethnicity)' were taboo, being reminders of wartime ultra-nationalism. For those such as the Physics Section who believed in the universality of science, it was virtually impossible to introduce a phrase such as 'Kokuminteki Kagaku' into their daily research activity or world-view. It has been said that the anti-universalistic approach of 'Kokuminteki Kagaku' is a literal translation of the Chinese Communist Party's wartime strategy. It was not appropriate to the history and reality of the Japanese scientific community of the time.

However, considering the differences in research fields and methods, it is not surprising that the 'Kokuminteki Kagaku' was advocated by practitioners of 'local science' in the Historical Section and the Geology Section (Chidanken). That is, in the fields of sciences dealing directly with local history and geology, workers were in close contact with, and saw the real life of the people in the lower social strata of Japan. In contrast to the Technology Section, for whom the 'Kokumin' – society – was de facto the industrial sector, workers in the local sciences had regular contact with grassroots people. Thus, the Chidanken was at one time the main actor of the 'Kokuminteki Kagaku.'

The Chidanken financially supported Minka for some time with the membership fees from their research groups, but they did not get along well with the Social Science Section, which they claimed 'argues but doesn't pay,' and finally withdrew from Minka in 1957.[5] Eventually the only Section Bulletins that officially acknowledged an association with Minka were *Rekishi kyōiku* (History and education) from the History Section and *Seibutsu kagaku* from the Biology Section. The Biology Section derived its methodology from the Michurin-Lysenco School. The Japan Michurin Society was formed in 1954 within this Section as a farmer's movement seeking progress in *Yarovizatsiya* technology. *Seibutsu kagaku* aimed to provide the Society's scientific and theoretical resources.

Minka as a Political Movement

Following their liberation from wartime oppression, some scientists reflected on their lack of resistance to the military domination of Japanese society. They were eager to organize a movement to speak up on day-to-day political affairs as scientists and apply pressure to the government. One of Minka's organizational objectives was to transmit politically-oriented peace statements from the JSC internationally, but when the Minka cadres stepped way-out on the political movement, average scientists could not follow them. And because Minka was essentially a voluntary association, the central secretariat could not tighten its control over scientists in the way that a political party might.[6]

Relationship with the Japan Communist Party

The aim of GHQ's ESS/ST was to develop a science and technology research structure that would contribute to Japan's economic recovery. To this end, they expected research to be shared with Minka, as well as the average Japanese citizenry. But the economic recovery GHQ sought was the recovery of a capitalist economy, and this was not the cadres' aim. The ultimate goal for the cadres of Minka was to create a communist society,

There was a firmly established view among the discussants of science and technology groups that under a capitalist regime, capitalists manipulated and distorted the 'true' course of S&T development. For instance, capitalists bought the patents of new inventions and then shelved them if it proved advantageous to their vested interests. From this perspective, S&T could 'truly' develop only under a socialist regime.

Minka was initially a loose and fuzzy organization. Around 1947 it became involved in the cultural policy of the Japan Communist Party.[7] In 1948 the Science and Technology Section of the Japan Communist Party was formed with a negotiation between the Minshu Gijutsu Kyōkai (The Society of Democratic Technology, Technology Section of Minka) and Sanzō Nosaka, Chairman of the Japan Communist Party.[8] It soon became clear that the Japan Communist Party intended to use Minka as an agent for their cultural activity.[9] In 1951, the 'Science and Technology Thesis' of the Japan Communist Party appeared,[10] providing a guideline for Party scientists. Around this time, the Japan Communist Party was outlawed by GHQ. The Party then went underground, but continued to make public statements through their affiliated cultural organizations. Minka's affiliations with the Party are clearly discernible in the organizational chart of the Japan Communist Party which includes the Nihon Minshu Shugi Bunka Renmei (The Japan League of Democratic Culture) as the upper structure of Minka.[11] Through this League, Minka was urged to issue many political statements.

Finally, Minka functioned simply as a mouthpiece for the underground Japan Communist Party to make statements about cultural affairs, in spite of the fact that the Minka organization was hollowed out in those days.

The Purge of Minka

ESS/ST was rather late in recognizing Minka's political affiliations. GHQ had sent a congratulatory message upon the founding of Minka, but subsequently there was no contact between them until 1948, when Shun'ichi Ono and his colleagues at the Research Recovery Conference petitioned ESS/ST (cf. ch. 9). ESS/ST then began to investigate Ono's group and discovered Minka behind them. Harry C. Kelly himself visited Minka during the investigation but was refused entry to their offices.[12]

This prompted GHQ to tighten its monitoring of Minka. A rumor circulated in the scientific community that anyone with connections to Minka would be denied visas to visit the US. Some foundation for this rumor can be found in Occupation documents; in 1950 Hajime Matsuura and in 1951 Shōichi Sakata and Yasuo Kondō were each denied overseas travel. They were all members of Minka. The effects of McCarthy's Red Purge in the US appears to have reached as far as the outpost offices in Japan.

Many scientists, especially in the physical sciences, wished to study in the US; so, because of the risk of being denied a visa, many withdrew from their Minka affiliations.[13] Others, like the Chidanken, concentrated on local sciences and did not consider it necessary to visit the US. They argued, instead, for the creation of 'Kokuminteki Kagaku' and opposed the trend of studying abroad.[14]

In 1955, the Japan Communist Party abandoned the radical line modeled after the Chinese Communist strategy. This change of position brought further confusion to the cadres of Minka. Moreover, Minka was in financial trouble. Organizational reform was discussed at the General Assembly in 1956, but they actually ceased activity then.[15] The Chidanken continued to support Minka activity for a while, but they found Minka did not give them any merit and, as previously mentioned, formally withdrew in 1957.

Factors in the Decline of Minka

Tetsu Hiroshige attributed the decline of Minka to the political maneuvering by its cadres.[16] For a short period after WWII, the Japan Communist Party was open-minded and more democratically oriented, but as the Cold War intensified in the international arena, they resumed their underground style of top-down directives from a quasi-militaristic organization in a state of emergency without the freedom to criticize the principles established as Party line. Secretivism and sectionalism prevailed and the organization remained closed to the outside world. Hideomi Tsuge, the Secretary-General of Minka, later confessed that he was misunderstood as being a secret member of the JCP, though he was not.[17]

Under Pax Americana – the protection of the Occupation Forces – the postwar Japanese world was, at least on the surface, an open and peaceful society, unlike the wartime society had been. The Minka activists' failure to acknowledge and adapt to this new environment, their anti-American motivated actions and their sense of political crisis created a gulf between Minka and the ordinary JSC members and general scientists. It led to Minka's alienation from the Japanese scientific community.

In the early 1960s, the History and Biology Sections still preserved the title of Minka branches but among those who had once participated in Minka activity and then withdrew there was a sense of rejection and antipathy. It became a liability to use Minka's title for conducting organizational activities.

In conclusion, though, we should consider Minka's history in a wider context. At the time of Minka's founding, scientists were eager to find a new direction for S&T and gathered together across disciplines to explore and develop the possibilities for the future. Then, as the postwar system became established and its direction (somewhat) set, the scientists returned to their laboratories and disciplines. They had no more need for an organization such as Minka.

Notes

1 According to Yoshirō Hoshino, Minka was organized by the Union of Proletariat Scientists rather than the Society of Materialism, and thus it had no liberal atmosphere from the outset. Hoshino, 1976 [7]

2 Shigeru Nakayama, 'Ijiri Shōji Ron (On Shōji Ijiri)', in *Shisō no kagaku* (The science of thought) (1966); also Shigeru Nakayama, 'Grassroots Geology,' in S. Nakayama et. al. (eds.), *Science and Society in Modern Japan*, 1974.

3 Hiroshige, 1960 [8] p. 153.

4 Hoshino, 1976 [7]

5 Ijiri, 1977 [7] pp. 105 ff.

6 Cf. 'Minka Seibutsu Bukai no Seibutsugaku Shisō (The Biology Section of Minka and its biological thought)' in Nakamura, 1970 [9]

7 Watanabe, 1974, [16] p. 321.

8 Nobuo Ōhashi, 'Sengo no kagaku gijutsusha undō (1), Ishii Kinnosuke ni kiku, (Postwar movements of scientists and engineers, no.1, Interview with

Kinnosuke Ishii),' in *Gijutsu to ningen* (Technology and humanity), April 1976, pp. 141–51.
9 Tsuge, 1980 [3]
10 *Zen'ei* (Vanguard), no. 58.
11 Torii and others, 1976 [7]
12 Tsuge, 1980 [3]
13 Hiroshige, 1960 [8] pp. 68–9.
14 Hiroshige, 1960 [8] p. 135.
15 Hiroshige, 1960 [8] p. 131.
16 Hiroshige, 1960 [8] p. 175.
17 Tsuge, 1980 [3]

Bibliographical Guide

[1] Minshushugi Kagakusha Kyōkai (Association of Democratic Scientists) (ed.), *Kagaku nenkan* (Yearbook of science).
Vol. 1 (1947) & vol. 2 (1948) were published by Nihon Kagakusha. Vol. 3 (1949) was published by Hokuryūkan. Each volume has a different editorial format.

[2] Minshushugi Kagakusha Kyōkai (ed.), *Warerano kagaku* (Our science).
Edited by Takeo Konno. A magazine started on 1 March 1946 and continued until June 1947. Yōichi Fukushima, Kazuto Matsumura, Minoru Tanaka, Ryūichi Yasugi and Yoshinori Kaneseki were regular contributors.

[3] Hideomi Tsuge, *Minka to Watashi: Ichikagakusha no ayumi* (Minka and I: Footsteps of a postwar scientist), Keisō, 1980.
The reminiscing of a biologist and Ex-Secretary-General of Minka based on document sources. Attached is a chronology of Minka and its affiliations, its constitution and a list of executives.

[4] Hideomi Tsuge, *Kagaku wa heiwa o motomete* (Science seeks peace), 1954.
A collection of the author's fragments. Informative about Minka and its relation with the World Federation of Scientific Workers.

[5] Nihon Kagakushi Gakkai (History of Science Society of Japan) (ed.), *Nihon kagaku gijutsu-shi taikei 5, Tsūshi 5* (A compendium of Japanese history of science and technology, vol.5, Historical Overview 5), 1964.
Chapter 3: 'Kagaku gijutsu bun'ya no minshuka undō (Democratization movement of science and technology fields),' includes conference records from the foundation of Minka through its peak period.

[6] Nihon Kagakusha Kaigi (Scientists Society of Japan) (ed.), *Kagakusha undō no shōgen* (Testimony of the scientists movement), 1978.
Fourteen members recall the activities surrounding the creation of Minka in various geographical areas and specialties. Attached is Minshushugi Kagakusha Kyōkai Kiyaku (The Rules of Minka), revised at the Third

General Assembly on 24–25 January 1948. This includes self-criticism of their past political bias. Places more emphasis on research activity and agrees to the founding of the Research Section. Documents and a simple chronology.

[7] Noboru Takahashi (ed.), 'Sengo no kagaku gijutsusha undō (Postwar movements of scientists and engineers)', in *Gijutsu to Ningen* (Technology and humanity).
Takahashi conducted interviews and panel discussions on the activities of scientists and engineers in postwar movements like Minka and ran a series of feature articles in the magazine. It provides a critical review from the perspective of the new left since the late 1960s. The following are directly related to Minka activity:

- Yoshirō Hoshino, 'Minshushugi Kagakusha Kyōkai no gijutsusha undō (Engineers' movement in the Association of Democratic Scientists)', in *Gijutsu to ningen* (Technology and humanity), May 1976, pp. 141–53.
- Hiroshi Torii et. al., 'Minka undō no rekishiteki haikei (Historical background of the Minka movement),' in *Gijutsu to ningen* (Technology and humanity), June 1976, pp. 141–53.
- Hisashi Koyama et. al., 'Minka, Mingikyō to genba gijutsuha (Minka, Mingikyō (Association of Democratic Engineers) and engineers on the shop floor),' in *Gijutsu to ningen* (Technology and humanity), July 1976, pp. 146–51.
- Hideomi Tsuge, 'Minshushugi Kagakusha Kyōkai to watakushi (The Association of Democratic Scientists and I),' in *Gijutsu to ningen* (Technology and humanity), September 1976, p. 144ff.
- Kiyoshi Mori, 'Soshiki toshiteno Minka (Minka as an organization),' in *Gijutsu to ningen* (Technology and humanity), October 1976, pp. 125–35.
- Shōji Ijiri and Masae Ōmori, 'Aru kaiko: Minka wa naze suitai shitaka (A retrospective: Why Minka declined),' in *Gijutsu to ningen* (Technology and humanity), February & March 1977. This dialogue was later included in Shōji Ijiri, *Hiteiteki seishin* (Mentality of antithesis), Tsukiji Shokan, 1986.

[8] Tetsu Hiroshige, *Sengo Nihon no kagaku undō* (The science movements in postwar Japan), Chūō Kōronsha, 1960.
The author was a member of Minka. In Chapter 5 'Minshushugi Kagakusha Kyōkai (Minka),' he followed the history of Minka and concluded that its political maneuverings were the reason for its decline. His thesis invited reactions and repulsion in contemporaneous movements. Critics said that Hiroshige's analysis was only criticism and provided no guidelines for the future of the movement. However, his work deserves to be respected as an orthodox history of the scientists'

movement during the postwar democracy period. It also provides an outline history of postwar science.

[9] Teiri Nakamura, *Seibutsugaku to shakai* (Biology and society), 1970. Includes 'Minka Seibutsu Bukai no seibutsugaku shisō (The Biology Section of Minka and its biological thought).'

Bulletins of Minka

[10] *Minshushugi kagaku.*
March to September 1946; later renamed *Shakai Kagaku.*

[11] *Shizen kagaku.*
Launched in July 1946.

[12] Quarterly *Seibutsu kagaku.*
Started in March 1949. Bulletin of the Biology Section. Its editorial principle was the introduction of the Soviet Michurin-Lysenco line and pioneering new fields like biochemistry and ecology.

Also, *Rekishi kyōiku* of the History Section and *Riron* of the Philosophy Section.

Miscellaneous

[13] Nihon Kyōsantō Kagaku Gijutsubu (Japan Communist Party, Science and Technology Section) (ed.), *Sovieto no kagaku to gijutsu* (Science and technology of the Soviet Union).

[14] Yasoji Kazahaya, 'Minka Sōritsu no koro (When Minka was established),' in Nihon Kagakusha Kaigi (ed.), *Kagakusha undō no shōgen* (Testimony of the scientists' movement), 1978, p. 35.
The author was the chairman of the Founding Assembly of Minka and was elected as the first Secretary-General. According to his testimony, he evaluated GHQ as a comrade in 1945 on the basis of their policy of liberating political criminals and the various democratization policies announced one after another. Toward the end of 1945, the young officers of GHQ, mainly the left-wing group from the State Department, with whom he had contact for discussing and reporting on their policy returned to the US and reactionary policy became prevalent. He was irritated by this trend and appealed to his fellow Japanese, such as members of the former Yuibutsuron Kenkyūkai (Materialism Society). They organized Minshushugi Kagakusha Kyōkai (Minka) with the support of many intellectuals. However, this claim is a little confusing: at the end of 1945, the left-wing New Dealers in GHQ had not yet departed. They returned to the US about two years later when the Cold War intensified.

In contrast, Tokunosuke Nakajima 'Sengo Nihon no kagaku kenkyū no hōkō (The direction of scientific research in postwar Japan),' in *Zen'ei*

(Vanguard), no. 612, 1991, p. 226, locates the right-turn of GHQ considerably later. He argues that with the establishment of the People's Republic of China in October 1949, US Occupation policy changed radically.
[15] Mononobe Nagaoki, 'Gen'ei no minshushugi kagaku (Democratic science as illusion),' in *Asahi Journal*, 26 October 1969, pp. 111–16.
[16] Watanabe, Yoshimitsu (dictated), Hiyaringu Gurūpu (Hearing Group) (ed.), *Shisō to gakumon no jiden* (An autobiography of thinking and learning), Kawade Shobō Shinsha, 1974.
Includes a chapter on Minka.

26 Democratization Movements in the Scientific World and the 'Red Purge'

Toshifumi Yatsumimi

From Cultural Nation to Democratic Nation

Although Japan had accepted the Potsdam Declaration, the country's future direction remained unclear. With the Allied Occupation imminent, Japan seemed unable to envisage a new national image for itself. The Prime Minister, Kantarō Suzuki, provided some leadership in this regard when he spoke of the way forward in a broadcast made on the night of 15 August 1945. Addressing the nation, he advocated 'retaining [the] national polity,' the 'promotion of science and technology, which proved to be our greatest shortcoming in the war' and the creation of 'a culture positioned to make a contribution to the civilization of mankind.'[1]

Communicating through both newspapers and radio, Tamon Maeda, Minister of Education in both the Naruhiko Higashikuni and Kijūrō Shidehara Cabinets, was another politician who emphasized the importance of science for rebuilding the Japanese culture, calling for the nation to foster 'pure scientific thinking founded upon a search for truth.'[2] Such thoughts formed the basis of the 'Education Policy for the Rebuilding of a New Japan'[3] which was released to the public on 15 September 1945. The central pillars of this policy were: the retention of a national polity, the construction of a peaceful nation, improvements in education and the cultivation of scientific thinking. The new policy was drawn up without instruction from the General Headquarters, Supreme Commander for the Allied Powers (GHQ) and can therefore be read as broadly reflective of the Japanese Ministry of Education's immediate postwar attitudes and assumptions.

While this idea of a national culture that promotes science appeared to symbolize the new, postwar Japan, it was in fact little more than a reworking of a theme often voiced during the war, when the emphasis was upon winning the 'scientific war.' Even the scientists themselves were amazed with the postwar resurfacing of this terminology. The University of Tokyo's geophysicist, Professor Chūji Tsuboi, gave expression to a more general sentiment when he posed the ironic question, 'Why is it that science must always be treated so specially?'[4] In thinking about these connections we must not discount nor downplay the possibility that

Japan's loss of the 'scientific war,' its failure to develop its own nuclear bomb, acted as an unstated, perhaps unconscious, motivation for its postwar promotion of science.

As to the question of 'culture,' this term had become popular in the Taishō period of Japanese history. Since then a search for and emphasis upon a notion of 'Japanese culture' had been a constant in Japanese public discourse. One need only recall that, in Japan, World War II was portrayed as a 'war for the sake of culture.'[5] Minister of Education Maeda stated that, 'What is more important than anything else in terms of national recovery after the war is creating a means of getting out of the despondency being felt, setting a new goal and arousing a new-found courage to move towards it. There is no other way for us to go.'[6] The new goal that Maeda and others set was the idea of a 'Cultural Nation,' a national ideology which would unify the populace.[7]

In September 1945 the Occupation Forces began implementing their basic policies of demilitarization and democratization, based on the Potsdam Declaration and the American 'initial basic policy.' Hence GHQ did not approve the Japanese Ministry of Education policy, which advocated the retention of a national polity as the first consideration of education,[8] but instead resolutely enacted General Douglas MacArthur's Five Major Reform Orders (11 October) and the Four Major Education Orders (22 October – 31 December), which focused on abolishing militarism and ultranationalism while encouraging liberalism and democracy. The GHQ edict relating to demilitarization and democratization covered the areas of politics, economics and society. As such it marked the point in time when the ideology of promoting science ceased to be a central theme for reconstruction and the concept of 'Cultural Japan' was transformed into that of 'Democratic Japan.'

The term 'democracy' is open to a multitude of interpretations. What version of democracy, then, were the Allied Powers, particularly the US, seeking for Japan? One way of determining this is to study the ways and contexts in which the term 'democratic' was used in Occupation documents and the reports of Allied delegations to Japan. In doing this one discovers that 'US democracy' involved: 1) the concept and system of opposing ultranationalism, imperialism, militarism, feudalism and autocracy, 2) ideas intimately associated with personal freedom and basic human rights and, 3) a welcoming of diversity and the principle of living by which minorities are respected.[9] One also notices that there was much discussion about where authority would reside in a Japanese democracy, in contrast to the US democracy where the emphasis was placed on the guarantee of individual rights. Overwhelmed by the constant release of democratization orders, there was little scope for Japan to question the underlying assumptions of the broader democratization project. In this way the term 'democracy' came to prominence in Japanese public discourse without ever being clearly defined.

In January 1946 an organization was born which for the first time combined 'democracy' and 'science' in its name – a symbol of the times. It was called Minshushugi Kagakusha Kyōkai (abbreviated as 'Minka': the Association of Democratic Scientists). This was an organization made up of those involved in the social sciences who, prior to the war, had been involved in the proletarian Science Research Institute and the Materialism Research Society as well as other academics from the arts and sciences, including natural scientists like Hideomi Tsuge and Mitsuo Taketani (cf. ch. 25; 'The Association of Democratic Scientists').

The Elementary Particles Theory Group

Within Minka's organizational structure it was left to sub-groups or newly formed researcher groups to enact the Association's aim to promote 'the democratization and expansion of science facilities and organizations.' Two particularly active groups within the democratization movement were the Elementary Particles Theory researchers and Chigaku Dantai Kenkyūkai (abbreviated as Chidanken; Association for Geological Collaboration Japan), the activities of which will now become our focus.

In Japan in the 1930s the Physics Departments of Tokyo, Kyoto and Tōhoku Universities were based on a hierarchical, European model, one not attuned to developments in the new field of Elementary Particles Theory. The breakthrough research in this field, undertaken by people like Hideki Yukawa and Shin'ichirō Tomonaga, was instead occurring in new facilities at the Osaka University and the privately funded Nishina Research Laboratory at Rikagaku Kenkyūsho (abbreviated as Riken; the Institute of Physical and Chemical Research). Yoshio Nishina – who had studied at Niels Bohr's Institute for Theoretical Physics at the University of Copenhagen – brought experience to Riken and at the same time encouraged new ideas and robust debate amongst his younger researchers. This more democratic approach came to be known as 'joint research,' and saw young researchers sharing and contesting ideas, as well as dividing the mathematical workload between them. In this way everyone developed and contributed their own theories about, and to, the central research goal. The old style of self-contained research in which individuals set themes according to their own areas of interest and completed the research without discussion was deemed unworkable in the field of elementary particle theory. Hence, there was a succession of jointly authored theses produced in this research area.

These collaborative groups were generally referred to by the name of the person, affiliated institutions, or place associated with the cutting edge research being done, such as the 'Yukawa Group' and the 'Tokyo Group.' However, in Japan the elementary particle theory researcher groups were collectively referred to as simply the 'Elementary Particle Theory Group.' The origin of the Elementary Particle Theory Group as a national organization can be traced to the Meson Association and the Mesotron Discussion Group, which existed between 1941 and 1944.[10] The central figures in these early associations were Hideki Yukawa, Shin'ichirō Tomonaga, Shōichi Sakata and Mitsuo Taketani. These men formed the nucleus of a cohesive, well functioning group that sought to overcome differences in institutional affiliations and actively encouraged free and open debate. Having experienced the benefits of joint research during the war, they set about democratizing the postwar world of physics.

One example of this was the 'laboratory democracy' espoused by Shōichi Sakata (cf. ch. 25). Sakata had come to prominence as the joint research partner of Yukawa. Following this, and with the expectation of the construction of new research laboratories, he was made a professor at Nagoya University in 1942. Much of his research interest was put on hold until after the war, at which time he began actively developing his new research unit. His basic approach was to create a system by which researchers could, of their own accord, conduct joint research on a common theme. This, though, flew in the face of tradition – where researchers were organized around individual courses, with the professors of the courses wielding enormous influence – and therefore met with considerable resistance. Despite opposition from these conservative forces Sakata managed to revolutionize Nagoya University's research culture, replacing the conventional courses with laboratories (the unit of joint research) and the practice of having a small number of people make decisions with a consultation between the joint parties. Under Sakata's guidance the Physics Department at Nagoya University staged the first Elementary Particle Physics Laboratory Conference on 24 January 1946. On 13 June of the same year it set down a charter for the department and instigated the new system (see Fig. 26-1) whereby the overall decision making body was the Kyōshitsu Kaigi (Department Council), which constituted all of the researchers affiliated with the department.[11]

The influence of the democratization of research practice, symbolized by this Department Council (referred to as 'laboratory democracy'), extended to universities across the country and brought about the adoption of Department Councils by researchers in disciplines other than physics. The influence though did not stop there. The research standards of Yukawa and his fellow elementary particle theory researchers was high and stood out in the low-key world of postwar Japanese science. This prominence enabled high profile members of the Group to enter the public debate and make pronouncements regarding Japan's social and political direction. As

Figure 26-1: Structure of the Department of Physics at Nagoya University (1946)

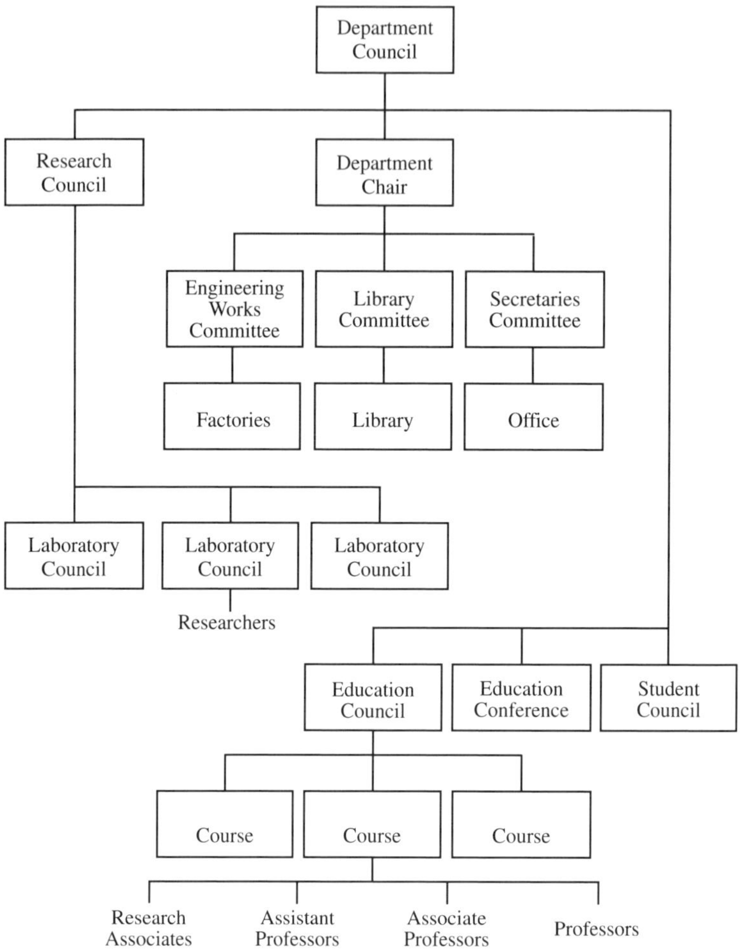

Source: Shōichi Sakata, *Kagakusha to shakai ronshū* (Scientists and society), vol 2, 1972, p. 33.

founding members of Minka, controversial figures like Shōichi Sakata and Mitsuo Taketani were able to voice their strong support for Japanese democratization. While Taketani emphasized the importance of technological research for the rebuilding of Japanese industry,[12] Sakata pressed for the rationalization of research organizations. Both understood the history of science to be the history of the development of science and both adopted dialectical materialism as a methodological theory for anticipating the future and interpreting the scientific past.[13] By virtue of their status as acclaimed scientists in the field of advanced nuclear physics Taketani and Sakata's advocacy of dialectical materialism carried great clout within the scientific and broader academic communities and contributed to a heightened interest in theoretical Marxism in general and dialectical materialism in particular (cf. ch. 28).[14]

Chidanken: The Association for Geological Collaboration Japan

Whereas the Elementary Particles Theory Group was greatly influenced by European and American academic trends and was active in the newly expanding field of physics, in the more traditional field of geology the democratization of the discipline by the Association for Geological Collaboration Japan (Chidanken) was unprecedented.

Chidanken was inaugurated on 2 February 1947 by twenty or so young geology researchers from the University of Tokyo, the Tokyo University of Literature and Science and the Tokyo Science Museum. This group was characterized by three defining features. First, its emphasis upon youth, encapsulated in its 'forty year old clause' which limited membership to those aged forty or under (those over forty wishing to join needed the group's approval). Second, its reformist agenda to democratize the academic world. And third, its promotion of group research as a preferred research method.

While there was a general trend towards youth in academic research in the immediate postwar period,[15] Chidanken, by stipulating a membership cut-off age of forty, was not merely following this trend but making a political point more specific to its own discipline. During the war a considerable number of young geologists were sent to the Asian mainland and South East Asia to conduct research into natural resources. Many of this number lost their lives in the conflict. In recognition of this fact the authors of *Nihon Chishitsu Gakkai-shi* (A History of the Geological Society of Japan) state that, 'It was sad but unavoidable that the young geologists who had graduated from 1940 onwards would become direct victims of the war.'[16] It was the geologists of that generation who survived the war (aged approximately 30 in 1945) that founded Chidanken and, in barring older members, began to transform the entrenched hierarchical culture of the old state universities.

One example of how Chidanken assumed authority and reformed the prevailing academic culture can be found in its modernization and democratization of Nihon Chishitsu Gakkai (the Geological Society of Japan). Prior to and during the war the Geological Society was more-or-less run by its president and a standing committee. While in theory the Board of Trustees were, according to a 1926 society regulation, to 'be elected from members at a plenary session,' in practice – as evidenced by reports of such plenary sessions in the society's journal, *Chishitsu Gakkai zasshi* – the matter of running the society was left to the discretion of the president. As such, the president unilaterally nominated the next president, the Board of Trustees and the standing committee. This inevitably created the impression if not the reality that the society, which during the war had grown into a national organization with over 1000 members, was being run by one elite section of its membership. Challenging this undemocratic practice at the May 1947 executive meeting, Chidanken called for elections to be held.[17] The result was that at the Kyoto plenary session later that month a ballot of members was carried out and for the first time the society obeyed its own democratic rules and elected its Board of Trustees.[18]

However it was not until the April and May plenary meetings of 1948 that Chidanken was able to initiate actual reforms to the society's rules. Key amongst these was the demand that along with the Board of Trustees, the president should be elected from and by the general membership. In response to this the old Board of Trustees proposed that forty members were needed to nominate a presidential candidate. This, though, was defeated by a Chidanken amendment that set the minimum nominating number at five.[19] From this point onwards the regulations guiding the day to day running of the society began to more accurately reflect the popular will of its membership.

Another cause taken up by Chidanken was the unfair distribution of science research funds. At the time the amount of such funds for geology was determined by Section 3 of the National Research Council, which came under the jurisdiction of the Ministry of Education. Under this scheme funds allotted to the universities and their different courses varied, but the University of Tokyo was disproportionately rewarded, as were professors. Chidanken exposed these inequalities and proposed to the National Research Council that all researchers be treated as independent scholars and that funds be distributed accordingly, i.e. fairly.[20]

Chidanken's push to reform the rules of the Geological Society and to seek a just distribution of research funds gained them prominence in the broader scientific community, so that when members of the Science Council of Japan (JSC; cf. ch. 11) came to elect office bearers in December 1948 two inaugural and influential members of Chidanken, Shoji Ijiri and Masao Gorai, were put forward as joint candidates from Minka and Chidanken. Reflecting the reformist mood of the time, Ijiri and Gorai – both in their early-to-mid 30s – defeated the Geological Society-backed candidates. In 1950 both men were selected to join the Geological Society Board of Trustees, where they quickly made their Chidanken affiliations felt. For example, on 3 April 1950 seventy-four members of the Elementary Particles Theory Group, including Kōji Fushimi, Mitsuo Taketani and

Shōichi Sakata, announced a Peace Declaration and the following day, at the Geological Society plenary meeting, a 'Resolution to Protect Science from the Interests of War' was adopted at the JSC. This was in large part due to the role played by Chidanken. On 28 April 1950 the JSC adopted a declaration stating that 'they would not be party to scientific research for the purpose of waging war.' Key proponents of this were Taketani and Ijiri.[21] As such initiatives attest, both the Elementary Particles Theory Group and Chidanken led the way for scientists to become active in the social and political debates current in postwar Japan.

Chidanken's commitment to democracy extended beyond reforming organizational structures amongst scientific societies to the very practice of scientific research. With his experience as one of a group of five members drawn from different affiliations to carry out the Transverse Geological Survey of the Tertiary in North-East Japan from 1939 to 1940, Shōji Ijiri became a keen advocate for 'collaborative research.'[22] In 1943 he carried out collaborative research with the employees of the geology section of the Tokyo Science Museum where he worked. This reinforced his conviction regarding the effectiveness of collaborative research.[23] This survey was concerned with the relationship between the siliceous rocks and the green rocks found beneath it in the area from the middle slopes to the top of Mount Dōdairayama in Saitama Prefecture. It contested Haruyoshi Fujimoto's 1937 theory that the geological structure was over thrust, pointing out that the fault plane did not exist in the area and concluding that these rocks were intrusive relations, a point that recent research appears to confirm. Drawing upon this collaborative experience, Ijiri concluded that while the ability of individuals or the amount of research funds available to individuals might prove insufficient, if a research theme were clearly identified and collaborative research carried out the result would invariably exceed that achieved through individual research.[25]

Collaborative research was born of necessity during the war as austerity budgets imposed restrictions on free, individual research. Developing research strategies in the postwar climate, though, people like Ijiri believed that even such restrictions had produced positive results. In his *Koseibutsugaku-ron* (Theory of paleontology) (1949), Ijiri claims that 'controls of research and research planning represent one of the most important things in terms of future research methodology.'[26] He contends that firm unity is necessary for collaborative research. In this sense, he suggests, collaborative research tends to be closely intertwined with research control.[27] One benefit that he claims for collaborative research is that in the group context the subjective element intrinsic to individual research is counteracted and precedence is given to collective rather than personal understanding. As such, he asserts that the group method leads to a higher order of scientific research.[28]

It is ironic that while Ijiri argued that one of the strengths of research was its ability to accommodate dissenting opinion and foster open discussion, he was unable to countenance any criticism of the collaborative research method itself. For Ijiri, a common view of the world, political awareness and a shared scientific ideology were considered necessary conditions for effective research.[29] Despite this glaring contradiction, Ijiri's theory about the efficacy of collaborative research went largely unchallenged and continued to wield great influence over the organizational management of Chidanken for many years to come.[30] In the postscript to a history of Chidanken, *Tomo ni manabu yorokobi…*(The Joy of Studying Together; 1953), Shō Ishimoda writes, 'They are all similar, just like a row of eggs.' What is telling about this assessment is that it was intended and received as a compliment rather than a criticism.

The objectives and actions of Chidanken reflected the gathering momentum of the movement towards democratization sweeping the country. Its membership increased from 20 at the time of its inauguration, to 200 in 1948, then to 300 in 1949.[31] Riding this wave of popularity, Chidanken succeeded in reforming the Geological Society in 1948 and, as already mentioned, managed to have its own nominees elected to the Science Council of Japan. In research terms, the collaborative method was first employed in a comprehensive study of the central southern part of the Miura Peninsula and became increasingly common as the preferred mode for geological research.

The Red Purge

In April 1949 Chidanken became a research chapter of Minka. Minka argued on political grounds for a joint organizational front and many key members of Chidanken, who supported Minka's policy platform, thought it only natural for the two organizations to join forces. Shōji Ijiri and Masae Ōmori recall that amalgamation with Minka heightened the Chidanken members' awareness of philosophical and socio-political issues and increased the possibility of extending the collaborative research methodology to other academic disciplines.[32] Not all of the Chidanken members, however, supported the merger. In ideological terms, Minka was to the left of Chidanken, with most of its activists being either members of the Communist Party or fellow travelers.[33] To join with Minka, then, was to align Chidanken with the Communist Party, a decision that caused considerable division and debate within the organization. The March 1949 edition of the *Chigaku Dantai Kenkyūkai kaishi* (Journal of the Association for Geological Collaboration) reports that the Kyūshū Chapter concluded it was premature to join Minka, while the Kyoto Chapter held a hearing to which members of Minka were invited and questioned in regard to Minka's ideological orientation.[34] Despite this disquiet amongst Chidanken's rank and file membership the April plenary meeting of Chidanken decided to amalgamate with Minka. Moreover, the presentation of a researcher of Tōhoku University drew severe criticism from the members of Chidanken. This led to a growing dissatisfaction with the Chidanken leadership and the subsequent disbandment of the Sendai branch of Chidanken, the host of the occasion.[35] In fact, Chidanken paid a heavy price for merging with Minka, almost immediately halving its national membership from 300 to 150.[36]

This radical reduction in the Chidanken membership is a specific example of a broader phenomenon.[37] While the push for the democratization of academia began as a broad-based movement, one that encompassed a wide range of political viewpoints, over time its activities and agenda became dominated by members and/or affiliates of the Communist Party of Japan. As this trend developed, the non-Communist members of the various democratization organizations withdrew their support and membership. This withdrawal was actively encouraged by the growing backlash against the radicalization of the democratization movement known as the Red Purge.[38]

The Red Purge refers to official efforts from 1949–1950 to rid government, government agencies and private enterprises of Communist Party members and fellow travelers. The purge is considered to have had three distinct stages. First, as part of the 1949 administrative restructuring and the curtailment of the business operations of private enterprises, in which approximately 600,000 workers were retrenched, communist factions were targeted for dismissal.[39] Secondly, from September 1949 to March 1950 the Ministry of Education attempted to expel communists and communist sympathizers from teaching posts across the country (with the exception of Kochi Prefecture). And thirdly, in response to MacArthur's letters to Japan's Prime Minister from June–July 1950 and the government decision of September 1950, communist activists were removed from Government, government agencies, municipal offices, public offices and private enterprises. According to recent research by Tetsuo Hirata, the total number of suspected communists or communist sympathizers removed from office exceeded 27,300. (See Table 26-1 for a breakdown of the numbers of those purged at each stage of the process.)[40]

The Red Purge occurred against the backdrop of the Cold War, in which there was an international surge in the size and power of the communist movement and a countervailing increase in anti-communist politics in the US. Within Japan the purges involved five distinct yet inter-related sectors of society: GHQ, the Government, business enterprises, the labor movement and the Japanese Communist Party. The degree of involvement of each sector altered in each individual case, so that an overall picture of the process can only be gained through further detailed research. This already complex task is made more difficult by the fact that many of the people

Table 26-1: No. of victims of the Red Purge 1949–50

Stage	Those affected	No. of Victims	Survey organization (person)
1	Administrative Restructuring and the Curtailment of the Business Operations	10,930	Special Investigations Bureau
	of Private Enterprises (main 45 companies)	3145	Ministry of Labor
2	Public School Teaching Staff	1113*	Isao Myōjin
	University Related	30-40**	Isao Myōjin
3	Private Enterprise	10,972	Ministry of Labor
	National Public Servants etc	1177	National Personnel Authority
Total		27,367–27,377	

* Isao Myōjin adds a fixed number to this, making it approx 1200.
** includes the numbers up until 1951.
Source: Tetsuo Hirata, *Rekishi hyōron* (Historical review), no. 472, August 1989, p. 10.

involved remain reluctant to talk about their role in these events. Hence, the Red Purge remains a poorly illuminated aspect of Japanese history.

Administrative Restructuring and the Curtailment of the Business Operations of Private Enterprises

By 1948 the USA had supported the Japanese economy to the tune of 1.25 billion dollars. While Japan strove hard to revive its economy, there was a growing sense in the US that these vast sums of financial aid were not being put to effective use.[41] Responding to domestic public opinion, in December 1948 the US government ordered Douglas MacArthur to implement the 'Nine Principles of Economic Stability,' which called for arresting inflation in the Japanese economy. In February 1949 Joseph Dodge arrived in Japan and in March he announced the 'Dodge Line,' central to which was an 'ultra balanced budget.' These external initiatives resulted in both the public and private sectors in Japan being put under extreme pressure to rationalize and restructure their organizations. A large part of this restructuring involved the reorganization of personnel. While the retrenchment of the thousands that followed were primarily motivated by economic factors, the restructuring provided a convenient pretext for the purging of communists and fellow travelers from the workforce.

The Administrative Bodies Employee Numbers Law (ABENL), which in principle reduced *gengyō* sections (ministries and agencies with public work operations) by twenty percent and *higengyō* sections (those without) by thirty percent, was proclaimed in May 1949. The result of this was that 50,000 people employed in government offices and another 95,000 employed in the Japan National Railway were dismissed.[42] Restructuring was also carried out in regional public bodies, affecting another 10,000 people.[43] In business enterprises, too, restructuring was employed as a means of curtailing the business operations of private enterprises. In 1949 alone more than 430,000 people had their employment terminated.[44] The impact of this was particularly great on the Electric Telecommunications Industries and the Steel and Chemical Enterprises. This policy of staff cutbacks attacked national, public and private scientific research bodies, all of which appeared to be facing a lean future.

While there was a succession of industrial disputes over the scale of the retrenchments, both labor and industry were reconciled to the fact that, faced with the stringent demands of the Nine Principles and the Dodge Line, personnel restructuring was unavoidable. The scientific community joined in the chorus of those voicing concern about these economic reforms. Both the March 1949 proposal of the JSC and the joint statement made in June by the members of the Physics Society of Japan and the physics section of the JSC called into question the uniform restructuring of personnel, and the inclusion of science technologists among those being dismissed. It proved difficult however to convince the Prime Minister, Shigeru Yoshida, that the scientists' should receive special treatment. Yoshida maintained that the only fair way to implement the restructuring of personnel was on the basis of the principal of uniformity.[45]

Despite their protests, government and private scientific research bodies were restructured and a number of science technologists were relieved of their duties. Two such examples were Yōichi Fukushima of the Agricultural Improvement Bureau of the Ministry of Agriculture and Forestry and Shōji Ijiri of the Tokyo Science Museum over which the Ministry of Education had jurisdiction. Fukushima was only one of the 4,301 employees of the Agriculture and Forestries Ministry subject to personnel reductions. In 1949, the Tokyo Science Museum was abolished, and the National Science Museum was organized anew. In response to the restructuring pressures, Fukushima retired in September 1949. According to the Ministry of Education Employee Fixed Numbers Regulation (Ministry of Education Ordinance No. 33), the number of the National Science Museum employees was reduced from 91 to 78 in 1949. Ijiri was unlucky to be retrenched amidst temporary employee reduction measures; by October 1950 the number of people employed by the National Science Museum had been restored to 91. This figure steadily increased in the following year, to 93 in January and 98 in April 1951.[46] Ijiri officially retired in November 1949 and all four of the researchers in the Geological Section with which he was affiliated were changed in 1950.[47]

Whether or not there was a genuine need for the 1949 restructuring of nearly 600,000 personnel in Japan's public and private sectors is beyond the scope of this discussion. Instead we must limit ourselves to questioning the criteria for this restructuring and the way it was implemented. Let us look first at the restructuring criteria. In March 1949 the government held talks with Chief Secretary Kaneshichi Masuda and suggested the following categories of people as candidates for retrenchment: 1) those who had reached the pension age; 2) those whose length of service had exceeded, for example, 25 years; 3) those whose technical skills were undeveloped, and; 4) those who had conspicuously underachieved in their job.[48] In addition to these criteria, when the ABENL was implemented in June, the government announced that the restructuring would attempt to eliminate elements that were either not conducive to the democratic rebuilding of Japan or were not in line with the country's economic reconstruction.[49] In the same month, at the Cabinet meeting organized to set down the government's restructuring criteria, a proposal outlined the following bases for retrenchment: 1) those with low levels of efficiency, and; 2) those lacking in dedication and who thereby detracted from the overall level of efficiency.[50] The restructuring criteria issued by the Ministry of Posts and the Ministry of Telecommunications in August was even more explicit in outlining the covert political motive behind the retrenchment program, underlining as it did, the importance for a worker to have a co-operative attitude over and above his or her abilities or level of work-related knowledge.[51]

When Ijiri was pressed into retiring from the Science Museum he received advice from a Japanese person affiliated with GHQ to the effect that, 'GHQ is considering exceptions [to the retrenchment criteria] in the case of excellent scientists, so I hope to have this apply to you.'[52] Ijiri declined this offer of assistance and instead became a victim of this administrative restructuring. Ijiri claims that before his enforced retirement he received a warning that he would 'get fired' unless he stopped his involvement with the union movement.[53]

There is no doubt that union activists with connections to the Communist Party were targeted in the administrative restructuring and curtailment of the business operations of private enterprises that occurred in 1949. Prime Minister Yoshida nostalgically recalls in his 'memoirs' that:

> As for the sackings, of course special recompense was provided and it proceeded by calling for volunteers. But we simultaneously followed GHQ's suggestion, which was in line with its rationale, that we set up a policy of terminating the employment of communist elements along with such people as it was feared would be detrimental to the normal running of government office operations. We urged each government office to run a wholesale check on those in question. We dealt with these elements along with the other regular retrenchees from July through to September of that year.[54]

The Government was unable to investigate all those under suspicion in the limited time available, hence its list of those to be retrenched drew heavily on information acquired by GHQ and the lists of names submitted by each branch of the Japanese Communist Party demanded by the Group Regulation Ordinance (GRO), which came into effect in April 1949. Thus, the GRO served the purpose for which it was designed, i.e., to effectively keep track of members of the Communist Party.[55]

The GRO was in accord with the GHQ Government Section's (GS) suggestion that the highest priority for termination of employment should be given to the communist element in the workforce.[56] Eiji Takemae claims that GHQ, and in particular the Public Service Qualifications Division (with Jack Napier as Division Chief), played a leading role in manipulating the restructuring of personnel to purge the workforce of communists. According to Takemae, the Public Service Qualifications Division's instructions were treated as 'input' into the drafting of the ABENL.[57] These instructions were considered top secret in GHQ, which meant that the Economic and Scientific Section (ESS) Labor Division and other divisions of GHQ were obliged to ratify the Japanese government's unidirectional pursuit of administrative restructuring, which were themselves based on the 'internal secrets' of the GS.[58]

Fearing criticism from the Soviet Union – one of the Allied Powers that 'officially' occupied Japan (cf. ch. 1) – GHQ never gave explicit instructions for there to be a purge of communists in the 1949 administrative restructuring or the curtailment of the business operations of private enterprises.[59] The secrecy surrounding the political criteria for retrenchment led to the implementation of devious means of enforcing retirement, whereby people were transferred from one department to another in order to make them retire. For example in August 1949 thirteen people were transferred from the Agriculture Research Institute of the Ministry of Agriculture and Forestry (six from the Kōnosu Research Institute and seven from the Nishigahara Central Institute) to the administrative bureau and made to retire on the same day.[60] Such practices reveal the political nature of these administrative reforms and illustrate the impediment they represented to genuine, effective economic restructuring.

The research of Tetsuo Hirata indicates that there were exceptions to the policy of purging communists as part of the administrative re-structuring. Hirata points to the Special Investigations Bureau of the Attorney General's Office (SIBAGO; the investigative organ of the GRO) estimate that while the vast majority of members of the Communist Party and their sympath-izers were removed during the restructuring of SIBAGO personnel, a considerable number remained. This assessment was shared by GHQ and the Government and was a contributing factor in the development of the third stage of the Red Purge, which occurred in 1950.[61]

The Red Purge in the Education Sector

The Japan Teachers' Union was formed in June 1947 and the All-Japan Students Autonomy Confederation was established in September 1948. These two organizations became the focus of communist activities in the education sector and as such came under close scrutiny from the Civil Information and Education Section (CIE) of GHQ.[62] To counter the growing influence of communism within education, CIE changed its policy from one of passive restraint through the promotion of democracy, to one of actively purging communists from the profession.[63] This policy shift was enthusiastically supported by the Civil Intelligence Section's Counter Intelligence Corps (CIC) and the regional military government teams. It provided direction for GHQ policy and its influence extended to the Ministry of Education.[64]

In September 1949 the Ministry of Education convened an informal, nationwide meeting of the heads of education and instructed each Prefectural Board of Education to take responsibility for conducting a purge of left-leaning teachers.[65] As a result, from October 1949 through to March 1950 approximately 1200 members of the teaching profession were purged from public schools across the country; the only exception being the Kōchi Prefecture.[66] Thus, the second stage of the Red Purge was enacted.

Teachers were officially dismissed on the grounds of being surplus staff or because of their individual incompetence or unsuitability to the profession. Very few of those purged were told directly that they were being dismissed because their membership in the Communist Party rendered them unsuitable for teaching. Rather they were included in a broad category of teachers considered unsuitable because they hindered the development of a new style of education. The criteria for purging were subjective and not limited to communists and their sympathizers. Victims of the purges included non-communist unionists and a number of people dismissed by opportunistic principals who disliked the keenness of some teachers' interest in education.[67]

The Red Purge also affected Japan's university sector, although not on a particularly large scale. One of its key instigators in this sector was the CIE Education Division High School Education Advisor, Walter Eells. With a long history as an educator and education administrator in the US behind him, Eells arrived in Japan in April 1947. Employed as a CIE advisor he quickly became involved in the reformation of the Japanese higher education system. His first impact on Japanese university life came when, in a commemorative address given at the opening of Niigata University in July 1949, he stated that members of the Communist Party were not fit to be educators.[68] In the US the intensification of the Cold War and the resultant increase in anti-communist sentiment gave rise to a report written in June 1949 by the Education Policy Committee of the largest education organization in America, the National Education Association, in which it was baldly stated that, 'Members of the Communist Party must not be employed as teachers.'[69] Eells, then, had merely repeated the National Education Association's assertion and unthinkingly transposed it to the Japanese context. The fact was, though, that trends in the US education system were all but unknown in Japan, so Eells' nakedly anti-Communist speech came as a great shock to his audience of Japanese academics and university administrators.

From November 1949 to May 1950 Eells, together with other members of the CIE Education Division, conducted a lecture tour of the country, visiting 27 national universities and spending two days in each. For example, on 17 and 18 November Eells, along with Donald Typer, the CIE Youth Organization Student Activities Officer, visited Yamaguchi University. According to the records of Yamaguchi University they spent the first morning delivering speeches on academic freedom and student autonomy, and the afternoon engaging lecturers in discussions of the same topic. On the second day Eells held seminars on the general subject of university education, while Typer's sessions focused once again on the issue of student autonomy.

In his morning address on 17 November Eells repeated his argument that being a university professor is a privilege and should be reserved for people who are independent thinkers. Hence, professors who are members of the Communist Party and who thereby discard their freedom of thought by submitting to Party directives are not worthy of that privilege, and as such should not be university professors. In the afternoon discussion he went further, suggesting that a statement be issued to the university rector calling on the university to announce a ban on communism and urging faculty heads to 'get the various faculties to report on the numbers of communists and what sort of things they were doing.' Unlike the university rector, Motonori Matsuyama, who avoided giving an immediate answer, the heads of the Economics, Engineering, Education, Literature and Science and Agriculture faculties all answered Eells' questions by providing him with statistics regarding the number of communists teaching in their disciplines.

Yamaguchi University was the fourth such institution on Eells' lecture tour. By the time he reached there his anti-communist addresses had already become the focus of considerable debate, resulting in the All-Japan University Professors League adopting a resolution on 22 October that stated: 'We cannot condone the dismissal of professors based simply on the fact that they are members of a political party. To do so would be to condone an injustice.'[70] Given his presentation at Yamaguchi University, we can assume that this kind of Japanese opposition failed to alter Eells' resolve. Undaunted, he continued his crusade, though he was not always received as calmly as he was at Yamaguchi. At both Tōhoku and Hokkaidō Universities Eells' addresses and discussion meetings were cancelled halfway through due to a storm of student protests.

As mentioned earlier, Eells' argument was attuned to the increasing popularity of anti-communism in other parts of the world, particularly the USA. This international movement gained considerable momentum from the publication of George Orwell's *1984*, a book that depicts an absolutist social system in which 2+2=5 by government decree and individuals are forbidden to think otherwise. Orwell intended this work as a warning against the appearance of such a society. Eells cited this point when speaking at Hokkaidō University and criticized the absolutist tendencies of nation states where the Communist Party monopolized political power. Despite being published in translation in Japan in April 1950 and selling out multiple editions within the month, *1984* was still relatively unknown at the time of Eells' address at Hokkaidō University in the middle of May. Hence, his reference to the text was lost on most of the audience, some of whom responded by rhetorically asking how, if 2+2=5 in the Soviet Union, could the Soviets ever have developed aeroplanes, let alone the atomic bomb? An irrelevant and increasingly heated exchange resulted and it soon became clear that the discussion would have to be abandoned.[71]

Eells believed that just as university students should apply themselves to their studies, so their professors should strive to increase their knowledge not only of the subjects they teach but also of teaching methodologies.[72] He advocated a teaching method in which the content of a subject was in part guided by the needs of those being taught. His controversial political pronouncements, however, precluded many in his audience from taking his quite progressive pedagogical theories seriously. To cite the example of Hokkaidō University again, when Eells commenced his lecture on methodology the audience interpreted it as being 'provocative,' 'an escapists' strategy' and 'nothing more than instruction which employs techniques which very skillfully foster a colonial education.'[73] In effect both Eell's political pronouncements and his statements in relation to teaching methodologies incited opposition.

Eells was a propagandist for the anti-communist cause, but there is no evidence to suggest he was directly involved with the purging of communists from the teaching staffs of universities. Speaking at Yamaguchi University, Typer admitted as much when he said, 'Here we debate for two days but what we say is certainly not binding. My opinions have found favor with the Supreme Headquarters but I was not forced or made to hold these views under duress.'[74] In reality though Eells' opinion was influential at the Board level of several universities, where, soon after his visit, resignation notices were issued to teaching staff who were Communist Party members.

Unlike their counterparts in other branches of the public service, teachers at national universities were afforded some protection from the Red Purge by the Education Public Service Particular Case Law. This meant that notification of dismissal did not lead automatically to the termination of employment. The issue was further complicated by the fact that these political dismissals occurred in the transition period from the old university system to the new. Hence, some people were pressed into resigning because universities following the new system did not employ teachers from the old. Teachers dismissed in this way, combined with those whose jobs were terminated for political reasons, numbered several dozen.[75]

In both cases the final decision regarding termination was made via the university administrative body's screening process, which meant that a certain number of university administrators actively participated in the Red Purge. To continue with the example of Yamaguchi University, the university management committee – made up of the heads of the various departments, two professors from each faculty and with the rector as Committee Chairman – rejected re-appointment applications by Hidekazu Aizawa, professor of the Yamaguchi College of Economics (economics) and Michikazu Uruma, professor of Yamaguchi Normal School (history). Aizawa was known as a Marxist economist who had, in the 1947 Lower House elections, stood as a Communist Party endorsed candidate. By contrast, Uruma's political affiliations were less overt, though he was involved in union activities. The fact that some of his past students had become prominent in the union movement was only thought as a reason for his re-application to be rejected.[76] It is important to remember that those responsible for the administration of the Red Purge in the university sector were not employees of GHQ or the Japanese Government, but were themselves university personnel.

The Korean War and the Red Purge

In contrast to the 1949 phase of the Red Purge which was carried out covertly under the guise of administrative restructuring and the curtailment of the business operations of private enterprises, the 1950 phase was explicit in its attempt to expel Communist Party members and communist sympathizers from positions of power and/or influence in Japanese society.

Early in the Occupation GHQ had released all political prisoners, including members of the Communist Party, and adopted a general policy of tolerance towards political dissidents. This, however, changed once the Party began exerting influence over various social movements, particularly the labor movement. When GHQ shifted its focus from demilitarization and democratization to economic revival it became increasingly important to try to control the labor movement, which was itself controlled by the

Communist Party. One expression of this power struggle was the already discussed undisclosed purges of 1949. Another was GHQ's low-level surveillance of the Party's activities. While MacArthur's Independence Day speech of July 1949 made it clear that Japan would not be swayed by communist propaganda, it also raised the possibility of future tolerance of communism. One year later though, his tune had changed and he announced strong measures against the Communist Party.

The reasons for this shift in policy were both international and domestic. After his speech in July 1949 international developments such as the Soviet Union's announcement that it possessed a nuclear arsenal and the October establishment of the People's Republic of China significantly increased Cold War tensions. On the domestic front, in January 1950 the Cominform criticized the Japanese Communist Party's theory of peaceful revolution. This gave rise to internal conflict in the Party within Japan as it struggled to determine how to respond to this policy correction. The result was a deepening of the Communist Party's opposition to the US Occupation. This heightened attitude of confrontation between GHQ and the communists was a deciding factor in MacArthur adopting a hard line policy against the Party.

In his Constitution Day address of May 1950 MacArthur declared that the Communist Party 'has thrown away the mask of legitimacy and in its place we find it has overtly become the minion of an international plundering force and has taken on a role of achieving a policy relating to overseas authority and imperialistic and destructive propaganda.' On 30 May a rally organized by the communists, but held under the auspices of the Tokyo Preparatory Group for the National Democratic Front, ended in violent clashes between the demonstrators and the American forces. Eight protestors were arrested. In response, on 2 June MacArthur prohibited public gatherings and demonstrations in the capital and proceeded to write a series of letters to Prime Minister Shigeru Yoshida ordering that all twenty-four members of the Japanese Communist Party Central Committee (letter of 6 June) and seventeen editorial committee members and other workers on the *Red Flag* (letter of 7 June) were to be dismissed from public office. Given that these measures did nothing to alter the editorial policy of the *Red Flag*, MacArthur then instructed Yoshida to suspend the paper's publication for thirty days (letter of 26 June). This edict became redundant when, in July, *Red Flag* and its successor, along with other affiliated publications, were banned indefinitely (letter of 18 July). As a result, the bulletins of some 1700 organizations sub-ordinate to the Party were suppressed.[77] In addition to these harsh measures, approximately 4,400 Communist Party members and their supporters were arrested for infringing Occupation policy.[78] For example, in August 1950 Kunio Oka, a then member of the Communist Party and committee member of the board of the History of Science Society of

Japan, followed Party instructions and pasted in excess of 100 posters defaming GHQ around the town of Karuizawa, where he resided. In September he was arrested on the grounds of infringing Occupation policy.[79]

On 25 June, amidst the implementation of a series of repressive policies aimed at the Japanese Communist Party on the Korean peninsula, the North Korean army, commanded by Kim Il Sung, conducted an artillery attack on the 38th parallel, thereby providing the spark for the Korean War. Japan immediately became a departure base for US forces, thus making Japanese intelligence management a vital war issue, one that prompted GHQ to regulate newspaper reporting and to oversee the purging of communists from public news reporting bodies. In accordance with these policies, on 15 July in Osaka, the 8th Army Foreign Radio Section (AFRS) ordered the Osaka Central Broadcasting Station to prohibit Communist Party members or their sympathizers from entering the facility.[80] Soon all managers of radio broadcasts, newspapers and telecommunications sections were directed to terminate the employment of communists and fellow travelers. According to a Ministry of Labor survey, a total of 704 persons in 50 companies were summarily dismissed. Unsurprisingly, it was Napier, the GS Public Office Qualifications Division Chief who had carried out much of the behind-the-scenes work during the first stage of the Red Purge, who gave the dismissal instructions to eight companies in the capital.[81] By September the purge had reached the movie and theatre industries, with a total of 136 communists and their sympathizers being sacked from Shochiku, Daiei, Tōhō, Nichiei and Riken Eiga (the science movie section of the Institute of Physical and Chemical Research).[82]

Japan's role as a supply base for the US forces during the Korean conflict and the resultant need to safeguard production of vital goods provided a pretext for GHQ to assist private enterprises, often informally and tacitly, in purging communists from their workforce.[83] Due to its interconnectedness with all other industries and its highly unionized workers, the electrical industry was among the first to receive such assistance.[84] Densan (Council of Electrical Industries Union) was formed as an independent labor union but fell victim to a confrontation between two rival factions that had developed within it – the communist and Mindō (Democratization Alliance) factions. The Mindō faction, which enjoyed the support of the majority of members, moved to expel the communist faction as a result of the chaos that erupted at Densan's national congress in May 1950. Hence, even the union was trying to purge itself of communists. While at the same time the management side of the electric power industry dismissed 2,100 workers, it did so ostensibly on the basis of their inefficiency rather than their political affiliations. It may or may not have been a coincidence though that most of those dismissed had also been expelled from Densan for being communists.[85]

In September Robert Amis, the GHQ Labor Division Chief, warned the mining, ship-building, steel, car, private railways, electrical engineering, heavy machinery, banking and chemical industries against dismissals of convenience and at the same time advised them to complete the Red Purge before the end of October. During that month similar notifications were given to a variety of other industries.[86] Amis believed that while communism would destroy normal union activities, the Red Purge needed to be instigated and implemented by the Japanese and, therefore, GHQ should neither initiate nor encourage it.[87]

It was against this backdrop of tacit GHQ support that the Japan Federation of Employers' Associations produced the 'Outline for Managing the Removal of the Red Element' on 2 October 1950.[88] Following this blueprint, the managers of the different industries began actively purging communists from their ranks. On 12 October Mitsui Mining, a coal mining company, notified the union of its intention to sack workers who were either communists or communist sympathizers. On 14 October 1950 *Mainichi Shimbun* published the company's criteria for dismissing suspected communists. It is interesting to note that, for example, the criteria for judging Communist Party sympathizers closely resembles the Japan Federation of Employers' Associations 'Outline.' This suggests that each manager took his direction not from an order issued from GHQ but rather from the Association's 'Outline.' One of Mitsui Mining's criteria for being a sympathizer is, 'those who are enrolled, who have been enrolled or who have encouraged enrolment as members in the following agencies: the Democratic Protection Alliance, the Alliance of Soviet Returnees, the Democratic Scientists' Alliance, the Youth Homeland Alliance, the Industrial Protection Council, the Sino-Japanese Friendship Society, the Russo-Japanese Friendship Society, the Workers and Farmers Relief Association or other agencies which come under the direct command of the Party.'

With the exception of the Democratic Scientists' Alliance, the agencies cited above appear in exactly the same order as those given in the Association's 'Outline.' In the one instance where Mitsui Mining deviated from the 'Outline' the Democratic Scientists' Alliance did not exist and was thought to be mistaken for Minka (the Association of Democratic Scientists). This meant that members of the Association were liable to be purged as communists even though not all members had such sympathies. The Association's Secretary-General, Hideomi Tsuge, the Secretary, Shōsaku Nishimura, protested to Mitsui Mining, forcing them to apologize and admit that, 'Minka is certainly not an object of the Red Purge.'[89] Further problems arose with the Mitsui Mining restructuring criteria when it was decided they were too broad and were eventually revoked in mid October.[90] Labor Division Chief Amis judged that the inclusion of Minka amongst the blacklisted agencies was evidence that things had gone too far. Tsuge recalls that, 'It was reported that Minka had been removed from Mitsui Mining's Article 2 sympathizer criteria.'[91]

In October the Red Purge reached into almost every field of private industry. In fact, in the latter half of 1950 the number of people dismissed on the basis of their political affiliations was in excess of 10,000. In November, in accordance with a Cabinet decision made in September, communist workers were purged from the public service and public utility enterprises. Just as politically based dismissals from the public service were made legal under the National Public Servant Law, Article 78, No. 3 (lack of eligibility), so the purging of 'reds' from public utilities was legalized by the Japanese National Railways Law, Article 29, No. 3 and the Japan Tobacco and Salt Public Corporation Law, Article 22, No. 2. Under these laws any communist or communist sympathizer could be sacked if they were potentially in a position to leak secret information or in any way hamper normal business operations.[92] From 2–17 November 1,171 people suspected of communist sympathies were purged from government agencies. Included in this number were eleven from the Ministry of Transportation's Meteorological Observatory and two from the Ministry of International Trade and Industry's Geological Survey of Japan.[93] Hence, what began in 1949 as a covert policy to remove communists from government positions was completed as an explicit campaign in 1950.

While the Japanese were responsible for the domestic implementation of the Red Purge, matters concerning international appointments came under the direct control of GHQ. As such, scientists wishing to apply for positions overseas had to receive GHQ approval, which was withheld from those critical of GHQ and the political policies of the US. Hidcomi Tsuge recalls many past Minka members seeking to study in the US asking him to vouch that they had never been involved with the organization.[94] When Minka members Hajime Matsuura and Shōichi Sakata attempted to represent Japan at an international conference they were harassed by GHQ and refused passports. In this way the democratization movement within the scientific community was politicized, resulting in many scientists becoming disenchanted with political ideologies *per se*, rather than becoming active anti-communists.

Future Directions in Red Purge Research

While there was a wide range of political opinion held by the people involved in the postwar democratization of Japan's academia, almost all of the central figures in this movement shared an interest in, if not a commitment to, Marxism. This was in part due to the fact that Marxism had been successfully repressed during the war, which in effect only earned it sympathy amongst many intellectuals. Another factor was that amidst the chaos of postwar Japan it was Marxism that provided the most coherent and systematic vision for the future. On a more psychological level, by embracing Marxism these intellectuals asserted their independence from US political philosophies and systems being imposed on them under the Occupation. In this way many scientists saw their affiliation with Marxism as an expression of their freedom of thought and as affirming both Minka's pledge to have 'a sincere conscience and unflagging courage' (statement of October 1949),[95] and the JSC's 'resolution never to accede to conducting scientific research which has war as its objective' (plenary statement of April 1950).

Even after the war, few Japanese scientists felt ashamed that their lack of social awareness had allowed their work to be co-opted by the war effort. While Minka expressed the view that 'scientists had been obliged to serve in an inhumane war' (Fifth National Conference),[96] it also claimed that many scientists had only assisted the military's research agenda because they were ignorant of the important role science plays in social and political life. Yet, these scientists failed to be fully aware of the process of the 'institutionalization' of science and to accept their social responsibility as scientists. The absence of such social awareness and consciousness made them insensitive to the importance of the internal freedom of individuals as well as its limitations. This tendency was combined with the totalistic orientations of socialist thinking and diverted the public opinion away from focusing upon the need for creating diverse, free-thinking individuals.

In the January 1952 edition of *Shizen* (Nature) a scholar of French literature who was also a member of the JSC's 'Committee to Guarantee the Freedom of Learning and Ideology' concluded his Science Council Plenary Report by stating that, 'Academic freedom is nothing more than a negative freedom which shares its fate with capitalist societies.' The most remarkable aspect of this statement is that it was made by someone who was arrested during the war for breaching the Maintenance of Public Order Law. As this suggests, most of Japan's postwar scientists had an underdeveloped understanding of the need for intellectual freedom and the 'freedom of truth.' Reflecting on that generation of scientists, Yōichi Fukushima writes, 'For the natural scientists and technicians, who represented the majority of Science Council members, I wonder just how seriously freedom of learning and ideology was taken.'[97]

The Red Purge was enacted primarily because GHQ believed that communists were unable to think or carry out their responsibilities independently, without recourse to the policies of the Party, and because the tenets of communism were contrary to the principles of a US-style democracy, and as such were considered dangerous. To more closely examine the freedom of thought of those who were purged and the importance of freedom of thought in the wider Japanese society is a task to be undertaken in future research on the Red Purge.

Notes

1 *Asahi Shimbun*, 17 August 1945.
2 *Asahi Shimbun*, 19 August 1945.
3 Ministry of Education, Secretariat Archives Division, *Shūsengo kyōiku jimu shori teiyō* (Summary of education administration management following the war),' collection 1, November 1945, pp. 67–9.
4 Chūji Tusboi, 'Iwayuru kagaku shinkō ni tsuite (On the so-called science promotion),' in *Sekai* (World), June 1946, p. 70.
5 Tsuyoshi Matsumoto, *Ryakudatsu shita bunka* (A culture which plundered), Iwanami Shoten, 1993, pp. 47–50.
6 Tamon Maeda, 'Shūsen chokugo gokagetsu zainin no kiroku (A record of five months in office immediately following the war),' in *Monbu jihō* (The monthly journal of the Ministry of Education), no. 824, January 1946, p. 3.
7 Nishikawa, 1993 [5]
8 Eiichi Suzuki, *Nihon no senryō to kyōiku kaikaku* (The Allied Occupation of Japan and education reforms), Keisō Shobō, 1983, pp. 62–4.
9 Extracts Nos 1) and 2) are from *Nihon senryō oyobi kanri jūyō bunshoshū* (Documents concerning the Allied Occupation and control of Japan and its administration), edited by the Division of Special Records, Ministry of Foreign Affairs, Japanese Government, vol. 1, 1949. No 3) is from *Amerika kyōiku shisetsudan hōkokusho* (Report of the United States Education Mission to Japan) by Minoru Murai (Complete translation and expository comments), Kodansha, 1952, and from *Nihon ni okeru kagaku to gijutsu no saihensei* (Reorganization of science and technology in Japan) edited by Nihon Gakujutsu Shinkōkai (Japan Society for the Promotion of Science), Kyōiku Shuppan, 1948.
10 Rokuo Kawabe and Michiji Konuma, 'Chūseishi-ron no tanjō (Birth of the meson theory),' in *Butsuri* (Physics), vol. 37, no. 4, April 1982, pp. 265–75.
11 Shōichi Sakata, *Kagakusha to Shakai ronshū* (Scientists and society), vol. 2, Iwanami Shoten, 1972, pp. 1–45.
12 Nihon Kagakushi Gakkai (History of Science Society of Japan), 1964 [1], pp. 96–101.
13 See Mitsuo Taketani, *Benshōhō no shomondai* (Various issues on dialectics), Rigakusha, 1946; Shōichi Sakata, *Butsurigaku to hōhō* (Physics and methodology), Hakutō Shokan, 1947.
14 Tetsu Hiroshige, *Kagaku to rekishi* (Science and history), Misuzu Shobō,

1965, pp. 1–53; Kenji Itō, 'Review,' in *Historia Scientiarum*, vol. 3, no. 3, 1994, pp. 240–2.

15 Shōhei Miyahara, 'Kagaku taisei sasshin ni tsuite (Science renewal),' in *Shizen kagaku* (Natural science), no. 10, January 1948, pp. 44; Gorai, 1976 [16], p. 7.

16 Nihon Chishitsu Gakkai (The Geological Society of Japan), 1953 [20] p. 53.

17 Gorai, 1976 [16] p. 143; *Chishitsu Gakkai zasshi* (Journal of the Geological Society of Japan), vol. 53, nos. 622–7, July–December 1947, p. 34.

18 *Chishitsu Gakkai zasshi*, vol. 53, nos. 622–7, July–December 1947 p. 40.

19 Ijiri, 1953 [12] p. 290; *Chishitsu Gakkai zasshi*, vol. 54, no. 634, July 1948, pp. 70, 82–88.

20 Ijiri,1953 [12] pp. 67–70; Ijiri, 1982 [15] vol. 3, pp. 181–2.

21 Yōichi Fukushima, *'Gakusha no mori' no yonjūnen* (Forty years of the 'forest of scholars'), vol. 1, Nihon Hyōronsha 1986, pp. 111–44.

22 Ijiri, 1983 [15] vol. 15, *Nenpu* (Personal chronological history), p. 5.

23 Kobayashi, 1966 [13] pp. 44–5.

24 Shōji Ijiri, Ryūji Sugiyama, Kennosuke Ogawa, Shirō Iwai, Makoto Wada, Yoshio Watanabe, and Tadashi Kimura, 'Kantōsanchi ni okeru oshikabuse kōzō no saikentō (A re-examination of the overthrust structure in mountain areas in the Kantō region),' in *Tokyo Kagaku Hakubutsukan kenkyū hōkoku* (Bulletin of the Tokyo Science Museum), no. 14, 1944, pp. 1–13.

25 Ijiri, 1982 [15] vol. 3, pp. 86–97.

26 Ijiri, 1982 [15] vol. 1, p. 130.

27 Ijiri, 1982 [15] vol. 1, p. 129.

28 Ijiri, 1982 [15] vol. 1, pp. 129–30.

29 Ijiri, 1982 [15] vol. 1, p. 128; Ijiri 1982 [15] vol. 3, p. 96.

30 Note that in the new edition of *Kagaku-ron* (Science theory; 1977), Ijiri softens his stance with regards to the conditions necessary for collaborative research to be effective.

31 Gijutsu to Ningen Henshū-bu (Editorial staff of Science and Humanity), 'Chigaku dantai kenkyū-kai to Minka (jō): Ijiri Shōichi Ōmori Masae ryōshi ni kiku (The Chidanken and Minka (pt 1) – Interviews with Shōji Ijiri and Masae Ōmori)' [4] February 1977, p. 138. Kobayashi, 1966 [13] p. 131.

32 Gijutsu to Ningen Henshū-bu (Editorial staff of Science and Humanity), 'Chigaku dantai kenkyū-kai to Minka (jō): Ijiri Shōichi Ōmori Masae ryōshi ni kiku (The Chidanken and Minka (pt 1) – Interviews with Shōji Ijiri and Masae Ōmori)' [4] February 1977, pp. 139–140; Ijiri, 1982 [15] vol.3, p. 183.

33 Gijutsu to Ningen Henshū-bu, 'Minka undō no rekishi-teki haikei Torii Hiroshi hoka ni kiku (Historical background of the Minka movement – Interviews with Hiroshi Torii and others)' [4] June 1976, p. 146.

34 *Chigaku dantai kenkyū-kai kaishi* (The journal of Chidanken), vol. 3, no. 2, March 1949, pp. 30–4.

35 *Chikyū kagaku* (Earth science), no. 12, April 1953, p. 6.

36 Takahashi [4] February 1977, p. 138.
37 Minka suffered a similar fate with its popularity peaking in 1950 and declining as it became more overtly political. See, Hideomi Tsuge, *Minka to watashi* (Minka and I), Keisō, 1980, pp. 179–80; Takahashi, June 1976 [4] p. 143; Hiroshige, 1960 [10] p. 153.
38 The origins of the term the 'Red Purge' are contested. Hirata, 1989 [21] claims it originates from protesting students, citing the written resolution adopted at the Second Metropolitan and Prefectural Student Council Representatives' Conference (the National New Structure Universities Autonomous Group Representatives' Conference), 11 March 1950. Kenjirō Takahashi, however, asserts the term was coined by Katsuzō Kimura, a student of the Faculty of Economics at the University of Tokyo. See, *Kaisō no Tōdai Komaba-ryō* (The Komaba Dormitory of the University of Tokyo in retrospection), Nesco, 1994, p. 108.
39 Takemae, 1982 [23] pp. 347–51; Hirata, 1989 [24]; Shigeru Yoshida, *Kaisō jūnen* (Reminiscing ten years), Shinchōsha, vol. 2, 1957, pp. 284–7.
40 Hirata 1989 [21] p. 10; Referring to the third stage of the Red Purge, Miyake, 1994 [29] claims that, 'It is accurate to say that in excess of 13,070 were expelled from office' (p. 12).
41 Dodge Ministerial Statement, 7 March 1949.
42 Ministry of Labor (ed.), *Shiryō rōdō undō-shi* (Reference data on the history of the labor movement), 1949, p. 243; Nihon Kokuyū Tetsudō (The Japan National Railways) (ed.), *Nihon Kokuyū Tetsudō hyakunen-shi* (Hundred year history of the Japan National Railways), vol. 12, 1973, p. 478.
43 Tadao Amatatsu (ed.), *Jichitai rōdō-sha* (Municipality workers), San'ichi Shobō, 1962, p. 75.
44 Ministry of Labor (ed.), see footnote 42, 1949, p. 86.
45 Shigeru Yoshida, *Kaisō jūnen* (Reminiscing ten years), Shinchōsha, vol. 3, 1957, pp. 260–261.
46 Kokuritsu Kagaku Hakubutsu-kan (National Science Museum) (ed.), *Kokuritsu Kagaku Hakubutsu-kan* (One hundred year history of the National Science Museum), 1977, pp. 380, 878.
47 Kokuritsu Kagaku Hakubutsu-kan (National Science Museum) (ed.), *Kokuritsu Kagaku Hakubutsu-kan* (One hundred year history of the National Science Museum), 1977, p. 441.
48 *Asahi Shimbun*, 19 March 1949.
49 *Asahi Shimbun*, 5 June 1949.
50 *Asahi Shimbun*, 8 June 1949.
51 *Asahi Shimbun*, 12 August 1949.
52 Shoji Ijiri, *Hiteiteki seishin* (Mentality of antithesis), Tsukiji Shokan, 1986, p. 56.
53 Shoji Ijiri, *Hiteiteki seishin* (Mentality of antithesis), Tsukiji Shokan, 1986, pp. 56–7.

54 Shigeru Yoshida, *Kaisō jūnen* (Reminiscing ten years), Shinchōsha, vol. 2, 1957, p. 286.
55 Shigeru Yoshida, *Kaisō jūnen* (Reminiscing ten years), Shinchōsha, vol. 2, 1957, p. 272.
56 Toshio Yoshihashi, 'Dantai-tō kiseirei ni tsuite (On the group regulation ordinance),' in Izutarō Hirosue (ed.), *Hōritsu jihō* (Law review), vol. 22, no. 10, October. 1950, p. 773.
57 Takemae, 1982 [23] p. 350.
58 Takemae, 1982 [23] p. 350.
59 Takemae, 1982 [23] p. 351.
60 Kyōjirō Nakagawa, 'Gekidō no ichinen (A year of turmoil),' in Ministry of Agriculture, Forestry and Fisheries, Agriculture Research Institute (ed.), *Kōnosu no omoide* (Recollections of Kōnosu), 1981, p. 85; Chidanken [13] 1969, p. 132.
61 Hirata, 1989 [24] p. 43.
62 Mark T. Orr, *Education Reform Policy in Occupied Japan*, translated into Japanese as *Senryō-ka Nihon no kyōiku kaikaku seisaku* by Gery Hōichi Tsuchimochi, Tamagawa University Press, 1993, pp. 32, 140–41.
63 Myōjin, 1989 [22] p. 28.
64 Myōjin, 1989 [22] p. 29; Takemae, 1982 [23] p. 347.
65 Myōjin, 1989 [22] p. 32.
66 Isao Myōjin, 'Kyōin reddo paaji no hi-tsuihōsha-sū o megutte (On the number of teaching personnel who were victims of the Red Purge),' in *Hokkaidō Kyōiku Daigaku kihō (Dai 1-bu C)* (Journal of Hokkaidō University of Education, Part 1, C:, Education science), vol. 38, no. 2, March 1988, pp. 33–47.
67 Myōjin, 1989 [22] pp. 33–4; Takemae, 1982 [23] p. 348.
68 Tokiomi Kaigo and Masao Terasaki, *Daigaku kyōiku* (University education), University of Tokyo Press, 1969, pp. 50–5.
69 Myōjin, 1989 [22] pp. 30–1; Takemae, 1982 [23] p. 347; *Asahi Shimbun,* 10 June 1949.
70 Tokiomi Kaigo and Masao Terasaki, *Daigaku kyōiku* (University education), University of Tokyo Press, 1969, p. 56.
71 Isao Myōjin, 'Hokudai Eells jiken no shōgen (1) (Testimony to the Eells Incident at Hokkaidō University (1),' in *Kushiro ronshū* (Journal of Hokkaidō University of Education at Kushiro), no. 14, 1982, pp. 122–7.
72 Eells [26] no. 1, 1949.
73 Isao Myōjin, 'Hokudai Eells jiken no shōgen (1) (Testimony to the Eells Incident at Hokkaidō University (1))', in *Kushiro ronshū* (Journal of Hokkaidō University of Education at Kushiro), no. 14, 1982, p. 128 & no.15, 1983, p. 113.
74 Eells [26] no. 2, 1949, p. 1.
75 Myōjin, 1989 [22] pp. 37–8.

76 Isao Myōjin, 'Hokudai Eells jiken no shōgen (1) (Testimony to the Eells Incident at Hokkaidō University (1))', in *Kushiro ronshū* (Journal of Hokkaidō University of Education at Kushiro), no. 14, 1982, p. 144.
77 Nihon Kyōsan-tō Chūō Iinkai (The Japanese Communist Party Central Committee) (ed.), *Nihon Kyōsan-tō no 70-nen* (Seventy years of the Japan Communist Party), vol. 1, Shin-Nihon Shuppansha, 1994, p. 232.
78 Nihon Kyōsan-tō Chūō Iinkai (The Japanese Communist Party Central Committee) (ed.), *Nihon Kyōsan-tō no shichijūnen* (Seventy years of the Japan Communist Party), vol. 1, Shin-Nihon Shuppansha, 1994, p. 232.
79 *Asahi Shimbun*, 7 September 1950; Setsu Matsumoto, 'Oka Kunio no isshō (The life of Kunio Oka)', in *Shisō no kagaku* (Science of thought), December 1971, p. 101.
80 Miyake, 1994 [29] pp. 55–6.
81 Asahi Shimbunsha Reddo Paaji Shōgen-roku Kankō Iinkai (Asahi Shimbun, Red Purge Testimonial Records Publishing Committee), *1950-nen 7-gatsu 28-nichi* (28 July 1950), Bansaisha, 1981, pp. 40, 44.
82 Kaneto Shindō, *Tsuihōsha tachi* (The purged), Iwanami Shoten, 1983, pp. 17, 61.
83 Takemae, 1982 [23] p. 355.
84 Takemae, 1982 [23] pp. 355–7.
85 Eiji Takemae, 'Reddo Paaji (The Red Purge),' in Shisō no Kagaku Kenkyū-kai (Research group of the Science of Thought) (ed.), *Kyōdō kenkyū/Nihon senryō-gun: Sono hikari to kage* (Joint research/the Occupation Army in Japan: Its light and shade), vol. 1, Gendaishi Shuppankai, 1978, pp. 285–7; Shōbei Shioda, *Reddo Paaji* (The Red Purge), Shin-Nihon Shuppansha, 1984, pp. 64–88.
86 Takemae, 1982 [23] pp. 357–8.
87 Eiji Takemae, *Shōgen Nihon senryō-shi* (A Testimonial to the history of the Occupation of Japan), Iwanami Shoten, 1983, p. 292.
88 Eiji Takemae, Akimasa Miyake, and Kōshi Endō (eds.), *Rōdō kaikaku to rōdō undō* (Labor reforms and the labor movement), Ōtsuki Shoten, 1992, pp. 338–44.
89 Hideomi Tsuge, *Minka to watashi* (Minka and I), Keisō Shobō, 1980, p. 171.
90 Miyake, 1994 [29], p. 88.
91 Hideomi Tsuge, *Minka to watashi* (Minka and I), Keisō Shobō, 1980, p. 171.
92 Shōbei Shioda, *Reddo Paaji* (The Red Purge), Shin-Nihon Shuppansha, 1984, p. 47; Nihon Kokuyū Tetsudō (The Japan National Railways) (ed.), *Nihon Kokuyū Tetsudō hyakunen-shi* (Hundred year history of the Japan National Railways), vol. 12, 1973, p. 480.
93 Meteorological Agency (ed.) *Kishō hyakunen-shi: Shiryō-hen* (A Hundred Year History of meteorological observations in Japan, volume on historical material), Nihon Kishō Kyōkai (Japan Meteorological Association), 1985,

p. 283. Chishitsu Chōsasho Bunkai Nijūgonen-shi Henshū Iinkai (Editorial Committee on the Twenty-five Year History of Geological Surveys in Japan) (ed.), *Daichi o Kizamu* (Recording the earth), Zen Shōkō Rōdō Kumiai Kanshin Shibu Chishitsu Chōsasho Bunkai (Geological Survey Team, Kantō and Shin-etsu Region Branch, All Commerce and Industry Labor Union), 1976, p. 96.

94 Hideomi Tsuge, *Minka to watashi* (Minka and I), Keisō Shobō, 1980, p. 172.

95 Hideomi Tsuge, *Minka to watashi* (Minka and I), Keisō Shobō, 1980, pp. 145–8.

96 Hideomi Tsuge, *Minka to watashi* (Minka and I), Keisō Shobō, 1980, p. 165.

97 Yōichi Fukushima, *'Gakusha no mori' no yonjū-nen* (Forty years of the 'forest of scholars'), vol. 2, 1988, pp. vii–viii.

Bibliographic Guide

From Cultural Nation to Democratic Nation

[1] Nihon Kagakushi Gakkai (History of Science Society of Japan) (ed.), *Nihon kagaku gijutsu-shi taikei: Tsūshi 5* (A compendium of the Japanese history of science and technology: vol. 5, Historical overview), Daiichi Hōki Shuppan, 1964.

[2] 'Zadankai: Kenkyū minshuka ni kansuru Zenkoku Butsuri Gakusha no kondankai – minshu sensen ni taisuru kagakusha no taido ni tsuite (Round-table discussion: The All Japan Physicists' Talkfest on research democratization: On the attitude of physicists in relation to the democratic front),' in *Shizen kagaku* (Natural science) vol. 1, no. 2, July 1946, pp. 22–33.

[3] Yoshimitsu Watanabe (dictated), Hiyaringu Gurūpu (Hearing Group) (ed.), *Shisō to gakumon no jiden* (An autobiography of thinking and learning), Kawade Shobō Shinsha, 1974.

[4] Noboru Takahashi (ed.), 'Sengo no kagaku gijutsusha undō (1) – (12) (Postwar movements of scientists and engineers (1)–(12)),' in *Gijutsu to ningen* (Technology and humanity), April 1976–June 1977.

[5] Nagao Nishikawa, 'Kokka ideorogii to shite no bunmei to bunka (Civilization and culture as a national ideology),' in *Shisō* (Philosophical thought), May 1993, pp. 4–33.

The production of a comprehensive and systematic compilation of materials of the immediate postwar writings that scientists and technologists published is desired but at the present time no such resource exists. As things stand the only thing we have that approximates this is a summary of [1]. I cite [2] in particular, in order to demonstrate that there was debate about public opinion concerning postwar science and the scientists' responsibility in the war. This is partially included in [1] but the historical significance is throughout the entire work. In terms of postwar democracy there are a great variety of documentary records but [3], which indicates a lack of historical reflection on the part of those concerned, is also important in terms of the research of Minka. [4] is a written record of what was said by people who played a part in the postwar science and technology movement. Summaries of each different field are included and one should read the entire 12 installations. [5] points out that 'culture' is a new term in the European languages which was born in the latter half of the 18th century as a national unifying ideology. [5] is an essay that abounds with suggestions. In terms of things indicating the cultural sterility of a cultural nation, see Marc Fumaroli's Bunka kokka – Kindai no shūkyō (L'Etat culturel: Essai sur une religion moderne), Tsuneo Amano (tr.), Misuzu Shobō, 1993.

The Elementary Particles Theory Group

[6] Nihon Butsuri Gakkai (Japanese Association of Physics), *Soryūshiron kenkyū* (Research in Elementary Particles Theory), vol. 1, no. 1, 5 August 1949.

[7] Yoshinori Kaneseki, 'Soryūshiron gurūpu (The elementary particles theory group),' in *Shizen* (Nature), April 1950, pp. 48–55.

[8] Hideki Yukawa, Shōichi Sakata, and Mitsuo Taketani, *Soryūshi no kenkyu* (Research into elementary particles theory), Keisō Shobō, 1965.

[9] L. M. Brown, R. Kawabe, M. Konuma and Z. Maki. (eds.), *Proceedings of the Japan-USA Collaborative Workshops on the History of Particles Theory in Japan, 1935–1960*, Kyoto: Yukawa Hall Archival Library, Research Institute for Fundamental Physics, Kyoto University, 1988.

[10] Tetsu Hiroshige, *Sengo Nihon no kagaku undō* (The science movements in postwar Japan), Chūō Kōronsha, 1960.

[11] Hitoshi Yoshioka, *Kagakusha wa kawaru ka – Kagaku to shakai no shisō-shi* (Will scientists change? A history of the ideology of science and society), Shakai Shisōsha, 1984.

In terms of elementary particle physics in Japan, serious research into the history of science became possible with the successive establishment of the Yukawa Hall Archival Library at Kyoto University, the Tomonaga Memorial Hall at Tsukuba University, and the Sakata Memorial Room at Nagoya University between 1979 and 1986. As for the history of the elementary particles theory in Japan centering on the war years, research continued as a Japanese-American Science Co-operation Enterprise from the end of the 1970s and is published as [9].

The history of the Elementary Particles Group was actively recorded in the 1950s. This is still instructive today, though it is best now to consider it as historical data. [7] ties together the group's network from the external position of a journalist. [8] is an enlarged and revised edition of *Shinri no ba ni tachite* (Standing in the space of truth), Mainichi Shimbunsha, 1951. It covers Shōichi Sakata, 'Chūkanshi riron kenkyū no kaiko (Reflections on research into meson theory)' and Mitsuo Taketani 'Soryūshi gurūpu no keisei (The formation of the elementary particles theory group).' In contrast to these stories, which belong to the internal science history, [10] is a critical narration of the scientists' groups from the perspective of the sociology of science. [11] further discusses the tendencies involved in Taketani's and Sakata's scientism with their theory as the subject of inquiry. In terms of the Elementary Particles Theory Group we have the agency bulletin [6]. In addition to the historical archives referred to above, there is also the Physics Historical Data Committee which was inaugurated by the Physical Society of Japan in 1984. As this consciously collects data, we can look forward to a development of research in the area of the sociology of science.

The Association for Geological Collaboration Japan

[12] Shōji Ijiri (ed.), *Tomo ni manabu yorokobi: Dantai kenkyū no hōhō* (The joy of studying together: Methodology of collaborative research), University of Tokyo Press, 1953 (2nd edn., 1969).
[13] Hideo Kobayashi (ed.), *Kagaku undō* (The science movements), Chigaku Dantai Kenkyūkai (The Association for Geological Collaboration Japan), Tsukiji Shokan, 1966.
[14] Chigaku Dantai Kenkyūkai (The Association for Geological Collaboration Japan), *Minna de kagaku o – Chidanken sanjūnen no ayumi* (Science together: Thirty years of the Association for Geological Collaboration Japan), Ōtsuki Shoten, 1978.
[15] Shōji Ijiri, *Ijiri Shōji senshū* (Selected work of Shōji Ijiri), 10 vols, Ōtsuki Shoten, 1981–1983.
[16] Masao Gorai, *Kaseiron e no michi – Sono henreki no jidai* (The road to the igneous theory: The period of intellectual journey), vol. 2, Kankōkai, 1976.
[17] Shigeru Nakayama 'Ijiri Shōji ron (On Shōji Ijiri),' in *Shisō no kagaku* (The science of thought), May 1966, pp. 100–6. This item was included in Shigeru Nakayama *Tenkan-ki no kagaku-kan* (Views on science in the period of transition), Nihon Keizai Shimbunsha, 1980, pp. 83–96.
[18] L. B. Halstead, *'Imanishi shinka-ron' hihan no tabi* (An intellectual journey for the critical analysis of Kinji Imanishi's 'theory of evolution'), Akiko Nakayama (tr.), Tsukiji Shokan, 1988.
[19] Miyashirō Akiho, 'Chikyū kagaku no rekishi to genjō (1) – (15) (The past and the present of geosciences (1)–(15)),' in *Shizen* (Nature), September 1965–November 1966.

[20] Nihon Chishitsu Gakkai (The Geological Society of Japan) (ed.), *Nihon Chishitsu Gakkai-shi – Nihon Chishitsu Gakkai rokujū shūnen kinen* (A History of the Geological Society of Japan: In commemoration of sixty years), 1953.

The Association for Geological Collaboration Japan published a record of its activities about every 10 years as it was a self-based organization. [12] is the record of its first seven years, [13] is the record of its first twenty and [14] is the record of its first thirty years. A new version of [12] was produced in 1969. The 1969 version is called the summation re-print. It does away with certain information included in the 1953 version which had become outdated or which lacked practical application as well as 33 pages of Shō Ishimoda's postscript 'Kokumin no tame no kagaku (Science for the people).' Because of this, those wishing to learn of the early period of the Association need to refer to the 1953 (original) version. However, the 1969 version of the part entitled 'Kagayaku ibara no michi (The path of the shining wild rose),' which makes reference to the history of the Association, contains an account not found in the 1953 version. There is thus value in utilizing the twin versions. The selected collection of Shōji Ijiri [15] is useful for an overall understanding of his thoughts. [16] is the edited version of Masao Gorai's diary and research diary. These are interesting by virtue of the fact that they contain information relating to the time when the Society was established. The abovementioned are the literary records of the Association of Physical Geography Groups. [17] and [18] offer a third party's view of the Society and Shōji Ijiri's theory. [19] is a literary document that reveals what presented as problems to geosciences around about 1965. [20] is an authentic history of the Geology Society which was made the subject of democratization by the Association for Geological Collaboration Japan. As this was the first history of a society edited in relation to geology, it contains not only the history of the Geology Society but also a short history of associated bodies as well. Nowadays it is recognized as documenting the historical role played by geology in Japan.

The Red Purge

[21] Tetsuo Hirata, 'Naze reddo paaji-shi no kenkyū ka (Why should we study the Red Purge?),' in *Rekishi hyōron* (Historical review), no. 472, August 1989, pp. 2–23.

[22] Isao Myōjin, 'Kyōiku-kai ni okeru reddo paaji (The Red Purge in educational institutions),' in *Rekishi hyōron* (Historical review), no. 472, August 1989, pp. 24–41.

[23] Eiji Takemae, *Sengo rōdō kaikaku – GHQ rōdō seisaku-shi* (Post-war labor reforms: A history of the GHQ labor policies), University of Tokyo Press, 1982.

[24] Tetsuo Hirata, '1949-nen no gyōsei seiri kigyō seibi ni okeru reddo paaji no hōshin to jittai (The direction and actual situation concerning the Red Purge during the administrative and enterprise re-adjustments of 1949),' in *Ōhara shakai mondai kenkyūsho zasshi* (The journal of the Ōhara Social Issue Research Institute), no. 367, June 1989, pp. 38–57.
[25] Tetsuo Hirata (ed.), *Daigaku jichi no kiki – Kōbe Daigaku reddo paaji jiken no kaimei* (The crisis of university autonomy: An analysis of the Red Purge incident at Kobe University), Shiraishi Shoten, 1993.
[26] 'Rengōgun sōshireibu minkan jōhō kyōiku-kyoku komon Eells-hakushi oyobi Typer-shi kōen kiroku (A record of the addresses given by Dr Eells and Mr Typer, advisors at the SCAP Civil Information and Education Section (17 November 1949 in the Department of Economics Theatre),' Yamaguchi University, mimeographs. 'Eells-hakushi, Typer-shi kondankai kiroku (A record of talks with Dr Eells and Mr Typer (17–18 November 1949, the Department of Economics Conference Room, the Library Reading Room),' nos. 1–3, Yamaguchi University, mimeographs. Both documents are held in the Library of the Research Institute for Social Research at Hiroshima University.
[27] Walter C. Eells, *Communism in Education in Asia, Africa and the Far Pacific*, Washington D.C., American Council on Education, 1954.
[28] Eiichi Suzuki, 'Walter Eells bunsho (The Walter C. Eells Documents),' in Hideo Satō (chief researcher), *[Kaigai gakujutsu kenkyū hōkokusho] Senryō-ki Nihon kyōiku ni kansuru zaibei shiryō no kenkyū chōsa* ([Overseas academic research report] Survey report on historical data in America relating to Japanese education during the Occupation period), Kokuritsu Kyōiku Kenkyūsho (National Institute for Education Research), 1988, pp. 97–107.
[29] Akimasa Miyake, *Reddo paaji to wa nani ka – Nihon senryō no kage* (What was the Red Purge? The dark shadow of the Occupation of Japan), Ōtsuki Shoten, 1994.

[21] and [22] appeared in Rekishi hyōron in 1989, the fortieth anniversary of the Red Purge. The overviews are particularly significant studies which indicate that Red Purge research had developed to the point where it was fully-fledged research. However, these theories together did not properly express the future direction of research. At the present time, with the collapse of the Soviet Union, it is the inclusion in one's field of vision of the very Red Purge history research which critiques the Japanese socialist movement and ideology that is desirable. In this sense it is indeed the research of Eiji Takemae, which makes full use of GHQ data and Japanese data, which provides the starting point for research into the history of the Red Purge. [23] is one such example (pp. 348–86 is on 'The Red Purge'). Based on Japanese materials, [24] explores the view that the Red Purge

started in 1949. There are a series of studies on the Red Purge in education by Isao Myōjin. [25] concerns itself mainly with a record of things unpublished by Setsurō Komatsu, himself a victim of the Red Purge. It also has been edited to contain a record of documents which discuss the Red Purge of university employees. Myōjin's outcomes have also been incorporated. The addresses given by Eells are well-known for their anti-communist statements, but in order to understand his way of thinking it is necessary to be aware of the contents of his speeches overall. Hirata's chapter [25] introduces numerous records in print but [26] provides a complete record of the events at Yamaguchi University. The Yamaguchi University records of Nobuyoshi Harima Kenpō o ikasu doryoku (Efforts to make the most of the Constitution), Shiki Shuppan, 1987, which [25] quotes, do not go beyond the Eells address schedule and are not a record of the addresses themselves. [27] was written by Eells himself. The Japanese section is Chapter 2 (pp. 21–36). [28] is a list of Eells' papers and documents. Of these, the materials which relate to Japan have been reproduced by Hideo Satō's team and are kept in the National Institute for Education Research, together with reproductions of historical data that are in the United States. [29] is the most recent research work on the Red Purge. It is a solid reference work which makes full use of the Valery Burati papers, the man who served as head of the Labor Relations Branch and Deputy Chief of the Labor Division at GHQ, while examining Japanese materials. These papers are held in the Reuther Library of Wayne State University. They are useful for developing an overall perspective of the situation.

27 The Mushrooming of Popular Science Magazines

Yukio Wakamatsu

Popular Science Magazines in the Publishing Boom

In September 1945, soon after the end of the war, when Japan lacked printing paper and its printing capacity was paralyzed, *Nichi-Bei kaiwa techō* (Japanese-American conversation notes, Seibundō Shinkōsha) was published and became a runaway best seller with as many as 3.6 million copies sold. After this, any reading materials that were published could be sold because they were welcomed by those with a thirst for reading. This was the beginning of the postwar publishing boom.

Freedom of speech and publication were permitted (in spite of the fact that all publications were submitted for censorship under the Occupation). Not only publishers and newspaper publishing companies, but various groups and organizations (including political organizations) started to republish or launch new newspapers, magazines and periodicals. In this publishing boom, science magazines were no exception. As the following discussion explains, there never has been and never again will be a situation where so many science magazines appeared. 'Science magazines' in this article primarily refers to general educational science magazines that are not written for professionals. In addition to natural science magazines, some magazines about technology and the social sciences are also included.

Tetsu Hiroshige remarks on the science enlightenment movement that occurred shortly after the war in *Sengo Nihon no kagaku undō* (Japan's science movement after the war):[1] 'due to the opportunity for science enlightenment, publishing companies that were about to be established immediately took interest in science magazines one after another. Easy explanatory science magazines were launched in rapid succession, and a variety of science magazines flooded the market.' Hiroshige mentioned that many of these science magazines were hopelessly rootless magazines. He discussed an argument often heard during this period as an explanation for the publishing boom after the war: 'Japan was defeated by scientific warfare. Promoting the development of science is the key to the rehabilitation of Japan and to the foundation of democracy.' However, we must ask whether, as Hiroshige interpreted it, this boom in science magazine publishing was simply an abortive flower blooming at the corner of the postwar publishing boom.

I shall survey the overall situation of science magazines from 1945 until the first half of the 1950s. Many science magazines published in 1946 and 1947, springing up like mushrooms after the rain, quickly disappeared. The number of general science magazines that continued to be published in the 1960s and 1970s was few. Technical magazines aimed at those interested in particular areas of science got out of this magazine boom, and survived.

It is very difficult to grasp the situation of the publishing industry immediately after the war. Jūrō Fukushima, who researched a massive number of magazines that made their first appearance or were republished from the end of the war until 30 July 1946, pointed out that there were many errors and much omitted information even in yearbooks of publication.[2] Table 27-1 is a list of science magazines published after the war, compiled from various resources.[3] However, this list leaves much to be desired.

Newly Published Science Magazines in 1946 and 1947

Science Magazines Appearing in Succession

Kagaku Asahi (Science Asahi) was published throughout the war years without missing any issues. *Kagaku* (Science) was launched in October 1945 (September issue). Science magazines republished in 1945 included *Kagaku gahō* (Pictorial magazine of science: November), and *Kodomo no kagaku* (Science for children: December). In December 1945, as one of the forerunners of the publication boom, Sankaidō launched a new magazine under the title *Kagaku no tomo* (Friends of science). This was a combination of the former *Nainen kikan* (Internal-combustion engine) and *Hōdō* (Report). The same quality of paper was used for the cover page and main text pages. This magazine was a thin book of only 16 pages. The contents of this magazine, such as home diagnosis of radio (written by Toshio Murata), electric portable cooking stove (Masao Ikeuchi), bulldozer and scraper (Yukio Onouchi), square-shaped nails (Junji Yamada), science of sweetness (Takeshi Mizukoshi) and an essay on the current situation (Kaku Arai), reflected the immediate postwar society of Japan.

In the first half of 1946, more than nine general science magazines such as *Kokumin no kagaku*, *Kagaku kōron*, *Minshushugi kagaku*, *Warera no kagaku*, *Kagaku no sekai*, *Kagaku to geijutsu*, *Shizen*, *Bunka jin no kagaku*, *Shizen kagaku*, etc. were published. The following are the contents of the inaugural issue of *Shizen* (Nature: May 1946) published on the basis of 'the need to establish science for the general public': penicillin (written by Hamao Umesawa), radar [1] (Shirō Negami), chronological table of the history of science and culture (I) (Mitsutomo Yuasa), increase of quantity

Table 27-1: New and republished science and technology magazines after the war

September 1945
Kagaku (Science), republished in October, beginning with vol. 15, no. 1 (September edition).
Kinzoku (Metal), monthly, Agune Kōgakusha, later Agune Shuppansha, Agune: first published in October 1931.
December 1945
Kagaku no tomo (Friends of science), monthly, Sankaidō; from April 1950: *Chūgakusei* (Junior high school students).
Magazines that are thought to have made their first appearance in 1946:
Nihon shuppan nenkan 1944, 1945 and 1946 (Japanese publications yearbook 1944, 1945 and 1946), published in July 1947.
Kokumin kagaku (Science for the people), Kagakusha, previously titled *Seisan seinen* (Productive youth).
Gijutsu bunka (Technological culture), Ten'nensha.
Kagaku to seikatsu (Science and life), Kikai Seisaku Shiryōsha.
Saiensu (Science), previously titled *Seinen gijutsushi* (Young technicians), Nishi Nihon Shimbunsha.
Kagaku shugi (Scientism).
Kagaku to mokei (Science and models).
Kagaku shō kokumin (Science for youth). This magazine might have been published in 1947, probably after *Ugoku jikken shitsu* (Moving laboratory).
Atomu (Atom), Nihon Kagaku Gijutsu Renmei (Japanese League of Science and Technology).
January 1946
Kokumin no kagaku (Science for national people), Shin Shōsetsusha.
February 1946
Yoron kagaku (Public opinion science), Yoron Kagaku Chōsa Kenkyūjo.
March 1946
Kagaku kōron (Science public opinion), Zen Nihon Kagaku Gijutsu Dantai Rengōkai.
Minshushugi kagaku (Democratic science), Shōkō Shoin; from November 1946: *Shakai Kagaku* (Social science).
Warera no kagaku (Our science), Shōkō Shoin, Minshushugi Kagakusha Kyōkai keimōteki kikanshi (Educative bulletin for the Association of Democratic Scientists).
From July 1947: *Warera no sekai* (Our world).
Igaku no ayumi (Progress in medical science), monthly later weekly, Nihon Igaku Zasshi, later Ishiyaku Shuppan.
Kagaku no sekai (World of science), monthly, Takayama Shoin.
April 1946
Kagaku to geijutsu (Science and art), Nihon Kagaku Geijutsu Kyōkai.
Journal of the Physical Society of Japan.
May 1946
Shizen (Nature), monthly, Chūō Kōronsha.
Bunka jin no kagaku (Science for civilized people), Kōsōsha.
June 1946
Shizen kagaku (Natural science), Minshushugi Kagakusha Kyōkai, Shizen kagaku-bu kikanshi (Bulletin of Minka's Department of Natural Science).
Igaku (Medical science), monthly, Nihon Isho Shuppan [final issue: vol. 14, no. 5 (May 1953)].
First half of 1946
Kagaku sekai (Science world), monthly, Hōbun Shorin: *Hikō Shōnen* (Flying boy).
July 1946
Hideki Yukawa, et al., *Progress of Theoretical Physics.*
October 1946
Ugoku jikkenshitu (Moving laboratory) [final issue: August 1950].

November 1946
Kagaku ken (Science circle).
Chikyū no kagaku (Earth science), monthly, Meguro Shoten, Nihon Chishitsu Gakkai (ed.) [final issue: vol. 5, no. 1 (1950)].

February 1947
Shōnen kōsaku (Engineering work for boys), monthly, Kagaku Kyōzaisha. This magazine was merged with *Kodomo no kagaku* (Science for children) in July 1950.

March 1947
Kagaku to minsei (Science and people's livelihood), Minsei Kagaku Kyōkai. This magazine was published under the new title of *Josei no kagaku* (Science for women) in January 1948.
Kodomo kagaku kyōshitsu (Science classroom for children), later *Watashitachi no kagaku kyōshitsu* (Our science classroom), monthly, Daiichi Kagakusha [final issue: vol. 4 no. 1 (1950)].

May 1947
Saiensu (Science).

July 1947
Kagaku no ryōiki (Territory of science), monthly, Nankōdō.

August 1947
Denpa (Radio wave), monthly, Shūkyōsha. Musen Gijutsusha Rennmei Kikanshi (Bulletin of the Federation of Radio Technicians) October 1947.
Kiso kagaku (Basic science), quarterly, Kōbundō, no. 31 published in 1953.

October 1947
Niji (Rainbow), monthly, Jitsugyō no Nihonsha; November 1948 = *Kagaku shōnen* (Science boys).

November 1947
Popular Science (Japanese edition), monthly, Ibuningu Staasha [final issue: no. 81 (September 1952)]. A different publisher later republished this magazine.

November 1947
Iden (Heredity), monthly, Nihon Iden Gakkai (ed.), Hokuryūkan, later Shōkabō.

As of the end of December 1947
Shuppan nenkan 1947 and 1948 (Yearbook of publications 1947 & 1948).
Zasshi mokuroku (Magazine catalogue), inaugural issue.
Kagaku kyōshitsu (Science classroom).
Nichi-Bei kagaku tsūshin (Japan and America science correspondences), weekly.
Popyurā saiensu (Popular science).

Year of publication uncertain
Shōnen shōjo kagaku gurafu (Illustrated magazine of science for boys and girls), seems to have been published between 1947 and 1948.
Kagaku to gijutsu (Science and technology), Communist Party, final issue: March 1961.

January 1948
Nihon kagaku zasshi (Chemistry magazine of Japan), monthly, Nihon Kagakukai (= January) 1921, *Nihon kagaku kaishi* (Bulletin of the Chemistry Society of Japan), under new title.

November 1948
Kagaku shōnen (Science boys), monthly, Jitsugyō no Nihonsha, previously published under the title of *Niji* (Rainbow) [final issue: March 1949].

January 1949
Tenmon to kishō (Astronomy and atmospheric phenomena), monthly, Chijin Shokan = *Tenmon to kikō* (Astoronomy and weather), 1934.

March 1949
Seibutsu kagaku (Biochemistry), quarterly, Iwanami Shoten, Minshushugi Kagakusha Kyōkai Seibutsu Gakubukai (ed.).

April 1949
Kagaku Yomiuri (Science Yomiuri), monthly, Yomiuri Shimbunsha [final issue: January 1966].
Shigen (Natural resources), quarterly later monthly, Shigen Kyōkai.
Seitai no kagaku (Body science), bimonthly, Nihon Igaku Zasshi, later Igaku Shoin.

May 1950
Kagaku gurafu (Illustrated magazine of science), monthly, Dai Nihon Tosho [final issue: no. 64 (October 1953)].

October 1950
Kagaku no jikken (Scientific experimentation), monthly, Kyōritsu Shuppan.

February 1951
Rikō tosho nyūsu (News on science and technology books), monthly, Kōgakusho Kyōkai.

April 1951
Kagaku (Chemistry), Kagaku dōjin, Kagaku Henshū Iinkai (ed.).

January 1952
Saishū to shiiku (Collecting and breeding), republished, monthly, Uchida Rōkakuhoshinsha, later Saishū to Shiiku no Kai = January 1939.

September 1952
Rika no kyōiku (Education of science), monthly, Tōyōkan Shuppansha.

October 1954
Kagaku kisoron kenkyū (Research on basic science theory), quarterly, later biannually, Sōbunsha, Kagaku Kisoron Gakkai (ed.).

June 1956
Kagaku gahō (Pictorial magazine of science), republished, monthly, Seibundō Shinkōsha [final issue: May 1961].

August 1956
Genshi enerugii (Nuclear energy), bimonthly, Atomusha [final issue: January 1961].

October 1956
Erekutoronikusu (Electronics), monthly, Ōmusha.
Tanpakushitsu, kakusan, kōso (Protein, nucleic acid and enzyme), monthly, Kyōritsu Shuppan.

January 1957
Genshiryoku hatsuden (Nuclear power generation), quarterly, Koronasha, final issue: vol. 8, no. 4 (December 1964).

October 1957
Kagaku no kyōshitsu (Science classroom), monthly, Gakushū Kenkyūsha.

June 1958
Igaku no ayumi (Progress of medical science), weekly, Ishiyaku Shuppan.

October 1959
Gijutsu jaanaru (Journal of technology), weekly, Nikkan Kōgyō Shimbunsha.

December 1960
Bussei (Physical properties), weekly, Shin Shoten.

July 1961
Seibutsu butsuri (Biophysics), quarterly, later bimonthly, Yoshioka Shoten (sold by Maruzen).
Mokei to kōsaku (Model and engineering work), monthly, Gijutsu Shuppan.

August 1961
Gakusei no kagaku (Science for students), republished = October 1940, monthly, Seibundō Shinkōsha [final issue: May 1962].

April 1962
Sūgaku seminaa (Mathematics seminar), monthly, Nihon Hyōron Shinsha, later Nihon Hyōronsha.

November 1962
Kagaku to hōdō (Science and information) annually, Osaka Nihon Kagaku Kisha Kyōkai.
Kagaku to seibutsu (Science and biology), bimonthly, later monthly, Tōdai Shuppankai, Nihon Nōgei Kagakukai (ed.).

June 1963
Sūri kagaku (Mathematical science), monthly, Daiyamondosha.

July 1963
Ningen no kagaku (Human science), monthly, Seishin Shobō. Ningen no Kagaku Kenkyū (ed.) [final issue: no.18 (December 1964)].

January 1965
Karada no kagaku (Body science), bimonthly, Nihon Hyōronsha.

April 1965
Kōgai to taisaku (Pollution and its measure), monthly, Kōgai Taisaku Gijutsu Dōyūkai.
Jōhō kagaku (Information science), monthly, Keiyūsha, later Jōhō Kagaku Kenkyūjo.
October 1965
Butsuri to kagaku (Physics and science), monthly, Seibunsha.
March 1966
Nihon no kagakusha (Japanese scientists), bimonthly, later monthly, Nihon Kagakusha Kaigi (Bulletin of Nihon Kagakusha Kaigi).
April 1966
Konchū to shizen (Insects and nature), monthly, Nyū Saiensusha.
May 1966
Denshi igaku (Electronic medical science), quarterly, Rajikon Gijutsusha
June 1966
Ishoku (Transplant), quarterly, Kyōdō Isho Shuppan.
October 1966
Kōgai (Pollution), quarterly, Hakua shobō [final issue: no. 5 (April 1967)].

Source: Until the end of 1967: *Nihon shuppan hyakunenshi nenpyō*, Nihon Shoseki Shuppan Kyōkai; confirmed by the 1968 publication.

of nutrition (Kaneo Inoue), science and the reconstruction of Japan (Yoshio Nishina), speed (Kōji Fushimi), bath (Tomio Ogata).

Table 27-2 is a list of general science magazines up to approximately the middle of 1946, based on the science literature catalogue published in *Shizen kagaku* (Natural science).[4] This catalogue was based on license applications to Nihon Shuppan Kyōkai (the Japan Publication Association). Publishers' names are listed in parentheses, except in the case where the publication was the initiative of some founding group other than publishers; then the name of the founding groups is listed. (The classification follows the catalogue.)

There is very little data available indicating the number of copies of each magazine published. However, in the data compiled by the Occupation Forces, there were lists of science magazines published immediately after the war, and their analyses includes estimates of the number of copies.[5] This document was dated 2 July 1946. Information sources are from Nihon Shuppan Kyōkai and the list provided by the Metropolitan Police Department. The estimated number of copies is based on the quota of paper. According to this information, there were forty-three science magazines and five science-related magazines. The number of copies of these forty-three magazines totaled approximately 700,000, and ten of these forty-three magazines achieved more than 20,000 copies each.

The following is a list of magazines and their estimated number of copies: *Kagaku no tomo* (Friends of science) [80,000], *Kagaku Asahi* (Science Asahi) [75,000], *Kagaku no sekai* (The century of science) [50,000], *Kagaku sekai* (The world of science) [50,000], *Kagaku gahō* (Science illustrated) [50,000], *Warera no kagaku* (Our science) [50,000], *Yoron kagaku* (Science of opinion surveys) [50,000], *Shashin kagaku* (Photographic magazine of science) [30,000], *Kokumin kagaku* (National

Table 27-2 : General science magazines published in the middle of 1946

1	Science Review (17 magazines) *Minshushugi kagaku/Warera no kagaku* (Minshushugi Kagakusha Kyōkai), *Seisan gijutsusha* (Minshushugi Seisan Gijutsusha Dōmei), *Kagaku shugi* (Kagakusha), *Kagaku kōron* (Zen Nihon Kagaku Gijutsu Dantai Rengōkai), *Bunkajin no kagaku* (Kōsōsha), *Gijutsu bunka* (Gijutsu Bunka Kenkyūkai), *Kagaku shisō* (Kagaku Shisō Kenkyūkai), *Kagaku chishiki* (Nihon Kagaku Kyōkai), *Kagaku to geijutsu/Seisan kagaku* (Nihon Kagaku Geijutsu Kyōkai), *Shisō no kagaku* (Senkusha), *Nihon kagaku bunka kyōkaishi* (Nihon Kagaku Bunka Kyōkai), *Kagakushi kenkyū* (Nihon Kagakushi Gakkai), *Gijutsushi kenkyū* (Hokuryūkan), *Kagaku seisaku* (Kagaku Seisaku Kyōkai), *Nihon kagaku* (Nihon Kagakusha Kyōkai).
2	Science Commentary [for adults and students] (14 magazines) *Shizen kagaku* (Minshushugi Kagakusha Kyōkai), *Shizen* (Chūō Kōronsha), *Kagaku Asahi* (Asahi Shimbunsha), *Kokumin no kagaku* (Shin Shōsetsusha), *Kokumin kagaku* (Kokumin Kagaku Kyōkai), *Kagaku sekai* (Hōbun Shorin), *Kagaku gahō* (Seibundō), *Kagaku no tomo* (Sankaidō), *Mura no kagaku* (Mura no Kagaku Kankōkai), *Seikatsu to kagaku* (Hokkaidō Kagaku Fukyūkai), *Kagaku to seikatsu* (Kikai Seisaku Shiryōsha), *Seikatsu kagaku* (Seikatsu Kagakuka Kyōkai), *Saiensu* (Nishi Nihon Shimbunsha), *Shizen to shiki* (Nisshin Shoin).
3	Science Commentary [for children] (6 magazines) *Kodomo no kagaku* (Seibundō), *Shōnen kagaku* (Shōnen Kagakusha), *Shōnen kōsaku* (Kagaku Kyōzaisha), *Shōnen bunko* (Hōbun Shorin), *Kagaku no sekai* (Takayama Shoin), *Shinri no kuni* (Mikensha).

Source: Minshushugi Kagakusha Kyōgikai (ed.) 2 July 1946, *Shizen kagaku*, vol. 1, no., pp. 44–5.

science) [30,000], *Kodomo no kagaku* (Children's science) [23,000], *Kagaku zukai* (Pictorial magazine of science) [20,000], *Kokumin no kagaku* (Science for the national people) [15,000], *Kōgyō kagaku* (Industrial chemistry) [14,000], *Rōdō to kagaku* (Labor and science) [12,000], *Kagaku shugi* (Scientific orientations) [10,000], *Kagaku kōron* (Public discourse on science) [10,000], *Kagaku* (Chemistry) [10,000], *Gijutsu shinron* (The new theory of technology) [10,000], *Makino shokubutsu konkon roku* (Makino's copious record on plant) [10,000], *Shinsei sūgaku to busshō* (Newly-born mathematics and material phenomena) [10,000], *Kagaku chishiki* (The knowledge of science) [9,400], *Kagaku* (Science) [8,000], *Kagaku to geijutsu* (Science and technology) [5,000], *Bunkajin no kagaku* (Science for cultured people) [3,000] and *Shizen* (Nature) [2,000].

Magazines were successively launched in 1946 and 1947. No documents that accurately recorded magazine publications for this period could be found, but from *Nihon shuppan nenkan 1947* and *1948* (Yearbook of publications in Japan) we can ascertain what other general science magazines were published.

The publication of specialized magazines after the war started from the republishing of *Kinzoku* (Metal) in September 1945, or from thereabouts. *Igaku no ayumi* (History of medicine) and *Chikyū no kagaku* (Global science) made their first appearance in 1946 and both *Kagaku no ryōiki* (Sphere of science) and *Iden* (Heredity) in 1947.

Science magazines aimed at young people included *Ugoku jikkenshitsu* (Moving laboratory), *Shōnen kōsaku* (Boys' science), *Kodomo kagaku kyōshitsu* (Science classroom for children), *Niji* (The rainbow), *Kagaku shō kokumin* (National children for science), etc as well as *Kodomo no kagaku* (Children's science), which had been published since before the war.

General science magazines: In November 1946 *Kagaku ken* (Scientific sphere), edited by Kiyotsune Inoue and Haruo Numanoi, made its first appearance. *Saiensu*, edited by scientists of Kyoto University and written in romaji (romanisation of Japanese scripts), was also newly published in May 1947. In October 1947 a quarterly magazine *Kiso kagaku* (Basic science) was launched and in the same year *Atomu* was brought out by Nihon Kagaku Gijutsu Renmei (Japanese Science and Technology League). In November 1947 *Popyuraa saiensu*, a Japanese version of the American *Popular Science* magazine began publication. *Kagaku zuihitsu amiiba* (Science essays ameba) [Tōsui Shobō] also seems to have been published around this time. The magazines listed above were aimed at educating the masses in general scientific knowledge. There were also various magazines that had emphasis on technology.

The Initiative for Magazine Publication

The science and technology magazines published after the war were often successors to, and supported by, magazines published during the war and republished under new titles.[6] Publishing groups and publishers were reorganized and renamed, both for a fresh start and as a survival strategy.[7] For instance, Dai Nihon Hikō Kyōkai, who had published *Hikō nihon*, was renamed Hōbun Shorin, and launched *Kagaku sekai*. Kagaku Shugi Kōgyōsha, who had originally published *Kagaku shugi kōgyō* and *Seisan seinen* was renamed Kagakusha, and produced *Kagaku shugi* (February 1946) and *Kokumin kagaku*, in place of the previous magazines. *Kōkū shōnen* from Seibundō Shinkōsha was transformed into a new magazine called *Nōkō to engei* (Agriculture and horticulture). *Kōkū Asahi* published from Asahi Newspaper was merged into *Kagaku Asahi* (Asahi science). *Zukai kagaku* (Illustrated magazine of science), initially published in December 1941, was taken over by Asahi Newspaper from 1944, vol. 3, no. 8 because of the dissolution of Chūō Kōronsha. After the war, though, Chūō Kōronsha was restored and *Zukai kagaku* returned to it, but the magazine was no longer published. Instead, the style was adopted by *Shizen*.

Various science organizations also took an initiative to publish science magazines. Minshushugi Kagakusha Kyōkai (the Association of Democratic Scientists, also known as Minka; cf. ch. 25) took the most active role in the science movement at this time and published *Minshushugi kagaku*, *Warera no kagaku* and *Shizen kagaku*. Kagaku Chishiki Fukyūkai, the publisher of

Kagaku chishiki from 1921 changed its organizational name to Nihon Kagaku Kyōkai during this time. Table 27-2 includes the names of various organizations that founded new science and technology magazines. This list, however, does not include magazines published outside of the major cities. In most cases, these magazines seem to be the initiative of scientists and engineers, rather than publishers. Scientists at Kyūshū University were the editors, contributors and publishers of *Mura no kagaku* (Village science; published by Mura no Kagaku Kankōkai) and scientists at Kyoto University who published a romanised magazine entitled *Saiensu* fall under this category. *Kagaku shō kokumin* [Teikoku Hatsumei Kyōkai Niigata Shibu (Niigata Branch of the Society of Imperial Invention)], *Gijutsu to shakai* (Technology and science) [Hokkaidō Sōgō Gijutsu Renmei (Federation of Hokkaidō Synthetic Technology)], *Shinsei kagaku* (Newborn science) [Hokkaidō Kagaku Fukyū Kyōkai (Society of Hokkaidō Science Promotion)], and *Seito jidō kagaku to hatsumei* (Science and invention for pupils) [Teikoku Hatsumei Kyōkai Hokkaidō Shibu (Imperial Society for Invention, Hokkaidō Chapter)] were all initiated by scientists living in areas other than large cities.

Some magazines specified the names of scientists as editors. Hideomi Tsuge, who played an active role in Minka, was responsible for *Kokumin no kagaku* (Science for the people). *Kagaku ken* also belongs to this type of magazine. Many of the new science and technology magazines during this period seemed to be either directly published by scientists or engineers, or by publishers with the help of science and technology organizations.

Of course, some publishers took the initiative in the publication of science magazines. *Shizen* and *Popyūraa saiensu* are examples of this category. Magazines aimed at young people, such as *Ugoku jikkenshitsu*, *Shōnen kōsaku* and *Niji*, probably fall into this category as well. There might be publishers who wanted to make a fortune during the publishing boom around this period. However, the publishing of educational science magazines might be considered to be a fairly conservative business and therefore not profitable. We can see 'aspirations' for science magazines by publishers and editors.[8]

Contents of Essays and Articles

Depending on the magazines, subjects taken up are diversified. Subjects on food, contagious diseases and other diseases, vectors, etc., that reflected the postwar chaotic state of the society were taken up. The inaugural issue of *Shizen* (May 1946) carried articles on penicillin written by Hamao Umesawa and increased nutrition by Kaneo Inoue. In the following issue Masami Kitaoka wrote on typhus, and in the August 1946 issue Manabu

Sassa and Yoshio Kawakita wrote articles on mosquitoes and epidemic encephalitis, respectively.

As a reflection of this time, there were critiques and arguments about how to reestablish science. For example, in *Kokumin no kagaku* Hiroto Saigusa and Yoshishige Kozai wrote essays on 'peace and science' (inaugural issue) and 'scientific mind and the establishment of a new Japan' (March 1946), respectively. Beginning with Yoshio Nishina's debate on the reconstruction of science and Japan in the inaugural issue of *Shizen*, these topics were continually discussed. In the editor's postscript in *Shizen kagaku* (July 1946), Y. H. (probably Yoshirō Hoshino, the editor) observes that 'general statements are numerous. These are only sermons. We may have needed sermons for the first three or four months after the war...' We can assume from these comments that there were many discussions on the reconstruction of Japan and science.

The topics taken up by these magazines are not limited to natural science. Not only science, agriculture or technology, but also the politics and education systems of various European countries were discussed in *Kokumin no kagaku*. In *Kagaku no tomo* (June/July 1946), Yoshio Nakano contributed an essay on 'science education and problems associated with Japanese language and characters.'

On the other hand, a wide variety of topics were dealt with using 'science' as a keyword. As an example, let's look at *Bunka jin no kagaku*. Initially published in May 1946, its initial issue was a special edition on women and science. The contributors were Toshiko Yuasa who wrote on women and science and Toshie Fujinaga on women's liberation and medicine. In this initial issue, there was an essay on 'development of electronic technology' written by Jun'ichi Takagi, but after this article, there were few articles on natural science. Instead, articles were more inclined towards medicine and especially psychiatry. From the fourth issue of *Bunka jin no kagaku*, the sub-title *Ningen kagaku no zasshi* (Magazine of human science) was added. The publication of this magazine ceased with the March 1947 issue (vol. 2, no. 1 [total of six issues]). Judging from the style in the editor's postscript, this editor didn't seem to be a man of the pen. In *Bunka jin no kagaku* (May 1946), Ranpo Edogawa contributed an essay of three pages on 'the criminal psychology that appeared in detective stories.' There were numerous magazines that used 'science' in the title, in spite of the fact that natural science was not discussed; for example, *Shisō no kagaku*.

The editor's objective for *Bunka jin no kagaku* does not seem to have been clear and the targeted readers not well defined. Some magazines also changed the nature of magazines. The editor's postscript in *Kagaku no tomo* (combined issue nos. 6 and 7, July 1946) described the objective as 'a magazine for the family and favored by people at home.' This magazine appeared to be aimed at students as well as adults. In 1948 it refocused and became a magazine exclusively written for young people, especially students.

Scientists as contributors: Tomio Ogata mentioned, in the preface of *Shizen* (June 1947) that science magazines for the general public in those days were 'not interesting' because the contributors were poor at explaining science. He cautioned that 'if we ourselves can't write interesting essays,...we should be discreet in suggesting that the general public has a low level of interest in science.' This problem with the quality of articles will be further discussed in the following section. But we should note here that this problem was exacerbated by the fact that, rather than a wide variety of scientists writing essays, the same few scientists seem to have contributed essays to various magazines. The list of contributors for *Shizen* is a good example of this. Except for Mitsutomo Yuasa who ran a series of chronological tables of science and culture from the inaugural issue over a long period of time, Koji Fushimi, Takanobu Suzuki, Hamao Umesawa, Yoshio Nishina and Ukichirō Nakaya were frequent contributors for the first few years of the magazine. It appears that the number of good writers were few. There were many instances of the same writer appearing in different magazines. Magazines do not get support from readers unless they secure good writers who can provide attractive essays and articles.

Criticism against Science Magazines

Publishers were surprised that so many science magazines appeared, so magazines themselves started to carry reviews and criticism of other science magazines. Yoshinori Kanaseki wrote an essay entitled 'reading overflowing numbers of science magazines'[9] in which he says:

> Science magazines are continuously published and republished, shooting up like mushrooms after the rain. However, the time appears to be near when shops with unsold magazines will be in a quandary. Due to the lack of paper and stagnation of printing, the rising prices of magazines is unavoidable. With readers on the verge of starvation, paying for magazines – even with the life of the new-yen – is the last thing they can think of. It is a matter of regret that there is no progress worthy of special mention in the publication of science magazines.

Kanaseki already predicted the future of science magazines when the boom of science magazines had just begun. *Nihon shuppan nenkan 1944, 1945 and 1946* commented on the booming of science magazines: 'science magazines are still no better than the training of pernicious learned men with fragmentary knowledge of science who argue about shallow scientific methodology.'[10] In 1946 Gundayū Aida wrote a 'criticism against science magazines' in the November issue of *Kagaku chishiki*.[11] He outlined what science magazines 'should' be, and classified the science magazines being published at that time. But his criticism was somewhat abstract and didn't seem to provide a guide for editors and writers.

The yearbook *Kagaku nenkan*, edited by Minka, explains the postwar situation of science in Japan. In its 1947 edition (Period of data: August 1945 until April 1947), Yoshinori Kanaseki repeatedly mentioned the 'overflowing numbers of science magazines.'[12]

When academic magazines were stagnant, so-called science magazines (which were used as media for commentary and criticism) overflowed. Kanaseki referred to this situation as 'the impression that in poor families debauchees are rampant.' He continuously and severely attacked science magazines, saying that 'worthless practical articles, cynical scientific news and hasty scientific talk mixed with unclear pictures and drawings filled up page space. It is like a shop front in a black market.' Regarding magazines whose primary objective was profit, he argued that 'from the beginning the writers didn't understand science and they didn't have a sound knowledge of science. This soon made readers distrustful of science magazines. Combined with a lack of paper and inflation, many magazines are on the verge of ruin.' Immediately after this statement, he said: 'the reason that science magazines ended in failure was that entrepreneurs, editors and writers did not grasp a close connection between the construction of science and the liberation of people.' This, however, was also a reflection of a typical contemporary ideology, and did not clarify how or what to do. He concluded that 'regardless of difficulty or ease and hardness and softness, it is difficult to tell which readers were targeted for the many science magazines. And even though magazines were repeatedly read, readers do not understand the articles due to insufficient explanations and mistakes in the articles...[U]nattractive science magazines are useless.'

Kokumin no kagaku (June/July 1946, pp. 24–5 and August 1946, pp. 54–5) also carried a review of science magazines, as did the weekly newspaper *Kagaku Bunka Shimbun* (from issue no. 64, 15 July 1947, until no. 147, September 1948).

The Softening of the Publishing Boom

I have already mentioned that the publishing of science magazines was stagnant already in 1946 when the publishing boom began. A shortage of paper significantly restrained magazine publication, and many monthly magazines could not be published regularly. Magazines frequently complained about the paper and printing situations in the editor's postscript. When I actually felt the magazines, the quality of paper of magazines published in 1946 and 1947 was worse than the quality of paper of those published immediately after the war.

Popular science, much discussed before its first appearance, carried translated articles of US science magazines, so the contents did not correspond to the situation in Japan. Furthermore, the paper and printing were of such poor quality that the magazine was shocking to look at.

Nihon shuppan nenkan commented that contrary to the increasing number of science magazines for the general public, purchasing power had actually declined. This decline was especially noticeable around the spring of 1947 and seemed to reach its peak that summer. However, the publication of new science magazines continued. The situation of the publication of general science magazines remained unstable.

Science magazines published from 1948 until 1950: *Kagaku Yomiuri* made its first appearance in April 1949, but the inauguration of new general science magazines peaked in 1947 and began to decline. Besides *Kagaku Yomiuri*, specialised magazines such as *Tenmon to kishō* (January 1949), *Seibutsu kagaku* (March 1949), *Seitai no kagaku* (April 1949), and *Kagaku no jikken* (October 1950) made their first appearance.

Many magazines born during the postwar publishing boom disappeared in a few years. Science magazines were included in this trend. We trace the number of magazines overall and its change. The number of magazines at the end of each year from 1946 to 1952 was 2904, 7249, 6778, 5243, 1537, 1572 and 1190, respectively.

With the depression of the economy at the end of 1949, many magazines were suspended and discontinued as rapidly as they had blossomed only a few years earlier. Many newly established publishers became bankrupt and were forced to cease their business activities. May 1950 appears to have been the 'peak' period for suspending and discontinuing magazines, and October 1950 records the highest number of publishers declaring bankruptcy and closing their businesses.

Data indicating how readers reacted to these science magazines is scarce. The result was that many magazines went out of existence without being able to obtain support from readers.

Science Magazines after 1950

When the situation of 1950 is examined by *Shuppan nenkan*'s 1951 edition, the 'science' section of the magazine catalogue listed thirty-two magazines, of which four magazines, *Kagaku gahō*, *Kagaku ken*, *Kagaku no tomo* and *Watashitachi no kagaku kyōshitsu* were discontinued, and two others – *Ugoku jikkenshitu* and *Kodomo no kagaku* – were merged. The total number of science magazines for children as well as general public and specialized magazines was seventeen: *Kagaku*, *Kagaku Asahi*, *Kagaku kyōshitsu*, *Kagaku gurafu* (Science graphics), *Kagaku ken bessatsu* (Science circle: a supplementary issue), *Kagakusha*, *Kagaku shō kokumin*, *Kagaku chishiki* (Scientific knowledge), *Kagaku no jikken* (Experiments in science), *Kiso kagaku*, *Kagaku Yomiuri* (Science Yomiuri), *Gijutsu to shakai* (Technology and science), *Kodomo to kagaku* (Children and science), *Kodomo no kagaku*, *Saiensu daijesuto* (Science digest), *Shizen* and *Popyūraa saiensu nihongo ban* (Popular science: Japanese version).

In 1951 there were fifteen magazines listed in the 'science' section, including *Kagaku shi kenkyū* (The study of science history) and magazines designed for children. There were ten magazines in 1952, still including *Kagaku shi kenkyū*, but the magazines for children have been listed in a different section. In 1953, there were twelve magazines, including not only the study of science history, but also magazines other than general science magazines such as science education news and reports from science research institutes. In this year *Kiso kagaku* published its thirty-first issue and then shut-down.

In 1954, the science section was changed to the natural science section. Forty-one magazines are listed under this classification, but this includes professional journals and the English language journals of various science societies. There are only about seven magazines for the general public. Categorised in the same manner, forty-one magazines are listed in 1955, and forty-five magazines in 1959, among which *Kagaku*, *Kagaku Asahi*, *Kagaku gahō* (relaunched in June 1956, folded in 1961), *Kagaku Yomiuri* (ceased to publish in 1966), and *Shizen* are included under the general science magazine category.

Science Magazines that Outlived the Publishing Boom

Science magazines sprang up like mushrooms after the rain in 1946 and 1947 immediately after the war. These magazines fall into three categories: (1) survivors of magazines published during the war; (2) magazines published by scientists as part of the science movements; (3) magazines republished and newly published on the initiative of publishers themselves. Regarding the increasing number of science magazines, categories (1) and (2) are more significant than Hiroshige suggests when he argues that these magazines were mainly produced by publishers wishing to make a profit.

Many of these magazines disappeared within a few years. This may have been directly caused by the paper and printing situation, the state of confusion after the war, and the economic recession. People were struggling simply to support themselves and science magazines were at the bottom of their priorities. Together with the confusion and stagnation of the science movement by Minka and others, the publication of the magazines that they initiated could not be maintained. Some magazines revised editorial objectives and targeted different readers.

In addition to this, there was the previously mentioned situation that good science writers (mainly scientists) did not exist. Although many scientists placed great significance on the concept of scientific enlightenment during those few years after the war, they were not trained to write general science articles, but rather scientific articles for other scientists. The significance of educational science articles and the training required to enlighten the general public about and through science remained a matter of grave concern, but many scientists went back to their research, turning away from the general public. While on the surface it may seem possible to do both – i.e., scientific research and public education – most scientists are not motivated to write for popular magazines because these articles do not count as academic achievements and therefore do not advance the scientist's professional standing.

Regardless of the explanations for the postwar publishing boom and the subsequent collapse, it is important to recognize that some of the science magazines born in the boom survived the collapse. *Iden*, *Konchū to shiiku*, *Denpa* and *Kagaku no ryōiki* belong to this group. The readers who supported these particular magazines were junior and senior high school (especially science) teachers, students and researchers. As a reflection of the increasing number of interests in various academic fields and especially newly established fields, wide varieties of technical and specialized magazines soon began to appear.

The number of general science magazines that went through the publishing booming and survived into the 1970s are few, including magazines such as *Kagaku*, *Shizen*, and *Kagaku Asahi*. The shift from science magazines to newspapers as the primary medium of science journalism and reporting from the latter 1950s will be discussed in volume 2.

Notes

1 Tetsu Hiroshige (1960), *Sengo Nihon no kagaku undō* (The science movement in postwar Japan), Chūō Kōronsha, p. 21.

2 Fukushima, 1985 [1] p. 634. Five magazines – *Kokumin no kagaku*, *Kagaku kōron*, *Warera no kagaku*, *Kagaku to geijutsu* and *Kagaku* – were introduced in the bibliographical essays on the inaugural publication and republication of major magazines (Section 2).

3 The four main sources used were: Nihon Shoseki Shuppan Kyōkai, 1968 [2]; Fukushima, 1985 [1]; Shuppan Kyōdo Kabushikigaisha [3]; and Shuppan Kyōdo Kabushikigaisha [4]. In addition to these, actual science magazines published after the war and related articles from newspapers were referred to. There are many points where confirmation could not be obtained.

4 Minshushugi Kagakusha Kyōkai, Shizen Kagaku Bukai (ed.), *Shizen kagaku* (Natural science), vol. 1, no. 2, July 1946, pp. 44-5.
5 Occupation Forces Document ESS(A)09906.
6 According to Fukushima (1985 [1] pp. 543–4) there were 33 science and technology magazines as of September 1944. The publication of many magazines was reorganized due to the 1943 readjustment of enterprises, but science and technology magazines were preserved and promoted rather than abolished. The number of copies of magazines printed (especially *Kōkū Asahi*, *Hikō Nihon*, *Kōkū shōnen*, *Hikō shōnen*, *Mokei kōkū*, etc.) proudly reached a high of over 100,000 copies each. Refer to Shuppan Kyōdo Kabushikigaisha [3] p. 41 ('Science and physical science' from 'Three years history' in the magazine section).
7 Notes written in the science publications catalogue (refer to note 4) attached in *Shizen kagaku* explained this change.
8 'Aspirations in the publishing of science magazines' were written in the remarks of inaugural issues or editors' notes of each magazine, and their records were not kept. Let's look at two boxed items from *Kokumin no kagaku* (Science for the national people), August issue, 1946. One article was titled 'Kagaku jaanarisuto sunden – Ozeki Takematsu no koto (Short biography of a science journalist – Takematsu Ozeki)' [p. 17], and the other article was 'Kagaku jaanarisuto (Science journalist)' [p. 63]. The latter was an account of Toshio Aijima, then assistant director general of Shuppan Kyōkai (Association of Publication), later an active science journalist, and the former was an account of the death of a nameless character who had edited science magazines such as *Kagaku pen* (Science pen), *Kagaku shichō* (Science thought), *Kagaku Nihon* (Science Japan), *Kagaku chishiki* (Knowledge of science), etc. before and during the war. It appears that the columnist (MSP) who wrote on Ozeki wanted to record the 'aspiration' of nameless editors of science magazines.
9 *Shizen kagaku*, vol 1, no. 2, July 1946, p. 19.
10 Shuppan Kyōdo Kabushikigaisha [3] 1947, p. 41.
11 *Kagaku chishiki*, vol. 6, no. 11, November 1946, p. 26.
12 Minshushugi Kagakusha Kyōkai, 1947, [5], p. 291.

Bibliographic Guide

Records regarding the situation of publication after the war for a few years are rather disordered. It is correct to say that actual publications should be referred, however it is very difficult. The following laborious work clearly explains the situation of magazines immediately after the war and comments on science magazines.

[1] Jūrō Fukushima (ed.), *Shinpan sengo zasshi hakkutsu* (New edition of excavation of magazines after the war), Yōsensha, 1985.

Information on the inaugural issues of magazines can be found in the following chronological table that contains publications until the end of 1967, but the information on the republication and new publication after the war is full of omissions.

[2] Nihon Shoseki Shuppan Kyōkai, *Nihon shuppan hyakunen-shi nenpyō* (Chronological table of one-hundred years of publication in Japan), Nihon Shoseki Shuppan Kyōkai, 1968.

Information on publication during the war and after the war can be found in the next two *Shuppan nenkan* (Yearbook of publication). These yearbooks classify information on books and magazines, but the classification differs from year to year.

The distributing company of publications that provided publication data closed down in August 1950 because it was designated as part of the concentration of excess economic power (cf. ch. 17). Due to this, the 1949 and 1950 editions were not published, so information on magazines published in 1948 and 1949 can't be put together.

[3] Shuppan Kyōdō Kabushikigaisha, *Nihon shuppan nenkan 1944, 1945 and 1946* (Yearbook of publication in Japan 1944, 1945 and 1946), Shuppan Kyōdō Kabushikigaisha, 25 July 1947.

[4] Shuppan Kyōdō Kabushikigaisha, *Nihon shuppan nenkan 1947 and 1948*, (Yearbook of publication in Japan 1947 and 1948), Shuppan Kyōdō Kabushikigaisha, 5 November 1948.

Minshushugi Kagakusha Kyōkai (Minka, cf. ch. 25) published the next three *Kagaku nenkan* (Yearbook of science). These yearbooks contain catalogues of books. Together with other data given, these catalogues provide information on the situation of publication.

Kinnosuke Ogura, president of Minka, comments, in the preface to the 1947 edition, on the concrete view regarding a variety of fields in the natural and social sciences. The contents of these yearbooks include general views on, and trends in, various academic fields, information guides to academic circles, research institutions and libraries, and catalogues of books, etc.

[5] Minshushugi Kagakusha Kyōkai (Association of Democratic Scientists) (ed.), *Kagaku nenkan 1947* (Yearbook of science 1947), Nihon Kagakusha, 15 November 1947.

[6] Minshushugi Kagakusha Kyōkai (Association of Democratic Scientists) (ed.), *Kagaku nenkan* (Yearbook of science), 2nd series (1947.7.5 – 1948.4), Nihon Kagakusha, 25 August1948.

[7] Minshushugi Kagakusha Kyōkai (Association of Democratic Scientists) (ed.), *Kagaku nenkan* (Yearbook of science), 3rd series, 1949 edition, Hokuryūkan.

The publication date is not written on the imprint, but the preface was dated 10 June 1949.

28 Marxism and Postwar Science in Japan

Kunio Gotō

Japan-Specific Features of Marxism and Science/ Technology

Diverse Interpretations of Marx and their Political Implications

Marxism exerted significant influence on the thought and ideology of many scientific movements in postwar Japan. For example, Minka, the Association of Democratic Scientists, initially represented a wide range of progressive political views, including non-socialist liberalism. Nevertheless, it is evident that orthodox Marxists assumed leadership of Minka from its inception. This was indicative of a more general phenomenon, namely that all of Japan's progressive postwar scientists' groups had Marxist orientations. Stated baldly, however, at that time in Japan most Marxists who were scientists were also scientific Marxists who took for granted the equation: Marxism = materialism – science. Given this, it is surprising that so few in-depth studies have been conducted into the relationship between science and technology and Marxism – an oversight which this chapter intends to at least partially correct through its investigation of the importance of Marxist thought to Japanese scientists' movements, especially during the second half of the 1940s.

The majority of Karl Marx's vast writings consist of unfinished manuscripts, notes, letters, journalistic pieces and other works written in the formative years of his intellectual life. One might go so far as to say that the first volume of *Das Kapital* and its French version are his only rigorously systematic, scholarly publications. The massive amount of unfinished manuscripts and sketchy writings has generated diverse interpretations of 'Marxist thought,' often leaving 'political orthodoxy' to determine 'theoretical orthodoxy.' The most influential 'political orthodoxy' in this regard was, of course, the Stalinist system that governed the Soviet Union and cemented Soviet-style Marxism as the 'theoretical orthodoxy.' As far as the natural sciences are concerned, orthodox Marxism is characterized by endless regression into Enlightenment rationalism, as the writings of Joseph Stalin and Mao Zedong demonstrate.

Orthodox Marxism was generally more influential in the field of science and technology than in other areas of intellectual endeavor.

Ironically, this is in part attributable to the relative paucity of references that Marx himself made to science, making it difficult for people to rebut claims of orthodoxy by reference to his works. Furthermore, orthodox Marxism seems to closely resemble the Enlightenment rationalism that governs the thinking of most people concerned with science and technology.

A few points should be made about the long-standing issue of 'Real Marx versus Marxism.' Critiques of modern science and modern civil society constituted one of the central themes of German classical philosophy ranging from Immanuel Kant to Georg Hegel. The same can be said of David Ricardo's economic theory, which was regarded as a 'dismal science.' Marx aimed to go beyond these criticisms, a task which made his vocabulary and reasoning extremely complex. One of the outcomes is the density of his argumentation and the opacity of his prose in *Das Kapital* and many of his earliest writings. By comparison, the 'philosophy of the Enlightenment' and 'modern rational thinking' were far more lucidly expressed, and therefore more accessible. 'Orthodox Marxism' attempted to rescue Marx's thought from its own complexity and convert it into a simple and precise scientific style of writing. In this way, the clear and explicit statements of Stalin and Mao were easily absorbed by their followers.[1]

It is undeniable that grounds can be found within both Hegelian and Marxist thinking for arguing that 'political orthodoxy' is equivalent to 'scholarly orthodoxy.' In his *Theses on Feuerbach* Marx states, 'The philosophers have interpreted the world in various ways; the point, however, is to change it.' This lies at the core of Marxism.[2]

Hegel, Marx's philosophical predecessor, thought the aim of philosophy was the 'realization of the Idea.' This led him to advocate the transformation of Germany from small, fragmented federations to a 'nation state' unified under a single set of constitutional principles. Thus, he became a strong defender of the monarchy and the German imperial system. Hegel's philosophy implied a kind of transcendence of modern civil society and modern thought, but once generalized beyond a philosophical context it was interpreted as advocating the 'establishment of a socialist state' with the authoritarian power of a monarchy. Thus, the 'realization of the Idea' became equated to a specific political outcome. This political distortion of Hegel's work set a precedent by which Marx's writings were later taken from their original context and used as weapons of ideological control by Communist Party leaders.

One of the key political justifications cited for the authoritarian dominance of technocrats in the Soviet Union was the unquestionable verity of the State's scientific achievements. This, the leaders and officials claimed, bestowed legitimacy upon both the Party and the State. The notion that 'being scientific' was an expression of political legitimacy reassured many scientists, enabling them to devote all their time to scientific research that was by definition progressive and 'free of ideology.' Hence it would be many years before the kind of critical analysis that Marx applied to 'political economy' in *Das Kapital* was brought to bear on modern science and technology.

Problems of Marxism and Science in Japan

One of the reasons Marx's social theory has come to be seen as an heir of Enlightenment philosophy is that some of its most influential interpreters, people like Friedrich Engels and Karl Kautsky, were far more indebted to Enlightenment ideals (e.g. rationality, democracy) than Marx himself; their interpretations of his work reflect this. The same can be said for his most significant Russian interpreters, Georgii Plekhanov and Vladimir Lenin, who both faced the challenge of re-interpreting Marx's post-capitalist revolutionary prescriptions in the context of Russia's pre-capitalist, feudal economy. Near the end of his life, when he was confronted with the problem of the modernization of rural Russia, Lenin became interested in Taylorism, just as American Taylorists, who preferred technocracy, began taking an interest in the socialist plans of the Soviet Union.[3]

As with pre-Revolutionary Russia, prewar Japan was governed by a despotic Imperial system which supported its development as a capitalist economy. Republicanism, which was advocated by moderate liberals in the West, was considered to be 'dangerous thought' in Imperial Japan. In this context, liberals and others imbued with the modernist spirit came to see Marxism, suppressed by the police, as embodying the esoteric justice of scientific truth. Support for Soviet-supervised revolutionary strategies became widespread among Japan's progressive thinkers. In 1932 the Comintern outlined a program for Japan's revolution in which the first priority was given to a 'bourgeois revolution which should overthrow the absolutist emperor system.' This directive remained the ultimate goal for many Japanese socialist activists and intellectuals, even after the war.

During this period some Japanese Marxists developed their own revolutionary plans. At the level of theories of tactical organization, they included Yamakawa's 'social democratic popularization line,' Fukumoto's 'radical intellectuals' leadership line,' Japan's own socialist revolution platform (a draft of the 1931 program), and the so-called Rōnō-ha faction, for whom bourgeois revolution was already achieved by the Meiji Restoration. The Communist Party of Japan, however, held firm to Soviet directives and rejected these indigenous revolutionary initiatives as heterodoxy.

Japanese Modernization and Marxism during the Fifteen Years' War

Although the organizational activities of Marxist parties suffered irreparable damage at the hands of government suppression during the so-called Fifteen Years' War of 1931–1945, the philosophical and academic influence of Marxism lingered on. The Japanese establishment at that time embraced both Imperial ideology and the maintenance of existing social institutions, as well as the introduction and promotion of Western science and technology in order to prepare for modern warfare. This military agenda made the expansion of higher education, particularly tertiary studies in science and technology, an imperative. Hence, the importance of science was emphasized in primary and secondary schools and the number of university graduates in science and engineering dramatically increased during the war years.[4]

Kunihiko Hashida, Education Minister in the Third Fumimaro Konoe Cabinet as well as in the Hideki Tojo Cabinet, actively promoted the 'spirit of science' among primary school children. Thus, the seemingly contradictory forces of Imperial ideology and scientific rationalism were united in a militarist cause. Marxism, suppressed by the authorities, could only exert its influence as the 'true agent of scientific rationalism' covertly. What follows are some illustrations of this covert activity.

Voluntary Obscurity

It is often the case that even while political activism is being successfully suppressed, 'dissident thought' survives in academia under the guise of refined and highbrow scholarship or as the facilitator of intellectual debate. This was certainly true of Japanese Marxist studies, where meticulous textual analysis became predominant and was far removed from the political and social realities of Japanese life. For example, the 'debate over Japanese capitalism,' which began as a discussion about revolutionary strategies, gradually turned into a series of esoteric academic arguments about 'manufacture as proto-industrialization', 'absolutism' and the 'theory of ground rent.'

Retreat into Compensatory Research Areas

In contrast, a kind of compensatory activity was carried out in spheres more closely aligned with Japan's social realities. In the case of Japanese Marxism, the 'theory on production force' and the 'theory on technology' represent a uniquely Japanese contribution to Marxist scholarship. Researchers in these fields were able to masquerade their studies as contributions to Japan's modern industrial development because, in keeping with Soviet orthodoxy, they advocated rather than criticized the modernization that comes with science and technology and the development of industry. Further, the 'history of science' and 'folklore studies' provided scope for covert Marxist research.[5]

Participation in the Wartime 'Planned Economy'

A more practical example of the influence of Marxism can be found in the 'wartime products mobilization program,' which strongly resembled the Soviet planned economy. The Japanese 'reformist forces' who executed this program in the 1930s were nationalist, right-wing, totalitarian groups whose politics came close to fascism. Their core membership was composed of 'reformist bureaucrats,' 'colonial bureaucrats' and 'young army officers.' Yet, the Stalinist system in the Soviet Union that had so rapidly established a heavy industry sector and developed a modern military by sheer force provided a powerful example for these Japanese reformists.[6]

Soviet Studies

Despite strict Imperial censorship, some pro-Soviet publications managed to find an audience in Japan. Two of the most influential of these were E. Varga's prediction of the Great Depression in the *World Economy and Economic Policy: A Quarterly Report*, and P. Dutt's Marxist study of fascism and Nazism. Participation in Soviet studies was compatible with the promotion of 'national policy' because the Soviet Union was the most likely potential enemy of the Japanese Army.[7]

Many of the intellectuals who took part in these four realms of Marxist influence were not themselves Marxists. A good number of them pinned their vague hopes on the possibility that industrial development based on scientific rationalism would 'trigger the intensification of contradictions between the mode of production and the relations of production and strengthen the prospect for revolution and modernization of Japan.' Underlying this belief was the esoteric influence of Enlightenment-oriented Marxism upon Japanese culture from the Taishō to the Shōwa periods. It is claimed that some believers in this line of thinking were more loyal to the Comintern's assessment of and prescriptions for Japanese society than the Japan Communist Party's own members.[8]

Marxism and Scientists' Movements Immediately after the Defeat in the War

Starting Point of Postwar Marxism

As a consequence of Japan's defeat in World War II the country's productive capacities and research facilities were largely destroyed. Like everyone else in postwar Japan, scientists and engineers endured impoverished conditions. For Marxists though the end of the war represented the overthrow of the authoritarian government and the legalization of the Communist Party by the Allied Occupation Forces. Thus, Marxist theory and activism were re-animated by Japan's defeat.

Most of this renewed activity was conducted under the auspices of the Communist Party of Japan. The domination of Soviet orthodoxy can be partly explained by the fact that the Japanese Government's repression of the Party before and during the war, and the Occupation Force's control of copyright after it, meant that the Japanese public was largely unaware of the Stalinist purges and the harsh realities of political life in the Soviet Union during the late 1930s and the early 1940s. Hence, most post-1945 Japanese Marxists had not become disillusioned with the Party in the way many of their Western comrades had and, therefore, they continued to derive their philosophical and political perspectives from a 1930s Party line.

Despite their relative ignorance of the realities of global politics, Japan's left-wing academics still managed to publish a significant body of scholarly work that was informed by Marxist theory but avoided calling itself by that name. Many of these social and philosophical works were originally published in the early 1940s and re-published to great acclaim immediately after the war.

Understanding 'Postwar Japan' and the Communist Party's 'Science and Technology Thesis'

Attempting to define the nature of postwar Japanese society was a pivotal issue, not only for Marxists engaged in a debate about revolutionary strategies, but also for the Japanese community at large. The Comintern's 1932 program, which became orthodoxy, regarded the prewar Japanese power structure as absolutist, with the Emperor at the apex of the hierarchy and privileged bureaucrats, landlords and capitalists forming the foundation. Accordingly, the forthcoming revolution was envisaged as a bourgeois concern which, although a mere stepping stone towards a socialist revolution, would in the first instance overthrow the Imperialist system and thereby rid the society of its feudal legacy. In the face of postwar realities, this framework underwent remarkably little change. Certainly, the Rōnō-ha groups who had defined the Meiji Restoration as a bourgeois revolution had a different perspective, but they had little philosophical influence over the scientists' movements at the time. Many of the issues that had dominated Marxist discourse before the war re-emerged after it, yet a small minority of Marxists considered these discussions out of date and irrelevant to contemporary problems.

In November 1946, soon after Japan's defeat in World War II, the Science and Technology Department of the Japanese Communist Party published a document entitled, 'The defects in Japanese science and technology and the duties of Communists.'[9] The paper attributes the underdevelopment of Japanese science and technology to the country's social and economic structure and interprets modern Japanese history in the following way:

> Having started as a *capitalist country* after the Meiji Restoration, Japan exploited workers and peasants endlessly on the basis of agricultural feudalism and serfdom and with the power of the absolutist emperor system. The nation waged a series of colonial plundering wars, expanded the capitalist class rapidly and entered into the *imperialist* stage. (*emphasis added*)

The document identifies three key dominating elements in Japanese society: (1) the absolutist emperor system; (2) the system of parasitic and feudal land ownership; and (3) monopoly capitalism. The paper then nominates Japan's semi-feudal underdevelopment as the greatest impediment to the growth of modern science and technology and asserts that Japanese capitalism was essentially colluding with feudalist forces and was therefore unable to act as the engine of Japan's modernization. The thesis also itemizes and criticizes the defects of Japanese science and technology under this system. Its argument is that anti-Imperial and anti-feudal struggles will inevitably lead to a split between the interests of capital and the interests of the old order, and that this will facilitate the modernization of science and technology. This unquestionably represents the Kōza-ha position, but it assumes that Japan was a capitalist country after the Meiji Restoration.

While recognizing that prewar and wartime Japan bore all the characteristics of a modern monopoly capitalist, imperialist society, the Kōza-ha based JCP had to reconcile this empirical reality with the Comintern's official definition of Japan as an absolutist state. This apparent contradiction caused heated debate within the Party in 1946–47, with its most famous opponents being Yoshio Shiga and Shigeo Kamiyama,[10] both of whom vigorously argued their position in the pages of the Party's official newspapers and magazines as well as in more general non-Party publications. Kamiyama maintained that Japan had reached what Lenin called the 'military and feudal imperialism' stage by the 1890s and that the absolutist emperor system had substituted or complemented the functions of the imperialist ruling class up until the defeat in World War II. Shiga (1947a and 1947b) however, contended that the concept of military and feudal imperialism that Lenin had used in his analysis of Tsarist Russia did not apply because Japan was already a modern imperial state. He pointed out that the characteristics of Japan's political power and its socio-economic foundation were mutually independent and argued that the absolutist emperor system controlled all of the political power.

The debate was politically settled in Shiga's favor at the Sixth Conference of the JCP in December 1947. Here Kamiyama and his fellow intellectuals were criticized for their 'modernism,' even though the Party's official document, 'The defects in Japanese science and technology and the duties of Communists,' closely resembled Kamiyama's argument. Some of these intellectuals supported the Taketani theory (which we will discuss later). It is also interesting to note that the so-called modernists attacked Japan's feudal and pre-modern characteristics far more openly and straightforwardly than their orthodox critics.

Beyond these philosophical discussions, one can observe how Japan underwent some fundamental changes during the war. These included the expansion of military industry, the growth of higher education in science and engineering and the reinforcement of government economic control. The impact of postwar changes also requires analysis.

Takuichi Ikumi, a Marxist based in Nihon Keizai Kikō Kenkyūsho (the Japan Research Institute of Economic Organization) at the time, addressed these issues and, as early as 1947, established a position that can be called a neo-Rōnō-ha school.[11] In 1930, as a Party member, Ikumi edited a series entitled, 'A History of the Development of Japanese Capitalism.' But, he had come to think that Japanese absolutism had gradually become monopoly capitalism during the twentieth century.

Harumaru Inoue and Seijirō Usami, also began, in 1948, to contribute painstaking articles to the journal *Chōryū* (The Trend). Inoue and Usami both published articles in the January 1949 special edition of *Chōryū* entitled, 'Japanese fascism and the resistance to it.' The issue was widely read and went into a second printing. Both articles describe and analyze how the large sums of money invested by the Japanese state into military industry led to changes in the industrial structures and the capital accumulation of large monopoly corporations. Despite being poorly argued, the articles convincingly assert that Japan maintained a system of state monopoly capitalism wherein the state was the major economic agent, despite the fact that Japanese industry had been largely destroyed during the war.[12]

The orthodox school criticized this analysis as 'economism' and for confusing the emperor system's political power with the socio-economic structure of the state, a version of the 'production force theory.' Yoshio Shiga once again led the orthodox campaign, publishing an introductory text entitled, *Kokka-ron* (Treatise on the State; 1949) and giving two lectures on this issue at the independent series organized by student activists at the University of Tokyo in autumn 1949. While the issue of state monopoly capitalism resurfaced later in various forms, the orthodox school invariably countered it with the argument that every economic problem was reducible to an issue of political power. When Usami and Inoue later published *Kiki ni okeru kokka dokusen shihonshugi no kōzō* (Structure of state monopoly capitalism under crisis; 1952), they accepted the criticism of the orthodox school.

The Comintern's 1932 program remained influential well into the 1950s. The Party viewed the US Occupation Forces as having assumed the role once played by the Emperor. Working from this premise, the 1951 Party platform espoused the 'national independence of Japan' as a strategic response to what it saw as the United States' colonial control over the country, a control aided and abetted by Japan's semi-feudal forces. As late as the 1970s, the JCP still argued that Japan was a non-sovereign country subordinate to the United States.

Marxist Theory of Science Immediately after the War

Oversimplified theories gained popularity among activists during the late 1940s and 1950s when social movements were at their height. Typically, the 'mass line' (which we will examine later) vulgarized, belittled and was even directly hostile to theoretical activities. Despite this, directly after the war, some detailed theoretical ideas did develop, some of which we will now discuss.

Kōza-ha and the Hani School of History

Gorō Hani was the Kōza-ha school's chief ideologue. His historical emphasis on Japan's social backwardness and feudality influenced a number of thinkers including Osamu Kuno and Mitsuo Taketani. Hani's work (1946), together with that of Kinnosuke Ogura (which we will discuss later), provided a theoretical framework for the postwar science movements.[13]

According to the official Soviet line, only the proletariat was able to discover scientific truth. Hence, scientists who politically deviated from Communist Party orthodoxy and did not see themselves as working in proxy for the proletariat were assumed to lack scientific judgment and were expelled from the Party. Despite this, modern science and technology was unquestionably understood to be a product of modern capitalism. In attempting to address this apparent contradiction, Hani argued – with illustrations from Florence, a Renaissance city state – that the Western bourgeoisie did in fact embody scientific truth because they were fighting for civic liberty and against feudalism. This account explained the discrepancy between the rapid development of science and technology in the United States, with its War of Independence in the eighteenth century right up to its fight against fascism in World War II, and the stunted scientific development of Japan, with its remnant feudalism.

Hani and others affirmed Lenin's theory that contemporary capitalism had entered its final, imperialist phase, one in which its degeneration precluded it from discovering further scientific and technological truths. They argued that now only socialism, in its struggle against capitalism, could make such discoveries. This view was conducive to the privileging of science and technology in the Soviet Union, the criticism of pre-modern Japanese science and the proposition that 'scientific progress would not be possible under monopoly capitalism' – a theory closely resembling the above mentioned 'science and technology' thesis of the Communist Party.

Japan's rapid economic development from the late 1950s onward effectively silenced this line of argument. In the 1960s and 1970s, however, Hani regained some of his former influence, leading a movement critical of 'science supported by the ruling power.'

The legacy of Yuibutsuron Kenkyūkai (the Materialism Study Group)

The prewar Materialism Study Group, which was suppressed and dissolved by the authorities during the war, passed on an important legacy to the postwar generation of Marxists. The most notable members of the group were Jun Tosaka and Tadashi Katō. Tosaka, who was the last leader of the Materialism Study Group, gained a wide readership when part of *Yuibutsuron zenshō* (The corpus of materialism) was republished immediately after the war. Consequently, two other books published prior to the war, *Kagakuron* (Science studies) and *Kagaku hōhōron* (Methodology of science), belatedly acquired a large number of readers. Numerous book reviews and essays written between 1936 and 1937 were collected in the book, *Dokusho-hō* (Methods of reading), which was posthumously published in 1946 after Tosaka's death in prison in June 1945.[14]

Katō, famous as the first translator of Engels' *Dialectics of Nature*, survived the war despite suffering a long illness. Defining himself as an 'objectivist,' he maintained a belief in scientific rationalism which he regarded as a reflection of the true structure of nature. Katō was also known for his prewar translation of Friedrich Dannemann's *Naturwissenchaften in ihrer Entwieldung und ihrem Zusammenhang*. The tradition of the Materialism Study Group was handed over to Minka, along with its monthly journal, *Riron* (Theory), though this did not survive long in its new incarnation.

Published Works on the History of Science and Science Studies

Many Marxists played active roles in Nihon Kagakushi Gakkai (History of Science Society of Japan), established in 1941. Their wartime scholarly achievements were also published immediately after the war.

Kinnosuke Ogura, for example, contributed to *Nihon shihonshugi hatatsu-shi kōza* (Series on the history of the development of Japanese capitalism) before the war and published the first volume of *Sūgaku-shi kenkyū* (Study on the history of mathematics) in 1936. Here he argued that historical developments in mathematics correlated with changes to social relations and fluctuations in democracy. He published volume 2 and republished volume 1 soon after the war.[15] Originally delivered as a lecture in 1943, a chapter in volume 2 which examined L'Ecole Polytechnique Superieure during the French Revolution stirred much debate.[16]

Mitsuo Hara published *Shizen benshōhō no kenkyū* (Studies in the dialectics of nature) at the beginning of 1947. Similar to Ogura's work, this book built upon the achievements of prewar and wartime Marxist studies in the history of chemistry and includes a review of Hessen and Borkenau's critiques of modern science, both of which caused much debate amongst Marxists in prewar Japan.[17] These debates were originally ignited with the publication of two Japanese translations of Hessen's *Social and Economic Roots of Newton's Principia* in 1931, and the partial translation of Borkenau's *Der Ubergang von feudalen zum būgerlichen Weltbild* in 1934 (the complete translation was published after World War II). Hessen argued that the nature of scientific development was determined by the prevailing socio-economic structure, while Borkenau asserted that class consciousness conditions the way one thinks.

Yōitsu Kondō published *Sūgaku shisō-shi* (A history of ideas in mathematics) and *Kikagaku shisō-shi* (A history of ideas in geometry) soon after the war. The latter of these was a history of nineteenth-century non-Euclidian geometry, and attracted attention because, while heavily indebted to Marxist thought, it avoided using Marxist terminology.[18] Other notable prewar works re-published soon after the war were Kunio Oka's histories of scientific ideas and Takeo Konno's studies of mathematics.

Each volume of *Kagaku koten sōsho* (A series on science classics), which was planned and began publication during the war, re-published the most significant works of the history of science in their original form, along with translations and commentaries on the texts. Each commentary is, itself, an excellent article in the history of science. Kiyoshi Amano's *Netsu-fukusharon to ryūshiron no kigen* (The thermal radiation studies that led to the genesis of the quantum theory) and Yoshito Shinotō's *Shokubutsu zasshu* (Botanical hybrid) were published during the war years, while Mitsuo Hara's *Sanso no hakken* (The discovery of oxygen) and Minoru Tanaka's *Busshitsu to denki* (Matter and electricity) appeared after the war.[19]

It should be noted that Hara, Tanaka and many other Marxists had carefully studied science classics during the war.

Classic Marxist Works on Science and Other Overseas Literature

Marxism has never produced a theoretical work concerned with the natural sciences that in influence and importance can compare with the enormous impact of *Das Kapital* upon the study of economics and the social sciences.

This is in large part due to the fact that so few classical Marxist texts actually address issues concerning the natural sciences, with Engels' *Dialectics of Nature* and the latter part of Lenin's *Materialism and Empirico-Criticism* being the only obvious exceptions. The few Marxist books about science that did exist were rarely seen in postwar Japan because of the restrictions placed upon the importation, translation and publication of foreign books (cf. ch. 12).

In the Soviet Union, the Stalinist regime exercised increasing control over theoretical and academic studies. The journal, *Under the Flag of Marxism*, virtually stopped publication at this time. Many scholars were purged, and those that survived toed the Party line and produced only textbook-type publications. Among the purged were: D. Ryazanov, the first to have published a Russian translation of *Dialectics of Nature* and *German Ideology* and to argue their importance; Nikolai Bukharin, who had played an active role in the second International Conference of the History of Science held in London in 1931; and B. Hessen, who at the same conference had delivered an influential paper entitled, 'Social and Economic Roots of Newton's Principia.' Soviet orthodoxy went so far as to question the legitimacy of the *Economic and Philosophical Manuscripts*, known then as the *Paris Manuscripts*, and criticized the Frankfurt School as revisionist. As a result of the repressive measures in the Soviet Union and Japan's own censorship laws, Japanese Marxist studies remained mired in the discourse of the early 1930s.

The Significance of the Debate between Hara and Ogawa in 1947

In this context the so-called Hara-Ogawa debate, which was conducted through the Japan Communist Party's official journal *Zen'ei* (Vanguard) took on special significance. Essentially, the two authors disagreed about the nature and theoretical importance of the relationship between the dialectics of nature and historical materialism.

In an article in issue no. 21 of *Zen'ei*, Mitsuo Hara criticizes Kunio Oka's contention that the dialectics of nature are predicated upon historical materialism. Quoting Stalin in support of his position, Hara maintains that the dialectics of nature govern the laws of nature. He goes on to argue that dialectical laws of nature also apply to the social world, so that historical materialism is in fact predicated upon the dialectics of nature and not the reverse.[20]

Responding in *Zen'ei* no. 24, Sadahisa Ogawa (a pseudonym for Shintarō Tanabe) criticized Hara's view as too simplistic, its connection of theory to its object as too mechanistic. Ogawa suggests that theory does not merely reflect its object but is filtered through the theorist's epistemological assumptions, and that greater emphasis should be given to the interrelations between theory and its object. This article was one in a series that demonstrated the results of Ogawa's study of dialectics as epistemology.[21] Ogawa's argument is deeply indebted to Lenin's *On the Problems of Dialectics*, written after Lenin had read Hegel in Geneva and realized the true significance of dialectics – a realization that most Japanese Marxists came to immediately after the war.

To more fully understand the Hara-Ogawa debate of 1947 we must give further consideration to four key points. First, both writers had produced excellent translations of Engels' *Dialectics of Nature*, and as such were fully conversant with that text. Both had also specialized in chemistry and possessed a sophisticated understanding of the field of natural science. Hence, they were equally qualified in both philosophical and scientific terms to lead such a debate.

Secondly, while citing Party documents written by Stalin and others, both Hara and Ogawa presented their respective arguments articulately and clearly outlined their theoretical differences. Although Hara's argument appeared to rely on Stalin's *Dialectical Materialism*, it was in fact constructed in conjunction with Katō (who called himself an 'objectivist') long before the publication of Stalin's text. Hara was also the translator of John Read's *Ten Days that Shook the World*, which provides an accurate account of Trotsky's role in the Russian Revolution.

Thirdly, as early as the late 1940s, both Hara and Ogawa exhibited reflexive tendencies, a methodology which was to become increasingly popular in and after the 1960s. The points of disagreement between the two men persisted, however, and can still be traced today in the ongoing debate between positivism and social constructionism.

Fourthly, it is noteworthy that this debate took place in the Communist Party's official journal, with both sides being treated with equanimity. This suggests that the Party was, at that stage, open and independent enough to facilitate discussions that challenged Soviet orthodoxy.

This debate was only one among many discussions about Marxist science occurring within the Japan Communist Party at the time. Although specific studies dedicated to this area of endeavor appear to have increased in quantity and improved in quality since the 1960s, it is a telling fact that Hara and Ogawa's dispute represents the last time that the Party encouraged open debate about the fundamentals of Marxist science.

The Taketani Theory and its Influence

The Rise of the Taketani Theory

It cannot be denied that the so-called Taketani theory – or, to be more accurate, the Taketani-Sakata theory – formed a brand of Marxism that gained widespread support among scientists and engineers in the late

1940s and the 1950s and as such proved quite influential in postwar Marxist science studies in Japan.

Mitsuo Taketani and Shōichi Sakata contributed to the journal, *Shizen kagaku* (Natural Sciences), which was published by Minka's natural science group and lasted for only a short period of time. The journal was quite an attractive publication, and their contributions in 1946 and 1947 commanded a large readership. These were later released as books, with Taketani publishing, *Benshōhō no shomondai* (Problems of dialectics) in 1946,[22] and Sakata publishing *Butsurigaku to hōhō* (Physics and methodology) in 1947.[23] Their philosophical status was further enhanced when they collaborated with Hideki Yukawa – who was awarded Japan's first Nobel prize in 1949 – in the development of meson theory when Yukawa was working at Osaka University and then later at Kyoto University.

In fact, the genesis of Taketani's theory can be traced back to an earlier, prewar association between he, Sakata and Yukawa. In 1936, Taketani published his first Marxist work in the journal, *Sekai bunka* (World culture) where he used a pseudonym Kazuo Tani. *Sekai bunka* (February 1935 – October 1937) and its student counterpart, *Gakusei hyoron* (Student review) (May 1936 – July 1937) were published as liberal journals in Kyoto during the war. His 'three stage theory' was published in 1942 in the journal *Kagaku* (Science).[24] After the war he provided a detailed account of a Special Security Police report written when he was twice arrested – in the 1930s and in the early 1940s – for his leftwing activities.

Speaking as Japan's foremost specialists in elementary particle physics, Taketani and Sakata's claims that their great success was achieved by their adherence to Marxist methodology carried considerable weight amongst natural scientists in general and physicists in particular. Their emphasis on the 'introduction of substance' and the 'substantial stages' was compatible with the research style of many natural scientists who tended not to examine complex natural phenomena in a direct manner but to study them through theoretical models that represent aspects of their characteristics. The central themes of Taketani's framework comprise the 'three stage theory' and the 'technology theory.' We will concentrate here on the former and deal with the latter of these in Volume 2.

'Three Stage Theory'

Taketani's theory is premised on what he regards as the validity of methodology and builds upon the already mentioned historiography of Gorō Hani. Taketani claims that the philosophies and methods necessary for the continuation of scientific progress reside in the social class emerging from its conflict with the old order. Knowledge of these philosophies and methodologies cannot be acquired through reading books, but must be absorbed through the activities of correctly deliberated and implemented scientific research. Taketani contends that the source of the progress of science is not to be found in socio-economic factors, à la Hessen, nor in epistemological considerations, à la Borkenau – both of which he regards as rather supplementary variables – but essentially in the logical structure of science itself. Taketani identified three stages in the logic of scientific development – phenomenal, substantial and essential – and attempted to match them with Hegel's dialectical developmental stages: being-in-itself, being-for-itself and being-in-and-for-itself. He considered the birth of Newtonian mechanics as exemplifying his scheme, where Tycho-Brahe's precise observation of data was *phenomenal*, Kepler's three laws of planetary motion was *substantial* and the Newtonian system was *essential*. Sakata extolled Taketani's theory as a 'discovery comparable to Marx's social development law in *Das Kapital*.'

Taketani emphasized the 'substantial stage,' which resides between the two conventional categories of philosophy – phenomenon and essence. Many regarded this approach as an original contribution to the theory of dialectics. But Taketani ignores the conventional dichotomy between accident and substance, the latter sometimes denoting the divine subject. In traditional Western philosophies of 'being,' 'substance' (being-in-itself) and 'accident' (being-in-another) constitute two mutually opposing categories. Some observers attributed Yukawa's success to his adherence to this model, with the introduction of the meson as 'substance' in his research on nuclear force, although Yukawa himself did not think so. Similarly, Sakata's 'two meson theory' – in which he posited two separate 'substances,' the pi meson as a 'substance' of nuclear force and the mu meson as the 'substance' in the cosmic ray – was understood by many in terms of Taketani's theoretical framework. But when Sakata proposed this idea in 1942, Taketani was surprised by this simple resolution of what was, at the time, a very difficult problem.

Sakata's own theoretical contribution to particle physics came in the form of his identification of 'stratification' in nature. In studying microscopic quantum theory, many physical scientists sensed that a kind of stratification existed in nature, suggesting the possibility of 'transformation from quantity to quality.' Sakata wanted to produce some generalizations in this area and to extend Engels' taxonomy of sciences into the social sciences.[25]

The Taketani theory was readily accepted by most Marxist natural scientists in Japan. Underpinning this theory was the influence of the above-mentioned Hani school of history and Taketani's naïve confidence in science and rationalism. He believed that the development of nuclear weaponry could be a great accomplishment in a democratic nation. He also espoused the theory that politics and economics would be best left to physical scientists in constant contact with scientific truth – a view widely held by many scientists immediately after the war.

'Institute Charter' and The Modernization of Physical Science Research Organizations

Sakata made various attempts to modernize the research organizations of physical science. He modeled his thinking after, *Social Functions of Science* (1939), written by the British Marxist physical scientist J. D. Bernal. This work was later seen to have set an influential precedent for the study of the sociology of science, but at the time of its publication Marxism was based on a mixture of economic determinism and scientific rationalism. Bernal claimed that science could only be fully developed in a socialist society and that it was retarded by the demands of capitalism. In other words, he advocated the transition from capitalism to socialism for the sake of scientific progress.[26] While overly simplistic, Bernal persuasively argues that the revolution in physical sciences emanating from the birth of quantum mechanics was the product of Weimar democracy and that Hitler's seizure of government led to a decline in German physics. Beyond this, Bernal's major contribution to the sociology of science is his concrete descriptions of British case studies, through which he demonstrates the practical necessity of developing science according to clearly designed national plans. Here he emphasizes the importance of 'laboratory democracy' in modern scientific research, which was by then based on organizational rather than individual activities.

As soon as the war was over Sakata proceeded to reform the Physics Institute of Nagoya University according to Bernal's findings. One of Sakata's most successful reforms was the implementation of the 'institute system' as a replacement for the old 'chair system,' so it then became possible to manage an institute through institute meetings. These reforms provided a model compatible with the expansion of the organizational scale of basic research and the necessity for scientific collaboration.[27]

While it might be argued that these reforms owed little to Marxist theory, but instead reflected the changing organizational demands of basic scientific research, it should be noted that Bernal, Sakata and Taketani were all self-declared Marxists, and all claimed that the changes they initiated were inspired by their Marxist beliefs.

The Split of the Party and the Korean War

The Cominform Criticism and the Split of the Communist Party which spread across the Marxist Movements in Japan

In January 1950, the Cominform – the Communist Information Bureau, which was based in Europe – published a commentary entitled, 'On the Situation of Japan' in its bulletin *For Lasting Peace and People's*

Democracy. The article sharply criticized the 'peaceful revolution lines' that the Japan Communist Party was taking during the Occupation years and directed much of its criticism towards Sanzō Nosaka, a leader of the JCP. While not the direct concern of science movements or theoreticians of science, the ramifications of this ideological rebuff were felt throughout Japan's Marxist social movements and theoretical activities, including science and technology studies.[28] As a result, Marxist theories of science and Marxist science movements lost momentum in the years immediately after the Cominform's criticism. Many journals and magazines ceased publication during this period. Three reasons can be given for this general decline in Marxist activity in postwar Japan.

The Cominform's criticism of the JCP caused an organizational split in the Party between the Politbureau and the Internationalist group, both of whom claimed to represent ideological orthodoxy. Each competed with the other for recognition of its faithfulness to either the Soviet or Chinese line, and both sought to demonstrate how close they were to authoritative, 'orthodox Marxism.' Even the anti-authoritarian leadership of the internationalist faction of the Zengakuren never failed to defend the Soviet Union and follow the Comintern's ideology of 'internationalism.' This factional conflict took a political, organizational form, which precluded any chance of a fair debate over scholarly, theoretical issues.

Economic Recovery under Capitalism and the Revival of the Science and Technology 'Community' in the 1950s

The Korean War and the special procurement boom it produced represented a turning point in Japan's economic recovery. While the deflationary measures imposed by the Dodge Plan inflicted pain and hardship on many Japanese, the special procurement boom caused by the US use of Japan as a supply base for the Korean War revitalized the nation's economy and paved the way for a phase of sustained economic growth that began in the mid-1950s. It was in this economic climate that the research and production facilities that had been under-funded since the end of WWII were gradually restored. Scientists and engineers began to return to the 'communities' to which they had once belonged. One consequence of this was that it devalued the importance of interdisciplinary communities such as Minka. Instead, vertically structured, scholarly research communities – unique to Japan and efficient in their own way – were revived as many scientists and engineers returned to their old work routines. Thus, Japan's economic recovery from 1949 onwards was potent enough to facilitate the 'restoration of science.'

Until the 1950s, Marxist S&T theories assumed that the pre-modern characteristics of Japanese society were preserved in the postwar society, which thereby precluded the possibility of real scientific and technological development. This theory, however, did not tally with the socio-economic reality. Hence, Marxist organizations were thrown into confusion as they witnessed the steady recovery of the Japanese economy, the smooth revitalization of scholarly research and the increasing importation of overseas technology. Until the mid-1950s most Japanese Marxist groups were in active denial about the glaring disparity between their theory and the economy's practice. Eventually, though, by the late 1950s Marxist scholars had begun to re-think how they interpreted Japanese society and constructed theories of science and technology.

National People's Science and the Thought of Mao Zedong

The Chinese Revolution of 1949 greatly impressed the progressive Japanese thinkers who had done much soul-searching about their country's wartime aggression towards China. There was little doubt that the Revolution was a great development, even though it established a rival source of orthodox authority and its overall influence was not entirely positive.

Regardless of what one thinks of his political ends, there can be little argument about the effectiveness of Mao's means – he was, undoubtedly, a master political and logistical strategist. His analysis of the conditions of Chinese society in the 1930s and the 1940s remains persuasive even today. Nonetheless, the theories he expounded in what were to rapidly become classic Marxist texts, *On Practice* and *On Contradictions* were based more on Enlightenment rationalism than on the early Lenin and Stalin. Like Stalin, though, as the supreme leader of China, Mao's philosophical and political directives exerted great influence on the thinking of Marxist intellectuals around the world.[29]

In terms of theories of science and technology, Mao's influence was felt in two main areas: (1) the emphasis placed on 'mass lines,' and (2) the rise of the 'science for national people' line.

The 'Mass Lines' in Science

During the Cultural Revolution, Mao expelled Lu Shao-chi from the Party for being a revisionist power holder. He once published a small pamphlet entitled, 'Discussing Mass Lines' in the Yanan period. The notion of 'mass lines' remained a core concept in Maoist thought. Lu's original text discusses the implementation of tactics and the operation of organizations, as well as moral issues concerning routine, daily activities. However, a different doctrine on the 'Mass Line', the one that came to influence much scholarly research and be adopted by Japan's postwar science movements, was that the theory of Marxism should be directly accessible to the masses and should serve their interests.

The concept of 'mass lines' remains ambiguous because the notion of the 'masses' is itself so ambiguous. The 'masses' has been employed to describe a variety of entities, ranging from 'people who happen to be around you' to 'people who are led by the Communist Party' or even 'by an individual leader who represents it.' Whatever the definition of 'masses,' one consequence of the 'mass lines' approach was that the elaborate Marxist theories that had inherited the complex logical structures of Hegel and Ricardo were regarded as incomprehensible to the masses and therefore afforded less and less attention. These theories were often replaced by a kind of pragmatism that pursued immediate outcomes for social movements.

'Science for the National People'

The Chinese Revolution liberated China from the control of imperialist powers and demolished the semi-feudal system that had supported them for so long. In Japan, this gave rise to the push for 'national independence,' as spelled out in the 1951 Japan Communist Party platform.[30] According to this perspective, in 1950 Japan was under the 'total control' of the United States, despite the Occupation having been converted to a security treaty between the US and Japan. Marxist science movements were expected to give priority to disrupting this arrangement and developing a nationalist brand of science. Minka thus started publication of a short-lived monthly journal, *Kokumin no kagaku* (Science for the national people), which published a number of articles stressing the importance of the restoration of Japanese traditional culture and the promotion of a brand of science closely connected to the nation's people.[31]

As with 'the masses,' the notion of '*kokumin* (national people)' was extremely vague and open to numerous interpretations. The concept and political reality of Japan's modern nation state and a concomitant national consciousness had rapidly developed during the Meiji period. As such, the Japanese concept of *kokumin* grew out of the reactionary elements of Hegelian thought and aided the establishment of the nation. Many Marxists, though, began to equate *kokumin* with other Asian manifestations of post-colonial nationalism.

While influential for a time, the 'mass lines' and 'science for national people' trends failed to evolve into fully-fledged science movements like the residents' and anti-pollution movements of the 1970s. This was not unusual, for many theoretical orientations of the 1940s were either abandoned or ignored in the 1950s.[32]

Antecedents to the 1960s

It would be unfair, though, to think of the 1950s as a totally futile period in terms of Japanese Marxism. The following two developments should be noted.

1. With the ending of the Occupation and the stabilization of the economy, the 1950s saw the systematic translation and publication of many

important Marxist texts. Amongst these were selected writings from Marx and Engels, so that works such as *Dialectics of Nature* and the *Economic and Philosophical Manuscripts* were for the first time widely available in Japanese. This enabled members of the public to study Marxism from its primary sources, rather than through the filter of Party-approved 'textbooks.'

2 The 1956 criticism of Stalin dramatically lowered the authority of Soviet-led orthodoxy and thereby relaxed constraints on Marxist theoretical research. Thus, independent researchers outside the Party organization began to freely study theories and doctrines that had been considered heretical in the recent past. An increasing number of 'Marxists' also began to take it for granted that the achievements of Marx and Lenin needed to be investigated objectively, without preconceptions.

Both of these developments aided the gradual reconstruction of Marxist theories of science; though it must be said that it was also in the 1950s that the science movements became less Marxist and more diverse in their philosophical and theoretical perspectives.

The orthodox Party line remained influential in some quarters because of its plain, simple theoretical structure, but it was constantly challenged by the development of rival forms of Marxist thought. Some of these focused on theories of alienation, others employed a structuralist approach, while still others attempted to rigorously systematize *Das Kapital*. All of these versions of Marxism faced the same challenging question of whether or not they could provide a methodology that adequately addressed the central, contemporary issues of science and technology.

Notes

1 A typical example is 'Benshōhōteki yuibutsu-ron to shiteki yuibutsu-ron (Dialectical materialism and historical materialism)' which Stalin himself purportedly wrote as a chapter in *Soren Kyōsan-tō-shi* (History of the Communist Party of the Soviet Union; 1938). The introduction clearly states that dialectical materialism developed from the study of nature and historical materialism is the application of the former to society.

2 This was originally part of an unfinished manuscript from Marx's youth. It was put into *Rūdovihhi Foierubahha to Doitsu koten tetsugaku no shūketsu* (Ludwig Feuerbach and the End of German Classical Philosophy), an enlightening work by Engels that became quite popular.

3 A lineage from Saint-Simonism and Taylorism to the Technocrat movement was eventually assimilated into 'scientific' socialism despite the denial from the party concerned. Refer to W. H. G. Armytage, *Tekunokuraato no bokkō* (The Rise of the Technocrats), Akio Akagi (tr.), Chikuma Shobō, 1972.

Chapter 12, 'Roshia no gijutsu interigenchia' (Russian technological intellectuals; pp. 227–49) describes the background of Lenin's activities. Also refer to Chapter 14, 'Amerika o ou Sobieto' (Soviet who chases America; pp. 270–91) for the relationship between the movement of technocrats in the United States and the planned economy of the Soviet Union.

4 Hiroshige, 1973 [16]; chapters 5–8 give appropriate descriptions of the science mobilization plan during the period.

5 For example, Nihon Kagakushi Gakkai (History of Science Society of Japan) was inaugurated in 1941.

6 There were a large number of former communists around these power groups. Among them, Mantetsu Chōsa-bu (South Manchuria Railway Co. Inc.) is one of the more famous.

7 These are legal publications. Varga's *Quarterly Reports* were re-edited and published as a 30 volume set in Japanese; Palm R. Dutt, *Social Revolution: A Study of the Economics and Politics of the Extreme Stages of Capitalism in Decay,* 1935, New York. The Japanese version was distributed by the Ministry of Foreign Affairs and then published.

8 Testimonies given by Mr. Takuichi Ikumi in Takuichi Ikumi and Kazuteru Okiura, '1930-nen dai ni okeru Marukusu-shugi undō (Marxist movements in the 1930s),' in *Gendai no riron* (Theory of the present day),' May 1977, pp. 49ff. Within the editorial activities for 'Nihon shihon-shugi hattatsu shi kōza (Seminar on the history of development of Japanese capitalism),' which reportedly aimed at theoretically strengthening the 1932 program of the Japanese Revolution by the Comintern, Eitarō Noro and Ikumi who were party members at the time, were reported to be unexpectedly flexible.

9 *Zen'ei* [8] nos. 10 and 11, 1946. These publications are reproduced in appendices to Mitsuo Taketani, *Kagaku to gijutsu – Sono riron to hōhō* (Science and technology: Its theory and method), Rironsha, 1950.

10 Shigeo Kamiyama, *Tennō-sei ni kansuru rironteki shomondai* (Theoretical problems in the Emperor system), Minshu Hyōronsha, 1947; Yoshio Shiga, 'Tō-shi no mikata (How to think the history of the Communist Party),' in *Zen'ei* [8] no. 19, 1947; Yoshio Shiga, 'Sekai-shi no ichi kōsatsu (A study in world history),' in *Zen'ei* [8] nos. 20 and 21, 1947.

11 Takuichi Ikumi, *Nihon shihon-shugi-ron* (Theory of Japanese capitalism), Kureha Shobō, 1948.

12 Harumaru Inoue and Seijirō Usami, 'Nihon ni okeru kokka dokusen shihon-shugi no tenkai (Development of national monopoly capitalism in Japan),' in *Chōryū* [9] October 1948.

13 Gorō Hani, *Jokonda no bishō* (Smile of Gioconda), Sanichi Shobō, papers published in 1946; 'Kagaku to shihon-shugi (Science and capitalism).'

14 Jun Tosaka, *Dokusho-hō* (Method of reading), Mikasa Shobō, October, 1946. A collection of book reviews and essays written between 1935 and 1937.

15 Ogura, 1948 [11]

16 Tetsu Hiroshige, 'Ogura Kinnosuke to kindai Nihon kagaku-shi (Kinnosuke Ogura and the history of modern Japanese science),' in *Kagaku kenkyū-shi* (Studies in the history of science), no. 65, 1962; Tetsu Hiroshige, *Kagaku to rekishi* (Science and history), Misuzu Shobō, 1965.

17 Hara, 1947 [12]

18 Yōitsu Kondō, *Kikagaku shisō-shi* (History of ideas in geometry), Itō Shoten, 1946.

19 Minoru Tanaka, *Busshitsu to denki* (Matter and electricity), Dainihon Shuppan Insatsu, 1946; Mitsuo Hara, *Sanso no hakken* (Discovery of oxygen), Dainihon Shuppan Insatsu, 1946.

20 Mitsuo Hara, 'Shizen benshō-hō to shiteki yuibutsu-ron (Dialectics of nature and historical materialism),' in *Zen'ei* [8] no. 21, 1947.

21 Sadahisa Ogawa, 'Shizen benshō-hō to shiteki yuibutsu-ron no iwayuru jōge kankei (The so-called hierarchy between dialectics of nature and historical materialism),' in *Zen'ei* [8] no. 24, 1948; Takashi Sugawara in Minshushugi Kagakusha Kyōkai, 1948 [4] p. 329. Sugawara introduces this debate both concisely and precisely with inclinations towards Ogawa's view. Hara and Tanabe had a number of debates on the classification of science and other matters beginning immediately after the war.

22 Taketani, 1946 [13]

23 Sakata, 1947 [14]

24 *Sekai bunka* (World culture) and its student version *Gakusei hyōron* (Student review) were published in Kyoto and were two of a handful of liberal speech movements. Behind them was Kyōsan-shugi-sha dan (The Communist League) in the Kansai (western Japan) region. This group understood the importance of 'Jinmin sensen' (People's front in France) and pursued it consciously.

25 The three-stage theory was explained using the studies by Tycho-Brahe, Kepler and Newton, and exhibited considerable persuasive power. There is a reference to the 'Taketani philosophy, which Japan proudly presents to the world' in the foreword of *Kagaku no tetsugaku* (Philosophy of science), Shōichi Sakata and Yōitsu Kondō (ed.) (Iwanami Kōza, *Tetsugaku* [Philosophy] VI), Iwanami Shoten, 1968. However, it is noteworthy that in Sakata's 'Gendai kagaku no gendaisei (Modernity of contemporary science)' which starts from p. 375 in the afore-mentioned book, his harsh criticism of modern science is emphasized along with the necessity of dialectical unity with the social sciences and an appraisal of early Marx. In this regard, however, the basic principle is the stratified structure of nature and science.

26 Bernal, 1939 [15]

27 'Nagoya Daigaku butsurigaku kyōshitsu kenshō (Nagoya University Physics Institute Charter)' was set up in June 1946. The contents are

explained in detail in Sakata, 1947 [14]. The charter contained notable ideas such as 're-organization of the chair system [kōza-sei] into a laboratory system,' 'the Institute Conference' which means high regard for 'democracy by involvement of all the researchers,' 'relative independence of research and education' and 'student participation in curriculum and its management.'

28 For example, there was a huge campaign staged by the Communist Party against the Kamiyama faction between 1953 and 1954 after Shigeo Kamiyama was expelled from the party. Those social scientists and philosophers who were attacked by name were primarily supporters of the Taketani theory. However, Taketani and natural scientists were not among the scholars attacked.

29 There was a brief honeymoon period between the Party renowned for 'free debate' and the intellectuals, but it did not last, and it is a well-known fact that the situation eventually turned to 'anti-right-wing struggle' and 'Cultural Revolution.' However, Sakata often praised 'free debate.'

30 This is a policy line of the Communist Party under the so-called '1951 Platform.' The book *Gendai Nihon shihon-shugi kōza* (A series on contemporary Japanese capitalism), Iwanami Shoten, 1953–54, edited following this policy line, was not a success.

31 A typical work is, Tadashi Ishimoda, *Rekishi to minzoku no hakken* (Discovery of history and a nation), Iwanami Shoten, 1952.

32 Tetsu Hiroshige, *Sengo Nihon no kagaku undō* (The science movements in postwar Japan), Chūō Kōronsha, 1960; presents a critical summary of the science movement at that time.

Bibliographic Guide

Classics

[1] Engels, *Shizen benshō-hō* (Dialectics of nature).
This was published in one volume by combining Engels' hand-written manuscripts on natural science (which are presumed to have been written between 1873 and 1886) and other related small works and letters. The original hand-written manuscripts remain in four bundles and in his letter to Marx dated 1873, Engels hinted at writing a book. This book has therefore been considered to be the original form of a 'book.' There are different editions as follows:

1 *Marukusu Engerusu Aruhiifu* (Marx-Engels Archive), vol. 2, 1925. All the hand-written manuscripts have been arranged in chronological order. The German text has a Russian translation printed on the opposite page. Quotations from other sources are cited in their original languages. Ryazanov was the chief editor. After some alterations such

as decoding the hand-written manuscripts, they were published as a single volume in German in 1926 and in Russian in 1929.

2 In 1935, this was published as a part of Old MEGA (MEGA is an abbreviation of Marx-Engels Gesammeltausgabe. It is referred to as 'Old' because the publication of 'New MEGA' was started in East Germany after the war. However, the publication has been in a state of confusion since 1989 and is yet to be completed). Old MEGA is a coherent book, but the original order of manuscripts has been completely altered (Adratskij edition).

3 Dietz edition published in 1952 (East Germany). This is said to have vastly improved the Old MEGA edition.

4 New MEGA edition, although the series is incomplete, became useful for critical-textual analysis.

5 Japanese translations include:

a Iwanami Bunko in 1929 and 1932 (translated by Tadashi Katō and Yūjirō Kako, based on the Ryazonov edition, 1926).

b Ōdosha, 1948. The translation was supervised by Hiromi Arisawa (based on the Old MEGA edition by Adratskij). However, there are many errors in the translation.

c Translated by Mitsuo Hara and published by San'ichi Shobō in 1949.

d *Marukusu Engerusu senshū* (Selected works of Marx and Engels), (translated by Takashi Sugawara and Tsunenobu Terasawa, based on the Adratskij edition), Ōtsuki Shoten, 1950.

e Kokumin Bunko edition in 1954 (translated by Takashi Sugawara and Tsunenobu Terasawa, based on the Dietz edition).

f Iwanami Bunko Shinpan (new edition) in 1956 and 1957, (translated by Shintarō Tanabe, from the Dietz edition).

[2] Lenin, *Yuibutsu-ron to keikenhihan-ron* (Materialism and Empirio-Criticism).

In 1909, Lenin wrote this while in political exile in London, criticizing Ernst Mach's philosophy concerning the internal power struggle within the Russian Social Democratic Party. In order to criticize Mach, topics in physics, such as radioactivity and the atomic theory became widely discussed at the end of the 19th century. As well as their philosophical interpretations, these are extensively covered in the latter sections. This book came to be recognized around the time of the 25th anniversary of its publication in 1934 and became the most acclaimed theoretical work according to Stalin. It was written before Lenin started reading Hegel's works in earnest in Geneva and therefore, it is not held in high regards by Western Marxists. Japanese translations are as follows:

1 Iwanami Bunko edition (translated by Fumio Sano).

2 Kaizō Bunko edition (translated by Kanson Arahata), from a German translation.
3 *Reein zenshū-ban* (The complete works of Lenin), (Ōtsuki Shoten; translator not specified).

Catalogues

[3] Minshushugi Kagakusha Kyōkai (Association of Democratic Scientists) (ed.), *Kagaku nenkan* (Yearbook of science), 1947 edition, Nihon Kagakusha, 1947.

[4] Minshushugi Kagakusha Kyōkai (Association of Democratic Scientists) (ed.), *Kagaku nenkan* (Yearbook of science), collection 2, Nihon Kagakusha, 1948.

[5] Minshushugi Kagakusha Kyōkai (Association of Democratic Scientists) (ed.), *Kagaku nenkan* (Yearbook of science), collection 3, 1949, Hokuryūkan, 1949.

These are basic catalogues compiled soon after Japan lost the war. They record information not only on the activities of Minka, but also on scientific activities in general. In particular, [3] contains information such as *Arubaito sōran* (Complete catalogue of books and articles) and a list of academic societies. *Arubaito sōran* discusses scientific publishing activities soon after the war (cf. ch. 27). This information is not contained even in *Zasshi kiji sakuin* (Indices of journal articles of the National Diet). [4] is also a valuable resource.

Magazines and Journals

[6] *Riron* (Theory).

Riron was first published in May, 1947 as an organ for Minshushugi Kagakusha Kyōkai (Minka: The Association of Democratic Scientists) (issued monthly). The contents of this publication were often papers on philosophy and science theories. It had the characteristics of a successor to *Yuibutsuron kenkyū* (Studies on materialism) which had been published before the war (some issues were called *Tetsugaku-bu kaishi* (Bulletin of the Philosophy Group)). It was discontinued in June, 1950 and succeeded by *Kokumin no kagaku* (Science for the national people). The publisher was Nihon Hyōronsha.

[7] *Shizen kagaku* (Natural science).

Published by Kasumigaseki Shobō from June, 1946 as an organ for Minka Shizenkagaku Bukai (Natural science group), its publication discontinued at no. 14. It put forth arguments concerning the theories of science and technology and responsibility for the war, as well as introducing papers which became classics in the interpretation of quantum mechanics through translation.

[8] *Zen'ei* (Vanguard).
A well-known monthly magazine by Nihon Kyōsan-tō (Communist Party of Japan).

[9] *Chōryū* (Trend).
A monthly magazine first published in January 1946. It was discontinued in May 1950, when the publisher Chōryūsha ceased to operate. This is an important magazine for understanding movements in Marxist theory immediately after the war.

Books

[10] Mitsuo Taketani, *Kagaku to gijutsu – Sono riron to hōhō* (Science and technology: Its theory and method), Rironsha, 1950.
This is a reprint version of *Kagaku to gijutsu no kadai* (Problem in science and technology) which was published in 1947 by San'ichi Shobō as a collection of papers. Apart from additional comments on 'Gijutsu-ron ronsō (Controversy about the philosophy of technology)' at the end of the book, the contents remained almost the same. The addition of 'Kagaku gijutsu kōryō (Science and technology thesis) of the Communist Party of Japan has made this reprint more useful.

[11] Kinnosuke Ogura, *Sūgaku-shi kenkyū* (Study on the history of mathematics), vol. 2, Iwanami Shoten, 1948.
Many of the papers contained in volume 1, which was published in 1936, dealt with the history of mathematics in Europe such as 'Kaikyū shakai no sanjutsu (Arithmetic in a class society).' The contents of volume 2 are primarily about the history of mathematics in Japan during the Meiji period.

[12] Mitsuo Hara, *Shizen benshōhō no kenkyū* (Studies in the dialectics of nature), Taigadō, 1947.
A collection of papers that had appeared in 'Kagakushi-kenkyū (History of scientific research)' and 'Yuibutsu-ron kenkyū (Studies on materialism)' which had been published since before the war.

[13] Mitsuo Taketani, *Benshōhō no shomondai* (Problems of Dialectics), Rigakusha, 1946.
This was Taketani's main published work at the time and was supported by a large number of readers. Taketani's views on a wide range of subjects such as methodology, his three stage theory, the philosophy of technology and the problem of measurement in quantum mechanics are efficiently presented.

[14] Shōichi Sakata, *Butsurigaku to hōhō* (Physics and methodology), Hakutō Shokan, 1947.

The title is from a paper included at the beginning of this book that was first printed in *Chōryū* (Trend). It was republished by Iwanami Shoten, but this edition had many additional papers and is thus a completely different book from the Hakutō Shokan edition. In the later edition, each paper has been revised and corrected.

[15] J.D. Bernal, *Social function of science,* Routledge & Sons, 1939 (*Kagaku no shakaiteki kinō*, Shōichi Sakata, Makoto Tatsuoka and Yoshirō Hoshino (tr.), Sōgensha, 1951).
A well-known classic in the sociology of science and technology, it was influential internationally, even after the war. The Japanese version was published in two volumes.

[16] Tetsu Hiroshige, *Kagaku no shakaishi – Kindai Nihon no kagaku taisei* (A social history of science: The structure of modern Japanese science), Chūō Kōronsha, 1973.
This is not related to Marxism, but is important in order to understand the mobilization of science and technology in wartime Japan and the trend influenced by the Enlightenment view embraced by Japanese Marxism after the war.

Consolidated Bibliography

ABCC Rōdō Kumiai Honbu (The ABCC Labor Union Headquarters) (ed.), *Tōron shiryō: Amerika no genbaku shōgai chōsa kenkyū kikan ABCC saihensei kōsō – Sengo o shorishi hibakusha kyūen zenjinrui-teki kikan to suru tameni* (Discussion Data: the American Atomic Bomb Casualty Commission's restructuring concept – Proposal to address postwar issues and to make it an instrument for the aid of the bomb victims and for all human-kind), ABCC Rōdō Kumiai Honbu (ABCC Labor Union Headquarters), not-for-sale item, 1972.

Abe, Akira, *Kyōiku kankei hōrei mokuroku narabini sakuin: Shōwa-hen* (Educational ordinances: An inventory and index, Shōwa Period, III)', Kazama Shobō, 1988.

Abe, Yūkichi and Kōji Kondō, 'Jinkō ninshin chūzetsu mondō (Questions and answers on induced pregnancy terminations)', in *Nihon ishikai zasshi* (Journal of the Japan Medical Association), 1948, vol. 22, pp.263–6.

Abe, Yūkichi, 'Yūsei hogohō ni tsuite (On the Eugenics Protection Law)', in *Nihon iji shinpō* (Japan medical affairs bulletin), 1948, no. 1270, pp.1005–6.

Aida, Nagamune, 'Shizan to jinkō ninshin chūzetsu (Stillbirth and induced pregnancy termination)', in *Kōsei jihō* (Health and welfare bulletin), July 1949, pp.24–6.

Aizawa, Hideyuki, '29-nendo yosan o chūshin to shita kenkyū-hi no shomondai' (Problems with research funding centering on the 1954 Budget), in *Gakujutsu geppō,* [5] vol. 7, no. 1, 1954.

Akiho, Miyashirō, 'Chikyū kagaku no rekishi to genjō (1) – (15) (The past and the present of geosciences 1–15)', in *Shizen* (Nature), September 1965–November 1966.

Akimoto, Sueo, *I no ronri o tou: Dai 731-butai deno taiken kara* (Questioning the ethics of medicine: From the experiences of the 731 Unit), Keisō Shobō, 1983.

Amatatsu, Tadao (ed.), *Jichitai rōdō-sha* (Municipality workers), San'ichi Shobō, 1962, p. 75.

American Dental Association Mission to Japan, 'Report, July 1951', Trainor Collection, (preserved at Kokuritsu Kyōiku Kenkyūsho), microfilm roll, no.60.

Andō, Kakuichi (ed.), *Sanji seigen no kenkyū* (Studies of birth control), Nihon Rinshōsha, 1947.

Andō, Kakuichi, Sa Yamazaki, Nen Yokota, and Minoru Tachi, 1946, 'Zadankai: Sanji seigen ni tsuite (Roundtable discussion on birth control)', in *Nihon rinshō* (Japanese clinical medicine), 1946, no.82, pp.120–2, no. 83, pp.143–5, no. 88, pp.224–5, no. 89, pp.240–2.

Aoki, Giyū, *Dōjinkai shinryō bōeki-han* (Clinical group for the prevention of epidemics of Dōjinkai), Igaku-bu Saikingaku Kyōshitsu Suiyō-kai (Wednesday Meeting Group of the Department of Bacteriology in the Faculty of Medicine), 1975.

Ara, Kei and Kentarō Awaya, 'Nippon haisen to kyū-shihaitaisei no kaitai (Japan's defeat and the disbanding of the old system of domination)', in *Rekishigaku kenkyū* (Historical research), special edition, October 1979.

Ara, Kei, *Nihon senryō-shi josetsu* (An introduction to research into the history of the Occupation of Japan), Kashiwa Shobō, 1994.

Arai, Shōji, President of Nippon Hassōden, *Denryoku setsubi no baishō ni taisuru yōbō* (Pleas relating to electric power facility reparations), (date of publication unknown).

Arisue, Seizō, *Arisue kikan no shuki* (Notes on the Arisue Secret Service), Fuyō Shobō, 1983.

Asahi Shimbunsha (ed.), *Asahi Shimbun ni miru Nihon no ayumi: Shōdo ni kizuku minshushugi* (A history of Japan as found in the pages of Asahi Shimbun: Democracy built on burnt land), 1973, p. 85.

Asahi Shimbunsha Reddo Paaji Shōgen-roku Kankō Iinkai (Asahi Shimbun, Red Purge Testimonial Records Publishing Committee), *1950-nen 7-gatsu 28-nichi* (28 July 1950), Bansaisha, 1981.

Ashida, Jōji, 'Nihon Gakujutsu Kaigi no hansei: Dai 7-kai sōkai no hōkoku' (Reflections on the Science Council of Japan: A report on its seventh general meeting)', in *Shizen* (Nature), no. 6, January 1951, p.23.

Atomic Bomb Casuality Commission, National Research Council, 'General Report', Jan. 1947, in Paul Kesaris, ed., *A Guide to the Manhattan Project: Official History and Document,* A Microfilm Project of University Publications of America, Inc., 1977, held by the National Diet Library.

Awaya, Kentarō (ed.), *Shiryō Nihon gendaishi* (Data: History of modern Japan), vol. 2, Ōtsuki Shoten, 1980.

Bearwald, H., *Shidōsha tsuihō* (Purging leaders), Rinjirō Sodei (tr.), Keisō Shobō, 1970.

Beikoku Gakujutsu Komondan (The United States Scientific Advisory Group), *Nihon ni okeru kagaku to gijutsu no saihensei* (Reorganization of science and technology in Japan), 28 August 1947.

Bernal, J.D., *Social Function of Science,* Routledge & Sons, 1939 (*Kagaku no shakaiteki kinō*, Shōichi Sakata, Makoto Tatsuoka and Yoshirō Hoshino (tr.), Sōgensha, 1951).

Braw, Monica, *Ken'etsu 1945–1949: Kinjirareta genbaku hōdō* (Censorship 1945–1949: Forbidden reporting on atomic bombings), Seiitsu Tachibana (tr.), Jiji Tsūshinsha, 1988.

Brown, Laurie M., R. Kawabe, M. Konuma and Z. Maki. (eds.), *Proceedings of the Japan-USA Collaborative Workshops on the History of Particles Theory in Japan, 1935–1960*, Kyoto: Yukawa Hall Archival Library, Research Institute for Fundamental Physics, Kyoto University, 1988.

Chiba Daigaku Igaku-bu Sōritsu Hachijūgo-shūnen Kinen-kai (Chiba University Medical Faculty Eighty-fifth Anniversary Commemorative Society), *Chiba Daigaku igaku-bu hichijūgonen-shi* (The eighty-five year history of the Medical Faculty of Chiba University), 1964.

Chigaku Dantai Kenkyūkai (The Association for Geological Collaboration Japan), *Minna de kagaku o – Chidanken sanjūnen no ayumi* (Science together: Thirty years of the Association for Geological Collaboration Japan), Ōtsuki Shoten, 1978.

Chūgoku Hōsō (China Broadcasting), *Genbaku kiroku eiga no sugao* (Real face of atomic bomb documentary film), 1984.

Chūgoku Shimbun-sha (Chūgoku Newspaper) (ed.), *Nenpyō Hiroshima yonjūnen no kiroku* (A chronological table, a record of forty years of Hiroshima), Chūgoku Shimbun-sha, 1986, p.68.

Chūō Rōdō Saigai Bōshi Kyōkai (Central Work Disaster Prevention Association), *Nihon no rōdō anzen eisei undō: Gojūnen no kaiko to tenbō* (Japan's safety and hygiene movement: Reminiscing on fifty years and future prospects), 1971.

Compton, Karl T., 'Mission Tokyo', in *The Technology Review*, vol. 48, no. 2, December 1945.

Cumbers, M. P. (ed.), *Quality Management and Product Liability including the Lesson from Japan*, 1981.

Daigaku Kijun Kyōkai (University Accreditation Association), *Daigaku Kijun Kyōkai jūnenshi* (Ten year history of University Accreditation Association), 1957.

Date, Jirō, 'Ichinen no Kaiko (Retrospective on Last Year)', in *Shizen* (Nature), vol. 6, no. 12, December 1951, pp.30–1.

Dees, Bowen C., *The Allied Occupation & Japan's Economic Miracle: Building the Foundations of Japanese Science and Technology 1945–52*, Japan Library, 1997.

Deming, E. Edwards, *Out of Crisis*, MIT, 1982.

Denki Jigyō Minshuka Kyōgikai (The Electric Power Industry Democratization Council), *Denki Jigyō minshuka jōnin iinkai giji yōroku (shūroku)* (Proceedings digest of the standing committee on the electrical industry's democratization – Record), 1947.

Denki Jigyō Minshuka Kyōgikai (The Electric Power Industry Democratization Council), *Denki jigyō no minshuka wa dō susume rarete iruka – zōho* (How is the democratization of the electrical industry proceeding?), revised and enlarged edition, 1947.

Denki Jigyō Saihensei Shingikai (Electric Industry Reorganization Council), *Denki jigyō no saihensei tō ni kansuru tōshin* (Report on the electricity industry reorganization and other matters), 1950.

Denki Jigyō Saihensei-shi Kankō Iinkai (Editorial Committee for the History of the Reorganization of the Electrical Industry) (ed.) *Denki jigyō saihensei-shi* (A history of the reorganization of the electrical industry), 1952.

Densan Chūō Jōnin Tōsō Iinkai (Central Standing Struggle Committee of the Labour Unions of the Electric Industry), 'Densan-an ni tsuite (On the Densan plan)', in *Denkijigyō saiken keikaku* (Reconstruction plan of the electric industry), 1948.

Dōjinkai (Dōjinkai Foundation), *Dōjinkai gojūnen-shi* (A fifty-year history of Dōjinkai), 1943.

Economic Stabilization Board, Electric Power Bureau, Energy Division, *Saikin no denryoku jijō* (The recent energy situation), 1949.

Economic Stabilization Board, Production Bureau, Supply and Demand Division; Power Bureau, Electric Power Division; and Construction Bureau, Industrial Facilities Division, *Keizai fukkō keikaku dai 1-ji shian no urazuke to naritaru denryoku 5-kanen keikaku-an* (Five year initial draft plan for electric power which served as the basis for the first draft of the Economic Recovery Plan), 1948.

Eells, Walter C., *Communism in Education in Asia, Africa and the Far Pacific*, Washington, D.C., American Council on Education, 1954.

Eikoku Chōsadan (British Investigation Team) 'Eikoku chōsadan hōkokusho: Hiroshima oyobi Nagasaki ni tōkasareta genshi bakudan no kōka (The British investigation team report: The effects of the atomic bombs dropped on Hiroshima and Nagasaki)', 1946, in Yamagiwa, Akira and Seiitsu Tachibana (eds.), *Shiryō Manhattan keikaku* (Reference material – The Manhattan Project), Seiitsu Tachibana (tr.), Ōtsuki Shoten 1993.

Electric Distribution Company of each of the Hokkaidō, Tōhoku, Kantō, Chūbu, Hokuriku, Kansai, Chūgoku, Shikoku and Kyūshu Regions, *Saihensei keikakusho* (A re-organization plan), 1948.

Electric Distribution Company of each of the Hokkaidō, Tōhoku, Kantō, Chūbu, Hokuriku, Kansai, Chūgoku, Shikoku and Kyūshū Regions, *Denki jigyō saihen keikaku-an* (A draft plan for the re-organization of the electricity industry), 1948.

Endō, Jun (ed.), *Senryō shiroku 1 – Kōfuku bunsho chōin keii* (A record of the history of Occupation 1: Circumstances surrounding the signing of the surrender documents), Kōdansha Gakujutsu Bunko, 1989.

Endō, Saburō, *Nitchū jūgonen sensō to watashi* (The fifteen-year Sino-Japanese War and me), Nitchū Shorin, 1974.

Etō, Jun (editorial supervision), Ken Kurihara and Sumio Hatano (eds.), *Shūsen kōsaku no kiroku* (A record of moves towards ending the war), vol. 2, Kōdansha Bunko, 1986.

Etō, Jun '*Tozasareta gengo kūkan: Senryō-gun no ken'etsu to sengo Nihon* (A closed space of speech: the Occupation Forces' censorship and postwar Japan), Bungei Shunjū, 1989 (Bunshun Bunko, 1994).

Fee, Elizabeth, 'Rekishiteki kanten kara mita atarashii kōshū eisei no dōkō (Historical view of the new trends in public hygiene)', in Satoru Sugita (tr.), *Kagaku igaku shiryō kenkyū* (Studies of data on scientific medicine), no. 179, 1989, pp. 1–7.

Flexner, A., 'Medical Education in the United States and Canada', in *Bulletin*, no. 4, Carnegie Foundation for the Advancement of Teaching, New York, 1910.

Fort Detrick Report, *Brief Summary of New Information about Japanese BW Activities*, 20 June 1947 (005).

Frercks, Rudolf, *Nachisu no yūsei seisaku* (The Nazi eugenics policy), Fumio Hashimoto (tr.), Risōsha, 1942.

Fujin Kyōdō Hōritsu Jimusho (Female Collective Law Office) (ed.), *Ima naze yūsei hogohō kaiaku ka?* (Why is the Eugenics Protection Law amended in the wrong direction?), Rōdō Kyōiku Sentaa (Labor Education Center), 1983.

Fujioka, Yoshio, 'Kagaku tokubetsu kyōiku (iwayuru eisai kyōiku-ron) no kiroku (A record of science education [so-called elite education]),' in *Kagaku kyōiku-ron* (Essay on science education), Kawade Shobō, 1947.

Fujiwara, Akira, *Nihon gunji-shi* (A Japanese military history), vol. 2, Nihon Hyōronsha, 1987.

Fukushima, Jūrō (ed.), *Shinpan sengo zasshi hakkutsu* (New edition of excavation of magazines after the war), Yōsensha, 1985.

Fukushima, Jūrō, 'Senryō-ka ni okeru ken'etsu seisaku to sono jittai (The realities of censorship policies during the Occupation)', in Takahide Nakamura (ed.), *Senryō-ki Nihon no keizai to seiji* (Japanese economics and politics during the Occupation years), University of Tokyo Press, 1979, pp.339–82.

Fukushima, Jūrō, 'Senryō-ki shoki ni okeru shimbun ken'etsu (Newspaper censorship during the early period of the Occupation)', in Shisō no Kagaku Kenkyū-kai (Study Group of the Science of Thought) (ed.), *Nihon senryō-gun: Sono hikari to kage, jō-kan* (The Occupation Forces in Japan: Their bright and dark sides, vol. 1), Gendai-shi Shuppan-kai, 1978, pp.115–36.

Fukushima, Shingo, 'Genbaku jōhō no sogai (Neglect of the atomic bomb reporting)', in *Heiwa kenkyū* (Peace research), no. 9, November 1984, pp.8–28.

Fukushima, Yōichi, *'Gakusha no mori' no yonjūnen* (Forty years of the 'forest of scholars'), vol. 1, Nihon Hyōronsha 1986, pp. 111–44.

Fumaroli, Marc, *Bunka kokka – Kindai no shūkyō* (L'Etat culturel: Essai sur une religion moderne), Tsuneo Amano (tr.), Misuzu Shobō, 1993.

Furukawa, Jun, 'Nenpyō: Senryō-ka no shuppan, engei, hōsō ken'etsu (Chronological table: Censorship of publications, entertainment and broadcasting under the Occupation)', in *Tokyo Keizai Daigaku kaishi* (Journal of the Tokyo University of Economics), no. 118, December 1980, pp. 231–341.

Furukawa, Atsushi, 'Senryōto chōhō: "Genbaku eiga" fairu to kiroku eiga no yukue (The Occupation and intelligence: The files of "Atomic Bomb Movies" and the future of documentary films)', in *Senshū hōgaku ronshū* (Law Journal of SenshūUniversity), joint issues, nos. 55 & 56, February 1992, pp.527–47.

Furukawa, Atsushi, 'Senryō to shuppan ken'etsu: Joron (The Occupation and publication censorship: An introduction)', in *Kenpō no tenbō: Kobayashi Naoki sensei koki shukuga* (An outlook on constitutional studies: In celebration of seventy years of achievement by Professor Naoki Kobayashi) Yūhikaku, 1991, pp.281–97.

Furukawa, Atsushi, 'Senryō-gun no chōhō katsudō: "MacArthur Report" yori (Intelligence activities of the Occupation Forces – from the "MacArthur Report")', in *Tokyo Keizai Daigaku kaishi* (Journal of the Tokyo University of Economics), no. 119, Jan 1981, pp.165–75.

Furukawa, Atsushi, 'Zasshi "Kaizō" ni miru senryō-ka no kenetsu no jittai (1) (The actual censorship situation under the Occupation as evident in the Magazine "Kaizō" (1))', in *Tokyo Keizai Daigaku kaishi* (Journal of the Tokyo University of Economics), nos. 116–117 (joint issue), September 1980, pp. 133–43.

Gaikokugo Tosho Shuppan, *Saikinsen-yō heiki no junbi oyobi shiyō no kado de kiso sareta moto Nippon gunjin no jiken ni kansuru kōhan shorui* (Court proceedings of former Japanese soldiers charged with the preparation and use of bacteriological weapons), Gaikokugo Tosho Shuppan (Foreign Language Book Publishers), Moscow, 1950.

Gakujutsu Taisei Kenkyū-kai (Science System Research Group), *Gakujutsu kenkyū no haikei* (A background to science research), Nihon Gakujutsu Shinkō-kai (Japan Society for the Promotion of Science), 1953.

Gakujutsu Taisei Sasshin Iinkai (Renewal Committee for Scientific Organizations), 'Shin gakujutsu taisei no ritsuan ni tsuite (On the planning of new scientific organizations)', Report to the Prime Minister, 8 April 1948.

Genbaku Shōgai Chōsa Iinkai (ABCC) 1947–1975 (Atomic Bomb Casualty Commission 1947–1975), *ABCC-Yoken kyōdō kenkyū hōkokusho* (Joint research report of ABCC and the National Institute of Health), ABCC, 1975.

General Headquarters, Supreme Commander for the Allied Powers, CCS, *C.C.S. Keiei kōza* (CCS Management Course) *Vol. 1, Toppu maneejimento no hōshin to soshiki* (The policy and organization of the top management) and Vol. 2, *Toppu maneejimento no taisei to unei* (Control and operation of top management), Japan Electrical Telecommunications Industries Association, Management Research Group (tr.), Diamond-sha, 1952.

General Headquarters, Supreme Commander for the Allied Powers, *History of the Non-military Activities of the Occupation of Japan, Introduction*, 1945 through 1951. Japanese translation: *Nihon senryō – GHQ seishi* (The Occupation of Japan: GHQ's official history), vol. 1, Nihon Tosho Sentaa, 1990.

General Headquarters, United States Army Forces, Pacific, Scientific and Technological Advisory Section, *Report on Scientific Intelligence Survey in Japan, September and October 1945*: Volume I, 'Brief Summary Report, Main Report (Objectives, Organizations and Results of Survey); Volume II, Appendix (Interviews and Other Materials Relating to Japanese Army and Navy Research and Development on Instruments and Equipment); Volume III, Appendix (Interviews and other Materials Relating to Wartime Research and Development Outside of Military Institutions); Volume IV, Chemical Warfare; Volume V, Biological Warfare; GHQ/SCAP Records, National Diet Library Archives.

Gimbel, John, *Science, Technology and Reparation: Exploitation and Plunder in Postwar Germany*, Stanford University Press, 1990.

Gorai, Masao, 'Nihon Gakujutsu Kaigi no dōkō (Trends of the Science Council of Japan)', in *Shizen* (Nature), vol. 4, no. 8, August 1949, p.13.

Gorai, Masao, 'Nihon Gakujutsu Kaigi sōritsu no koro (On the founding of the JSC)', in *Nihon no Kagakusha* (Japanese scientists), vol. 5, no. 12, Dec. 1970, pp.34–5.

Gorai, Masao, *Kaseiron e no michi – Sono henreki no jidai* (The road to the igneous theory: The period of intellectual journey), vol. 2, Kankōkai, 1976.

Goutsmitt, S.A., *Nachi to genbaku – Arusosu: Kagaku jōhō chōsa-dan no hōkoku* (The Nazis and the atomic bombs – Alsos: Report of the Scientific Information Survey Group), Kazuo Yamazaki and Tsūji Konuma (tr.), Kaimeisha, 1977.

Groves, Leslie R., *Now It Can Be Told: The Story of The Manhattan Project*, 1964.

Groves, Leslie R., *Watashi ga genbaku keikaku o shiki shita* (I directed the atomic bomb plan: The inside story of the Manhattan Project) [The original title, *Now It Can Be Told: The Story of the Manhattan Project*], Kengo Tominaga and Yuzuru Sanematsu (tr.), Kōbunsha, 1964.

Gruef, S., *Manhattan keikaku – genbaku kaihatsu gurūpu no kiroku* (The Manhattan Project: Records of the group that developed the atomic bomb), Seitarō Nakamura (tr.), Hayakawa Shobō, 1970.

Guerlac, H.E., *Radar in World War II*, Tomash Publishers, 1987.

Gun'i gakkō bōeki kenkyū hōkoku (Report on prevention of epidemics by the Army Medical College), Rikugun Gun'igakkō Gunjin Bōekigaku Kyōshitsu (Epidemics Prevention Department of the Army Medical College), typed printing.

Hacker, Barton C., *The Dragon's Tail: Radiation Safety in the Manhattan Project, 1942–1946*, University of California Press, 1987.

Haiden Jigyō Zenkoku Todōfuken-ei Kisei Dōmeikai (All Japan Alliance for Prefectural Administration of Electricity Distribution), *Denki jigyō saihensei ni tomonau haiden jigyō todōfuken-ei mondai ni tsuite* (On the question of prefectural administration of electric distribution, accompanying the re-organization of the electric power industry), 1948.

Haiden Jigyō Zenkoku Todōfuken-ei Kisei Dōmeikai (All Japan Alliance for Prefectural Administration of Electricity Distribution), *Haiden jigyō todofuken-ei kihon hōshin setsumei-sho* (An explanation of the basic policy of the prefectural administration of electric distribution), 1948.

Halstead, L.B., *'Imanishi shinka-ron' hihan no tabi* (An intellectual journey for the critical analysis of Kinji Imanishi's 'theory of evolution'), Akiko Nakayama (tr.), Tsukiji Shokan, 1988.

Hara, Genkichi, 'Gakujutsu bunken no katsuyō (Utilization of scientific literature)' in *Gakujutsu geppō* (Science monthly), vol. 7, no. 4, July 1954, p. 24.

Hara, Genkichi, 'Nihon ni kagakugijutsusha wa donokurai iruka (How many scientists and engineers in Japan?)', in *Kokumin no kagaku* (Science for the national people), no. 3, May 1958, p. 4.

Hara, Genkichi, *Kagaku kenkyū-hi* (Science research funds), Kagaku Shimbunsha, 1982.

Hara, Mitsuo, 'Shizen benshō-hō to shiteki yuibutsu-ron (Dialectics of nature and historical materialism),' in *Zen'ei* (Vanguard), no. 21, 1947.

Hara, Mitsuo, *Sanso no hakken* (Discovery of oxygen), Dainihon Shuppan Insatsu, 1946.

Hara, Mitsuo, *Shizen benshōhō no kenkyū* (Study in the dialectics of nature), Taigadō, 1947.

Haruhara, Akihiko, 'Senryō ken'etsu no ito to jittai (The intent and realities of Occupation censorship)', in *Shimbun kenkyū* (Newspaper research), Parts 1–3, nos. 395, 396 and 397, June, July and August 1984.

Hata, Seizaburō, 'Jinkō ninshin chūzetsu ni tsuite (Induced pregnancy terminations)', in *Nihon ishikai zasshi* (Journal of the Japan Medical Association), 1948, vol. 22, pp.248–53.

Hata, Takashi, 'Sengo kyōiku kaikaku to kagaku gijutsu no shomondai (Postwar educational reform and various issues of science and technology)', in *Kōwa dokuritsu go no wagakuni kyōiku kaikaku ni kansuru chōsa kenkyū* (Research on Japan's educational reform after the conclusion of the Peace Treaty), Kokuritsu Kyōiku Kenkyūsho (National Institute for Educational Research), March 1992, pp.1–50.

Hatsumei Kōhō Kyōkai (Society for the Official Reporting of Inventions) (ed.), *Eibei sonota tekikoku shoyū no tokkyo-ken shūroku* (Recorded collection of the patent rights of the property of the other enemy countries of America and Britain), Hatsumei Kyōkai (Society for Inventions), 1942.

Hattori, Takushirō, *Daitōa sensō zenshi* (A complete history of the Great East Asia War), Hara Shobō, 1965.

Hewlett, Richard G. and Oscar E. Anderson Jr., *The New World, 1939/ 1946: A History of the United States Atomic Energy Commission*, University Park, 1962.

Hewlett, Richard G., and Francis Duncan, *Atomic Shield, 1947–1952* (A History of the United States Atomic Energy Commission, Vol. 2), University Park, Pennsylvania University Press, 1969.

Higashida, Iwao, 'Intaan hando bukku (The intern handbook (1))', in *Nihon iji shinpō* (Japan medical affairs bulletin), no. 1203, November 1946, p. 505.

Hikiage Engo-in Ken'eki-kyoku (Repatriation Assistance Agency, Quarantine Bureau), *Hikiage ken'eki-shi, dai ichibu* (History of repatriation and quarantine, Part 1), 1949.

Hirata, Tetsuo (ed.), *Daigaku jichi no kiki – Kōbe Daigaku reddo paaji jiken no kaimei* (The crisis of university autonomy: An analysis of the Red Purge incident at Kobe University), Shiraishi Shoten, 1993.

Hirata, Tetsuo, '1949-nen no gyōsei seiri kigyō seibi ni okeru reddo paaji no hōshin to jittai (The direction and actual situation concerning the Red Purge during the administrative and enterprise re-adjustments of 1949)', in *Ōhara shakai mondai kenkyūsho zasshi* (The Journal of the Ōhara Research Institute of Social Issues), no. 367, June 1989, pp.38–57.

Hirata, Tetsuo, 'Naze reddo paaji-shi no kenkyū ka (Why should we study the red purge?)', in *Rekishi hyōron* (Historical review), no. 472, August 1989. pp.2–23.

Hiroshige, Tetsu, *Kagaku no shakaishi – Kindai Nihon no kagaku taisei*

(A social history of science – The structure of modern Japanese science), Chūō Kōronsha, 1973.
Hiroshige, Tetsu, *Kagaku to rekishi* (Science and history), Misuzu Shobō, 1965.
Hiroshige, Tetsu, *Sengo Nihon no kagaku undō* (The science movements in postwar Japan), Chūō Kōronsha, 1960.
Hiroshima Genbaku Iryō-shi Henshū Iinkai (Editorial Committee on a History of Medical Treatment after the Hiroshima Atomic Bomb), *Hiroshima genbaku iryōshi* (A history of the medical treatment after the Hiroshima atom bomb), Zaidan Hōjin Hiroshima Genbaku Shōgai Taisaku Kyōgikai (Counter-Strategy Council on Hiroshima Atomic Bomb Damages, Incorporated), 1961.
Hiroshima Genbaku Iryōshi Henshū Iinkai (Editorial Committee on a History of Hiroshima Atomic Bomb Medical Treatment), *Hiroshima genbaku iryōshi* (A history of the atomic bomb medical treatment), Hiroshima Genbaku Shōgai Taisaku Kyōgikai (Counter-measures Committee on atomic bomb damage in Hiroshima), 1961.
Hiroshima-ken (Hiroshima Prefectural Government) (ed.), *Hiroshima kenshi: Genbaku shiryō-hen* (A history of Hiroshima Prefecture: Volume on the atomic bomb), 1972.
Hiroshima-ken Ishikai (Hiroshima Prefectural Medical Association), *Hiroshima igaku* (Hiroshima medicine), vol. 5, nos. 5 and 6, 1952, vol. 20, nos. 2 and 3, 1967.
Hiroshima-ken Ishikai (Hiroshima Prefectural Medical Association) (ed.), *Genbaku no ki* (A record of the atomic bomb), vol. 1, 1970.
Hiroshima-shi (Hiroshima City Office) (ed.), *Hiroshima shin-shi: Shiryō-hen (Tsuzuki shiryō)* ('The new Hiroshima City: volume on data, vol. 1 [Tsuzuki Data])', 1981.
Hiroshima-shi Nagasaki-shi Genbaku Saigai-shi Henshū Iinkai (The Hiroshima and Nagasaki Municipal Editorial Committee for Magazines on Atomic Bomb Damages), *Hiroshima Nagasaki no genbaku saigai* (Damages caused by the Hiroshima and Nagasaki atomic bombs), Iwanami Shoten, 1979.
Hishiyama, Tatsuichi, 'Warerano gikai o tsukurō (Let's make our own Parliament)', in *Warerano kagaku* (Our science), vol. 1, no. 1, March 1946, p.3.
Hokkaidō Daigaku (Hokkaidō University) *Hokudai Hyakunenshi* (History of Hokkaidō University), 1980.
Horiba, Kiyoko, 'Puresu kōdo ka no genbaku hōdō (The atomic bomb reports under the press code)', in *Tosho*, no. 494, 1990, p.8.
Horne, R.W., *Science on Service*, Manuscript, March 1987.
Hoshino, Yoshirō, 'Minshushugi Kagakusha Kyōkai no gijutsusha undō (Engineers' movement of the Association of Democratic Scientists)', in

Gijutsu to ningen (Technology and humanity), May 1976, pp.141–53.

Hōsōhō Seiritsu Katei Kenkyū-kai (Research Group on the Broadcasting Legislation Enactment Process) (ed.), *Shiryō: Senryō-ka no hōsō rippō* (Data: Broadcasting legislation under the Occupation), University of Tokyo Press, 1980.

Hutchings, David, 'The Historical Development of Quality Control in the West', in Naoto Sasaki and David Hutchings (eds.), *The Japanese Approach to Product Quality: Its Applicability to the West*, Pergamon, 1984, p.45.

Ichikawa, Hiroshi, 'Chōsen Sensō izen ni okeru hinshitsu kanri to kanri gijutsu no igi ni tsuite; GHQ no tai-nichi sangyō seisaku no shiten kara (On the significance of quality control and management technology before the Korean War: from the Perspective of GHQ's industrial policy for Japan)', in *Osaka Shiritsu Daigaku Keiei Kenkyūkai Keieigaku kenkyū* (Journal of management studies, Osaka Municipal University), vol. 38, no. 5, 1987.

Ichikawa, Hiroshi, 'GHQ kagaku gijutsuka no seisaku to katsudō ni tsuite (On the policy and activity of ESS/ST)', in *Osaka Shiritsu Daigaku Ronsō* (Journal of Osaka Municipal University), no. 54, 1987.

Ijiri, Shōji (ed.), *Tomo ni manabu yorokobi: Dantai kenkyū no hōhō* (The joy of studying together: Methodology of collaborative research), University of Tokyo Press, 1953 (2nd edn., 1969).

Ijiri, Shōji and Masae Ōmori, 'Aru kaiko: Minka wa naze suitai shitaka (A retrospective: Why Minka declined)', in *Gijutsu to ningen* (Technology and humanity), February & March 1977.

Ijiri, Shōji, *Hiteiteki seishin* (Mentality of antithesis), Tsukiji Shokan, 1986.

Ijiri, Shōji, *Ijiri Shōji senshū* (Selected work of Shōji Ijiri), 10 vols., Ōtsuki Shoten, 1981–83.

Ikeda, Naeo and Saburō Araki, *Hashōfū dokuso narabini sesshu ji ni okeru kin 'kuronakishii' ni tsuite (Shidō Nagayama chūsa)*, (About muscle 'chronaxie' upon the injection for tetanus toxin [Supervisor, Lieutenant Colonel Nagayama]). Date of publication unknown (around 1940).

Ikumi, Takuichi, *Nihon shihon-shugi-ron* (Theory of Japanese capitalism), Kureha Shobō, 1948.

Imabori, Seiji, *Gen-suibaku jidai – Gendai-shi no shōgen* (The age of atomic and hydrogen bombs – A testimony of modern history), vol. 1, San'ichi Shobō, 1959.

Imaoka, Kazuhiko, *Tokyo Daigaku Daini Kōgakubu* (Second School of Engineering, Tokyo University), 1987.

Imoto, Minoru, 'Haisen zengo no nikki (A diary before and after the end of the war)', in *Osaka Daigaku-shi kiyō* (Journal of the history of Osaka University), vol. 1, May 1981, pp.111–16.

Inoue, Harumaru, and Seijirō Usami, 'Nihon ni okeru kokka dokusen shihon-shugi no tenkai (Development of national monopoly capitalism in Japan),' in *Chōryū* [9] October 1948.

Institute of International Education, *Education for One World, Annual Census of Foreign Students in the United States*. Since 1954 the title has been *Open Doors*.

Ishigai, Seikan, 'Anzen gijutsu no shohōsoku to kihon mondai (Various laws and basic problems concerning safety techniques)', in Nihon Kikai Gakkai (Japan Machinery Society) (ed.), *Gijutsu no kokoro* (The heart of technology), vol. 2, Maruzen, 1985.

Ishigai, Seikan, *Kōgaku gairon* (Zōhoban) (Introduction to technology) [Enlarged edition], Corona-sha, 1977.

Ishige, Eizō, 'Kenkyūkikan no seiri (Clearance of research laboratories)', in *Nissankyō geppō* (Japan Industrial Association monthly), May 1949.

Ishii, Shintarō, 'DDT no sayō kijo (Function and operation of DDT)', in *Igaku no ayumi* (Path of development of medicine), vol. 1, 1946, pp. 333–42.

Ishikawa, Hiroyoshi, *Yokubō no sengoshi* (Postwar history of desire), Kōsaidō, 1989.

Ishikawa, Kaoru, *Nihonteki hinshitsu kanri: TQC to wa nanika* (Japanese style quality control; What is TQC?), 1981.

Ishikawa, Kaoru, *Quality Control Circles at Work*, Asia Productivity Organization, 1984.

Itō, Junkichi, 'Handai no kyū saikurotoron shimatsuki (Disposal of the Osaka University cyclotron)', in *Iwanami kōza gendai butsurigaku geppō* (Monthly bulletin of the Iwanami series on contemporary physics) no.9, 1965.

Iwasaki, Akira, *Senryō sareta sukuriin – Watashi no sengo-shi* (The occupied screen – My postwar history), Tōyō Keizai Shinpōsha, 1975.

Jones, Vincent C., *Manhattan, the Army and the Atomic Bomb* (United States Army in World War II, Special Studies), Center of Military History, United States Army, Wahington, D.C., 1985. Held by the National Diet Library.

Junod, Marcel, 'Hiroshima no sanka (Disaster in Hiroshima),' in *Sekijūji Shimbun*, 1 May, 1 June and 1 July 1983.

Kagaku Asahi, 'ABCC to wa nani ka: Genshi bakudan shōgai chōsa iinkai o tazunete (What is the ABCC? We visited the Atomic Bomb Casualty Commission)', in *Kagaku Asahi* (Science Asahi), February 1950.

Kagaku Gijutsu Seisakushi Kenkyūkai (Workshop on the History of Science and Technology Policy), *Nihon no kagaku gijutsu seisaku-shi* (History of Japanese policy on science and technology), Mitō, 1990.

Kagaku, 'Genbaku shōgai chōsa iinkai (ABCC) wa nani o shite kitaka (What has the Atomic Bomb Casualty Commission (ABCC) Been Doing?)' in *Kagaku* (Science), vol. 22, no. 6, June 1952.

Kaigai Haken Seido Kenkyūkai (Study Group of the Overseas Dispatch System) (ed.), *Kaigai haken, shisatsu sōran* (Survey of overseas studies and inspections), 1972.

Kaigo, Tokiomi and Masao Terasaki, *Daigaku kyōiku* (University education), University of Tokyo Press, 1969.

Kaigun Shisetsu-kei Gijutsukan no Kiroku Kankō Iinkai (Committee for the Publication of 'A Record of Naval Facility Engineers and Technicians') (ed.), *Kaigun shisetsu-kei gijutsukan no kiroku* (A record of naval facility engineers and technicians), limited press, May, 1972.

Kakinuma, Usaku, 'Gakujutsu Jōhōshitu no ninmu to shimei (The work and mission of the Scientific Information Branch),' in *Gakujutsu geppō* (Science monthly), vol. 5, no. 6, September 1952, p. 2.

Kamata, Kei, *Jidōsha ōkoku no kurayami* (Gloom of automobile empire), 1984.

Kamata, Kei, *Jidōsha zetsubō kōjō* (Desperate car factory), 1984.

Kamiyama, Shigeo, *Tennō-sei ni kansuru rironteki shomondai* (Theoretical problems in the Emperor system), Minshu Hyōronsha, 1947.

Kanamori, Tokujirō, 'PB Repōto no katsuyō o nozomu (Hoped-for utilization of PB reports)', in *Dokusho shunjū* (Reading chronicle), vol. 4, no. 10, 1953, p.1.

Kanayama, Muneo, *Kaisei denki jigyō-hō chikujō kaisetsu* (An article-by-article explanation of the amended Electric Industry Law), Nippon Hassōden Co. Ltd., Chūgoku Branch, 1947.

Kaneseki, Yoshinori, 'Sokoku saiken to kagakugijutsusha no mondai (Recovery of Japan and the problems of scientists and engineers)', in *Shizen kagaku* (Natural science), no. 9, August 1947.

Kaneseki, Yoshinori, 'Soryūshiron gurūpu (The elementary particles theory group)', in *Shizen* (Nature), April 1950, pp.48–55.

Kanō, Ryūichi and Hajime Mizuno, *Hiroshima nijūnen – Genbaku kiroku eiga seisakusha no shōgen* (Twenty years after Hiroshima: Testimonies of the atomic bomb documentary filmmakers), Kōbundō, 1965.

Katsumata Minoru Sensei Tsuitōroku Kankōkai (Editorial Committee on Professor Minoru Katsumata's Work in His Memory), *Kindai kōshū eisei no chichi, Katsumata Minoru* (Father of modern public hygiene, Minoru Katsumata), 1970.

Katsuta, Shūichi, 'Sengo ni okeru shakaika no shuppatsu (Beginning of social studies in the postwar period),' in *Katsuta Shūichi chosakushū* (Collection of literary works by Shuichi Katsuta), vol. 1, Kokudosha, 1972.

Kawabe, Rokuo and Michiji Konuma, 'Chūseishi-ron no tanjō (Birth of the meson theory)', in *Butsuri* (Physics), vol. 37, no. 4, April 1982, pp.265–75.

Kawakami, Kōichi, *Genshiryoku no seiji keizai-gaku* (The political economy of atomic power), Heibonsha, 1974.

Kawakami, Takeshi, *Gendai Nihon iryōshi* (A history of the medical treatment of modern Japan), Keisō Shobō, 1965.

Kawasaki-shi Nakahara Heiwa Kyōiku Gakkyū (Nakahara Peace Education Classes in Kawasaki City) (ed.), *Watashi no machi kara sensō ga mieta – Bōryaku himitsu kichi Noborito kenkyūsho no nazo o ou* (We could see the war from our town – Investigating the mysteries of the Noborito Research Institute, the secret base of warfare plots), Kyōiku Shiryō Shuppankai, 1989.

Kawashima, Kiyoshi, *Kushū: Ryōka* (Haiku collection: Lonely flowers), self-published, 1974.

Kawashima, Midori, 'Hitotsu no kango kyōiku-shi 1946–53: Tokyo kango kyōiku mohan gakkō de mananda hitobito, 1–3 (A nursing education history, 1946–53: People who learned at the Tokyo Model School for Nursing Education, 1–3), in *Kangofu kyōiku* (Nursing education), vol. 33, nos. 1–3, 1992.

Kazahaya, Yasoji, 'Minka sōritsu no koro (When Minka was established)', in Nihon Kagakusha Kaigi (ed.), *Kagakusha undō no shōgen* (Testimony of the scientists' movement), 1978.

Keidanren Jimukyoku (Secretariat of the Keidanren), 'Shintokuju no igi to kenzai kyōryoku no kōsō ni tsuite (On the significance of new special procurement and the idea of economic cooperation)', in *Keizai rengō* (Economic federations), August 1952.

Keiō University, *Keiō Gijuku Daigaku Kōgaku bu sanjūgonen-shi* (Thirty-five year history of the Faculty of Engineering, Keiō University), Keiō Gijuku Daigaku Kōgaku-bu, 1974.

Keizai Dantai Rengōkai, Keizai Kyōryoku Kondankai (Economic Cooperation Conversazione), 'Tokuju keiyaku jōken tō ni kansuru yōbō iken (Request and opinion about the conditions of special procurement contracts)', in *Keizai rengō* (Economic federations), no.7, 1953.

Kelly, Harry C., *Collection*, North Carolina State University Library, Raleigh, North Calorina.

Kesaris, Paul (ed.), *A Guide to the Manhattan Project: Official History and Document,* A Microfilm Project of University Publications of America, Inc., 1977, held by the National Diet Library.

Kikkawa, Takeo, 'Denki jigyō saihensei ni okeru min'ei chiiki-betsu 9-bunkatsu-an no keisei katei: Saihensei no shuyaku wa dareka (The formulation of the 9-block plan in the reorganization of Japan's electric power industry in 1951: Who was the opinion leader?)', in *Shakai keizai shigaku* (Socio-economic history), vol. 57, no. 6, March 1992, pp.735–61.

Kikkawa, Takeo, 'The Reorganization of Japan's Electric Power Industry and General Headquarters/Supreme Commander of the Allied Powers (1)–(4)', in *Aoyama keiei ronshū* (Aoyama Journal of Business), vol. 25, no. 3, Nov. 1990; vol. 25, no.4, Mar. 1991; vol.26, no. 1, June 1991; vol. 26, no. 3, Dec. 1991.

Kikkawa, Takeo, 'Denryoku-gyō (The power industry)', in The Ministry of International Trade and Industry Editorial Committee on International Trade and Industry Policy History (ed.), *Tsūshō sangyō seisaku-shi* (A history of international trade and industry policies), vol. 3, Tsūshō Sangyō Chōsa-kai, 1992, pp.317–95.

Kindai Nihon Kyōiku Seido Shiryō Henshū Iinkai (Editorial Committee for the Historical data on Modern Japan's Education System), *Kindai Nihon kyōiku seido shiryō* (Historical data on modern Japan's education system), 35 vols., Dai Nippon Yūbenkai Kōdansha, 1956–64.

Kitagawa, Toshio, 'Nihon Gakujutsu Kaigi Dai 12-kai Sōkai okaerimite (Reflecting on the twelfth general meeting of the Science Council of Japan)', in *Shizen* (Nature), vol. 7, no. 7, July 1952, p.34.

Kitano, Seiji, 'Bōeki hiwa' and 'Zoku bōeki hiwa (The secret story of the prevention of epidemics: parts 1 and 2)', in *Nihon iji shinpō* (Japan medical affairs bulletin), nos. 1947 (1961) and 2516 (1972).

Kitayama, Setsurō, *Rajio Tokyo: Senji taiseika Nippon no taigai hōsō – III Haiboku e no michi* (Radio Tokyo: Japanese wartime overseas broadcasts, vol. III: The road to defeat), Tabata Shoten, 1988.

Ko Itō Yōji-kun Kinen Ronbun-shū Kankō-kai (The Late Yōji Itō Commemorative Works Publication Committee), *Itō-san no omokage* (Memories of Mr Itō), 1956.

Kobayashi, Hideo (ed.), *Kagaku undō* (The science movements), Chigaku Dantai Kenkyūkai (The Association for Geological Collaboration Japan), Tsukiji Shokan, 1966.

Kōdansha, *Shōwa niman-nichi no zenkiroku 7, haikyo kara no shuppatsu, Shōwa 20nen-21nen* (Complete Records of 20,000 Days in Shōwa 7: Departure from ruin, 1945–1946), Kōdansha, 1989.

Kōei Denki Fukugen Undō-shi Henshū Iinkai (Editorial Committee on History of Movements to Restore the Local Government Management of Electric Power Industry), *Kōei denki fukugen undōshi* (A history of movements to restore the local government management of electric power industry), 1969.

Kōeki Jigyō Iinkai (Public Utilities Committee), (editorial supervision), *Shin denryoku jigyō meikan* (New directory of electric power enterprises), Nikkan Kōgyō Shimbunsha, 1951.

Kōeki Jigyō Iinkai (Public Utilities Committee), *Kigyō saihensei keikaku ni kansuru kettei shirei* (The Order relating to the enterprise re-organization plan), 1951.

Kōgō, Hideki, 'Tai-nichi ken'etsu seisaku no keisei katei to hōsō ken'etsu: Bei tōgō sanbō honbu bunsho o chūshin ni (The formation process of the censorship policies for Japanese broadcasting: Primarily based on the documents of the U.S. Joint Intelligence Headquarters)', in *Hōsō kenkyū to chōsa* (Research and survey on broadcasting), vol. 38, no. 1, January 1988, pp.22–31.

Kogure, Masao, *Nihon no TQC: sono saiginmi to shintenkai* (Japanese TQC: Its reevaluation and new developments), 1988.

Kōgyō Gijitsu-in (Agency of Industrial Science and Technology), *Kagaku gijitsu-in nenpō* (The Science and Technology Agency annual), from 1951.

Kōgyō Gijutsu-chō (Agency of Industrial Science and Technology), *Gijutsu hakusho: Wagakuni kō-kōgyō gijutsu no genjō* (Technology white paper: Mining and manufacturing technology in Japan), Nikkan Kōgyō Shimbunsha, 1949, 1951, 1956.

Kōgyō Gijutsu-chō(Agency of Industrial Science and Technology), *Kenkyū hakusho* (Research white paper), Nikkan Kōgyō Shimbunsha, 1951.

Kōgyō Gijutsu-in Hyōjunbu (Agency of Industrial Science and Technology, Standards Section), *Wagakuni no kōgyō hyōjunka; nijūnen no ayumi* (Japanese industrial standardization: A twenty year history), 1969.

Kōgyō Gijutsu-in Sōmu-bu Gijutsu Kenkyū Shinkō-ka (Agency of Industrial Science and Technology, General Affairs Department, Technical Research Promotion Division), *Shiken kenkyū josei seido no kaisetsu* (Explanation of the experimental research assistance system), Kagaku Gijutsusha, 1964.

Kōgyō Kyōiku Kenkyū Shūkai Chūō Un'ei Iinkai (Institute of Engineering Education, Central Coordination Committee of the General Meeting), *Kōgyō kyōiku kenkyū shukai hōkoku* (Report of the general meeting of the Institute for Engineering Education), 1951.

Koike, Shinzō, *Kaisei denki jigyō-hō kaisetsu* (An Explanation of the amended Electric Industry Law), Dai Nippon Denki-kai (The All-Japan Electric Association) (year of publication unknown).

Kokuritsu Kagaku Hakubutsu-kan (National Science Museum) (ed.), *Kokuritsu Kagaku Hakubutsu-kan* (One hundred year history of the National Science Museum), 1977.

Kokuritsu Kyōiku Kenkyūsho (National Institute of Educational Research, Japan), *Kaigai gakujutsu kenkyū: Hōkokusho, senryōki Nihon kyōiku ni kansuru zaibei shiryō no chōsa kenkyū* (Survey of sources preserved in the US on Japanese education during the Occupation period), March 1988.

Kokuritsu Yobō Eisei Kenkyūsho (National Institute of Health) (ed.), *Yoken nijūnen no ayumi* (Twenty years of the National Institute of Health), 1967.

Kokuritsu Yobō Eisei Kenkyūsho (National Institute of Health), *Kokuritsu yobō eisei kenkyūsho nenpō* (Annals of the National Institute of Health), no.1 (1947), no.2 (1948), no.3 (1949), no.4 (1950), no.5 (1951), no.6 (1952), Kokuritsu Yobō Eisei Kenkyūsho (National Institute of Health).

Kokuritsu Yobō Eisei Kenkyūsho/Genbaku Shōgai Chōsa Iinkai (National Institute of Health and the Atomic Bomb Casualty Commission) (eds.), *Kokuritsu Yobō Kenkyūsho/ABCC nijūnen no ayumi* (Twenty years of the National Institute of Health and the ABCC), Kokuritsu Yobō Eisei Kenkyūsho, 1966.

Kokuritsu Yoron Chōsasho (National Institute of Opinion Surveys), *Jinkō monai ni kansuru yoron chōsa* (Opinion surveys on population issues), 1950.

Kokuritsu Yoron Chōsasho (National Institute of Opinion Surveys), *Jutai chōsetsu ni kansuru yoron chōsa* (Opinion surveys on birth control), 1952.

Kondō, Yōitsu, *Kikagaku shisō-shi* (History of ideas in geometry), Itō Shoten, 1946.

Konno, Takeo (ed.), *Warera no kagaku* (Our science), Minshushugi Kagakusha Kyōkai (Association of Democratic Scientists).

Konuma, Tsūji and Yoshio Takata, 'Nihon no genshiryoku kenkyū ni tuiteno dainiji taisengo no senryō seisaku (Occupation policy on Japanese nuclear research after World War II)', in *Kagakushi kenkyū* (Studies of the history of science), Autumn 1992, pp.138–46.

Kōsei Mondai Kenkyū-kai (Welfare Research Foundation), *Kōseishō gojūnen-shi* (Fifty year history of the Ministry of Health and Welfare), 2 volumes, 1988.

Kotani, Masao, 'Kokusai Gakujutsu Rengō Kaigi to wa nanika (What is the International Council of Scientific Unions [ICSU]?)', in *Kagaku* (Science), vol. 22, no. 6, pp.47–50.

Koura, Kōzō, 'Hinshitsu kanri nenpyō (Chronological table of the quality control movement)', in *Hinshitsu kanri* (Quality control), vol. 41, 1990, pp. 391–5.

Koyama, Hisashi et. al., 'Minka, Mingikyō to genba gijutsuha (Minka, Mingikyō (Association of Democratic Engineers) and engineers on the shop floor)', in *Gijutsu to ningen* (Technology and humanity), July 1976, pp. 146–51.

Kuhn, Thomas S., *The Structure of Scientific Revolutions*, University of Chicago Press, 2nd edition, 1970.

Kuji, Naotarō, 'Sanji seigen-setsu no taitō to kokumin yūsei-hō no saikentō (The rise of the birth control argument and a reevaluation of the national eugenics law),' in *Sanka to Fujinka* (Obstetrics and gynecology), April 1946, pp.7–11.

Kurihara, Tōyō (ed.), *Gendai Nihon sangyō hattatsu-shi, III Denryoku* (A history of the development of modern Japanese industry: III Electric power), Kōjunsha, 1964.

Kuroda, Yasuo, 'Kenkyū-hi taikei to daigaku (The research funding system and universities)', in *Kōza Nihon no daigaku kaikaku, 4* (Series on Japanese university reform, vol. 4), Aoki Shoten, 1982.

Kusaba, Ikurō, 'Nihon no hinshitsu kanri (Japanese Quality Control),' in *Hinshitsu*, July 1992 [2] vol. 22, no. 3, p. 260.

Kusama, Yoshio, 'Amerika no igaku kyōiku (American medical education)', in *Nihon ishikai zasshi* (Journal of the Japan Medical Association), vol. 24, no. 5, May 1950, pp. 418–26.

Kusama, Yoshio, 'Amerika no senmon'i seido to wagakuni no seido shian (The American medical specialist system and Japan's draft plan)', in *Nihon ishikai zasshi* (Journal of the Japanese Medical Association), vol. 24, nos. 10–12, Oct–Dec 1950 (serialized).

Kusama, Yoshio, 'Kusama Yoshio sensei kikigaki (Yoshio Kusama: His verbatim account)', in *Igaku-shi kenkyū* (Medical history research), no. 57, 1983, pp. 43–8.

Kusama,Yoshio, 'Amerika no igaku kyōiku (American medical education)', in *Nihon iji shinpō* (Japan medical affairs bulletin), no. 1170, 15 October 1945, p.184.

Kyōiku Sasshin Iinkai (Education Reform Committee), *Kyōiku sasshin iinkai sōkai gijisokkiroku* (Records of the minutes of the general meeting of the Education Reform Committee), 1946–51, Noma Kyōiku Kenkyūjo (Noma Institute of Education).

Kyoto Daigaku Kōgakubu (Kyoto University, Faculty of Engineering), *Kyoto Daigaku Kōgakubu hachijūnenshi* (Eighty years history of the Faculty of Engineering, Kyoto University), 1980.

Kyoto Daigaku Shichijūnenshi Henshū Iinkai (Editorial Committee for a Seventy-year History of Kyoto University) (ed.), *Kyoto Daigaku shichijūnenshi* (Seventy years history of Kyoto University), 1967.

Kyūshū Daigaku (Kyūshū University), *Kyūshū Daigaku gojūnenshi, Tsūshi* (Fifty-year history of Kyūshū University: Outline history), 1967.

Kyūshū Daigaku (Kyūshū University), *Kyūshū Daigaku shichijūgonenshi, Shiryōhen (Ge)* (Seventy-five year history of Kyūshū University: Source materials [vol.2]), 1989.

Lindee, M. Susan, 'What is a Mutation? Identifying Heritable Change in the Offspring of Survivors at Hiroshima and Nagasaki', in *Journal of the History of Biology*, vol. 25, no. 2, Summer 1992.

Low, Morris S., 'Mosquito Coils, Ball Bearings and Batteries; The Australian Scientific Mission in Occupied Japan, 1945–47', Paper presented at Japanese Studies Association of Australia, Sixth Biennial Conference, 6–9 July 1980.

Maeda, Masao, 'Kagaku gijutsu gyōsei kikō no kakuritsu (Toward the establishment of S&T administration)', in *Nissankyō geppō* (Japan Industrial Association monthly), July 1952.

Maeda, Tamon, 'Shūsen chokugo gokagetsu zainin no kiroku (A record of five months in office immediately following the war)', in *Monbu jihō* (The monthly journal of the Ministry of Education), no. 824, January 1946.

Maema, Takanori, *Jetto enjin ni toritukareta otoko* (Men possessed by jet engines), Kōdansha Bunko, 1992.

Maeshiba, Kakuzō, 'Gakusha no ryōshin ha ikitteita (The conscience of scholars was still alive)', in *Shizen* (Nature), vol. 8, no. 7, July 1953, p.14.

Mainichi Shimbun, *Ichioku-nin no Shōwashi, Nihon senryō 1, kōfuku/shinchū/hikiage* (One hundred million people's history of Shōwa, Japan during the Occupation 1: surrender, occupation, repatriation), Mainichi Shimbunsha, 1980.

Manhattan Engineering Project, 'Final Report of Findings of the Manhattan Project Atomic Bomb Investigating Groups at Hiroshima and Nagasaki', 19 April 1946, in Paul Kesaris, ed., *A Guide to the Manhattan Project: Official History and Document,* A Microfilm Project of University Publications of America, Inc., 1977, held by the National Diet Library.

Manhattan Engineering Project, 'Photographs of the Atomic Bombings of Hiroshima and Nagasaki', 30 June 1946, in Paul Kesaris, ed., *A Guide to the Manhattan Project: Official History and Document,* A Microfilm Project of University Publications of America, Inc., 1977, held by the National Diet Library.

Manhattan Engineering Project, 'The Atomic Bombings of Hiroshima and Nagasaki', 30 June 1946, in Paul Kesaris, ed., *A Guide to the Manhattan Project: Official History and Document,* A Microfilm Project of University Publications of America, Inc., 1977, held by the National Diet Library.

Mann, Nancy R., *The Key to Excellence: the Deming Philosophy*, Study of Prestwick Book, 1989.

Masamura, Kimihiro, *Sengo-shi* (Postwar history), vol. 1, Chikuma Bunko, 1990.

Matsubara, Hiromichi, *Nihon nōyakugakushi nenpyō* (Chronological table of Japanese agricultural chemicals), Gakkai Shuppan Sentaa, 1984.

Matsumae, Shigeyoshi, *Waga Shōwa-shi* (My history of the Shōwa era), compiled by Hisaya Shirai, Asahi Shimbunsha, 1987.

Matsumoto, Setsu, 'Oka Kunio no isshō (The life of Kunio Oka)', in *Shisō no kagaku* (Science of thought), December 1971, p. 101.

Matsumoto, Tsuyoshi, *Ryakudatsu shita bunka* (A culture which plundered), Iwanami Shoten, 1993, pp. 47–50.

Matsumura, Toshio, *Daihon'ei Rikugun-bu 'tairikumei tairikushi sōshūsei'* (The Army of the Imperial Headquarters; the orders of the Army of the

Imperial Headquarters; the instructions of the Army of the Imperial Headquarters: Complete edition) 10 vols., MT Shuppan, 1994.

Matsunaga, Yasuzaemon, *Denryoku saihensei ni tsuite* (On the reorganization of electric power industry), 1950.

Matsunaga, Yasuzaemon, *Nippatsu o kokonotsu no genhaiden kuiki ni bunkatsu shita riyū o nobu* (Stating the reasons for dividing the Hassōden into the present nine local regions of power distribution).

Matsuura, Hajime, 'Gakumon no kiki (Science in crisis)', in *Shizen* (Nature), vol. 7, no. 10, October 1952, p.12.

Matsuura, Hajime, 'Nihon Gakujutsu Kaigi Dai10kai Sōkai kara (From the Tenth General Assembly of the JSC),' in *Shizen* (Nature), vol. 6, no. 7, July 1951.

Matsuura, Sōzō, *Matsuura Sōzō no shigoto 2: Senchū senryō-ka no masukomi* (Sōzō Matsuura's Work, volume 2: The media during the war and under the Occupation), Ōtsuki Shoten, 1984.

Matsuura, Sōzō, *Senryō-ka no genron dan'atsu* (Suppression of the freedom of speech during the Occupation), Gendai Jaanarizumu Shuppan-kai, 1969 (enlarged and revised, 1974).

Ministry of Commerce and Industry, Electric Power Bureau, *Denki jigyō shiryō* (Electric industry data), 1948.

Ministry of Commerce and Industry, Electric Power Bureau, *Denki jigyō minshuka iinkai gijiroku* (Minutes of the Electric Industry Democratization Committee), 1948.

Ministry of Commerce and Industry, Electric Power Bureau, *Hatsuden setsubi kinkyū fukkyū seibi yotei-hyō* (Schedule for the restoration and maintenance of power producing facilities), 1947.

Ministry of Commerce and Industry, Electric Power Bureau, *Hatsuden setsubi higai jōkyō meisai-hyō* (Detailed chart giving damage sustained by power producing facilities), 1945.

Ministry of Commerce and Industry, Electric Power Bureau, *Kūshū higai songai kingaku sōkatsu-hyō* (Comprehensive table of air raid damage), (year of publication unknown).

Ministry of Education, Bureau of Higher Education and Science, Technical Education Division (tr.), *Tainichi kōgyō kyōiku komondan hōkokusho* (Report by the Advisory Group for Industrial Technology Education in Japan), February, 1952.

Ministry of Education, Bureau of Higher Education and Science, Technical Education Division (tr.), *Tainichi kōgyō kyōiku komondan hōkokusho* (Report of the Engineering Education Mission to Japan), August 1951.

Ministry of Education, Records Unit, *Tsūtatsu-tsuzuri* (Files of official notices), held by the National Educational Research Institute.

Ministry of Education, Science and Library Bureau, UNESCO International

Department (Section of Students Abroad), *Wagakuni no ryūgakuseiseido no gaikan: Ukeire oyobi haken* (An Outline of Japanese students exchanges: Receiving and sending), 1977.

Ministry of Education, Science International Bureau, *Wagakuni no gakujutsu* (Science in Japan), Nihon Gakujutsu Shinkōkai (Japan Society for the Promotion of Science), 1975.

Ministry of Finance, Public Finance History Section (ed.), *Shōwa zaisei-shi*: *Shūsen kara kōwa made* (A history of public finances during the Shōwa period: From the end of the war to the peace treaty), Tōyō Keizai Shinpōsha, vol. 2 (1981), vol. 3 (1976).

Ministry of Foreign Affairs (ed.), *Nippon no sentaku: Dai 2-ji sekaitaisen shūsen shiroku* (Japan's choice: A history of the end of World War II), vol. 2, Yamate Shobō Shinsha, 1990.

Ministry of Foreign Affairs (ed.), *Shūsen shiroku* (A historical record of the end of the war), Shimbun Gekkan-sha, 1952.

Ministry of Foreign Affairs, Political Affairs Bureau, Special Data Division, *Nihon senryō oyobi kanri jūyō bunshoshū – Dai 1-kan: Kihon-hen, Dai 2-kan: Seiji, gunji bunka-hen* (The Occupation of Japan and a collection of important management documents – vol. 1: Basic; vol. 2: Political, military and cultural affairs), 1949. Reproduced as *Nihon senryō jūyō bunsho* (Important documents relating to the Occupation of Japan), Nihon Tosho Sentaa, 1989.

Ministry of Foreign Affairs, Special Material Section, (ed.), *Nihon senryō jūyō bunsho* (Important documents of the Occupation of Japan), vol. 1, 1949 (re-published in 1989).

Ministry of Health and Welfare, *Boshi eisei no omonaru tōkei* (Statistics on mother and child ante-natal health), 1992.

Ministry of Health and Welfare, Medical Affairs Bureau, *Isei hachijūnen-shi* (The eighty year history of the medical system), Ministry of Health and Welfare, 1955.

Ministry of Health and Welfare, Medical Affairs Bureau, *Isei hyakunen-shi* (The one hundred year history of the medical system), 1976.

Ministry of Health and Welfare, Population Research Institute, *Sanji seigen mondai no jinkō seisaku-teki kōsatsu* (Study of population policy in regard to the birth control question), July 1949.

Ministry of Health and Welfare, Population Research Institute, *Shogaikoku ni okeru sanji seigen no fukyū jōkyō* (The situation concerning the popularization of birth control in several overseas countries), 1949.

Ministry of International Trade and Industry (MITI) (ed.), *Shōkō seisaku-shi* (A history of the policies of commerce and industry), 24 vols., Shōkō Seisaku-shi Kankō-kai (Publication Group of History for the Commercial and Industrial Policies), 1965–85.

Ministry of International Trade and Industry, Agency of Natural Resources,

Electric Power Bureau, *Denki jigyō saihen no kon'nichi made no keika* (The course of the re-organization of the electric power industry to date)'. This is thought to have been published in 1950.

Ministry of International Trade and Industry, Public Utilities Bureau (later the Ministry of International Trade and Industry, Agency of Natural Resources and Energy, Public Utilities Department), *Denki jigyō no genjō* (The present situation concerning electric power industry), published annually.

Ministry of Labor, *Shiryō rōdō undō-shi* (Reference data on the history of the labour movement) published annually.

Minshushugi Kagakusha Kyōkai (Association of Democratic Scientists) (ed.), *Kagaku nenkan* (Yearbook of science), vol. 1 (1947) & vol. 2 (1948), Nihon Kagakusha; vol. 3 (1949), Hokuryūkan.

Minshushugi Kagakusha Kyōkai (Association of Democratic Scientists) (ed.), *Riron* (Theory), Nihon Hyōronsha.

Minshushugi Kagakusha Kyōkai, Shizen Kagaku Bukai (Association of Democratic Scientists, Natural Science Section), *Shizen kagaku* (Natural science), Kasumigaseki Shobō.

Miura, Masayuki, 'Sengo Nihon no kōshū eisei kaikaku to PHW – shokatsudō o gaikan suru shippingu risuto (Reform of public hygiene in postwar Japan and PHW: Shipping lists provide a general view of various activities)', in *Ritsumeikan Daigaku Jinbunkagaku Kenkyūjo kiyō* (Journal of the Humanities Institute at Ritsumeikan University), separate volume no. 5, 1985, pp. 123–48.

Miura, Yutaka, *Rōdō to kenkō no rekishi* (A history of work and health), vol. 1 – 4, Rōdō Kagaku Kenkyūsho (Research Institute of Labor Science), 1978.

Miyahara, Shōhei, 'Kagaku taisei sasshin ni tsuite (Science renewal)', in *Shizen kagaku* (Natural science), no. 10, January 1948, pp.44.

Miyake, Akimasa, *Reddo paaji to wa nani ka – Nihon senryō no kage* (What was the Red Purge? The dark shadow of the Occupation of Japan), Ōtsuki Shoten, 1994.

Miyatake, Tsuyoshi, *Shōgun no yuigon: Endō Saburō nikki* (The will of a general: The diary of Saburō Endō), Mainichi Shimbunsha, 1984.

Mizoguchi, Hazime, 'Seiyakugyō shinten ni kōken shita GHQ no "DDT kakumei" (GHQ's DDT revolution contributed to the development of the pharmaceutical industry)', in *Ekonomisuto* (Economist), 22 May, 1990, pp.92–3.

Mizuno, Shigeru, *Kaisha sōgō hinshitsu kanri: TQC no dōnyū to tenkai* (Company-wide quality control: Introduction and development of TQC), 1985.

Mori, Akihide, 'DDT to penishirin (DDT and Penicillin)', in *Nihon fūzoku jiten, Amerikan karuchaa 1 '45–50's* (Encyclopedia of Japanese

manners and customs: American culture 1, 1945–50's), Sanseidō, 1981, pp.18–20.
Mori, Kiyoshi, 'Soshiki toshiteno Minka (Minka as an organization)', in *Gijutsu to ningen* (Technology and humanity), October 1976, pp.125–35.
Morishima, Yutaka, 'Jutai chōsei no hōhō to hihan (Birth control methods and criticisms)', in *Sanka to fujinka* (Obstetrics and gynecology), August 1949, pp.330–40.
Moto Gunrei-bu Tsūshin Kachō no Kaisō Kankō-kai (Publications Committee on the Recollections of the Communication Chief of the Former Naval General Staff Division), *Moto gunrei-bu tsūshin kachō no kaisō* (Recollections of the Communications Chief of the Former Military Command Division), 1981.
Moulton, S.E., 'Nihon ni okeru igaku kyōiku kaizen-an (Proposals for medical education reform in Japan)', in *Nihon iji shinpō* (Japan medical affairs bulletin), no. 1191, July 1946.
Myōjin, Isao, 'Hokudai Eells jiken no shōgen (1) (Testimony to the Eells Incident at Hokkaidō University (1))', in *Kushiro ronshū* (Journal of Hokkaidō University of Education at Kushiro), no. 14, 1982.
Myōjin, Isao, 'Kyōiku-kai ni okeru reddo paaji (The Red Purge in educational institutions)', in *Rekishi hyōron* (Historical review), no. 472, August 1989, pp.24–41.
Myōjin, Isao, 'Kyōin reddo paaji no hi-tsuihōsha-sū o megutte (On the number of teaching personnel who were victims of the Red Purge),' in *Hokkaidō Kyōiku Daigaku kihō (Dai 1-bu C)* (Journal of Hokkaidō University of Education, Part 1, C:, Education science), vol. 38, no. 2, March 1988, pp. 33–47.
Nagai, Takashi, *Nagai Takashi zenshū* (The complete works of Takashi Nagai), Kodansha, 1971.
Nagai, Takashi, *Nagasaki no kane* (The bell tolls for Nagasaki), Hibiya Shuppan, 1949.
Nagamura, Jun'ichi, 'Kagakugijutsusha notameno Rosiago (Russian for scientists and engineers),' in *Gakkensha*, 1968, preface.
Nagaoki, Mononobe, 'Gen'ei no minshushugi kagaku (Democratic Science as illusion)', in *Asahi Journal*, 26 October 1969, pp.111–16.
Nagasaki Shiyakusho (Nagasaki City Office) (ed.), *Nagasaki genbaku higaishi* (A compilation of damages caused by the atomic bomb dropped on Nagasaki), vol. 1, Nagasaki Kokusai Bunka Kaikan, 1979.
Nagatoya, Yōji, 'Sengo Amerika no Nihon igaku ni oyoboshita eikyō (jō) – Kōshū eisei to igaku kyōiku o chūshin to shita shiron (American influence on Japanese medicine after the war, the first volume: An essay focused on public hygiene and medical education)', in *Nihon iji shinpō* (Japanese medical affairs bulletin), no. 1963, 1961, pp.53–7.

Nagatoya, Yōji, 'Senryōka eisei gyōsei-shi memo (Memorandum on the history of administration of public health during the Occupation)', *Igakushi kenkyū* (Studies of medical history), no. 20, 1966, pp.1047–9.

Nagoya Daigaku (Nagoya University), *Nagoya Daigaku gojūnenshi Bukyokushi* (Fifty-year history of Nagoya University, Section History), vol. 2, November 1989.

Nakagawa, Kyōjirō, 'Gekidō no ichinen (A year of turmoil),' in Ministry of Agriculture, Forestry and Fisheries, Agriculture Research Institute (ed.), *Kōnosu no omoide* (Recollections of Kōnosu), 1981, p. 85.

Nakagawa, Seizō, *Dokyumento: Kaigun gijutsu kenkyūsho – Erekutoronikusu ōkokuno senkusha-tachi* (Document: Navy Technology Institute – Pioneers in the kingdom of eletronics), Kōdansha Bunko, 1990.

Nakagawa, Yasuo, 'Hiroshima Nagasaki no genbaku hōshasen eikyō kenkyū: Kyūsei-shi kyūsei shōgai no kashō hyōka (Research into the effects of the Hiroshima and Nagasaki atomic bombs' radiation: The underrated sudden deaths, sudden injuries)'; 'Hōshasen ni yoru banhatsuteki eikyō no kashō hyōka (I) (The underrated late-developing effects of radiation (I))'; 'Hōshasen ni yoru banhatsuteki eikyō no kashō hyōka II (The underrated late-developing effects of radiation (II))', in *Kagakushi kenkyū* (Studies of science history), Period II, vol. 25, no. 157, 1986; vol. 26, no. 163, 1987; vol. 26, no. 164, 1987.

Nakagawa, Yasuo, 'Kokusai hōshasen bōgo kijun no hensen to genshiryoku kaihatsu (Changes to the international radiation protection criteria and the development of atomic energy)', in *Kobe Daigaku kyōyō-bu kiyō ronshū* (Journal of the College of Liberal Arts at Kobe University), no. 33, March 1984.

Nakagawa, Yasuo, 'Manhattan keikaku no hoshasen hibaku kanri to hōshasen eikyō kenkyū (The Research of the Manhattan Project on radiation exposure management and radiation effects)', in *Kobe Daigaku kyōyō-bu kiyō ronshū* (Journal of the College of Liberal Arts at Kobe University), no. 36, October 1985.

Nakagawa, Yasuo, *Hōshasen hibaku no rekishi* (A history of radiation exposure), Gijutsu to Ningen-sha, 1991.

Nakagawa, Yasuzō, 'Kaigun erekutoronikusu jinmyaku (Electronics: personal connections with former naval colleagues),' in *Will*, June 1985.

Nakagawa, Yasuzō, *Kaigun Gijutsu Kenkyujo* (Naval Technology Research Institute), Nihon Keizai Shimbun Co. 1987.

Nakajima, Shigeru, 'Japanese radar development prior to 1945', in *IEEE Antennas and Propagation Magazine*, vol. 34, no. 6, 1992, pp.17–22.

Nakajima, Shigeru, 'Magunetoron to reedaa no kenkyū: Taiheiyō sensō shūketsu made (Magnetron and radar research: Until the conclusion of the Pacific War)', in *Denki gakkai kenkyū-kai shiryō: Denki gijutsushi*

kenkyū-kai, HEE-94-1~9 (Research Group materials of the Electricity Society: Research group on electrical technology history/HEE-94-1~9), 1994, pp.49–58.
Nakajima, Shigeru, 'Reedaa kaihatsu no rekishi (The history of the development of radar)', in *Kōden Gihō* (The technological reports of Kōden Co. Ltd.), no. 4, 1988, pp. 19–34.
Nakajima, Shigeru, *Gijutsu to keiei – Tokuni benchaa bijinesu no igi ni tsuite* (On technology and management: Particularly the significance of nurturing venture businesses), record of address delivered to the Nittsū Shōsangyō Training Institute, 27 November 1979.
Nakajima, Shigeru, *Hi senryō-ka ni okeru seisan hinmoku sentei to sono go no seisan jōkyō (Nihon Musen no baai)* (Product selection under the Occupation and the subsequent production situation [The story of Japan Wireless]), record of the address given by the Diffusion of Technology Working Group of the 149th committee meeting of the Japan Society for the Promotion of Science, 19 February 1992.
Nakajima, Shigeru, *Mitō gijutsu to no tatakai* (The Fight with unexplored technology), record of the address delivered to the Tōhoku Branch of the Electronic Communications Society, at Akita University, 3 June 1983.
Nakajima, Shigeru, *The Study of the Microwave Magnetron*, Japan Radio Co., 1947.
Nakajima, Tatsumi, 'Hiroshima: Sono kageri wa fukaku (Hiroshima: Its deep and dark clouds)', in *Gekkan Shakai-tō* (The Socialist Party monthly), July 1984.
Nakajima, Tokunosuke, 'Sengo Nihon no kagaku kenkyū no hōkō (The direction of scientific research in postwar Japan)', in *Zen'ei* (Vanguard), no. 612, 1991, p.226.
Nakamura, Hatsuo, 'Shinjuku CIE Toshokan (CIE Library at Shinjuku),' in *Dokusho shunjū* (Reading chronicle), vol. 8, no. 2, 1951, pp. 36–8.
Nakamura, Humitaka, 'Sengo Nihon no gijutsu dōnyū seisaku no keisei to sono seisakuteki na kiketsu (The formation of technology importation policy in postwar Japan and its consequence),' in *Meiji Daigaku Shakai Kagaku Kenkyūsho Kiyō* (Bulletin of the Research Institute of Social Sciences, Meiji University), October 1989, p. 118.
Nakamura, Masao, 'Kagaku Akademii no shuyō na mondai (Main problems of Science Academy)', in *Kokumin no kagaku* (Science for the national people), vol. 2, no. 1, 1947.
Nakamura, Sachio, 'Shiryō to dōko: Kenkyū hukkō kaigi no kaidō (Sources and trends: Conference on the way to research recovery)', in *Shizen Kagaku* (Natural science). no. 9, August 1947.
Nakamura, Takahide (ed.), *Senryōki Nihon no keizai to seiji* (Economy and Politics of Occupied Japan), University of Tokyo Press, 1979.

Nakamura, Takahide, *Shōwa-shi 2* (History of the Shōwa Era, vol. 2), 1993.
Nakamura, Teiri, 'Minka Seibutsu Bukai no seibutsugaku shisō (The Biology Section of the Association of Democratic Scientists and its biological thought)', in *Seibutsugaku to Shakai* (Biology and society), 1970.
Nakamura, Teiri, *Kiki ni tatsu Kagaku* (Science in crisis), Kawade, 1976.
Nakamura, Teiri, *Seibutsugaku to shakai* (Biology and society), 1970.
Nakayama, Shigeru (ed.), *Nihon no gijutsuryoku: Sengoshi to tenbō* (Japan's technological power: Postwar history and perspective), Asahi Shimbun, 1986.
Nakayama, Shigeru and Hitoshi Yoshioka (eds.), *Sengo kagaku gijutsu no shakai-shi* (A social history of postwar science and technology), Asahi Shimbunsha (Asahi Sensho), 1994.
Nakayama, Shigeru and Hitoshi Yoshioka, 'Tekunohisutorii hanseiki (Half a century of techno-history)', in *Shūkan ekonomisuto* (Weekly Economist), Mainichi Shimbun, April 1990 to September 1991.
Nakayama, Shigeru, 'Grassroots geology', in Shigeru Nakayama, David Swin and Eri Yagi (eds.), *Science and Society in Modern Japan*, 1974.
Nakayama, Shigeru, 'Harii Kerii (Harry Kelly)', in Kagaku Asahi (ed.), *Sukyandaru no kagakushi* (History of scandalous science), Asahi Shimbunsha, 1988.
Nakayama, Shigeru, 'Ijiri Shōji ron (On Shōji Ijiri)', in *Shisō no kagaku* (The science of thought), May 1966, pp.100–6.
Nakayama, Shigeru, 'Joseph Needham, an Organic Philosopher', in S. Nakayama and N. Sivin (eds.), *Chinese Science*, MIT Press, 1973.
Nakayama, Shigeru, *Science, Technology and Society in Postwar Japan*, Kegan Paul International, 1991.
Nakayama, Shigeru, 'The American Occupation and the Science Council of Japan', in Everett Mendelsohn (ed.), *Transformation and Tradition in Sciences*, Cambridge University Press, 1984.
Nakayama, Shigeru, 'The Future of Research: A Call for A Service Science', in *Fundamenta Scientiae*, vol. 2, no. 1, 1981, pp.85–97.
Nakayama, Shigeru, *Academic and Scientific Traditions in China, Japan and the West*, University of Tokyo Press, 1984.
Nakayama, Shigeru, *Kagaku to shakai no gendai-shi* (Contemporary history of science and society), Iwanami, 1981.
Nakayama, Shigeru, *Tenkan-ki no kagaku-kan* (Views on science in the period of transition), Nihon Keizai Shimbunsha, 1980.
National Research Council, Division of Medical Sciences, Committee on Atomic Casualties, Field Agency, *Semi-Annual report, 1 July 1952 thru 31 December 1952*, Headquarters, Hiroshima, Japan.
Nawa, Takeshi, *Nawa Takeshi tsuisōroku* (Takeshi Nawa's memoirs), Nawa Takeshi Tsuisōroku Kankōkai (Takeshi Nawa Reminiscences Publication Committee), 1973.

Needham, Joseph and Dorothy, *Science Outpost*, 1948.

Neel, J. V. and W. J. Shull, *The Effect of Exposure to the Atomic Bombs on Pregnancy Termination in Hiroshima and Nagasaki*, National Academy of Sciences, National Research Council, USA, no. 491, 1956, held by the National Diet Library.

Neel, J. V., W. J. Shull, D. J. MacDonald, N. E. Morton, M. Kodani, K. Takeshima, R. C. Anderson, J. Wood, R. Brewer, S. Wright, J. Yamazaki, M. Suzuki and S. Kitamura, 'The Effect of Exposure to the Atomic Bombs on Pregnancy Termination in Hiroshima and Nagasaki: Preliminary Report', in *Science*, vol. 118, 6 November, 1953.

Neibel, F. and C. Bailey, *Mohaya kōchi nashi – Hiroshima genbaku tōka no himitsu* (Not even any high ground – Secrets concerning the dropping of the atomic bomb on Hiroshima), Masahiro Sasagawa and Reiko Sugibuchi (tr.), Iwanami Shoten, 1960.

Nihon Bosei Hogoi Kyōkai (Japanese Medical Association for the Protection of Mothers), *Nijūshūnen kinenshi* (Twentieth anniversary commemorative publication), 1970.

Nihon Butsuri Gakkai (Japanese Association of Physics) (ed.), 'Kagaku Renraku Shōgaikai no shōkai (Introducing the public relations coordination group of scientists)', in *Nihon Butsuri Gakkaishi* (Journal of the Japanese Association of Physics), vol. 1, no. 1, appendix 1946.

Nihon Butsuri Gakkai (Japanese Association of Physics), *Soryūshiron kenkyū* (Research in Elementary Particles Theory), vol. 1, no. 1, 5 August 1949.

Nihon Chishitsu Gakkai (The Geological Society of Japan) (ed.), *Nihon Chishitsu Gakkai-shi – Nihon Chishitsu Gakkai rokujū shūnen kinen* (A history of the Geological Society of Japan: In commemoration of sixty years), 1953.

Nihon Denki Sangyō Rōdō Kumiai (The Japan Electric Industries Labour Union), *Kado keizai-ryoku shūchū haijohō tekiyō ni yoru denki jigyō saiken keikaku* (A plan to rebuild the Japanese electric industry based on the application of the Elimination of Excessive Economic Concentration Law), 1948.

Nihon Denki Sangyō Rōdō Kumiai Kyōgikai, Chūō Kyōdō Tōsō Iinkai (Japan Labor Union Federation of Electrical Industries, Central Joint Struggle Committee) *Densan yōkyū kyūyo minshuka-an zen* (The complete draft of the electrical industry's salary democratization demands), 1946.

Nihon Gakujutsu Kaigi (Science Council of Japan), *Kagakusha seikatsu hakusho* (White paper on the living standard of scientists), 1959.

Nihon Gakujutsu Kaigi Genshi Bakudan Saigai Chōsa Hōkokusho Kankō Henshū Iinkai (Science Council of Japan, Editorial Committee on Atomic Bomb Damages) (ed.), *Genshi bakudan saigai hōkokushū* (A report on atomic bomb damages), 2 volumes, Nihon Gakujutsu Shinkōkai (Japan Society for the Promotion of Science), 1953.

Nihon Gakujutsu Kaigi Nijūgonen-shi Henshū Iinkai (Science Council of Japan Editorial Committee of the Twenty-five Year History) (ed.), *Nihon Gakujutsu Kaigi nijūgonen-shi* (Twenty-five year history of the Science Council of Japan), Nihon Gakujutsu Kaigi (Science Council of Japan), 1974.

Nihon Gakujutsu Shinkō-kai Kagaku Gijutsu Seido-shi Henshū Iinkai (Japan Society for the Promotion of Science, Editorial Committee on the History of Science and Technology System) (ed.), *Kagaku gijutsu seido-shi: Kagaku kenkyū-hi (Shōwa 21 – 34-nen), Shiryō-hen* (A history of the science and technology system: Science research funding (1946–1959); volume on data), Nihon Gakujutsu Shinkōkai (Japan Society for the Promotion of Science), 1960.

Nihon Gensuikyō Senmon Iinkai (The Japan Council against Atomic and Hydrogen Bombs, Specialist Committee) (ed.), *Gensuibaku Hakusho* (The white paper on atomic and hydrogen bombs – Hidden facts), Nihon Hyōronsha, 1961.

Nihon Igaku Kyōiku Gakkai (Japan Medical Education Society) (ed.), *Nihon igaku hakusho* (A white paper on medical education), Shinohara Shuppan, 1990.

Nihon Kagakusha Kaigi (Scientists Society of Japan) (ed.), 'Minshushugi Kagakusha Kyōkai Kiyaku (The Rules of the Association of Democratic Scientists, [as revised at the Third General Assembly on 24–25 January 1948])', in *Kagakusha undō no shōgen* (Testimony of scientists' movements), 1978.

Nihon Kagakusha Kaigi (Scientists Society of Japan) (ed.), *Kagakusha undō no shōgen* (Testimony of scientists' movements), 1978.

Nihon Kagakushi Gakkai (History of Science Society of Japan), *Nihon kagaku gijutsu-shi taikei* (A compendium of the Japanese history of science and technology), 25 volumes, Daiichi Hōki Shuppan, 1996.

Nihon Kazoku Keikaku Kyōkai (Japan Family Planning Association) (ed.), *Gendai no kazoku keikaku* (Contemporary family planning), Nihon Kazoku Keikaku Kyōkai, 1984.

Nihon Kazoku Keikaku Renmei (Japan Family Planning League) (ed.), *Chūzetsu kinshi e no hanmon: Kanashimi o sabake masuka* (Refutation of the prohibition of abortions: Can we adjudge one's grief?), Ningen no Kagakusha, 1983.

Nihon Ketsueki Gakkai (Japan Haematology Association) (ed.), *Hōshanō narabini genbaku shōgai* (Radiation and atomic bomb damage), (Ketsuekigaku Tōgi-kai hōkoku dai5-gō) (Haematology Conference report no. 5), Nagai Shoten, 1953.

Nihon Kikaku Kyōkai (Japanese Standards Association), *Kikaku to kijun* (Standard), later renamed *Hyōjunka to hinshitsu kanri* (Standardization and quality control).

Nihon Kokuyū Tetsudō (The Japan National Railways) (ed.), *Nihon Kokuyū Tetsudō hyakunen-shi* (Hundred year history of the Japan National Railways), vol. 12, 1973.

Nihon Kyōbunsha (ed.), *Taiji wa ningen dewa nainoka: Yūsei hogohō no gimonten* (Aren't fetuses human beings? Problems of the Eugenics Protection Law), Nihon Kyōbunsha, 1982.

Nihon Kyōsan-tō Chūō Iinkai (The Japanese Communist Party Central Committee) (ed.), *Nihon Kyōsan-tō no shichijūnen* (Seventy years of the Japan Communist Party), vol. 1, Shin-Nihon Shuppansha, 1994.

Nihon Kyōsantō Kagaku Gijutsubu (Japan Communist Party, Science and Technology Section) (ed.), *Sovieto no kagaku to gijutsu* (Science and technology of the Soviet Union).

Nihon Musenshi Henshū Iinkai (The Editorial Committee on the History of Japan Wireless), *Nihon musenshi* (A History of the wireless in Japan), 13 vols., Denpa Kanri Iinkai (Supervisory Committee of Electrical Waves), 1950–1951.

Nihon Shoseki Shuppan Kyōkai, *Nihon shuppan hyakunenshi nenpyō* (Chronological table of one-hundred year history of publication in Japan), Nihon Shoseki Shuppan Kyōkai, 1968.

Nikka Giren (Japanese Union of Scientists and Engineers) (ed.), *Nihon Kagaku Gijutsu Renmei jūgoshūnen kinen* (The first fifteen years of JUSE), May 1961.

Nikka Giren (Japanese Union of Scientists and Engineers) (tr. & ed.), *Demingu hakase kōgiroku: Tōkeiteki hinshitsu kanri no kiso riron to ōyō* (Lecture notes of Dr. Deming: Basic theory and the applications of statistical quality control).

Nikka Giren (Japanese Union of Scientists and Engineers) (tr. & ed.), *J. M. Juran jutsu; Hinshitsu kanri seikōhō* (Juran talked about how to succeed in quality control), 1956.

Nippon Denshin Denwa Kōsha Nijūgonen-shi Henshū Iinkai (Editorial Committee on the Twenty Five Year History of the Nippon Telegraph and Telephone Public Corporation) (ed.), *Nippon Denshin Denwa Kōsha nijugonen-shi, Jōkan* (1977), *Chūkan* (1978), *Gekan* (1977) *and Bekkan* (1978) (The Nippon Telegraph and Telephone Public Corporation: A twenty-five year history), vol. 1, (1977), vol. 2 (1978), vol. 3 (1977), Special vol. (1978)).

Nippon Hassōden Co. Ltd., Research Department, Planning Division, *Denryoku juyō no sūsei* (Trends in electric power demand), 1949.

Nippon Hassōden, Research Department, *Wagakuni denki jigyō no genjō* (The present situation of the electrical power industry in Japan), 1947.

Nippon Hōsō Kyōkai Hōsō Bunka Chōsa Kenkyūsho Jōhō Chōsa-bu (NHK Research Institute on Broadcasting Culture, Broadcast Information Research Division) (ed.), *GHQ bunsho ni yoru senryō-ki hōsōshi nenpyō, Shōwa 20-nen 8-gatsu 15-nichi – 12-gatsu 31-nichi* (A chronological table of a history of broadcasting during the Occupation compiled from GHQ documents, 15 August–31 December 1945), NHK Hōsō Bunka Chōsa Kenkyūsho Jōhō Chōsa-bu, 1987.

Nippon Hōsō Kyōkai Hōsōshi Henshū-shitsu (Japan Broadcasting Corporation, Editorial Office for Broadcasting History) (ed.), *Nippon hōsōshi* (History of Japanese broadcasting), vol. 1, Nippon Hōsō Kyōkai (Japan Broadcasting Corporation), 1966.

Nishikawa, Nagao, 'Kokka ideorogii to shite no bunmei to bunka (Civilization and culture as a national ideology)', in *Shisō* (Philosophical thought), May 1993, pp.4–33.

Nishimura, Keizō, 'Keizai no gōrika to Demingu hakase (Rationalization of economy and Dr. Deming)', in *Keizai rengō* (Economic federations), September 1951, p.257.

Nishimura, Kō, 'Kansai Chiku PB sentaa no zōkyō (Strengthening PB centers in the Kansai area)', in *Dokusho shunjū* (Reading chronicle), vol. 3, no. 2, 1952, p.17.

Nishimura, Kō, 'PB Repōto to sono riyō (PB report and its utilization)' in *Dokusho Shunjū* (Reading chronicle), vol. 7, no. 5, 1956, pp.8–10.

Nishina Kinen Zaidan (The Nishina Commemorative Foundation) (ed.), *Genshi bakudan – Hiroshima Nagasaki no shashin to kiroku* (The atomic bombs: Photographs and records of Hiroshima and Nagasaki), Kofūsha Shoten, 1979.

Nishinarida, Yutaka, 'Senryō seisaku no tenkan to Nihon keizai (Change of the Occupation policies and Japanese economy),' in *Nihon dōjidai-shi* (Contemporary history of Japan), vol. 2, Aoki Shoten, 1990.

Ōbayashi, Michiko, *Josan-fu no sengo-shi* (Postwar history of midwives), Keisō Shobō, 1989.

Odaka, Tomoo, 'Yunesuko kūpon seido ni tsuite (On the UNESCO coupon system)', in *Gakujutsu geppō* (Science monthly), vol. 3, no. 6, 1950, p.1.

Odake, Sokuichi, 'One-hundred year history of electricity, vol. 2,' 1980, Seikei-sha.

OECD, *Towards Mass Higher Education*, 1974.

Office of Military Government for Germany (US), *Military Government Regulations Title 27. Control of Scientific Research* (effective date 30 July 1946) Part 2, Prohibited Research, ESS(E)06455–7.

Ogishima, Hideo and Katsurō Shimomura, *Amerika no igaku kyōiku to iryō jijō* (America's medical education and medical situation), Igaku Shoin, 1970.

Ogura, Kinnosuke, *Sūgaku-shi kenkyū* (Study on the history of mathematics), vol. 2, Iwanami Shoten, 1948.

Ōhashi, Nobuo, 'Sengo no kagaku gijutsusha undō (1), Ishii Kinnosuke ni kiku, (Postwar movements of scientists and engineers, no.1, Interview with Kinnosuke Ishii),' in *Gijutsu to ningen* (Technology and humanity), April 1976, pp. 141–51.

Okano, Kiyoshi, 'Gakujutsu gyōsei (Science administration)', in *Gakujutsu geppō* (Science monthly), vol. 29, no.1, April 1976.

Okazaki, Fuminori, 'Shussan chōsei wa ikani okonau bekika (How should birth control be handled?)', in *Kōsei jihō* (Health and welfare bulletin), December 1949, pp.16–17.

Okuizumi, Eisaburō (ed.), *Maryland Daigaku Karejji Paaku-kō Makkerudin Toshokan Tōa Toshobu Gordon W. Prange bunko kaisetsu: Senryō-gun kenetsu zasshi mokuroku kaidai, Shōwa nijūnen – Shōwa nijūyonen* (Comments on Gordon W. Prange Collection, East Asia Section, McKeldin Library, College Park Campus, Maryland University: An inventory and synopsis of magazines censored by the Occupation Forces, 1945–1949), Yūshōdō Shoten, 1982.

Okuizumi, Eisaburō and Atsushi Furukawa, 'Nihon senryō-ki no kyokutō beigun jōhō shūshū katsudō to soshiki (The information gathering activities and organization of the Far East American military in Occupied Japan)', in *Tokyo Keizai Daigaku kaishi* (Journal of the Tokyo University of Economics), nos. 109–10, December 1978, pp.115–43.

Okuizumi, Eisaburō, '*Warera no kagaku* oyobi *Warera no Sekai* shishi: Minshushugi Kagakusha Kyōkai henshū-shi ni miru beigun no ken'etsu no jittai (The realities of American military censorship as evident in the magazines edited by the Association of Democratic Scientists, *Our science* and *Our world*)', in *Toshokan zasshi* (Library magazine), vol. 75, no. 8, August 1981, pp.461–5.

Ōmori, Masae, 'Kagaku kenkyūhi seido no henka to Nihon Gakujutsu Kaigi (Changes to the science research fund system and the JSC)', in *Nihon no kagakusha* (Japanese scientists), vol. 21, no. 8, August 1986, pp.443–9.

Onodera, Shinsaku, *Kensa: Seizōbutsu no sōgō-teki gijutsu tenkai* (Inspection: General technological development of products inspection), Nihon Kikaku Kyōkai (Japan Standards Association), 1989.

Orr, Mark T., *Education Reform Policy in Occupied Japan*, translated into Japanese as *Senryō-ka Nihon no kyōiku kaikaku seisaku* by Gery Hōichi Tsuchimochi, Tamagawa University Press, 1993, pp. 32, 140–41.

Osaka Daigaku (Osaka University), *Osaka Daigaku gojūnenshi, Tsūshi* (Fifty-year history of Osaka University; Outline history) 1985.

Ōshima, Jinpei, 'PB repōto (PB reports)', in *Dokusho shunjū* (Reading chronicle), vol. 3, no. 11, 1952, pp.8–9.

Ōta, Tenrei, 'Sanji seigen no hantai ni kotaeru (In response to birth control critics)', in *Kōshū eisei-gaku zasshi* (Journal of public health and hygiene), 1947, vol. 2, pp.215–19.

Ōta, Tenrei, *Chūzetsu wa satsujin dewa nai* (Abortions are not murders), Ningen no Kagakusha, 1983.

Ōta, Tenrei, *Datai kinshi to yūsei hogo-hō* (Abortion prohibition and the Eugenics Protection Law), Keieisha Kagaku Kyōkai, 1967.

Ōta, Tenrei, *Nihon sanji chōsei hyakunen-shi* (One-hundred years of birth control in Japan), Ningen no Kagakusha, 1976.

Ōta, Tenrei, *Sei no kenri: Datai kaihō no rekishi* (Sexual rights: a history of abortion liberation), San'ichi Shobō, 1970.

Oughterson, Ashley W. and Shield Warren (ed.), *Medical Effects of the Atomic Bombs in Japan*, National Nuclear Energy Series: Manhattan Project Technical Section, Division VIII – vol. 8, McGraw-Hill, 1955. Held by the National Diet Library.

Ōyodo, Shōichi, *Miyamoto Takenosuke to kagaku gijutsu gyōsei* (Takenosuke Miyamoto and science and technology administration), Tōkai Daigaku Shuppankai, 1989.

Patent Agency (compilation), *Kōgyō shoyūken seido hyakunen-shi* (One-hundred year history of the system of industrial property rights), First volume, Last volume, Separate volume, Hatsumei Kyōkai (Society for Inventions), published between 1984–85.

Patent Agency, *Tokkyo shichijūnen-shi* (Seventy year history of the patent system).

Peck, M. J. and M. J. Scherer, *The Weapons Acquision Process*, Manuscript submitted to Graduate School of Business Administration, Harvard University, 1962.

Putnam, David L., 'Reorganization of Science and Technology in Japan (1945 to September 1950)', in *History of the Non-military activities of the Occupation of Japan*, Kokkai Toshokan (National Diet Library), microfilm number CHS-04304.

Reimann, Günther, *Hittoraa no tokkyo senryaku* (Patents for Hitler), Hideo Tanaka, Masato Matsujō, and Toshimichi Komine (tr.), Daiyamondosha (Diamond Company), 1983.

Research Institute for the History of Industrial Policy, Toranomon Annex of the Research Institute of the Ministry of International Trade and Industry, *Kaitei-ban: Shōkōshō – Tsūshō sangyōshō gyōsei kikō oyobi kanbu shokuin no hensen* (Revised edition: The Ministry of Commerce and Industry/The Ministry of International Trade and Industry – Changes to the administrative structure and executive staff, 1925–1980), 1981.

Rikugun Gun'i gakkō (Army Medical College), *Rikugun Gun'i gakkō gojūnen-shi* (Fifty-year history of the Army Medical College). 1936.

Rōyama, Masamichi (ed.), *Daigaku seido no saikentō* (A re-examination of the university system), Minshu Kyōiku Kyōkai (Democratic Education Association), 1962.

Saitō, Hajime, *Furuburaito ikkisei* (First year Fulbright students), 1984.

Saitō, Jūkō, *Bōryaku-sen: Dokyumento rikugun Noborito kenkyūsho* (Plot warfare: Document – Noborito Research Institute of the Army), Jiji Tsūshinsha, 1987.

Sakata, Shōichi, 'Chūkanshi riron kenkyū no kaiko (Reflections on research into meson theory)', in Yukawa, et al., *Shinri no ba ni tachite* (Standing in the space of truth), Mainichi Shimbunsha, 1951.

Sakata, Shōichi, *Butsurigaku to hōhō* (Physics and methodology), Hakutō Shokan, 1947.

Sakata, Shōichi, *Kagakusha to Shakai ronshū* (Scientists and society), vol. 2, Iwanami Shoten, 1972.

Sakayanagi, Yoshimi, 'Sengo no ichimen: Rikagaku Kenkyūsho rokujūnen no ayumi (One aspect of the postwar period. The footsteps of Riken for sixty years),' in *Shizen* (Nature), vol. 33, no. 13, special issue, p. 117.

Sams, C. F., *DDT kakumei: Senryō-ki no iryō fukushi seisaku o kaisō suru* (The DDT revolution: Reflections on medical welfare policy during the Occupation), Eiji Takemae (tr.), Iwanami Shoten, 1986.

Sams, C.F., 'Igaku Kyōiku Iinkai no jigyō ni taisuru shishin (A guide to undertakings of the Medical Education Committee)', in *Nihon ishikai zasshi* (Journal of the Japan Medical Association), vol. 22, no. 11, December 1948, pp.439–41.

Sams, C.F., 'Igaku kyōiku kaikaku ni tsuite: Igaku kyōiku shingikai dai 1-kai kaigi ni okeru kunji (On medical education reforms: Address at the first meeting of the Deliberative Council on Medical Education, 17 February 1946)', in *Nihon iji shinpō* (Japan medical affairs bulletin), no. 1190, July 1946.

Sangar, Margaret, *Sanji chōsei-ron* (A treatise on birth control), Toshisada Oku (tr.), Seika Shoin, 1921.

Sangyō Seisaku-shi Kenkyūsho (Research Institute of Industrial Policy History), *Sangyō seisaku-shi kaisōroku* (Reflections on the history of industrial policy), 32 volumes, 1985.

Sasaki, Atsushi, Shunsuke Tsurumi, Ken'ichi Tominaga, Masanori Nakamura, Kimihiro Masamura, and Yōichirō Murakami (eds.), *Sengo-shi daijiten* (An encyclopedia on postwar history), Sanseidō, 1991.

Sasamoto, Yukuo, 'Genbaku shōdō chōsa ni okeru Nihon-gun no yakuwari (The role of the Japanese Military in the initial atomic bomb investigation)', in *Rekishi to shakai* (History and society), no. 9, 1989.

Satō, Kōji, *Bei gunsei kanri to heiwa undō* (The US military administration and peace movements), Seseragi Shuppan, 1991.

Sawamoto, Takahisa, 'Kagakugijutsusha no kenkyū jōhō chōsa (Survey of research information for scientists and engineers)', in *Library Science,* vol. 5, July 1967, pp.137–52.

Scherkenbach, William W., *The Deming Route to Quality and Productivity*, CEEP Press Books, 1986.

Sengo no Tsūshin Kōgyō Henshū Iinkai (Editorial Committee on the History of the Postwar Telecommunications Industry) (ed.), *Sengo no tsūshin kōgyō* (Postwar telecommunications industry), Nihon Denki Tsūshin Kōgyō Rengō-kai (Federation of Telecommunications Industries Association of Japan), 1959.

Shakai Hyōronsha Henshū-bu, *Josei no sei to chūzetsu: Yūsei hogohō no haikei* (Female sexuality and abortion: Background to the Eugenics Protection Law), Shakai Hyōronsha, 1983.

Sherwin, Martin J., *Hametsu e no dōtei – Genbaku to dai 2-ji sekai taisen* (The road to destruction – Atomic bombs and World War II), Mikio Katō (tr.), TBS Britannica, 1978.

Shiga, Yoshio, 'Sekai-shi no ichi kōsatsu (A study in world history),' in *Zen'ei* (Vanguard), nos. 20 and 21, 1947.

Shiga, Yoshio, 'Tō-shi no mikata (How to think the history of the Communist Party),' in *Zen'ei* (Vanguard), no. 19, 1947.

Shikiba, Ryūsaburō, *Sangar fujin-den to sanji chōsetsu tenbō* (Mrs. Sangar's biography and the future of birth control), Daigensha, 1947.

Shimazaki, Reiko Ryder, 'Hi-senryōka (1945–1951nen) ni okeru Nihon no kango seisaku 1–7 (Japanese nursing policies during the Occupation (1945–1951), 1–7)', in *Kangofu kyōiku* (Nursing education), vol. 31, nos. 2–8, 1990.

Shimizu, Katsuyoshi, *Zoku kōshū eisei no hattatsu* (Development of public hygiene: A sequel), Nihon Kōshū Eisei Kyōkai (Japan Public Health Association), 1983.

Shimomura, Kainan, *Shūsen-ki* (Notes on the end of the war), Kamakura Shoten, 1949, republished by Kōdansha Bunko, 1985.

Shimono, Katsumi, *Sengo Nihon kagaku kōgyō-shi* (History of the chemical industry in postwar Japan), 1987.

Shindō, Kaneto, *Tsuihōsha tachi* (The purged), Iwanami Shoten, 1983.

Shinkai, Kōsaku, 'Kenkyū kikan no kakujū kyōka (Expansion and empowerment of research organizations)', in *Nissankyō geppō* (Japan Industrial Association monthly), April 1949.

Shinobu, Seizaburō, *Sengo Nihon seiji-shi* (A history of politics in postwar Japan), vol. 1, 1965, vol. 2, 1996.

Shinozaki, Nobuo, 'Gendai seiseikatsu no jittai (Present day sex lives)', in *Kōsei jihō* (Health and welfare bulletin), December 1949, pp.6–14.

Shinozaki, Nobuo, 'Sanji seigen no jitsujō o saguru (Exploring the realities of birth control)', in *Kōsei jihō* (Health and welfare bulletin), May 1949, pp.29–35.

Shinozaki, Nobuo, *Sanji chōsei to fūfu sei-seikatsu no jittai hōkoku* (Report on birth control and the sex life of married couples), Muramatsu Shoten, 1949.

Shioda, Shōbei, *Reddo Paaji* (The Red Purge), Shin-Nihon Shuppansha, 1984.

Shisō no Kagaku Kenkyūkai (Study Group on the Science of Thought) (ed.), *Kyōdō kenkyū, Nihon senryōgun sono hikari to kage* (Joint research, the Occupation Forces in Japan, their bright and dark sides), Tokuma Shoten, 1987.

Shūkan Books Tokubetsu Shuzai-han (Special Investigation Team of the Weekly Books magazine) (ed.), *Naze ima yūsei hogohō 'kaisei' ka* (Why the proposed 'revision' of the Eugenics Protection Law now?), Gendai Shorin, 1983.

Shuppan Kyōdō Kabushikigaisha, *Nihon shuppan nenkan 1944, 1945 and 1946; 1947 and 1948* (Yearbook of publications in Japan 1944, 1945 and 1946; 1947–1948), Shuppan Kyōdō Kabushikigaisha, 1947 and 1948.

Singer, Charles (ed.), *The History of Technology* (five volumes), Oxford University Press, 1954–58.

Smith, Alice Kimball, *A Peril and a Hope: The Scientists' Movement in America, 1945–47*, MIT Press; Tetsu Hiroshige (tr.), *Kiki to kibō – Amerika no kagakusha undō* (Danger and hope – The American scientists' movements; 1945–1947), Misuzu Shobō, 1968.

Smythe, H.D., *Genshi bakudan no kansei: Smythe hōkoku* (The Completion of the atomic bomb: The Smythe Report), Yoshio Nishina, (editorial supervision), Asao Sugimoto, Eizō Tajima, and Eiichi Kawasaki (tr.), Iwanami Shoten, 1951.

Sodei, Rinjrō, 'Genbaku hōdō (Atomic bomb reporting)', in Fumio Etō, Shunsuke Tsurumi, and Akira Yamamoto (eds.), *Kōza komyunikeeshon 5: Jiken to hōdō* (Series on communication studies, vol. 5: Incidents and reporting), Kenkyūsha, 1972, pp.201–25.

Sugai, Shirō, *Shiryō, kokudo keikaku* (Materials, national land planning), Daimyōdō, 1975.

Sugihara, Yoshio, 'ABCC mondai ni tsuite (On the ABCC issue)', in *Nihon no kagakusha* (Japanese scientists), vol.2, no.3, December 1967.

Sugiyama, Akiko, *Senryōka no kōshū eisei – Shoki seisaku o megutte* (Public hygiene during the Occupation: Early policies), presented at Senryōshi Kenkyūkai (Study Group on the History of Occupation), 18 April 1992.

Sugiyama, Akiko, *Senryō-ki no iryō kaikaku* (Medical reforms during the Occupation period), Keisō Shobō, 1995.

Sumimoto, Toshio, *Senryō hiroku* (Secret records of the Occupation), Mainichi Shimbunsha, 1952, republished by Chūkō Bunko, 1988.

Suzue, Kōhei, 'Kagaku Gijutsu Gyōsei Kyōgikai o kaerimite (Reflecting on STAC)', in *Gakujutsu geppō* (Scientific monthly), vol. 5, no. 12, 1952, pp.784–7.

Suzuki, Eiichi, 'Walter Eells bunsho (The Walter C. Eells Documents)', in Hideo Satō (chief researcher), *[Kaigai gakujutsu kenkyū hōkokusho] Senryō-ki Nihon kyōiku ni kansuru zaibei shiryō no kenkyū chōsa* ([Overseas academic research report] Survey Report on historical data in America relating to Japanese education during the Occupation period), Kokuritsu Kyōiku Kenkyūsho (National Institute for Educational Policy Research), 1988, pp.97–107.

Suzuki, Eiichi, *Nihon no senryō to kyōiku kaikaku* (The Allied Occupation of Japan and education reforms), Keisō Shobō, 1983.

Suzuki, Zenji, *Gendai no yūseigaku* (Contemporary eugenics), Sankyō Shuppan, 1983.

Tachi, Minoru, 'Jinkō mondai kara mita sanji seigen (Birth control viewed from the perspective of the population problem),' in Kakuichi Andō (ed.), *Sanji seigen no kenkyū* (Birth control research), Nihon Rinshōsha, 1947, pp.27–154.

Tachi, Minoru, *Jinkō mondai setsuwa: Asahi jōshiki kōza* (Treatise on the population question: the Asahi common sense lecture), Bon'yōsha, 1943.

Takahashi, Bonsen, *Datai mabiki no kenkyū* (Abortion research), Daiichi Shobō, 1926, republished by Chūō Shakai Jigyō Kyōkai, 1981.

Takahashi, Noboru (ed.), 'Sengo no kagaku gijutsusha undō (Postwar movements of scientists and engineers)', in *Gijutsu to ningen* (Technology and humanity).

Takakuwa, Kōkichi, *MacArthur no shimbun ken'etsu: Keisai kinshi sakujo ni natta shimbun kiji* (MacArthur's newspaper censorship: Newspaper articles that were banned or deleted), Yomiuri Shimbunsha, 1984.

Takamatsu, Toichirō, 'Genshi bakudan no hōdō: Sono genkai to seigen (Atomic Bomb reporting: The limits and limitations)', in *Tokyo Daigaku Shimbun Kenkyū-sho kiyō* (Journal of the Newspaper Research Institute of the University of Tokyo), no. 3, 1954, pp.1–26.

Takano, Kazumoto, 'Senryō-gun no zasshi ken'etsu to *Minshu hyōron* (Occupation Forces' magazine censorship and *Democratic voice*)', in *Ōhara Shakai mondai kenkyūsho zasshi* (The Journal of the Ōhara Research Institute of Social Problems), no. 406, October 1992, pp.18–42.

Takeshima, Kazu, 'Senryō-ki no ken'etsu to hōsō: Rajio kōdo (Censorship and broadcasting during the Occupation: The radio code)', in *Hōsō kenkyū to chōsa* (Research and surveys on broadcasting), vol. 29, no. 12, December 1979, pp.24–9, 59.

Takemae, Eiji and Yukuo Sasamoto, 'Senryōka no iryō fukushi nenpyō (1) zenshi (Chronological table of medical welfare during the Occupation (1) The antecedent phase)', in *Tokyo Keizai Daigaku Kaishi* (Journal of the Tokyo University of Economics), no. 147, 1986, pp.259–70.

Takemae, Eiji and Yukuo Sasamoto, 'Senryōka no iryō fukushi nenpyō (2) (1945. 9–1946. 12) (Chronological table of medical welfare during the Occupation (2) (1945. 9 – 1946. 12)', in *Tokyo Keizai Daigaku kaishi* (Journal of the Tokyo University of Economics), no. 157, 1988, pp.91–123.

Takemae, Eiji and Yukuo Sasamoto, 'Senryōka no iryō fukushi nenpyō (3) (1947. 1–1948. 12) (Chronological table of medical welfare during the Occupation (3) (1947. 19 – 1948. 12)', in *Tokyo Keizai Daigaku kaishi* (Journal of the Tokyo University of Economics), no. 159, 1989, pp.135–74.

Takemae, Eiji and Yukuo Sasamoto, 'Senryōka no iryō fukushi nenpyō (4) (1949. 1 – 1952. 4) (Chronological table of medical welfare during the Occupation (4) (1949. 9 – 1952. 4)', in *Tokyo Keizai Daigaku kaishi* (Journal of the Tokyo University of Economics), no. 157, 1989, pp.131–62.

Takemae, Eiji and Yukuo Sasamoto, 'Senryōka no iryō to fukushi (2), GHQ/PHW no soshiki to jinji (Medical treatment and welfare (2), Organization and personnel of GHQ/PHW)', in *Tokyo Keizai Daigaku kaishi* (Journal of the Tokyo University of Economics), no. 156, 1988, pp.267–85.

Takemae, Eiji (editorial supervisor), Yukuo Sasamoto, Kazumoto Takano, and Yukuo Wakamatsu (eds.), *GHQ e no Nihon seifu taiō bunsho sō shūsei* (The Japanese Government's policy towards GHQ: Total collection of documents), MT Shuppan, 1994.

Takemae, Eiji, 'Reddo Paaji (The Red Purge)', in Shisō no Kagaku Kenkyū-kai (Research Group of the Science of Thought) (ed.), *Kyōdō kenkyū/Nihon senryō-gun: Sono hikari to kage* (Joint research/the Occupation Army in Japan: Its light and shade), vol. 1, Gendaishi Shuppankai, 1978.

Takemae, Eiji, Akimasa Miyake, and Kōshi Endō (eds.), *Rōdō kaikaku to rōdō undō* (Labor reforms and the labor movement), Ōtsuki Shoten, 1992.

Takemae, Eiji, *GHQ*, Iwanami Shinsho, 1983.

Takemae, Eiji, *Sengo rōdō kaikaku – GHQ rōdō seisaku-shi* (Postwar labor reforms: A history of the GHQ labor policies), University of Tokyo Press, 1982.

Takemae, Eiji, *Senyrō sengo-shi: Nihon kanri seisaku no zenyō* (A history of the Occupation after the War: A total picture of the management policies for Japan), 1980.

Takemae, Eiji, *Senryō to sengo kaikaku* (The Occupation and the postwar reforms), Iwanami Shoten.

Takemae, Eiji, *Shōgen Nihon senryō-shi* (A Testimonial to the history of the Occupation of Japan), Iwanami Shoten, 1983.

Takemae, Eiji (supervising editor), *GHQ shirei sōshūsei: SCAPIN* (A complete collection of GHQ directives: SCAPIN), 15 vols., MT Shuppan, 1994.

Takemura, Harumasa and Jirō Harada (ed.), *Bosei hogo* (Mother's welfare), Shin Nippon Igaku Shuppan, 1972.

Takeshima, Kazu, 'Senryō-ki no ken'etsu to hōsō: Rajio kōdo (Censorship and broadcasting during the Occupation: The radio code)', in *Hōsō kenkyū to chōsa* (Research surveys on broadcasting), vol. 29, no. 12, December 1979, pp. 24–9, 59.

Taketani, Mitsuo, 'Soryūshi gurūpu no keisei (The formation of the elementary particles theory group)', in Yukawa, et. al., *Shinri no ba ni tachite* (Standing in the space of truth), Mainichi Shimbunsha, 1951.

Taketani, Mitsuo, *Benshōhō no shomondai* (Problems of Dialectics), Rigakusha, 1946.

Taketani, Mitsuo, *Kagaku to gijutsu – Sono riron to hōhō* (Science and technology: Its theory and method), Rironsha, 1950.

Takeyama, Akiko, *Sensō to hōsō* (War and broadcasting), Shakai Shisōsha, 1994.

Tamaki, Hidehiko, 'Aru kenkyūsho no unmei – Eikō to kunan no Rikagaku Kenkyūsho (The fate of a research Institute: Glory and hardship at the Institute of Physical and Chemical Research)', in *Kagaku Asahi*, November 1961.

Tanaka, Masao, 'Daigaku Kijun Kyōkai no keisei to "daigaku kijun" no seiritsu (jō)(ge) (Formation of the Japan University Accreditation Association and 'university standards')', in *Bulletin* (The Japan University Accreditation Association), nos. 44–45, 1981.

Tanaka, Minoru, *Busshitsu to denki* (Matter and electricity), Dainihon Shuppan Insatsu, 1946.

Taniguchi, Yasaburō, 'Yūsei hogohō kaisetsu (Explanations of the Eugenics Protection Law)', in *Nihon ishikai zasshi* (Journal of the Japan Medical Association), 1948, vol. 22, pp.215–9.

Taniguchi, Yasaburō, 'Yūsei hogohō ni okeru shitei ishi no sentei hōhō ni tsuite (On the nominated doctor selection method under the Eugenics Protection Law)', in *Nihon iji Shinpō* (Japan medical affairs bulletin), 1948, no. 1272, pp.1076–7.

Temporary Administration Investigation Committee Secretariat, *Rinchō saishū teigen* (Final proposal of the Temporary Administration Investigation Committee), Gyōsei Kanri Kenkyū Sentaa (Administration Management Research Centre), 1983.

Terasaki, Masao (tr.), *Sengo kyōiku kaikaku shiryō 10: Beikoku tainichi kyōiku shisetsudan ni kansuru sōgōteki kenkyū* (Postwar education reform data 10: Comprehensive research on the United States Education Mission to Japan), March 1991.

Terasaki, Masao, 'Shiryō Tokyo Daigaku kyōiku seido kenkyū iinkai (Data on committee for Tokyo University education reform)', in *Tokyō Daigakushi kiyō* (Journal of the history of the University of Tokyo), no. 7, March 1989.

Terasaki, Masao, *Sengo Nihon no kyōiku kaikaku 9, 'Daigaku kyōiku'* (Education reform in postwar Japan 9: University education), University of Tokyo Press, 1970.

Teruya, Hiroyoshi, *Sengo Okinawa no iryō: Watashi no ayunda michi kara* (Postwar medical treatment in Okinawa: The path I took), Medikaru Furendosha (Medical Friend Co.), 1987.

Tetsudō Gijutsu Kenkyūsho (Railway Technology Research Institute of the Japan National Railways) (ed.), *Tetsudō gijutsu kenkyūsho sōritsu rokujūshūnen – jūnen no ayumi* (Sixtieth anniversary of the Railway Technology Research Institute – Ten years of progress), 1967.

Tetsudō Gijutsu Kenkyūsho (Railway Technology Research Institute of the Japan National Railway), *Jūnen no ayumi – sōritsu rokujūshūnen* (History of the last ten years at the sixtieth anniversary), 1967.

Tetsudō Gijutsu Kenkyūsho (Railway Technology Research Institute), *Jūnen no ayumi – sōritsu sichijūshūnen* (History of the last ten years of the Railway Technical Research Institute – at the seventieth anniversary), 1977.

Tetsudō Gijutsu Kenkyūsho Gojūnenshi Kankō Iinkai (Committee for Publication of the Fifty Year History of the Railway Technology Research Institute of the Japan National Railways) (ed.), *Nihon kokuyū tetsudō gojūnenshi* (The fifty year history of the Railway Technology Research Institute of the Japan National Railways), Zaidan Hōjin Ken'yūkai, 1957.

Thomas, Gordon and Max Morgan-Witts, *Enola Gay – Dokyumento genbaku tōka* (Enola Gay: Documenting the dropping of the atomic bomb), Sen Matsuda (tr.), TBS Britannica, 1980.

Tōhō Denryoku-shi Henshū Iinkai (Editorial Committee on History of Tōhō Denryoku), *Tohō denryoku-shi* (A history of Tōhō Denryoku), 1962.

Tōhoku Daigaku (Tōhoku University), *Tōhoku Daigaku Gojūnenshi* (Fifty-year history of Tōhoku University), Tōhoku University, 1960.

Tokkyo-chō Kōgyō Shoyūken Seidoshi Kenkyū-kai (Society for the Study of the History of the Industrial Property Rights System of the Patent Agency) (ed.), *Tokkyo seido no hassei to hensen* (Creation of and changes to the patent system), *Ōkura-shō Insatsu-kyoku* (Printing Bureau of the Ministry of Finance), 1982.

Tokyo Daigaku Hyakunenshi Henshū Iinkai (Editorial Committee for a One-Hundred Year History of the University of Tokyo) (ed.), *Tokyo Daigaku hyakunen-shi* (The one hundred-year history of the University of Tokyo), 10 vols., University of Tokyo Press, 1985–1987.

Tokyo Daigaku Seisan Gijutsu Kenkyūsho (Institute of Industrial Science,

University of Tokyo), 'Nijūshūnenshi (Twenty year history)', in *Seisan kenkyū* (Industrial science), vol. 21, no. 5, May 1969.

Tokyo Daigaku Seisan Gijutsu Kenkyūsho (Institute of Industrial Science, University of Tokyo) (ed.), *Tokyo Daigaku Daini Kōgakubushi* (History of the Second School of Engineering, University of Tokyo), November 1968.

Tokyo Daigaku Shakai Kagaku Kenkyūsho Sengo Kaikaku Kenkyū-kai (Research Committee on Postwar Reforms, Social Science Research Institute, University of Tokyo) (ed.), *Sengo zasshi mokuji* (Index of magazines published after the war), vols. 1 & 2, University of Tokyo Press, 1976, 1977.

Tokyo Kōgyō Daigaku (Tokyo Institute of Technology), *Tokyo Kōgyō Daigaku hyakunenshi* (One hundred-year history of the Tokyo Institute of Technology), 2 volumes, 1985.

Tokyo Kōgyō Daigaku (Tokyo Institute of Technology), *Tokyo Kōgyō Daigaku kyūjūnenshi* (Ninety-year history of the Tokyo Institute of Technology), 1975.

Tomita, Tetsuo, *Shijō kyōsō kara mita chiteki shoyūken* (Intellectual property rights from the perspective of market competition), Daiyamondosha, 1993.

Tomizuka, Kiyoshi, *Hachijūnen no shōgai no kiroku* (Records of my eighty years), 1975.

Tomonaga, Shin'ichirō and Hidehiko Tamaki (ed.), *Nishina Yoshio, Denki to kaisō* (Yoshio Nishina, Biography and recollections), Misuzu Shobō 1952.

Torii, Hiroshi, et. al., 'Minka undō no rekishiteki haikei (Historical background of the Minka movement)', in *Gijutsu to ningen* (Technology and humanity), June 1976, pp.141–53.

Tsuboi, Chūji, 'Iwayuru kagaku shinkō ni tsuite (On the so-called science promotion),' in *Sekai* (The World), June 1946, p. 70.

Tsuchimochi, Hōichi Gary, 'Beikoku Gakujutsu Komondan hōkokusho to sengo Nihon no kōtō kyōiku kaikakuan – Adams danchō bunsho o chūshin ni (The report of the United States Scientific Advisory Group and Japan's postwar reform of higher education with special reference to the Adams papers)', Kokuritsu Kyōiku Kenkyūsho (National Institute for Educational Research), Sōsuke Watanabe (ed.) *Kōwa dokuritsu go no wagakuni kyōiku kaikaku ni kansuru chōsa kenkyū* (Research on Japan's educational reform after the conclusion of the Peace Treaty), March 1992, pp.51–64. Also in *Daigaku ronshū* (Research in higher education), no. 20, 1991.

Tsuchimochi, Hōichi Gary, 'Senryōka no kyōiku kaikaku – Daiichiji beikoku tainichi kyōiku shisetsudan hōkokusho to kōtōkyōiku kaikaku (Education reform during the Occupation: The first report by the

United States Education Mission to Japan and Reform of Higher Education)', in *Daigaku ronshū* (Research in higher education), no. 18, 1989.

Tsuge, Hideomi, 'Minshushugi Kagakusha Kyōkai to watakushi (The Association of Democratic Scientists and I)', in *Gijutsu to ningen* (Technology and humanity), September 1976, pp.144ff.

Tsuge, Hideomi, *Kagaku wa heiwa o motomete* (Science seeks peace), 1954.

Tsuge, Hideomi, *Minka to Watashi: Ichikagakusha no Ayumi* (Minka and I: Footsteps of a postwar scientists), Keisō, 1980.

Tsuneishi, Keiichi (ed. and tr.), *Hyōteki Ishii – 731 butai to beigun no chōhō katsudō* (Target Ishii – 731 Unit and American military intelligence activities), Ōtsuki Shoten, 1984.

Tsuneishi, Keiichi and Tomizō Asano, *Saikinsen butai to jiketsu shita futari no igakusha* (The germ Strategy Unit and two doctors who committed suicide), Shinchōsha, 1982, pp.147–275.

Tsuneishi, Keiichi, 'Riken ni okeru uran bunri no kokoromi (The uranium splitting tests at the Institute of Physical and Chemical Research), in *Nihon butsuri gakkai-shi* (Journal of the Japan Physics Association), vol.45, no.11, Nov 1990, pp.820–5.

Tsuneishi, Keiichi, *Kieta saikin-sen butai: Kantō-gun dai 731-butai* (The lost germ warfare corp: The Kwantung Army No. 731 Unit), Kaimeisha 1981, republished by Chikuma Bunko, 1993.

Tsūshin Kōgyō Nijūnen no Ayumi Henshū Iinkai (Editorial Committee on the History of Twenty Years of Progress in the Telecommunications Industry) (ed.), *Tsūshin kōgyō nijūnen no ayumi* (Twenty years of progress in the telecommunications industries), Tsūshin Kikai Kōgyō-kai (Communications Industry Association of Japan), 1968.

Uemura, Shinsaku, Hiroshi Kawamura, Machiko Tsuji, Shigeyuki Tomita and Shizuo Maeda, *Nōyaku dokusei no jiten* (Encyclopaedia of toxicity of agricultural chemicals), Sanseidō, 1988.

Uki, Boku, *Kankoku no gijutsu hatten* (Technological development in South Korea), 1988.

Un'yushō Tetsudō Gijutsu Kenkyūshochō (Director of the Railway Technical Research Institute, Transportation Ministry), 'Baishō wariate yori kikai o jogaisuru ken (On the release of machineries from the reparation list) 6 September 1946', ESS(B)11723–34.

United States Department of Education, 'Development in Japanese Education in terms of the Report of the United States Education Mission to Japan,' Wanamaker Papers. Japanese translation, *Beikoku kyōiku shisetsudan hōkoku*, Kokusai Tokushinsha, 1946.

United States Department of Education, 'Report of the Second United States Education Mission to Japan.' Japanese translation, in Akio

Igasaki and Kōichirō Yoshihara (eds.), *Sengo kyōiku no genten 2* (The origin of postwar educational reform), Gendaishi Shuppankai, 1975.
United States Strategic Bombing Survey, *The Effects of Atomic Bombs on Hiroshima and Nagasaki*, US Government Printing Office, 1946. Microfilm held by the National Diet Library. Japanese translation: 'Amerika gasshūkoku senryaku bakugeki chōsadan hōkoku' in Yoshito Ōsumi and Sadao Kamata (tr.), *Nagasaki Zōsen Daigaku kenkyū hōkoku* (The Nagasaki Shipbuilding University research report), vol. 18, no. 1, 1977.
Usui, Masuo, 'Shizen kagaku, jiron, gakkaishi ni tsuite (On natural science, commentaries and scientific journals)' in *Shizen kagaku* (Natural science), no. 9, September 1947, p. 1.
Wakabayashi, Yoshitaka, 'Kagaku gijutsu kenkyūkikan no kakuritsu (The establishment of S&T research organization)', in *Nissankyō geppō* (Japan Industrial Association monthly), May 1952.
Wakabayashi,Yoshitaka, 'Minkan gakujutsu kikan no josei (Subsidy for private research organizations)', in *Nissankyō geppō* (Japan Industrial Association monthly), July 1951.
Walton, Mary, *Deming Management at Work*, Perigee Books, 1991.
Warren, S. L., *Preliminary Medical Report, Hiroshima-Nagasaki Bombings*, 27 Nov. 1947, in Paul Kesaris, (ed.), *A Guide to the Manhattan Project: Official History and Document,* A Microfilm Project of University Publications of America, Inc., 1977, held by the National Diet Library.
Watanabe, Yoshimitsu (dictated), Hiyaringu Gurūpu (Hearing Group) (ed.), *Shisō to gakumon no jiden* (An autobiography of thinking and learning), Kawade Shobō Shinsha, 1974.
Weiner, Charles , 'Japan's nuclear bomb project', in *Science*, no. 199, 17 February 1978.
Weiner, Charles, 'Cyclotrons and internationalism: Japan, Denmark and the United States, 1935–1945', in *Proceedings of XIVth International Congress of the History of Science, 1974* (Tokyo, 1975).
Winner, Langdon, *Autonomous Technology*, MIT Press, 1977.
Yakushiji, Taizō, 'Sengo kōgyō gijutsu-shi no shōgen (Testimonies on the history of postwar industrial technology),' in *Kogyo gijutsu*, July 1986.
Yamada, Keiichi and Shūichi Tsukahara (eds.), *Kagaku Kenkyū no raifu saikuru* (Life cycles of scientific research), 1986.
Yamagiwa, Akira and Seiitsu Tachibana (eds.), *Shiryō Manhattan keikaku* (Reference material – The Manhattan Project), translated by Seiitsu Tachibana, Ōtsuki Shoten 1993.
Yamamoto, Senji (chief ed.) *Sei to shakai* (Sex and society), no. 9, 1925–26, republished by Fuji Shuppan, 1983.
Yamamoto, Senji (chief ed.), *Sanji chōsei hyōron* (Review on birth control), nos. 1–8, 1925, republished by Fuji Shuppan, 1983.

Yamamoto, Senji, 'Sengo no hatsuden-sho no fukkyū ni tsuite (Postwar restoration of power plants)', paper submitted to the Fourth World Power Conference, Nippon Hassōden Co. Ltd., 1950, pp.2–21.

Yano, Kenzō, *Kōjō saigai fujo-ron* (Factory disaster support theory), Sanseidō, 1931.

Yasuda Sōgō Kenkyūsho (Yasuda Research Institute), *Seizōbutsu sekinin* (Products liability), Yūhikaku, 1989.

Yomiuri Shimbun, *Shōwa-shi no tennō* (Shōwa Emperor in history), 4 volumes, Yomiuri Shimbunsha, 1968.

Yoshida, Shigeru, *Kaisō jūnen* (Reminiscing ten years), Shinchōsha, vols. 2 and 3.

Yoshihashi, Toshio, 'Dantai-tō kiseirei ni tsuite (On the group regulation ordinance),' in Izutarō Hirosue (ed.), *Hōritsu jihō* (Law review), vol. 22, no. 10, October. 1950, p. 773.

Yoshimura, Toshio, *Kiju kaiko* (Reminiscing on my seventy-seventh birthday celebration), Yoshimura Sensei Kiju Kinen Gyōji-ka (Celebration Committee of Mr Yoshimura's Seventy-seventh Birthday), 1984.

Yoshioka, Hitoshi, *Kagakusha wa kawaru ka – Kagaku to shakai no shisō-shi* (Will scientists change? A history of the ideology of science and society), Shakai Shisōsha, 1984.

Yui, Daisaburō, *Mikan no senryō kaikaku* (Unfinished Occupation reforms), University of Tokyo Press, 1989.

Yukawa, Hideki, Shōichi Sakata, and Mitsuo Taketani, *Soryūshi no kenkyū* (Research into elementary particles theory), Keisō Shobō, 1965.

Appendix 1

Organization Chart: General Headquarters Supreme Commander for the Allied Powers (GHQ, SCAP), Tokyo, Japan, 5 June 1948

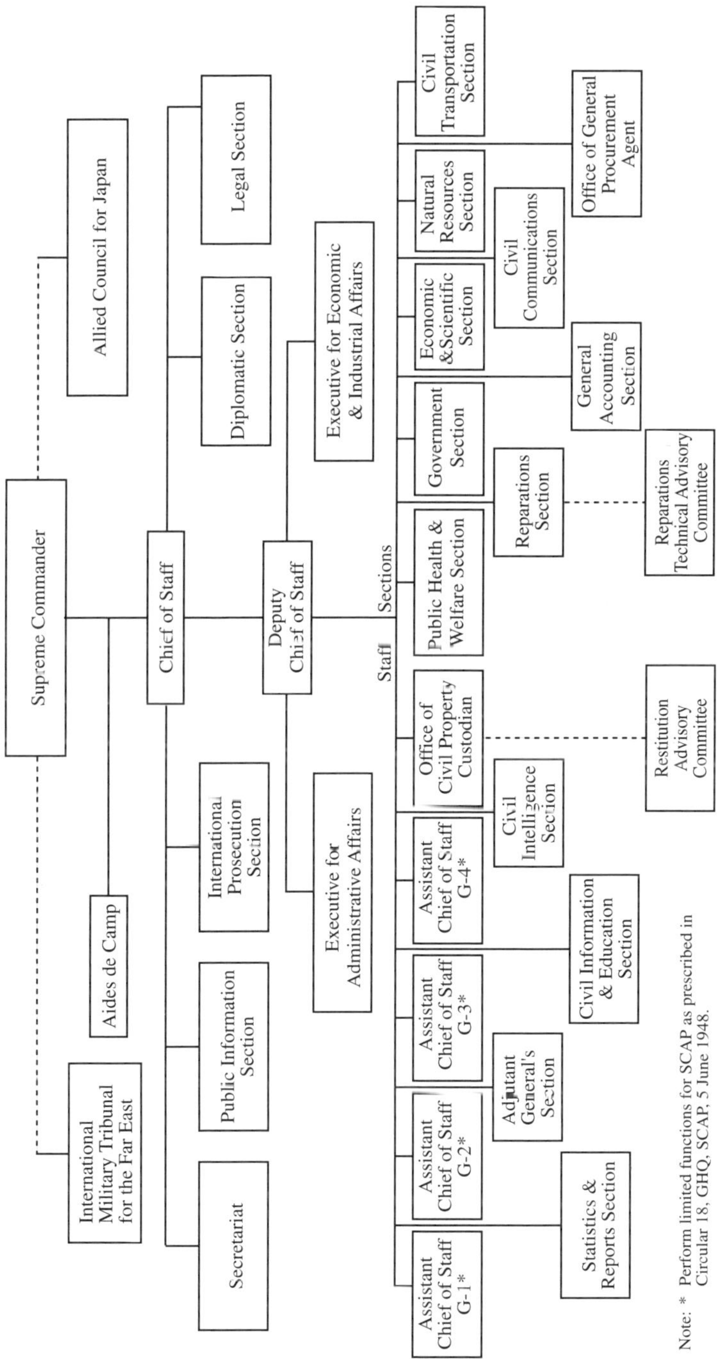

Note: * Perform limited functions for SCAP as prescribed in Circular 18, GHQ, SCAP, 5 June 1948.

Appendix 2

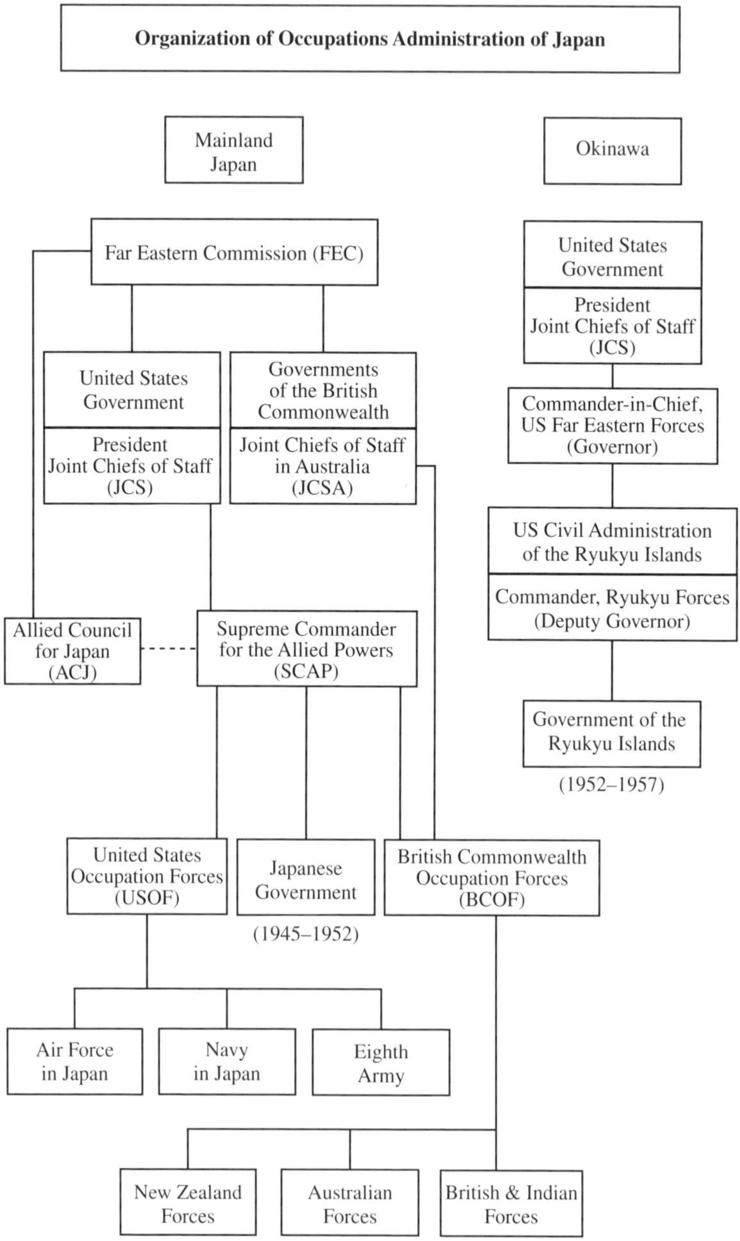
Organization of Occupations Administration of Japan
Mainland Japan
Okinawa
Far Eastern Commission (FEC)
United States Government
President Joint Chiefs of Staff (JCS)
Governments of the British Commonwealth
Joint Chiefs of Staff in Australia (JCSA)
Allied Council for Japan (ACJ)
Supreme Commander for the Allied Powers (SCAP)
United States Occupation Forces (USOF)
Japanese Government
(1945–1952)
British Commonwealth Occupation Forces (BCOF)
Air Force in Japan
Navy in Japan
Eighth Army
New Zealand Forces
Australian Forces
British & Indian Forces
United States Government
President Joint Chiefs of Staff (JCS)
Commander-in-Chief, US Far Eastern Forces (Governor)
US Civil Administration of the Ryukyu Islands
Commander, Ryukyu Forces (Deputy Governor)
Government of the Ryukyu Islands
(1952–1957)

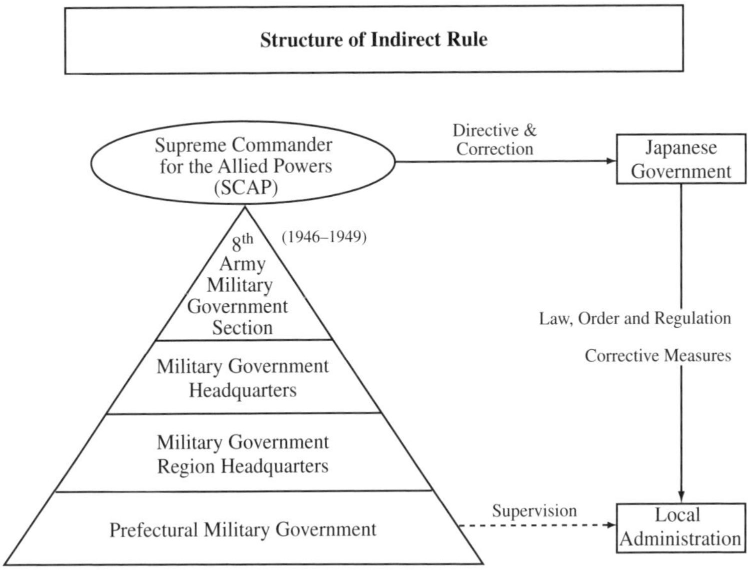

Source: Eiji Takemae, *Senryō to sengo kaikaku* (The Occupation and the postwar reforms), Iwanami Shoten.

Index